THE U.S.-JAPAN SCIENCE AND TECHNOLOGY AGREEMENT

The U.S.-Japan Science and Technology Agreement

A drama in five acts

CECIL H. UYEHARA

Ashgate

Aldershot • Burlington USA • Singapore • Sydney

Published by
Ashgate Publishing Limited
Gower House
Croft Road
Aldershot
Hampshire GU11 3HR

Ashgate Publishing Company
131 Main Street
Burlington, VT 05401-5600 USA

Ashgate website: http://www.ashgate.com

British Library Cataloguing in Publication Data
Uyehara, Cecil H.
The U.S.-Japan Science and Technology Agreement : a drama in five acts
1.Technology transfer - United States 2.Technology transfer - Japan 3. United States - Foreign relations - Japan 4.Japan - Foreign relations - United States
I.Title
327.7'3'052

Library of Congress Control Number: 00-133999

ISBN 0 7546 1414 X

Printed and Bound in Great Britain by
Antony Rowe Ltd, Chippenham, Wiltshire

Table of Contents

Appendices

Abbreviations Used

Abbreviation	Full Name
80STA	1980 U.S.-Japan Science and Technology Agreement
88STA	1988 U.S.-Japan Science and Technology Agreement
ABCC	Atomic Bomb Casualty Commission
AEC	Atomic Energy Commission (U.S.)
AIST	Agency for Industrial Science and Technology (Japan)
AmEmb	U.S. Embassy in Tokyo
ANRE	Agency of Natural Resources and Energy (Japan)
A/S	Assistant Secretary
AT&T	American Telephone and Telegraph
CA	California
CAD	Computer-aided Design
CENTO	Central Treaty Organization
CIA	Central Intelligence Agency
CIM	Computer Integrated Manufacturing
CISET	Committee on International Science, Engineering and Technology
COCOM	Coordinating Committee of Export Controls (NATO)
CRS	Congressional Reference Service, Library of Congress (U.S.)
DAS	Deputy Assistant Secretary of State
DC	District of Columbia
DCM	Deputy Chief of Mission
DOC	Department of Commerce (U.S.)
DOD	Department of Defense (U.S.)
DOE	Department of Energy (U.S.)
EA	Environmental Agency (Japan)
EAP	Bureau of East Asian & Pacific Affairs (State, U.S.)
EC	European Commission
EPA	Environmental Protection Agency (U.S.)
EPC	Economic Policy Council in the Reagan Administration
ESCAP	Economic and Social Council for Asia and the Pacific (U.N.)
ESS	Economic and Scientific Section in GHQ under SCAP in Tokyo, 1945-52

FCCSET	Federal Coordinating Committee on Science, Engineering & Technology
FOI	Freedom of Information
FOIA	Freedom of Information Act
FSX	Fighter Support Aircraft-Experimental (Japan)
FY	Fiscal Year
GAO	General Accounting Office (U.S.)
GD	General Dynamics Corporation
GE	General Electric
GHQ	General Headquarters under SCAP
GM	General Motors
GNP	Gross National Product
GOJ	Government of Japan
HHS	Department of Health and Human Services (U.S.)
IBM	International Business Machines
ICOT	Institute for New Generation Computer Technology
IPR	Intellectual Property Rights
ISA	Office of International Security Affairs, DOD (U.S.)
ISAS	Institute for Space and Astronautical Sciences (Japan)
ITA	International Trade Administration (DOC, U.S.)
JAERI	Japan Atomic Energy Research Institute
JDA	Japan Self-Defense Agency
JDC	Japan Documentation Center, Library of Congress (U.S.)
JEI	Japan Economic Institute (Japan)
JHLAP	Joint High Level Advisory Panel
JHLC	Joint High Level Committee
JOIS	Japan Online Information System
JSC	Japan Science Council
JSTI	Japanese Science and Technology Information
JTEC	Japanese Technology Evaluation Center, Loyola College, Baltimore, MD
JTECH	Japanese Technology (Science Applications International Corp)
JTF	Joint Task Force (U.S.)
JWLC	Joint Working Level Committee
LDP	Liberal Democratic Party of Japan
MAFF	Ministry of Agriculture, Forestry and Fisheries (Japan)
MHI	Mitsubishi Heavy Industries
MHW	Ministry of Health and Welfare (Japan)

MIT	Massachusetts Institute of Technology
MITI	Ministry of International Trade and Industry (Japan)
MOE	Ministry of Education, Science, Sports and Culture (Japan)
MOFA	Ministry of Foreign Affairs (Japan)
MOU	Memorandum of Understanding
MPT	Ministry of Post and Telecommunications (Japan)
MT	State of Montana
NAS	National Academy of Sciences
NASA	National Aeronautical and Space Administration (U.S.)
NASDA	National Aeronautics and Space Development Agency (Japan)
NBS	National Bureau of Standards (DOC, U.S.)
NCI	National Cancer Institute (Japan)
NEPL	National Public Employee Law (Japan)
NIH	National Institutes of Health (U.S.)
NIPR	National Institute of Polar Research (Japan)
NOAA	National Oceanic and Atmospheric Administration
NRC	National Research Council
NSF	National Science Foundation (U.S.)
NSC	National Security Council (U.S.)
NTIS	National Technical Information Service (U.S.)
NTT	Nippon Telegraph and Telegraph
OECD	Organization for Economic Cooperation and Development
OES	Bureau of Oceans, International Environmental and Scientific Affairs, Department of State (U.S.)
OMB	Office of Management and Budget (U.S.)
OPTI	Office of Production, Technology & Innovation (DOC, U.S.)
OSTP	Office of Science & Technology Policy (U.S.)
PRC	People's Republic of China
PSAC	President's Science Advisory Committee
R&D	Research and development
RERF	Radiation Effects Research Foundation (U.S./Japan)
S&Es	Scientists and Engineers
S&T	Science and Technology
S&TA	Science and Technology Agency (Japan)
SAC	Science Advisory Committee
SCAP	Supreme Commander for Allied Powers (Pacific)
SDI	Strategic Defense Initiative
SEATO	Southeast Asia Treaty Organization
SEMATECH	SEmiconductor MAnufacturing TECHnology, an R&D

	Consortium (U.S.)
SSC	Superconducting Supercollider
STA	Science and Technology Agreement
STAC	Scientific and Technical Administration Commission (Japan)
STH	Science, Technology and Health
TGH	Techno-Growth House
TFT	Technology-for-Technology (U.S./DOD Initiative)
TLG	Technology Liaison Group
TMDI	Theater Missile Defense Initiative (for Asia)
UJST	U.S.-Japan Science and Technology Agreement
USDA	U.S. Department of Agriculture
USG	U.S. Government
USJ	United States and Japan
USTR	U.S. Trade Representative
WWII	World War II
WTEC	World Technology Evaluation Center, Loyola College, Baltimore, MD
WVA	State of West Virginia

I. Introduction

The 1988 U.S.-Japan Science and Technology Cooperation Agreement (88STA) was hailed by the then science adviser to President Reagan, Dr. William Graham, as one of the major accomplishments of the Reagan Administration.[1] On the other hand, Glen Fukushima, a U.S. delegate who participanted in the process of creating this "achievement" described his experience as one of the worst, one of the most difficult.[2] While the agreement itself and the experience with creating it are different issues, it is worthwhile to note the highly divergent feelings about this Agreement.

The 88STA was indeed a departure from past patterns of science and technology (S&T) agreements, but whether a major accomplishment is another issue. It was a major "revision" -- a new agreement in reality -- of the then existing science and technology agreement between the U.S. and Japan signed by the U.S. President and the Prime Minister of Japan at a 1980 summit meeting. It was intended to put these relations on an even keel, on a basis of equality between the world's two major powers in science and technology and to set the stage as a model for negotiating similar agreements between the U.S. and other countries. It was the first major international S&T agreement concluded by the U.S. after the passage of significant federal legislation in 1986 on technology transfer and intellectual property rights.

There have been hundreds of articles on various aspects of Japanese S&T accomplishments, failures, potentialities, openness, obstacles which were looked upon variously as threats, challenges or just simply normality. There have also been a number of studies on bilateral negotiations concerning such thorny issues as textiles, automobiles, etc. There have not been many on U.S.-Japan S&T relations as a whole. Indeed there are none on how U.S. policy on U.S.-Japan S&T relations were formulated or on how such S&T agreements were negotiated between the two countries. It was for this reason, that this study was begun in 1988. The 88STA is a good example of how the S&T relations between the two countries have been affected by larger political and economic forces. This was one of the major factors influencing this entire process, in contrast to earlier less high level arrangements.

Over the past forty years, the U.S. and Japan have gradually developed probably the most intensive and extensive set of relations in science and technology between their respective governmental institutions of any two countries. In the beginning, the U.S. was interested in reviving

Japanese science and technology as part of the Free World. Gradually, the range and breadth of these relationships were increased to the point where, according to the 1987 report on *Science, Technology and American Diplomacy* "a new dimension in the bilateral relationship has emerged: the importance of the U.S. remaining abreast of Japanese S&T activities."[3]

Most of the S&T arrangements have been between U.S. Government Departments or Agencies and Japanese Ministries. The *public record* of these S&T relations has been excellent. This point cannot be over-emphasized. There have been, however, two agreements signed at the Presidential/Prime Ministerial level in 1980 and 1988. Although a 1961 cooperative science agreement stemmed directly from a 1961 summit meeting between the U.S. President and the Japanese Prime Minister, it was not signed by the principals. The 1961 program was the oldest of it's kind, and has been successfully executed to the satisfaction of both governments. So successfully that they have "served as a model for bilateral S&T agreements with many other countries."[4]

The 1980 agreement is quite generalized. It provides for the conduct of joint projects and programs, exchange of personnel, exchange of information through various kinds of meetings, information on policies, practices and legislation. In contrast, the 1988 Agreement sets forth in substantial detail the principles governing U.S.-Japan S&T cooperation, calls for both governments to provide "comparable access" to government-sponsored or government-supported research facilities and activities. The 1980 Science and Technology Agreement (STA) created a Joint Committee to oversee project activities; the 1988 STA created an elaborate management structure with a hierarchy of committees. It must be remembered that new U.S. and Japanese (USJ) S&T agreements in the postwar period were initiated by and at the behest of the USG (except one in 1979 on energy resources).

Because of my career experiences in the USG in the policymaking process, my intellectual interest in this process, and as my consulting centered on USJ S&T issues, and because there was no study on this subject, I began this study in 1988. The purpose was to find out through what process the two governments, USG and GOJ, arrived at a policy for bilateral S&T relations and how the bilateral negotiations were conducted and concluded. It was an intellectual adventure and search to find out what happened and why, unencumbered by an *a priori* overarching thesis or hypothesis. I started out with the idea of writing an academic journal article. But early in this endeavor, I was very fortunate to be given a substantial cache of unclassified USG documents – one could describe these documents as examples par excellence of "gray literature" – which immediately revealed the controversial nature of this topic at least inside the USG. It became utterly clear that the issue of future S&T relations centering around the 1980 STA "revision" was not simple and straightforward but loaded with semi-high drama involving emotional confrontation, suspicion, mutual distrust at high levels in the USG. This discovery gradually evolved into a book length manuscript and persuaded me to subtitle the study as "A Drama in Five Acts". That search for what occurred and why, became exciting and challenging. In explaining the process one could gloss over the micro aspects and present only the macro aspects, i.e. analyze the major new S&T agreement signed in 1988 by no less than the principals, the U.S. President and Prime Minister of Japan after some somewhat difficult negotiations and implemented without too much ado, and renewed not once but twice. Such a macro-oriented presentation would not reveal the raw emotions that surfaced inside the USG and between the USG and GOJ deletions. Only through a detailed exposition of the micro details of this process can one truly appreciate and properly gauge the depth of feelings which contributed, even though it was only one strand of many, to the increasingly long shadow over USJ relations. It is in laying out the mundane that we will learn how a goodly number of the participants felt. The 88STA was also a good example of how the S&T relations between the two countries have been affected by larger political and economic forces. This was one of the major factors influencing this entire process, in contrast to earlier less high level and less high profile USJ S&T arrangements.

It is hoped that this study will contribute to: 1) increasing our understanding of the bureaucratic process and negotiations within the U.S. and Japanese governments in drafting an agreement and the interaction of the negotiators in the outcome, 2) increasing our knowledge about how the U.S.-Japanese relationship in science and technology in the public sector is managed, 3) throwing some light on how domestic factors impact on preparing for and negotiating a new agreement between the U.S. and Japan on science and technology, 4) developing insights into the negotiating styles of each country, 5) assessing its role as a model agreement for negotiating similar agreements with other countries because most of the major issues were not related to "many unusual features" of the U.S.-Japanese S&T relationship, 6) learning some lessons for future negotiations with Japan in the science and technology area and with other countries if this Agreement is to be used as a model.

Research Methodology. The research methodology of this study is based on interviews and analysis of the relevant documents and articles augmented by an analysis of selected studies on U.S.-Japan S&T relations.

At this juncture, an important point must be emphasized: Based on my existing knowledge of the entire process, the part of this study on how the process was actually carried out, would be, in the first instance, an "insiders' or witnesses' story". From whom else can one obtain an interpretation of what transpired? Only insiders can provide an initial "witnesses's" presentation of what presumably transpired, what were the major issues, and what was the milieu. The term "witness" was used deliberately to suggest the witnesses in the movie *Rashōmon* where witnesses provided conflicting facts and interpretations of the same sets of developments. It must be remembered that there were many insider witnesses with several interpretations of what and why certain issues surfaced, insiders with many vigorously and tenaciously held beliefs, positions, determinations and suspicions. This researcher, as the "outsider", will act as the narrator and provide the analysis of what transpired.

Another point that must be emphasized is that the policymaking process and negotiations (the latter is normally a governmental function) were, in this instance, essentially concentrated in both governments with little input from industry or academe or from the U.S. Congress or the Japanese Diet.

1. Interviews. Starting in late spring 1988 through 1989, I interviewed officials in various departments and agencies of the U.S. government and a few members of the Embassy of Japan in Washington, DC. Interviews were resumed in 1990 through 1995 when I had opportunities to visit Tokyo. All in all, I interviewed over 50 people (listed in appendix 5) in Washington and Tokyo. The number of U.S. officials interviewed was much greater than that for the Japanese government. There was a much larger group of people involved in this policy debate about future U.S.-Japan S&T relations in the USG than there appeared to be in Tokyo. These interviews provided important information on interpreting the documents and helping to read between the lines of these documents. Since I was interviewing people about an event and experience which for almost all was not a pleasant recollection, I felt it necessary to take notes in confidence and not for attribution. While this is unfortunate from the scholarly standpoint this was unavoidable. Only a very few persons would not agree to meet with me. Of the very few who claimed this "privilege" of declining to meet with me were two science advisers to the U.S. President; they chose not to meet with me despite repeated requests and appeals.

Most officials, both U.S. and Japanese, were willing to meet with me to discuss their experience. Some officials in some U.S. departments, while they would meet with me, were basically not friendly, even antagonistic to my study -- Office of Science and Technology Policy (OSTP) and the U.S. Department of Commerce (DOC) for example -- and did not provide much information or proffer even unclassified materials; instead they explained that they could not, would not, reveal U.S. negotiating strategies which had led to the signing of the 1988 agreement. Some other officials provided me with important unclassified reports and other papers which were critical to the study. Through these interviews I learned of the bitterness, rancor and distrust that came to suffuse the atmosphere at USG interagency meetings. It was this atmosphere that Glen Fukushima was referring to in his book mentioned above. The depth of this suspicion among State, Commerce and OSTP was impressive and very sad to hear about. This antagonism was mirrored by a similar feeling which pervaded the U.S.-Japan negotiations between and among certain members of the U.S. delegation. The Japanese delegation was also particularly upset with the style and conduct of the negotiations by selected U.S. delegates.

The Japanese officials I interviewed were, in general, quite open and forthcoming. They were frank in their evaluations of both the Japanese and U.S. chairmen of the two delegations. Some volunteered information which threw further light on the processes in the GOJ. Two officials shared their personal notes of their negotiating experiences with some personal commentary on the process and on some delegates. These were useful in understanding some reported Japanese reactions to U.S. demands and expectations. Since I used a set of questions to ask each interviewee, I began to notice a similar pattern of reaction to certain parts of the U.S. proposal which they insisted Japan could not carry out even if it agreed to do so since the Japanese governmental arrangements would not permit this to happen. Because of the laws in Japan concerning disclosure of documents, all Japanese officials were meticulous in not giving me any documents, even unclassified materials.

2. Documentation. The second part of the research methodology was a study of documents provided by U.S. officials who were sympathetic to my study, and those obtained through the Freedom of Information Act.

The first group came from a variety of sources all of whom must remain anonymous. In this way, I feel that about 75% of the most important documents on the U.S. side were obtained early on in my study. These were augmented by materials received from the U.S. Department of State through a request made to the Department in late December 1988 in accordance with the Freedom of Information Act. This request produced between 2,000 and 3,000 pages of materials which, one might say, provided, in many instances, the glue or mortar binding together the various documents I had been given. Naturally, there were *probably* many justifiable deletions. I say probably justifiable because one does not know what has been deleted and therefore how justifiable these deletions actually were. I appealed the deletions, of course, to the State Department but, by definition, I did not know and could not know whether these deletions were justifiable or not. Some deletions were re-instated after review and released to me. As of this date, my FOIA request to the State Department for materials has been completed.

At the time my request was made, a 1989 GAO report on how Departments and Agencies were handling FOI requests described the Department of State's FOI program as "ill-managed and poorly prepared to handle the approximately 2,700 requests it receives each year. This number of requests is one of the lowest among federal agencies". I want to emphasize, however, that the staff in the FOIA office in the State Department was always courteous, helpful to the extent they could be, and patient with my queries and insistence on more meaningful reassessments and re-evaluations of deleted materials, although I must admit that little improvement has been noticeable in the amount of materials released. I did request the intervention of Congresswoman Connie Morella (Republican, Maryland) in this endeavor which resulted in some new releases.

Since my career was for a quarter century as a career civil servant in the U.S. Federal Government in military weapons system planning, military assistance policy and economic assistance, I am thoroughly aware of and appreciate the need for deletions for *genuine* privacy and *genuine* national security purposes before documents are released to the public. The FOIA allows a number of exemptions for releasing government documents. If the spirit of the FOIA is respected and fully carried out, then more materials could be released without compromising any privacy or national security issue. Even though years have elapsed since the 88STA was signed, even unclassified materials were subject to strange deletions. Courses of suggested USG actions are sometimes deleted and sometimes not deleted in the very same document. I have found several sets of the same documents all unclassified among the papers released but one set with large amounts of deletions in one, while the other was completely untouched.

Among the documents in the State Department files are some prepared by another agency. These were meticulously sent to that agency or department for their action and this involved almost all U.S government departments and agencies. In most cases, exemptions were invoked and precious little came from these other agencies. All appeals were turned down -- I suspect, without even a perfunctory review of the deleted documents. The other agencies and departments were obviously unsympathetic to the FOIA process. What few documents were released were of little significance.

While the problems with deletions that I have experienced were not crippling to the study, they are in stark contrast to the "special privileges given to ex-cabinet officers" when they leave office if they are inclined to write their memoirs. A General Accounting Office report in 1990 studied the release of highly classified materials to a few Cabinet officers under President Reagan, particularly Secretary of State George P. Shultz and Secretary of Defense Casper W. Weinberger, and found irregularities in the handling, i.e. release of government documents for their use in writing their memoirs. State Department released 75,000 documents (not pages, please note), of which 60,000 were classified and the top Secret documents were not even properly inventoried as required by the Department's own regulations. This led Senator David Pryor (Democrat, Arkansas) to comment:

> There appears to be an inverse relationship between the level one attains in the executive branch and one's obligation to comply with the law governing access to, and control of, classified information.[5]

The apparent discrepancy between the leniency with which a former Secretary of State is provided documents to write his memoirs and the ability of the ordinary researcher like myself who has to struggle to obtain unscathed even unclassified materials, let alone classified materials, is blatant and unfair.

While the researcher can obtain some documents from the USG by using the FOIA process, the GOJ only recently on May 7, 1999 enacted the Access to Government Information Act; it will become operative in about two years, too late, of course, for this study. This law affecting the national government was enacted after more than 20 years of domestic debate and 17 years after local prefectural governments started adopting similar laws; all 47 prefectures have had such information access laws for many years.[6]

As I mentioned earlier, the Japanese government officials with whom I met were open, helpful and courteous but they were clearly constrained by a 1978 Japanese Supreme Court decision. A newspaper reporter obtained secret cables from a lady friend in the Ministry of Foreign Affairs concerning the 1972 U.S.-Japan negotiations on the reversion of Okinawa and Japan's secret agreement to pay $5 million to compensate Okinawans in land damage claims, despite the government's claim not to have made any secret arrangements. Both parties were convicted of violating the National Public Employees Law (NEPL), article 100(1) which provides that "public employees shall not leak secrets that become known to them through their official duties." The Supreme Court held that 1) the courts have the authority to determine what constitutes a state secret under the NEPL and what is merely a political secret; 2) the government's secrecy when negotiating in this case was appropriate; 3) the government's failure to bring the facts before the Diet in this case did not violate the constitutional order or constitute illegal secrecy; and 4) the defendant violated the prohibition against inducement in his ethically questionable relations with the supplier of the materials (both were married) though free news gathering and reporting were of great importance to the people's right to know and freedom of expression. Japanese scholars sided with the defendants but the defendant's own newspaper, *The Mainichi Shimbun*, did not back its own reporter who was the defendant and remained silent despite their avowed assertion of the people's right to know.[7]

There is thus a strong incentive for Japanese government officials not to provide any materials to outsiders whoever they may be. There is, however, still a strong tradition among senior civil servants to take home collections of papers on their major accomplishments especially by those who have aspirations for writing their memoirs at some time. In conclusion, there is a built-in effective deterrent to the release of Japanese government documents.

Structure of the Study. It was basically written in a chronological fashion. Despite the possible tediousness, short accounts are inserted periodically about discussions with the GOJ, about Circular 175 authority actions, etc., in order to provide a sense of how events evolved or developed in a bilateral fashion as the future of USJ S&T relations are discussed within the USG and with the GOJ. The short fleeting accounts about meetings with GOJ officials indicate the degree to which the Japanese were basically passive participants -- if not onlookers peering through a frosted glass window on a murky and indecipherable image of U.S. officials and institutions struggling, pondering back and forth about S&T relations with Japan. The Circular 175 authority, normally a cut and dried bureaucratic process, to assure that the various agencies of the USG are agreed upon a course of action -- the U.S. equivalent of the Japanese *ringisei* -- sometimes revealed nuggets of interpretations and occasionally leads to a major intra-agency collision.

While there is a substantial body of literature on public policy decision-making processes and international negotiations, only occasional references are made to a few of these studies. When this writer negotiated agreements with a foreign government during his Federal career, academic studies as to how such negotiations should be or were conducted were not read nor considered; there is simply no opportunity nor time to do so. The immediate task at hand was to negotiate these agreements; the best approach in line with our mutual interests was devised, and an assessment of relative advantage was made in deciding on the degree of insistence we would hold to in adhering to our terms and conditions.

This study concludes with simple yet fundamental recommendations which Japan and the U.S. could follow to their mutual interest. In light of my own governmental experiences, I feel that U.S. or Japanese officials will not have the psychological and real time leeway to devote much time to the findings of this study. My objective was to re-create and throw some light on the internal USG policymaking processes and the conduct of bilateral negotiations.

As the decision-making process moved forward in the USG and as the bilateral negotiations inched forward toward an agreement, it devolved into plodding bureaucratic trench warfare. In this instance, "the truth" of the U.S.-Japan science and technology relationship in concept and management, lies in *studying the truly grubby, boring but often interesting details* of the numerous ideas and proposals, strategies and tactics described in documents, studies, cables and memoranda. The U.S. chairman told me that "we slogged it through to a new agreement". In a similar fashion and with equally dogged determination, we shall slog it through in this narrative to an agreement , it's implementation, renewal and a second renewal.

This study was conducted without financial support from either the governments of the United States or that of Japan, from any Foundation or research center. The U.S. Department of State did indirectly contribute to making this study possible by not charging me for the search and copying of materials supplied to me under FOI. The views, interpretations and mistakes in this study are, therefore, uniquely mine.

I wish to express my great appreciation and thanks to all the people who kindly agreed to meet with me – they are listed in appendix 7 – to share and discuss their experiences, knowledge and insights about the processes in the USG and the GOJ and in the bilateral negotiations about the USJ science and technology agreement and, in addition, many others who patiently responded to my many requests for materials, e.g. the staff in the Legal Office and the FOI offices in the U.S. Department of State. Then there is also, most importantly, the endless amount of patience and support given me by my wife, Allie Marie, throughout the ten plus years of research and writing and for putting up with my periodic crankiness in the writing process and with the seemingly interminable struggles with the computer.

Notes

1. Interview with Dr. Graham, 1988.
2. Glen Fukushima, *Nichi-Bei Keizai Masatsu no Seijigaku* (The politics of U.S.-Japan Economic Friction). Asahi Shimbunsha. 1992. p. 278. Fukushima was a member of the U.S. Delegation (from the U.S. Trade Representative's Office) which negotiated the 1998 STA with the Government of Japan.
3. Joint Committee Report of the House Committees on Science, Space and Technology and Foreign Affairs. U.S. House of Representatives. July 1987. p. 27. This year was chosen since that was the year that negotiations began between the U.S. and Japan concerning future S&T relations.
4. ibid. p. 28.
5. *The Washington Post*, 11/14/1990.
6. *The Nikkei Weekly* 5/17/1999, *Yomiuri* 5/8/1999, 5/9/1999, 5/11/1999.
7. John O. Haley, ed., *Law and Society in Contemporary Japan*. Dubuque, IA. Kendall-Hunt. 1988. p. 21-22; Lawrence Beer, *Law and Contemporary Problems*, 53:2 (Winter/Spring 1990): 56-59.

II. U.S.-Japan Postwar Science and Technology Relations

Since this study focuses on the science and technology (S&T) relations between the U.S. and Japan in the public sector, this chapter correspondingly will provide, in the main, a brief history emphasizing the interactions between the S&T public sectors of the two countries. While these interactions are not, of course, completely independent of S&T developments in the private sectors in both countries, they can be safely described independently without at the same time providing a detailed history of postwar Japanese S&T policies and development. It will begin with a summary of the early postwar expressions by Japanese leaders about the role they thought S&T should play in the reconstruction of their country. This will be followed by the first steps in restructuring postwar Japanese S&T together with and under the guidance of the U.S. Occupation, two cases of specialized USJ S&T relations, and concluded with an account of the formal USJ S&T relations up to the mid-1980s.[1]

1. Science and Technology, the Foundation for Reconstruction

"Starting from Zero" figuratively describes the point from whence in 1945 began the physical re-building of the Japanese science and technology enterprise to a technological superpower by 1990 -- much less than fifty years.[2] Yet due to a confluence of circumstances in the early postwar years, one could say that Japan was able to vigorously push her traditional and long standing policy of a special emphasis on science and technology development.

Perhaps because two atomic bombs, then the epitome of scientific achievement, had been dropped on Japan in the summer 1945, a series of statements, assessments, admonitions and exhortations to the Japanese people came in quick succession from the political and scientific leadership of the country to "build the nation [Japan] through science and technology" (*Kagaku Gijutsu Rikkoku*) between August 1945 and mid-1946. The purpose in citing these statements is to provide an example of the kind of thinking that prevailed among the top Japanese civilian leadership when the War ended, a clear symbol of their vision of the principal means for the reconstruction of Japan.

- August 15, 1945, evening. Emperor Hirohito had made a personal broadcast at noon ending the war. The then Prime Minister Suzuki Kantarō declared that:

> ... It is essential that the people should cultivate a new life spirit of self-reliance, creativity and diligence in order to begin the building of a new Japan, and in particular *should* (emphasis added) strive for the progress of science and technology which were our greatest deficiency in this war.

- August 16, 1945. Prime Minister Prince Higashikuni Naruhiko: S&T had been Japan's biggest shortcoming in the war.

- August 17 or 18, 1945. The outgoing Minister of Education, Maeda Tamon, thanked school children for their wartime efforts and urged them to dedicate themselves to elevating Japan's science and spiritual power to the highest possible levels, and urged that basic science be strongly emphasized together with sharpening the thinking ability of Japanese.

- August 20, 1945. The *Asahi Shimbun* declared that "we lost to the enemy's science."

- August 20, 1945. In the *Yomiuri Hōchi Shimbun*, Dr. Yoshio Nishina, a physicist and later very close friend of Harry Kelly, a physicist with the Occupation, declared that Japan must reassess its education of gifted children and called for the use of S&T in the rebuilding of Japan.

- the latter half of 1946. The Kagaku Gijutsu Seisaku Dōshikai (S&T Policy Association) under Dr. Yagi Hidetsugu's guidance, was created in June 1946. Dr. Yagi had been the President of the wartime Gijutsu-in (Technology Institute). In July 1946 a 70-member parliamentary group for S&T was created, met weekly and was able to pass a Diet resolution calling for the revitalization of S&T. A cabinet-level group was created to finalize the Yagi S&T plan for Japan but it was put on hold by the Occupation as being an inadequate plan -- probably more likely that the Occupation was supporting an alternative plan being created by a different set of scientists. Because of his wartime service, Yagi was purged from public life at the end of 1946. Yagi was well known for his original scientific work with antennas.

- soon after August 1945. Dr. Tomizuka Kiyoshi, professor of aeronautics at the University of Tokyo during the war, declared that since Japan can no longer rely on heavy industry it must turn, like Switzerland, to such precision industries as watches and clocks.

- 1946. The Ministry of Foreign Affairs. A special committee issued a report, *Nihon Keizei Saiken no Kihon Mondai* (The fundamental issues concerning the economic reconstruction of Japan): without democratization it will be difficult to achieve technological advance, and at the same time without technological advance we will be unable to achieve true economic democratization.[3]

Thus, there was obviously a strong expression of policy expectation for Japan to rebuild itself from the ashes of war through science and technology. This emphasis, however, was not new. According to one of Japan's outstanding science historians, Nakayama Shigeru, Japan was the first -- even ahead of the European powers -- in the 19th century to use militarily tilted (*gunji henkō*) science and technology as national policy through technology transfer and development of technology. This policy emphasis was obviously still further emphasized in the 20th century leading to the war in Asia. It was with this kind of philosophical approach to the revival of Japan that a basically new group of Japanese scientists encountered the Occupation, each with their anxieties and fears, expectations and obligations.

2. The U.S. Occupation and Japanese S&T Restructuring

The history of U.S.-Japan science and technology relationships since 1945 naturally begins with the Occupation of Japan. The initial objectives of the U.S. occupation of Japan were to demilitarize and democratize Japan. The administrative structure of the Occupation included numerous sections including one called Economic and Scientific Section (ESS), the largest in General MacArthur's Headquarters. In the beginning many of these sections were manned by military officers who may or may not have had knowledge or experience in the fields for which the various sections were responsible. There was a Scientific and Technical Division in ESS but it appears not to have had scientifically trained personnel at least in the early part of the Occupation. Soon after the beginning of the Occupation, this led to a major criticism of ESS.

The initial Post-Surrender Policy stated that "specialized research and instruction directed to the development of war-making power are to be prohibited". Then in November 1945 a more specific military directive instructed the occupation authorities to "insure that all laboratories, research institutes and similar technological organizations are closed immediately except those you [General MacArthur] deem necessary to the purposes of the occupation. ... You will provide for the maintenance and security of physical facilities thereof when you deem necessary, and the detention of such as are of interest to your technological or counter-intelligence investigations. You will at once investigate the character of the study and research that have an obviously peaceful purpose under appropriate regulations which 1) define the specific type of research permitted, 2) provide for frequent inspection, 3) require free disclosure to you of the results of the research, and 4) impose severe penalties, including permanent closure of the offending institution whenever the regulations are violated." The Occupation was particularly concerned with the status of Japanese atomic and nuclear research. For example, the Physical and Chemical Research Institute (*Rikagaku Kenkyūjo*, known popularly as RIKEN) had built a 23 ton cyclotron, the first to be built and operated successfully outside the U.S. before 1941. Based on the recommendation of the first postwar Scientific Intelligence Survey of Japan in September 1945 GHQ authorized the use of the cyclotron for experiments in biology. This permission was suddenly revoked just as the scientists were preparing the cyclotrons for experiments. Without proper knowledge about these cyclotrons and with eagerness to carry out Washington's directive, GHQ ordered that five of these cyclotrons be dismantled and destroyed.

This action prompted a substantial criticism of MacArthur's GHQ for this precipitous action, especially by U.S. scientists. This was immediately elevated from a mere error by the Occupational military authorities to the relation of scientists to society, to the military, especially in the new age of the atomic bomb and the internationalism of science in the new peacetime era.

It became obvious that GHQ needed a more coherent policy for science and technology on how to approach and dismantle military related R&D as distinct from R&D for civilian purposes and for the possible reconstruction of Japanese science and technology. Because of this urgent need to augment and strengthen the ESS with scientifically trained personnel, Gerald

Fox and Harry Kelly were recruited to go to Japan. They arrived in Japan in early January 1946 with the intention of staying for a short period, around three months and principally for intelligence gathering on Japanese S&T activities especially as these activities related to atomic research and military R&D. Fox stayed only eight months and returned to the U.S.; Kelly stayed on for four years and played a substantial and unusual role in laying the foundation for and reconstruction of the Japanese S&T establishment in the postwar years.[4]

Kelly was a physicist who had worked on defensive radar at the Radiation Laboratory and did not know about the Manhattan Project. He was born in 1908 in Wilkes-Barre, Pennsylvania and obtained his PhD from MIT in physics in 1936. Apparently for Kelly the loss of the cyclotron was not the core issue, rather the opportunity it provided for a dialogue over the role of science in society. He believed that science would play a major role in the reconstruction of Japan. When he arrived in Tokyo he was assigned to the Fundamental Research unit as an adviser. Since GHQ was not prepared to handle S&T issues, the question was what to do after their arrival. Kelly first issued a directive to the Japanese government for all reports, whether classified or not, on Japanese S&T; he was inundated! Based on his meetings with Japanese scientists and engineers, he concluded that the Japanese scientific enterprise did not pose any immediate military threat and that he would have to convince his Japanese counterparts about his credibility -- and also his superiors in the U.S. military chain of command -- and his motivations to help them in reconstructing Japanese science and technology. He chose to work with the Japanese on a "self-policing basis" and his confidence and trust was not violated by the Japanese.

This new relationship began in the frozen northern most island of Japan, Hokkaidō, in the dreary mid-winter months of early 1946. Kelly visited Hokkaidō University on the suspicion that some militaristic death-ray research had been conducted or was being conducted at the Microwave Laboratory. Initially he found nothing but was skeptical of the condition of the labs. He suggested that someone talk to him before he left; someone did and he was led to an obscure school in a village miles away with the missing equipment. But his gamble and the risk taken by a small delegation of Japanese scientists started a chain of events and the building of a confidence between this Occupation official and the Japanese scientific community. After revealing the existence of the equipment which was apparently effective only for 20 feet at most both discussed the future of Japanese science and its need for reconstruction and restructuring for its new role in postwar

Japan and realized they were thinking along similar wave lengths about future possibilities. In the meantime, a snow storm hit, the trains could not move; Kelly was "stranded" in a remote village with his new found Japanese scientists. They tried to reach the nearest train station from which trains were operating with two horse drawn sleighs with blankets and some food. But the sleighs overturned with Kelly and one of the Japanese scientists thrown into the swirling snow! The Japanese side asked Kelly to have U.S. troops removed from their labs; Kelly accepted. When he succeeded Japanese confidence in Kelly rose several notches. Thus the expedition which began as an intelligence gathering operation, through risk-taking on both sides, candor about the future and its possibilities, Kelly's philosophical disposition about the role of science in society, their common experiences in the snow blizzard had turned into a positive turning point for Kelly and the Japanese scientific community.

The subsequent dialogue continued in Tokyo between Kelly and various Japanese scientists about the reorganization of Japanese science. But Kelly, of course, had to work through the Japanese government which controlled the allocation of resources for science in education and R&D in general, and the more energetic, younger Japanese scientists who were interested in a wholesale re-shaping, re-structuring, re-assessment of Japan's scientific establishment. The existing Japanese scientific structure was regarded as an impediment, with an over-emphasis on basic research and with no one agency responsible for coordinating the whole scientific endeavor. By mid--July 1946, the Japan Association for Science Liaison had, through Kelly, sent a letter to the U.S. National Research Council about the state and future of Japanese science. Thus by summer, barely six months after Kelly had arrived for intelligence gathering and surveillance, his role now had evolved into one of mediation and friendly guidance toward the reconstruction of Japanese science and technology, and playing a major role in the country's economic recovery and possible contributions to human welfare and *to guide the Japanese shift from basic research to applied R&D to achieve this reconstruction*. It should be noted this change of direction antedated the so-called "reverse course" by almost one and a half years when the USG decided in January 1948 that it would emphasize the economic restructuring of Japan. Note an unusual confluence of factors in these human affairs, the eagerness of the involved Japanese S&Es to build a NEW S&T structure for a NEW Japan, the temperament, philosophy, inclinations, determination of one Occupation official who transcended his narrowly defined responsibilities in the Occupation hierarchy and bureaucracy and perhaps most importantly the trust that was created, fostered and apparently never violated on both sides, was critical in this vastly important reconstruction and restructuring endeavor.

There was a continual jockeying for power between the old guard Japanese S&T groups, and the younger scientists who were helped and assisted in various ways by the Occupation which was encouraging the Japanese S&Es to create their own plan of renewal. Eventually, a younger more energetic 108 member "Science Organization Renewal Committee" was created by late summer 1947. At about the same period, the "U.S. Scientific Advisory Group" spent six weeks traveling to many Japanese cities. While no specific plan for reorganization came out of this visit, their presence at the inauguration of the above-mentioned Renewal Committee lent prestige and support to the reconstruction efforts of this Committee and that of the Liaison committee. The validity of these efforts were further enhanced when the Renewal Committee met with Prime Minister Katayama Tetsu; the head of GHQ's S&T Division, and the chairman of the U.S. Advisory Group who also attended this auspicious occasion. The latter stressed the important link between science and industry and that through the cooperation of scientists with industry much could be done toward achieving a balanced economy. The Renewal Committee deliberated for seven months, prepared a report for the Prime Minister recommending the creation of a new Science Council of Japan and the Scientific and Technical Administration Commission (STAC) within the Prime Minister's Office; these two were created by law in June and December 1948, respectively. The Agency of Industrial Science and Technology was created on August 1, 1948 in the Ministry of Commerce and Industry (later reorganized into the Ministry of International Trade and Industry on May 25, 1949) and became the leading governmental facility in implementing the new policy of emphasizing practical R&D in contrast to basic research. The Ministry of Education continued to govern the distribution of government funds for scientific research at universities. Thus, by the end of 1948 a new institutional start had begun for Japanese science and technology.[5]

In the first two years of this bright and sunny beginning for the Japan Science Council it received about one inquiry per month from the government and offered 71 analyses of government issues between January 1949 and mid-1951; by 1955 this rate dropped to three per year, then down to one to two a year. The need for the Council was raised; it survived primarily as a vestige. This was probably due to the politicization of the Council by the leftwing and its confrontation with a conservative government; it could not compete with the rise of the strong bureaucracies which had been given the task of emphasizing industrial research in contrast to less emphasis on basic research in universities under the guidance of the Ministry of Education and Prime Minister Yoshida Shigeru, who himself seemed disinclined to increase research support for the MOE. Its counterpart coordinating body, the Scientific and Technical Administration Commission, created in March 1949, also did not fare well in the bureaucratic turf competition although it was chaired by the Prime Minister. It was a consultative body of 26, half from the government and half from the Science Council. In its first year, its major accomplishment is described as persuading government agencies that for the sake of Japan's economic recovery R&D resources should be shifted to R&D that will have more immediate payoff and benefits -- although this felt need had been recognized much earlier, this time, it now had official blessing and imprimatur to proceed full speed ahead in this direction. Because these institutions had not lived up to expectations, soon after the Occupation ended in 1952, the Government created by law the Science and Technology Agency (STA) in 1956 to promote, in the main basic research, and the Council for Science and Technology in 1959 chaired by the Prime Minister to establish the fundamental S&T policies to be followed by the Government. These agencies have continued to this day and have played constructive roles in the government. The S&T Council, for example, does indeed establish S&T policies which are gradually over the years reflected in changing shifts in S&T emphases and directions and in revised budgetary allocations.[6]

In addition to the role Kelly played in institutional building, he also had a persuasive role with GHQ in obtaining its approval to import radioisotopes to Japan for research purposes and a truly crucial role in convincing the group in GHQ's ESS -- which included Kelly's Science and Technology Division -- responsible for breaking up economic concentrations and holding companies not to dismantle RIKEN. The isotopes were important in reviving Japanese R&D in medicine, agriculture, and technological development.[7] Kelly was also persuasive in dissuading GHQ from dismantling RIKEN. As a result, RIKEN would play an important role in the revival of Japan's R&D and Japan's economic reconstruction, RIKEN's re-creation and re-establishment was approved by GHQ in April 1947 and in line with the prewar's successful endeavor embarked on manufacturing penicillin. Dr. Nishina Yoshio became the Director of RIKEN in its new reincarnation. Nishina and Kelly, both physicists, had met soon after the latter arrived in Japan in early 1946. Their personal chemistries immediately created a close working relationship. After four years in Japan, Kelly and family left for the U.S. in early 1950. In appreciation of Kelly's role in laying the foundations for the revival of Japanese S&T, he was awarded the Second Class of the Order of the Sacred Treasure, the highest awarded to foreigners.[8]

At the 25th anniversary (October 1974) of the establishment of the Japan Science Council, Kelly was invited to address the Council. It is instructive to select some passages from these remarks as indicative of his approach and assessment of Japanese S&T.

> The presence of civilian science advisers to the military was due to both guilt and fear on the part of the United States. There was a deep feeling of guilt for the use of atomic bombs at Hiroshima and Nagasaki ... Our fear resulted from our knowledge of Japanese scientific capabilities in nuclear research and our ignorance of how far the Japanese had progressed in developing these capabilities. Such were the reasons we first came together -- guilt and fear. ... It was particularly important that there were a large number of underlying convictions and postulations which were tacitly accepted by all of us. Perhaps most important of these was that the search for truth could be tampered with only at great peril. ... Japan, with its great economic difficulties, needed its scholars more than ever in identifying and overcoming obstacles to recovery. Japanese scholars were world leaders in the basic sciences, such as mathematics, genetics, and theoretical physics; but they were inactive in most applied fields. ... [We] all agreed that the present [1946-48 S&T] organizational structure needed a critical review and that it was a most opportune time for change; but all equally urged that whatever was done should be done solely by the Japanese. ... The first stage of the occupation was kind of an emergency period when Japanese scholars and American advisers learned to work together in complying with occupation directives and, more importantly in trying to get back to normal. The second stage began with the discussion or organization led to the formation of the Japan Science Council. ... All Japanese scholars who participated in the early stages ... showed great patience, vigor, wisdom and willingness to cooperate.[9]

While the Japanese were continuously admonished by their leaders immediately after the war ended in defeat that science and technology must be harnessed for the recovery of Japan, they did not know, of course, how the U.S. Occupation authorities would react to the use of science and technology in the reconstruction of Japan, if Japanese S&T facilities and personnel would be permitted to continue their research work even for peaceful purposes. The initial signs were indeed ominous: public discussion of the atomic bombing of Hiroshima and Nagasaki, the societal consequences of these bombings, economically, medically, socially and politically, were prohibited, and the dumping of RIKEN's cyclotron in the Bay of Tokyo in November 1945. When the Fox-Kelly combo of science advisers arrived at GHQ in January 1946 with the task of S&T intelligence gathering for an unknown purpose from the Japanese point of view this must not have been particularly reassuring. They, of course, did not know of Kelly's humanistic approach to the fostering of S&T and the potential contributions S&T could make to the enhancement of society and the quality of living. Kelly, himself, in light of his major assignment of intelligence gathering, probably was not conscious of how his sojourn in Japan as part of the Occupation bureaucracy would evolve.

An unusual confluence of circumstances in the first half of 1946 – at least two years BEFORE the so-called "reverse course" policy of the USG to emphasize the economic reconstruction of Japan was inaugurated – led to an unusual dialogue between an official of the victorious Occupation Headquarters and an accidental gathering of scientists of defeated Japan about the role of S&T in the reconstruction of Japan and the intelligence gathering function faded into invisibility. This dialogue was continuous and intense, gradually expanding to involve all segments of the Japanese S&T establishment and its S&Es. While Kelly maintained that the future institutional arrangements for Japanese S&T must emanate from the Japanese, his responsibilities as an Occupation official and his own humanistic philosophy of science would naturally assert themselves as "suggestions" which many of the Japanese S&Es involved in the restructuring effort wholly supported and endorsed. For example, Kelly, in striving for the democratization of science, aimed for an organization where all S&Es through representation could have a voice in future Japanese S&T and that for the future reconstruction of Japan, a much greater emphasis would need to be placed on applied R&D in contrast to the more traditional emphasis placed on basic fundamental research practiced by Japanese scientists in the past. To the Japanese scientists Kelly's trend of thinking must have appeared as a vindication of the emphasis placed on science and technology by the Japanese government immediately after the war. Since Kelly did not allude to or mention, at least in any prominent manner, the early Japanese governmental admonitions to the people to seek revival in S&T, he probably was unaware of this acceptance of science by the Japanese as a major means of salvation for Japan.

The extensive dialogue resulted in the creation of the Japan Science Council with its complicated election process, the Scientific and Technical Administration Commission, an S&T coordinating body in the Prime Minister's Office and the reorganized Japan Society for the Promotion of Science. While the Science Council was intended to be a forum through which S&E opinions could be reflected in government S&T policy, due to circumstances described above it soon lost its influence among the high political councils. The coordinating Commission also could not compete with other S&T governmental bureaucracies controlling the necessary financial resources. It evolved into the Science and Technology Council also chaired by the Prime Minister to establish basic S&T policies for the nation. While the institutional legacies from this period cannot be described as great long lasting successes, they were major achievements at a time of great chaos, uncertainty and dubious hopes in Japan about the future. In light of future developments, the outstanding phenomenon was that an unusual cohesion took place among the Japanese and U.S. players. In late 20th century terms, one might say that through this dialogue and mutual experiences "a strong bonding" took place between Kelly and the more energetic and forward looking Japanese S&Es. An uncommon trust, confidence and friendship had developed between Kelly -- and many other U.S. scientists who had participated in the process by being members of the two U.S. Science Advisory Missions to Japan in 1947 and 1948 -- and his Japanese S&E compatriots, one which apparently was not violated. One might wish to dismiss that as an opportunistic Japanese move of the defeated to utilize a sympathetic ear of a U.S. official who tended to think in their direction. But based on the available evidence, this was a relationship that transcended the outward trappings of those extraordinary times.

The JSC was manipulated by an energetic, dedicated and determined group of left leaning scientists, who apparently followed the lead, in the main, of the Japanese Communist Party. This dedicated core guided the JSC in opposition to the conservative government and politicized S&T policy. While initially the Conservative government under Yoshida Shigeru was not a stalwart supporter of funding science and technology, just when the left-wing dominated JSC was demanding greater governmental attention to science and technology. Ironically, the politicization of S&T did not take hold among and supported by S&Es, it was instead the technical agencies of the Japanese government, the MOE, MITI, S&TA, which later pushed for expanded commitments to the sciences and gradually eroded the power of the leftist group in the JSC to irrelevancy.[10] Thus, the cornerstone of Kelly's legacy, the JSC, was deprived of its power which slipped to the technical agencies of the Japanese government.

3. The Atomic Bomb Casualty Commission and Nuclear Energy

The S&T relations symbolized by these two cases are rather specialized and unusual; one stems from the dropping of the atomic bombs on Hiroshima and Nagasaki and the other from Japan's decision to rely on nuclear energy to develop a semblance of energy independence.

a. The Atomic Bomb Casualty Commission (ABCC)

President Truman created by executive order in October 1945, barely two months after the atomic bombs were dropped, the ABCC to investigate the biological effects of the bombings together with Japanese doctors, scientists, and related personnel. Because of the exemplary cooperation between the U.S. and Japanese personnel, about a year later this commission was transformed into a joint ABCC and a Japanese doctor from the University of Tokyo was appointed one of four chief scientists for ABCC. The ABCC was funded by the U.S. Atomic Energy Commission. As many as 50 U.S. scientists resided in Japan, and dozens of Japanese doctors, scientists, pathologists, and other medical personnel worked for the ABCC. This was an early example of the ability of the victors and defeated to work closely and cooperatively so soon after the end of a vicious, desperate and racial war in a project which hopefully would tell the world the after-effects of dropping an atomic bomb.

The ABCC was by agreement re-created into a binational organization, Radiation Effects Research Foundation (RERF) funded jointly and equally by the two governments. Management was provided by the Japanese Ministry of Health and Welfare and the U.S. National Academy of Sciences. Although the RERF had administrative and at times political issues to contend with, it conducted high quality and objective research and has continued its activities to this day, just over half a century.

The ABCC/RERF relationship is the oldest S&T project between the U.S. and Japan but it is a rather unusual program emanating directly from the last throes of the Pacific War. Despite its administrative and occasional political difficulties it is yet another example of model S&T cooperative efforts by both sides for a study with worldwide implications. Notwithstanding its scientific contributions it is hardly ever mentioned as part of USJ S&T relations. This oversight probably stems from its special character and is not regarded as part of the mainstream and conventional R&D cooperation.

b. Nuclear Energy

Japan was the first country under President Eisenhower's Atoms for Peace Program to sign an agreement in 1958 with the U.S. for cooperation in the peaceful uses of nuclear energy.[11] This was the beginning of a long standing lucrative relationship (in the billions of dollars) for U.S. companies who sold equipment and technology for nuclear power plants, transfer of fissionable materials to Japan, Japanese participation in and investment of at least $150 million in U.S. nuclear R&D programs -- a fact that is little known; it is also the largest foreign contribution for this purpose. Japan also paid substantial license fees, and substantial monies to the AEC and successor agencies primarily for purchase of nuclear fuel services, especially isotopic enrichment of uranium. As a result, with over 20+ nuclear power plants in Japan, providing one-third of its electricity needs, it has developed, despite restrictions placed on it by the U.S., one of the world's most advanced nuclear power programs.

The U.S. and Japan expanded their nuclear cooperation in 1969 (and extended it in 1979 for another ten years) to the fast breeder reactor field, in 1973 to nuclear safety issues and to business arrangements in medical uses of radio-isotopes. Despite numerous opportunities, the Japanese have not tried to sell their nuclear power technology and know-how to potential world recipients.

In order to avoid the appearance of giving Japan preferential treatment (stemming from Japan being a low-risk non-proliferation country) and the fear of setting a precedent which would apply to less-reliable nation-states, the U.S. tried repeatedly to impose restrictions on Japanese nuclear reprocessing in the context of nuclear non-proliferation and, in the process, tensions rose between the two countries on these issues. The U.S. refused to transfer technology to Japan for reprocessing spent fuel from nuclear power plants; Japan then turned to France to build a pilot plant in Japan. But when it was ready the U.S. still refused to allow spent fuel from reactors of U.S. origin to be reprocessed in this plant because it maintained that Japan could not safeguard against the diversion of plutonium from the plant. In the mid-1980s this major fracas still persisted. The diplomatic compromise was to allow the French contractor to process some fuel to fulfill its contractual obligations. Fuel has continued to be re-processed in Japan but Japan does not have control over its own plant. Japan agreed that the construction of a full scale reprocessing plant would not take place without USG approval; that approval had not been given by the mid-1980s and remained a contentious issue. But this problem did have a brighter side; it resulted in a new collaboration between the U.S. and Japan with France and the International Atomic Energy Agency on collaborative research on nuclear safeguards.[12]

The USG by law could not provide Japan with uranium enrichment technology since it is classified. That prohibition may be effective with less developed countries but not with Japan. It decided to develop its own indigenous enriched uranium technology without foreign assistance, thus negating USG's own restrictions. This kind of circumvention of USG restrictions probably will arise again.

Through their nuclear power plant investment and R&D in these fields, Japan has been able to develop its capabilities to the point of independence and little reliance on U.S. technology and materials. The U.S. tried unsuccessfully to impose certain proliferation restrictions on Japan but did not succeed. Japan gained enormously from this nuclear technology collaboration but the USG and U.S. companies also derived huge sums of payments for the technology, equipment and personnel provided. While from the overall point of view the nuclear S&T relationship was a distinctive plus for both sides, the USG has problems on its side with trying to enforce its strictures on scientifically and technologically developed countries like Japan. Their fluctuating policies with the recurrent changing of Administrations will, undoubtedly, raise questions in the minds of potential S&T collaborators about S&T cooperation with the U.S. in the nuclear field. In order for the bilateral program to expand further, the U.S. will have to reach a political decision to participate in Japanese nuclear power programs. Depending on U.S. policies and attitudes, Japan may feel it has no alternative but to go it alone or with other advanced nations in further development of nuclear technology. The value of the nuclear arrangement "is inestimable. ... It represents the strongest nuclear relationship that the United States enjoys with any country." (Bloom 3:36)

4. "Conventional" USJ S&T Collaboration since 1961

This period begins with the Summit Meeting in June 1961 between the newly elected President John F. Kennedy and Prime Minister Ikeda Hayato. Their communique said in part:

> The President and the Prime Minister also recognized the importance of broadening educational, cultural, and scientific exchanges between the two countries. They therefore agreed to form two United States-Japan committees, one to study expanded cultural and educational cooperation between the two countries, and the other to seek ways to strengthen scientific cooperation.

There is more than a little irony in this statement. This short statement resulted in one of the longest running S&T cooperation programs in the basic sciences between the U.S. which prides itself in the postwar era of its basic science achievements and contributions, and Japan which had a reputation for not being creative in the basic sciences. It was not a communique for convenience and window dressing. The subsequent USJ S&T cooperation program in the basic sciences became a model for other countries to follow and came to be regarded by the U.S. National Science Foundation as one its most important bilateral relationships in scope, depth and breadth. It has now been in effect for almost 40 years, the longest time period for any such arrangement between the U.S. and any other country. This subsection will trace briefly the S&T relationship between the USG and Japanese agencies.

Over the next twenty years, the USJ S&T relationship blossomed into almost every technical field: basic sciences, medical sciences, environmental protection, outer space, energy R&D, natural resources, transportation, building technology, urban affairs. It became the most intensive, broadest and deepest of any two countries in the post-1945 era. Interestingly, during this period, the initiative for specific projects basically came from the U.S. side. In addition to the extensive basic science cooperation managed by the NSF on the U.S. side, there were approximately 50 R&D projects by the mid-1980s.

The scope of the S&T relationship can also be gauged by listing the various bilateral cooperative R&D agreements in chronological order to the early 1980s:

- 1961 Cooperative Science Program.

- 1964 Natural Resources: diving physiology, marine electronics and communications, marine facilities, marine geology, marine mining, sea-bottom surveys, conservation, recreation and parks panel, earthquake prediction, fire research, forage germ plasm exchange and evaluation, forestry, mycoplasmosis, protein resources, toxic microorganisms, water research and technology, wind and seismic effects.
- 1965 Medical Science.
- 1969 Transportation.
- 1969 Space Cooperation.
- 1973 Cancer Research.
- 1975 Radiation Effects Research Foundation (mentioned in the previous section).
- 1975 Environmental Protection: sewage treatment, solid waste management, contained bottom sediments, air pollution-related meteorology, photochemical air pollution, automobile pollution, stationary source pollution control, environmental impact assessment, closed systemization of industrial waste liquid treatment, identification and control of toxic substances, environmental economics and incentives for pollution control, water conservation and flow reduction, controls in water quality, food additives.
- 1976 Vision Research.
- 1979 Energy: coal conversion, controlled thermonuclear fusion, high energy physics, photosynthesis and photoconversion, geothermal energy.
- 1980 R&D in Science and Technology: space sciences, environmental research, medical research.

This indeed is an impressive array of technologies where the U.S. and Japan have been cooperating on a governmental basis for decades. They involve almost every ministry and department and many independent agencies in both governments. This has naturally involved a substantial exchange of personnel at all technical levels and access to each other's R&D facilities. Since the U.S. was, in contrast to a later period, the initiator of these projects, there must have been an implied judgment that the Japanese S&Es were equal to or nearly equal to U.S. S&Es, or even in some cases ahead of the U.S. There have been no known major disagreements or conflicts; manageable irritations have occurred but nothing to mar or impede the ongoing R&D cooperation. Let us briefly comment on these various programs.

The Basic Science Program. NSF created its first overseas office in 1961 in the Tokyo Embassy to administer this program. By 1982 about 1,250 U.S. and Japanese S&Es were part of this program. The U.S. was spending about $5 million per year and the GOJ about half that amount but comparisons are difficult due to salary scales, accounting systems, etc. Originally it had been the only bilateral program between the U.S. and Japan but because of the mutual trust and confidence that had developed between the two parties over more than 20 years, the program was gradually expanded to include, in addition, three larger projects: Japanese funding and scientific staffing in the international phase of Ocean Drilling through the Glomar Challenger of the U.S., the earthquake shake-table in the seismic testing laboratory in Tsukuba city, and photoconversion and photosynthesis managed by the DOE under the Energy Agreement with Japan (discussed later). The ocean drilling and earthquake shake-table projects are unequaled in the world.

> ... no one has questioned the program's validity. From the American side, the large majority of scientists have said that their contacts with Japanese counterparts had led to new knowledge and insights and they considered themselves to be dealing with equals. ... the program remains at the highest level of priority ... in Tokyo and Washington. ... [It is] a program that has withstood the test of time better than most. (Bloom-1:93-94). ... the agreement is in the first rank among all bilateral S&T relationships under the purview of the [National Science] Foundation (Bloom-3:8).

Medical Sciences. There are several parts to this cooperative endeavor. 1) NIH chose many Japanese scientists for its research and training fellowships, more than any other country. This later became a source of great irritation, not at NIH which was well satisfied with its choices, but with the OSTP for political reasons. This program was outside the general framework mentioned above since the NIH program was an international training and research project; 2) NIH and the Ministry of Health and Welfare invested about $12 million per year in the study of diseases endemic to Southeast Asia and the results of this research were felt in this area, Africa and Latin America, not in Japan and the U.S.; 3) many aspects of cancer research; 4) smaller projects also exist for vision, shellfish sanitation, food product regulation, pharmaceuticals, biologicals and medical devices.

It has been "making first-class contributions to this most difficult of medical research fields" (Bloom-1:98). "Collaboration between American and Japanese biomedical scientists is exceptionally good and communication is relatively easy since most Japanese doctors speak English. Results from the joint research are disseminated freely to the rest of the world, thereby not only extending knowledge on the treatment of disease but demonstrating that the two countries can work together effectively on public health measures" (Bloom-3:11). [In the cancer area] ... "the relationship [is] one between equally qualified partners, and the best of its [NCI] bilateral programs ... [and] Japanese scientists have proven to be ahead of their American colleagues in a few disciplines such as immunotherapy, chemotherapy, and new drug testing" (Bloom-3:13).

Natural Resources Development. This cooperative program was started in 1964 as an offshoot of a bilateral cabinet-level committee on trade and economic affairs to augment the existing science agreement but in the more applied areas and was managed by the USDA and NOAA (DOC). As listed above it has a large array of panels (17 in mid-1980s). According to Bloom,

> ... the program has survived and thrived to a reasonable degree. ... [the U.S.] has gained far more from the Japanese experience than we gave up and the information we obtained is of great value in terms of public benefit ... [and] documentation of the results of the cooperation is extensive. (Bloom-1:96) ... While the program is not designed to do so, substantial economic and industrial benefits have been derived as well. The program has grown and matured without high-level political attention or involvement, primarily because of the dedication of individual scientists and engineers" (Bloom-3:29).

Environmental Protection. By the mid-1970s the U.S. and Japan were contributing the most in financial and intellectual investment in this area. Until 1981 (the first year of the Reagan Administration) when U.S. policy changes occurred and political problems within the U.S. EPA caused a reduction in commitment, it was "a very productive relationship" (Bloom-2:90); it has

> enabled the U.S. to acquire a great amount of Japan's sophisticated technology for sewage treatment, solid waste management and stationary source pollution control (Bloom-1:100). ... The program with Japan remains the best bilateral relationship in EPA's portfolio but it will be degraded further if budget reductions and lack of management support and attention continue to cause adverse effects (Bloom-2:44) ... [During the Nixon, Ford and Carter Administrations] special importance was attached to

> the Environment Agreement, but the [Reagan Administration] has not focused on it. ... the Japanese government was persuaded to facilitate access by EPA technical representatives to the pollution control technology employed by Japanese steel plants -- the best technology in the world. ... much of the technology has now been made available to American companies through private channels ... [thus demonstrating that] ... trade issues between the two countries can be solved or ameliorated by technical discussions out of the bright light of politics (Bloom-3:45).

Space collaboration. Japan was the only country to receive space technology in hardware and know-how in such amounts and sophisticated level; "there is no question that Japan has been given special treatment in this regard" (Bloom-1:102) and it was "able to establish a capability in space at a much lower cost than would have been necessitated by purely indigenous development" (Bloom-1:90). In the early 1980s, Japan decided to develop its own space capabilities with foreign, i.e. U.S., assistance, despite restrictions placed on Japan by the U.S. While Japan has largely succeeded in this effort, it still has major cooperative space projects with U.S. and U.S. companies. Space sciences research in Japan is the responsibility of the MOE where $50 million per year is a large amount; space applications is under the S&TA which is accustomed to large amounts of funds. Later, another space cooperation project with Japan was begun (after negotiations with Europe had been on-going for some time) for a space station. This project was "parked" under the 1980 STA for political convenience. It is still on-going with 15-20 active space projects.

Energy Research programs. Following the oil shock of 1973, the U.S. concluded its first bilateral agreement for non-nuclear energy – with Japan – covering geothermal, solar, synthetic fuels from coal, energy conservation measures, etc. Despite prodding from the U.S., GOJ did not invest sufficient funds in the non-nuclear energy research programs. "The agreement proved to be an absolute failure" (Bloom-1:104). Then quite surprisingly when Prime Minister Fukuda visited Washington in 1978 he suggested "the Fukuda Initiative" for the two countries to invest a billion dollars in a long-range program in controlled thermonuclear fusion and photosynthesis. The U.S. response was to accept if GOJ would invest in the USG's expanding coal conversion program. After much internal debate GOJ agreed and the 1979 energy agreement was born. The GOJ (and then the Federal Republic of Germany) agreed to furnish 25% (100s of millions of dollars over 10 years) of the cost of the U.S. SRC-II process for converting coal to synthetic crude oil. When the Reagan Administration came to power, it scrapped the SRC-II project in August 1981 and angered both Japan and Germany, and only fostered the image that the USG was "untrustworthy" in international projects. Another $70 million was committed by Japan for five years for the fusion project. In the beginning, the U.S. did not put forward its most advanced fusion projects because it thought it was ahead of Japan; but the latter "mostly by itself" (Bloom-2:91) has reached world class status in fusion, high energy physics, solar and geothermal energy and energy conservation technology.

This brings us to the mid-1980s when the future of the 80STA was being considered. Notwithstanding some difficulties and irritations, the USJ S&T relationship among S&Es was excellent; but because of the strains resulting from trade frictions, there was considerable apprehension and distrust of Japan in its acquisition of technology especially from the U.S. whether through exchange of information, or through purchase of technology. Bloom described the overall S&T relationship as "phenomenally successful when taken in large segments or as a whole. ... The conditions that led to the current level of cooperation probably were unique ... and ... no country at present ... appears to qualify." (Bloom-1:109-110) ... Although at first these [R&D] programs were politically motivated and designed to help Japan to recover from the devastation of war, they emerged later as a means for exchanges of advanced technical information approaching equilibrium in a two-way flow. Japan has recognized the intellectual debt it owed to the U.S. for the assistance it received by giving the U.S. a special place in its international technical relations and by investing heavily in U.S. government R&D programs" (Bloom-2:92).

Notes

1. This chapter will not include a discussion of the November 1983 Exchange of Notes between the USG and GOJ concerning the transfer of military technology from Japan to the United States and the highly controversial and tangled negotiations, again between the USG and GOJ with the intervention of the U.S. Congress, known as the FSX, a next generation fighter support aircraft for the Japanese Air Self-defense Force. Both governments and industry could not agree on the full implementation of the former arrangement so it was never fully implemented as originally conceived and planned. The latter was an exceedingly bitter controversy as to how and from whom the Japanese Air Self-defense Force would purchase its next generation fighter support aircraft. The rancor and bitterness engendered by especially the FSX controversy was in stark contrast to the general atmosphere of cordiality and cooperation in the sciences.

2. Starting from Zero was the title of a book in English by Taro Nakayama on the transformation of Japan by science and technology (Regional Centre for Technology Transfer. ESCAP, Bangalore, India. 1985. 140 p.). The term "technological superpower" was used by Justin Bloom in his *Japan as a Scientific and Technological Superpower*, (U.S. Department of Commerce, NTIS, 1990. 194 p.). In September 1988, Senator John D. Rockefeller commented in a speech at the sixth National Leadership Conference of the U.S.-Asia Institute that "the United States and Japan are in a race, a contest to be the No. 1 technological power".
3. The above information was selected from a number of sources as follows:
   Essays in Japanese: Nakayama Shigeru, Kagaku Gijutsi Rikkoku, Sengo Kaikaku to sono Isan, v. 6 of *Sengo Nihon: Senryō to Sengo Kaikaku.* 1995. p. 105-136;
   Books, articles in English: John W. Dower, The Bombed: Hiroshimas and Nagasakis in Japanese Memory, *Diplomatic History* 19:2(Spring 1995): 275-295; Tessa Morris-Suzuki, *The Technological Transformation of Japan*. Cambridge University Press. 1994. Chapter 7 "Technology and the Economic Miracle", 1945-1973. p. 161-208; Nakayama Shigeru, *Science, Technology and Society in Postwar Japan.* Kegan Paul International London/New York. 1991. Chapter 2, Democracy versus Technocracy in Science. p. 14-45.

   For those who wish to study postwar Japanese S&T policies and developments the following books, in addition to those cited above, are suggested: Moritani Masanori, *Gijutsu Kaihatsu no Showa-shi* (A Showa history of technology). Asahi Shimbunsha. 1990. Asahi Bunko. 291 p.; Nakayama Shigeru, *Kagaku Gijutsu no Sengo-shi* (Postwar Science and Technology [in Japan]). Tokyo. Iwanami Shoten. 1995. Iwanami Shinsho #395, 198 p.; Nakayama Tarō, *Starting from Zero: Transformation of Japan by Science and Technology*, ESCAP, Regional Centre for Technology Transfer. Bangalore, India. 1985. 140 p.; Nakayama Shigeru and Yoshioka Hitsoshi, eds. *Sengo Kagaku Gijutsu no Shakai-shi* (The social history of postwar science and technology [in Japan]). Tokyo. Asahi Shimbunsha. 1994. Asahi Sensho 511. 360 p.; Kagaku Gijutsu-chō (Science and Technology Agency). *Kagaku Gijutsu Hakusho* (S&T White Paper). 1995 ed. 468 p. Translated as Science and Technology: White Paper on Science and Technology 1995 – 50 years of Postwar Science and Technology in Japan. FBIS Report: FBIS:JST:96-021. May 1996. 289 p.
4. Since I am not writing a detailed history of the USJ S&T relations under the Occupation but merely trying to provide a brief summary I am focusing on Kelly's activities as symbolic of the milieu that had been created. After all, he was regarded by the GOJ as having made a major contribution in the reconstruction of Japanese S&T. In doing so, I am not implying any denigration of other U.S. actors in this area, nor the contributions in other S&T fields that other offices in the U.S. Occupation may also have made in reconstructing Japanese S&T. While there are a number of articles about Kelly's activities in Japan, I know of only one book on Kelly's role in Japan. This account is based on *Science has no National Borders: Harry C. Kelly and the Reconstruction of Science and Technology in Postwar Japan* (The MIT Press. 1994. 137 p.) by Hideo Yoshikawa and Joanne Kaufman. After this entire manuscript had been prepared, Justin Bloom brought to my attention, *The Allied Occupation and Japan's Economic Miracle,* by Bowen C. Dees, Japan Library (Curzon Press, Ltd, Surrey, UK, 1997). Dees was a participant in this effort and has devoted over 300 pages to "building the foundations of Japanese Science and Technology 1945-52" the seven year Occupation period, and naturally provided some differing emphases. It was my hope in this section to describe the milieu that was created among the working scientists and

engineers which set the stage for decades-long S&T collaboration between the S&Es across the Pacific. Since this account will be intentionally only a summary description, for those who wish to read about these events in detail should immerse themselves in these books, particularly the one by Dees.

5. Representatives to the new Council were chosen through a rather complicated procedure on December 20, 1948. There would be two major Departments, Cultural Sciences (with divisions for literature, philosophy, and history, law and politics, economics and commerce), and the Natural Sciences (with division for fundamental science, engineering, agriculture, medicine, dentistry, and pharmacy). Each division was given 30 members making for a total of 210, each elected for three year terms.
6. In a major administrative reform of the government in 1999, the STA became part of the MOE.
7. RIKEN was created in 1917 as a private foundation clearly emphasizing R&D for industrial development. It attracted many of Japan's best scientists, e.g. its Nobel prize winners. Over the years its ran into hard times, after World War I and during and after WWII. In order to survive, it created two companies in the mid-1920s to manufacture and market synthetically fermented rice wine (sake) and vitamin A. This unusual move was successful and its staff grew; during the 1930s it understandably became closely related to and dependent on the needs of the military resulting in only 20% emphasis on basic research by the late 1930s. During WWII, it fell on hard times again with air-raid damage, lack of electricity and staff; its Director was arrested for four months as a possible war criminal. In late 1945, it suddenly came into the news when its cyclotron was dumped in the Bay of Tokyo; this incident was one of the major factors which had led to Kelly (and Fox) to be assigned to GHQ as science advisers.
8. Said Kelly of this relationship with Dr. Nishina: When I would talk with anyone like Nishina, I would trust him like a brother. Their families became close friends. When Nishina suddenly died in 1951, the familial relationships continued on a very personal level; in the U.S. the Kelly's befriended the Nishina sons when they came to the U.S. to study and Kelly went to Japan for Nishina's son's wedding and even stood in for the bride's father when the younger son married in the U.S.

   Kelly returned to the U.S. to become the chief of the Scientific Section, Office of Naval Research in Chicago. In 1951 he became the Assistant Director for Scientific Personnel and Education in the newly created National Science Foundation. While Kelly was in Washington, the GOJ appointed its first science counselor, Mukaibō Takashi, to its Embassy in Washington. Kelly took considerable pains to introduce Mukaibō to the scientific establishment in Washington. In 1962 he moved to North Carolina State University as dean of the faculty and later as provost. He continued his close relations with Japan in many ways in supporting Japanese students study in the U.S. When the U.S.-Japan Joint Committee on Scientific Cooperation was created in 1961, he became the first U.S. co-chairman with his old colleague, Dr. Kaneshige Kankurō, from the early institution building years after 1945. Kelly retired in 1974; he died suddenly on February 2, 1976, at 67. Part of his ashes were brought by his wife, Irene, to the Tama cemetery for burial next to his friend, Nishina. A Harry C. Kelly Memorial Fund was created funded equally by Japanese and U.S. sources to support the Japan Center Program in Science and Technology at North Carolina State University.
9. *Science has no National Borders*, Hideo Yoshikawa and Joanne Kaufman. p. 107-116.

10. This activist group was known as Minka. In the mid-1980s the government was able to revise the election rules so that technical societies became the unit for election of representatives to the JSC.
11. The only person who has written consistently and systematically evaluating the technical value of almost all aspects of the USJ S&T relationship was Justin L. Bloom, a nuclear engineer, who was U.S. minister-counselor for science, 1976-81, at the U.S. Embassy in Tokyo. There is no need for me, a political economist, to try to duplicate his excellent work. This section, inevitably therefore, relies on the following of his selected writings for evaluative comments: 1) The U.S.-Japan Bilateral Science and Technology Relationship: A Personal Evaluation, *Scientific and Technological Cooperation Among Industrialized Countries: The Role of United States.* National Academy Press. 1984. p. 84-110 (Referred as to as Bloom-1:92, etc; 2) Bilateral Cooperative Programs: A Case Study – The United States and Japan, *Journal of the Washington Academy of Sciences,* 77:3(9/1987):87-92, (Referred to as Bloom-2:92, etc); 3) *An Evaluation of Major Scientific and Technical Agreements in Effect between the United States and Japan* (Final Contract Report to State OES). 1983. 66 p, (Referred to as Bloom-3:92, etc; 4) A New Era for U.S.-Japan Technical Relationship?: A Cloudy Vision, *U.S.-Japan Science and Technology Exchange: Patterns of Interdependence.* Westview Press. 1988. 219-256; Hyunjoo Byun's paper, *U.S.-Japan Cooperation in Science and Technology: Past and Present.* Unpublished paper prepared at the request of the Japan Technology Program, Department of Commerce and for the Asian Studies Certificate, Georgetown University. 9/92 was also consulted for this chapter.
12. Another little known fact is the Japanese Federation of Electric Power Companies agreed with DOE to invest $18 million in operations and R&D in connection with cleaning up the radioactive contamination at the Three Mile Island nuclear power plant. It is unusual in that it was an agreement between the USG and a private Japanese company.

III. The 1980 U.S.-Japan Science and Technology Agreement

1. Preliminary Dialogue

Japan's dependence on the importation of necessary fossil energy sources, e.g. oil, liquified natural gas and coal, was brought home vividly to the Japanese people and government through the two "oil shocks" of the 1970s. During the state visit to Washington of Japanese Prime Minister, Fukuda Takeo, in May 1978 – one of many regular pilgrimages by Japanese Prime Ministers -- he proposed to President Jimmy Carter that the U.S. and Japan combine their efforts to develop new energy resources for the 21st century. This proposal was popularly known as the "Fukuda Initiative". This was the first time since 1945 that the GOJ had made a major S&T proposal to the U.S. "The idea of broad new scientific cooperation with Japan began with the Fukuda Initiative."[1] Interestingly, it was the U.S. that had been the active partner in proposing joint research projects which the Japanese, in most instances, gladly supported. Since Japan during those years was regarded by the U.S. and by themselves as being substantially behind the U.S. in S&T, it is significant that Japan "as the catcher-upper" was not the party pursuing the more senior S&T partner for joint research projects. Obviously by 1978, the GOJ felt that Japanese R&D was sufficiently abreast of world S&T, if not the U.S. too, to feel confident enough to propose the "Fukuda Initiative" to the USG and, if such an initiative was successful, contribute to Japan's attempt to become more energy independent.

President Carter responded favorably but expressed the U.S. preference would be to broaden the proposed cooperation to include non-energy areas. According to Justin Bloom, the U.S. science counselor at the American Embassy at this time, "the concept of a broad based S&T agreement arose in the White House Office of Science and Technology Policy [then headed by Dr. Frank Press] apparently in reaction" to the Fukuda Initiative.[2] Simply stated, herein lies the genesis of the 1980 U.S.-Japan Science and Technology Agreement (hereafter referred to as the 80STA). President Carter continued to press the USG's interest at the Bonn Summit and in letters to Prime Minister Fukuda in the summer and fall 1978 and again to Prime Minister Ohira Masayoshi in December 1978 and again in June 1979 at the USJ bilateral summit. Apparently Ohira responded that while fully appreciating the ideas of broadening joint research he said that the GOJ needed to study the matter further -- a high level of saying, it's a great idea but we wish to proceed quite cautiously. In further discussions with Japanese diplomats, the State Department reported caution was the watchword in proceeding toward a non-energy S&T agreement.[3]

The U.S. motivation for asking the GOJ to consider a new agreement in research and development in areas other than energy appear to stem from several sources: 1) an increasing lack of R&D funds in the U.S., 2) use of a science and technology agreement for political purposes and 3) extrapolation of some assumptions stemming from GOJ actions under the USJ 1979 Energy Agreement.

In a memorandum of conversation with a Japanese MOFA official on April 2, 1979, in Washington, the justification for the President's proposal was amazingly based on a review by President Carter and the scientific community as to "what was happening in the United States in basic research" (i.e. it was being ignored and one could not look to industry for further funding) which led to the proposal by the President to the Japanese Prime Minister. It also noted that Japan's White Papers [presumably Japan's S&T White Papers without specifying which issues] had come to the same conclusion. During the discussion it also became clear that "it was advantageous not to limit research to 'basic research' but also to include applied research". That is to say, the United States was turning to Japan to help fund basic research since it was faltering in funding basic research and Japan appeared to have available resources. During this same meeting the USG explained that it had a list of 29 projects and that the cost to the GOJ "on an average basis would be about 150 million dollars." The cost to the USG would be "four or five times higher". There is a highly divergent report from this account provided by Justin Bloom who was then in the American Embassy as the science counselor.

The USG apparently assumed in 1979-80 that because the GOJ had agreed, under the 1979 Energy Agreement, to make substantial investments in U.S. projects, the GOJ could be persuaded to do the same in the non-energy areas. During the period from conceptualization to signing, OSTP looked upon the proposed new S&T agreement as a device to obtain financial compensation from Japan for the so-called "free" transfer of scientific and technical information from the U.S. since 1945. This was to be accomplished by proposing a series of bilateral projects originating in various technical agencies of the U.S. Government for which funding was not readily available through conventional U.S. budget and appropriations channels, and for which Japan would be expected to provide most of the necessary financial resources. The concept was resisted by the agencies, but they bowed to the will of OSTP and presented a collective list of some 40 projects meeting the criteria and involving investment by Japan in the order of $100 million.[4] This dovetailed with the generalized feeling in some circles in the U.S. that Japan should pay for or more adequately compensate the U.S. for technology given or sold to Japan since 1945.

The USG assured the GOJ that it wished to conduct these discussions in a "low-key manner so as to avoid a diplomatic 'blowup'" if the [proposed 29] projects proved unacceptable. The MOFA representative pointed out that while getting funds in the next fiscal year for the energy proposals would be difficult, obtaining funds from the Ministry of Finance for non-energy research would face "great difficulties" in most instances, a Japanese way of saying no. The USG said it hoped to include these proposals in a non-paper in about two weeks for the GOJ.[5] In another document two weeks later [it does not appear to be the non-paper] the USG's objectives were more bluntly stated: the primary objective in seeking this "enhanced cooperative effort" was to seek greater Japanese investment in U.S. research [basic and applied] to match our increasing investment in this area. The secondary objective was to inject through this effort and the Fukuda Initiative, a positive new note in our bilateral relations.[6]

The diplomatic position in negotiating a possible STA with Japan was that the two countries had reached parity in their technical prowess and that further cooperation at the government level in S&T should be conducted on the basis of equal sharing of information and costs. In light of this stance, the American Embassy proposed "as a minimum" that "there were unique programs going on in Japan that were worthy of U.S. participation and moderate investment." But this was "rejected in Washington."[7]

For the next 15 months from May 1978, while the emphasis on USJ discussions on S&T centered on completing the 1979 "energy" R&D agreement, slowly preparations were being made by the U.S. and Japan and meetings held between the two governments for a "non-energy agreement". The USG felt it "faced an unclear tactical situation" in making proposals to the GOJ in the non-energy area. On the other hand, while the GOJ recognized the value of expanded cooperation with the West, especially with the U.S., and was admonishing itself through its own S&T White Papers that Japan was at a turning point in S&T support by the government, i.e. the GOJ must change the 70/30 ratio of R&D supplied by industry (70%) versus the GOJ (30%) so that the government shoulders a higher percentage of R&D funds and also funds international scientific activities. Since the S&T gap between Japan and the West was narrowing, the opportunity for Japan to obtain new information and research data from the U.S. and Europe might substantially lessen.

On the other hand, there was a Japanese concern that the U.S. was seeking GOJ funding of secondary U.S. projects to help decrease the increasingly large trade surplus with Japan. A major aim of the GOJ was energy independence. To this mix was added budget constraints which is "without question the greatest inhibiting factor" in the USG estimate of persuading the GOJ to provide this support. While the Japanese S&T budget remained fairly constant to the total government budget in the seventies, i.e. 3.05% in JFY 77 and 3.19% in JFY75, the general account was running a substantial deficit: 37% in 1978 and an expected increase to 40% in 1979 to meet economic priorities. In addition, because of the Fukuda Initiative, the technical agencies had to reprogram their budgets for JFY79 to provide $3.0 million for coal conversion and $12.5 million for fusion. Thus, these agencies "having just swallowed this pill ... may well take a conservative approach toward the non-energy program." According to a USG document [Scope Paper, Part II Action Program], the Prime Minister had "personally ruled" against allocating a special budget even to fund the GOJ's own Fukuda Initiative energy program[8]: it could not be expected to treat the non-energy projects proposed by the USG in a different manner. The GOJ was expected to discuss these projects on the assumption that further cooperative research would be subject to the availability of funds -- no different from what the USG stance would be in similar circumstances. It concluded that the talks would be held "in an atmosphere that does not imply commitment by the GOJ."[9]

The USG even toyed 1) with the possibilities that the GOJ S&T budget might be increased to accommodate these USG projects although that did not look probable, and 2) with the possibility of some latitude for reprogramming even though "no answers" were available. The USG chose projects which it thought were responsive to GOJ national priorities as evidenced in S&T White Papers. The projects also had "to be of sufficient magnitude and importance to be included" but those which were being negotiated -- NSF's continental shelf drilling project and NASA's trans-Pacific Balloon project -- with good prospects of success were excluded. All in all, the USG felt they were "well-matched" to Japanese national priorities as exemplified in the space budget, basic research, and on the "qualitative improvement of the people's lives" as noted in a White Paper and recommended by the Prime

Minister's Council on Science and Technology. For example, in the health area where the USG suggested five medical research projects, the USG report commented that the "Japanese have demonstrated excellence in the biomedical sciences ... [and] one of the advantages of cooperation with Japan in the medical sciences is the availability of three discrete cultural universes for comparative study: ethnic Japanese in Japan, continental United States, and Hawaii (the half-way house)." These are projects in which the USG was interested and were grouped to appear to fit in with Japanese national interests; they were not Japanese projects. These projects could not be described as of such scientific merit as to be "Presidential" in stature. Years later, the USG criticized these kinds of projects which itself had proposed as being not of Presidential stature and not in line with the high level nature of the bilateral agreement. The proposed projects are listed in the following table.

The above indicates that the USG entertained hopes that it could persuade the GOJ to increase its R&D budget to accommodate USG projects, or reprogram other funds to accommodate USG projects. Since the GOJ had agreed to contribute funds to selected USG projects in the past, this was not totally unreasonable; however, in light of the problems with the projects under the energy STA and feeling among some Japanese officials that the USG was trying to lessen the trade deficit through this means, and was pushing a concept and emphasis, "non-energy R&D projects", which obviously the GOJ was less than enthusiastic about, should have been an indicator that this approach would not work.

According to a USG Talking Paper for presentation to the Japanese, the USG is "currently undertaking or is planning to undertake subject to its legislative process" -- i.e. not yet funded nor approved by the U.S. Congress -- all of these projects except for the Saturn Orbiter project which is scheduled for initial funding in USFY83. The total cost for five years was estimated to be about $2.66B; the total annual average costs for both countries would be about $370 million of which the Japanese contribution would be about $90M per year.[10]

While the USG presented the 23 projects which seemingly responded to or were in line with Japanese S&T priorities, the underlying objective, according to the science counselor in the Tokyo Embassy at that time, "was to try to exact from the Japanese a substantially increased financial investment in U.S. government R&D programs -- beyond the $150 million to $200 million already committed by Japan."[11]

These proposals were given to the GOJ on April 18, 1979 to give the GOJ time to "staff out" the proposals. In early June 1979, the GOJ asked three questions: 1) what are the relations between these proposals and on--going programs in the U.S.? Answer: They are additional. 2) Will they be carried out within the framework of existing programs or under a new umbrella agreement. Answer: "any mutually acceptable arrangement that will expedite cooperation." This is an unusual response since ultimately we pushed for a new agreement. 3) Is the USG engaged in similar type initiatives with other countries? Answer: USG had approached several advanced countries. Actually Joint U.S.-European Community working groups also were exploring cooperation in several areas.[12] At the June 1979 Summit, President Carter reiterated the USG desire to move ahead with its non-energy proposals, and proposed sending Dr. Frank Press, his science adviser, to Tokyo for this purpose. In mid-July 1979, the GOJ agreed. "The Japanese felt obliged to give us [the USG] an opportunity to fully explain all [sic] 24 projects".[13] Despite the dialogue with the GOJ since April 1979, "we still know very little about what the Japanese really think about our specific proposals."[14] As to the form of the new arrangement the GOJ said it would prefer "to carry out any cooperation emerging from the Press visit within the framework of existing programs". However, they also expressed their willingness to conclude an agreement "if we [the USG] attach political significance to setting up a separate program under Dr. Press' direction." Thus we sighed, "the question is still open."[15]

The first formal step or major milestone in these discussions took place when President Carter's science adviser, Dr. Frank Press, visited Tokyo on September 20-21, 1979, at the invitation of the GOJ. Press met with the Japanese Prime Minister Ohira, Foreign Minister Sonoda and the Minister for the Science and Technology Agency in addition to the formal discussions with the GOJ whose delegation was headed by the Deputy Minister of Foreign Affairs Miyazaki. The U.S. delegation included senior representatives from The White House, the Departments of State, Energy, Health, Education & Welfare, Environmental Protection Agency, National Aeronautics and Space Administration, and The National Science Foundation. The GOJ delegation consisted of representatives from the Ministries of Foreign Affairs, Health and Welfare, Agriculture, Forestry and Fisheries, International Trade and Industry, Transport, Posts and Telecommunications, Construction, and the Science and Technology Agency. While the discussions were held at two levels at the suggestion of the GOJ, policy level and expert technical working groups, the primarily purpose of the visit was for the USG to present detailed technical information on the various R&D projects.[16]

As usual, a joint communique was issued but more interesting was the adoption of a "policy paper" by both governments. The communique noted that the USG "stressed, in particular the importance of such cooperation with Japan in a number of areas of advanced science and technology where *a new partnership* [emphasis added] showed promise of producing greater results than could be made by each side proceeding separately." This theme of the importance of Japanese S&T is repeatedly discussed, stressed and emphasized as the rationale for closer USJ cooperation. Citing the Carter-Ohira communique that "the two governments will study seriously the prospects for cooperative efforts in areas of basic and applied research" agreed that promotion of Japan-U.S. cooperation in these areas is also important in building up a productive partnership between the two countries for the 1980s.

The unusual policy paper was titled "A Comparative Analysis of U.S. and Japanese Science Policies" and was described as "major document" produced by these meetings.[17] While the writing of such reports, let alone a bilateral policy paper and comparative analysis, are normally left to institutional anonymity, the authors are clearly identified as Dr. Frank Press and Mr. Yamano Masato, Deputy Minister, Science and Technology Agency, September 20, 1979. The main points can be summarized as follows:

1. As two of the most technologically advanced nations of the world, the United States and Japan share common aspirations and problems in planning their scientific future.

2. Japan realizes the need to increase its investment in R&D in today's austere economic situation, especially where the risk factor is such that the incentive for industry to invest is slight and that cooperative projects with the U.S. would multiply resources available, particularly in the case of globally important high-risk projects.

3. The costs of research in many areas of common USJ interest, such as health, environmental protection, resource and energy development surpass the ability of one nation to accommodate. "A desirable solution to this dilemma" is the development of mechanisms for nations to join their efforts and to share the necessary costs of these programs of acknowledged value. The different governmental administrative procedures "must be skillfully managed" to provide the research resources to initiate and carry out "joint research projects of significant magnitude."

4. The U.S. and Japan benefit greatly from a long and highly successful scientific and technical relationship which can be the foundation of a new era of cooperation characterized by recognition of equality between the US and Japan and by a shared perception of research needs.

5. Cooperation on *new jointly funded* research projects will contribute to good political relations between the U.S. and Japan. It mentioned several joint research projects which involved the expenditure of "large sums of money in ways that have been mutually agreed". This is described as a turning point which lends confidence to the premise that the U.S. and Japan can now identify those other projects for "this new era of cooperation [and joint funding which was implied and unstated] and to move aggressively toward the realization of these commonly desired goals."

6. Policy level dialogue between both parties will "serve to expose problems, suggest new avenues for cooperation and provide increased understanding of the two different systems." The hidden agenda or underlying assumption behind these USG ideas, was its presumption that it would be able to use successfully these analyses and discussions to modify Japanese behavior, practices, laws, regulations, legislation and institutions, and her willingness to make substantial financial contributions.

7. At the same time, the technical experts were to "pursue with equal determination the definition of joint research programs", the outline of resources required for their conduct, and the responsibilities of each of the parties and "evaluate concrete proposals for expanded cooperation with the objective of implementing a large number of them."[18]

The U.S. side felt these meetings to have been a substantial success. As the above summary implies, the communique and joint policy paper raised U.S. expectations that 1) since substantial joint funding of some projects has already taken place others would follow (despite the informal GOJ response was obviously far from enthusiastic), 2) policy review mechanisms will allow future discussion of S&T policy issues and expose problems, and 3) joint projects generate political goodwill in the U.S., 4) the USG had sensitized Japanese officials at the highest level about the benefits of cooperative projects which are likely to require major inputs of funds from both sides. The austere budget situation in both Japan and the U.S. would make joint financing much more difficult if not improbable. Only with the highest level of Japanese support would this level of funding be possible; even that was unrealistic and also ignored the difficulties on the U.S. side of obtaining similar funding. These formalized expectations by the USG will come to haunt and sour the future negotiations on the renewal of the 1980 agreement. In the discussions so far, there was no evidence that the issues of access to Japanese R&D facilities, scientific information, intellectual property rights -- which later became such contentious issues surfaced in the bilateral discussions. This would seem to underscore the real USG intention in initiating these discussions was to try to persuade the GOJ to make financial and other contributions to USG research programs -- in line with "A prominent aspect of our [USG's] historical approach in our scientific relationship was to seek Japan's contributions, including financial, to U.S. research programs."[19]

2. The Draft 1980 STA

The first draft of the 80STA was prepared by OSTP, not OES, and given to the Japanese Embassy in Washington on December 10, 1979 and described as a "draft umbrella agreement". OSTP, in explaining the draft "stressed the political value of our two countries working together on such a broad initiative." Since it was "patterned" on the already signed 1979 Energy Agreement and other standard phraseology USG hoped it "could be formalized at the next meeting in February [1980]. OSTP commented that it was the USG agent "since this demonstrates continued high-level interest and added "weight to the overall proposal." In this same meeting the USG gave a status report on the afore-mentioned 24 [sic] projects it was proposing as joint projects and, by implication, for inclusion in this umbrella agreement. In light of future developments, this draft is of particular interest in what it describes as the principal objective of the agreement and what is left out of this draft.

Article I declares unequivocally that the principal objective of this agreement was

> To promote and facilitate joint research and development activities in areas of mutual interest and of benefit to both countries and to mankind.

Table 3.1 Recommended Non-Energy Projects
In millions of dollars, 9/1979, periods in US fiscal years

Project (Agency)	Average Total Cost ($) per year	Period
Space (NASA)		
1. Geodynamics	40.0	79-82
2. Open Program	35.2	81-89
3. Halley Flyby/Temple 2 Comet	67.1	82-88
4. Saturn Orbiter	36.9	83-96
Basic Research		
5. Neutron scattering (DOE)	4.5	80-88
6. Accelerator Research (DOE)	1.3	80-84
Environmental Protection		
7. Effects of Carbon Dioxide (DOE)	6.0	current
8. Effects of Diesel Particulates (DOE)	6.0	79-84
9. Liquified Gas Safety & Control (DOE)	5.0	80-90
10. Effects of Electric/Magnetic Fields (DOE)	2.0	current
11. Hazardous Material Handling (EPA)	.66	80-82
12. Nitrogen Oxide Control (EPA)	.5	80-82
13. Study of Environmental Diseases (EPA)	1.0	80-82
14. Ocean Dumping Studies (EPA)	1.0	80-82
15. Fish Bioconcentration (EPA)	.75	81-82
16. National Toxicology Plan (HEW)	40.0	80-84
Health (HEW)		
17. Center for Alcohol Drug Abuse & Mental Studies	25.0	81-85
18. Expanded Program of Immunization	4.6	80-83
19. Animals, including Primates for biomedical research	50.0	81-86
20. Recombinant DNA Research	30.0	80-90
21. Development of Antivirals	12.0	80-90
Disaster Prevention & Agriculture (USDA)		
22. Snow and Avalanche Management & Landslide Prediction and Control	.66	79-84
23. Food Health Regulations & Animal Plant/Diseases	.2	79-83

Areas and kinds of cooperation are specified as follows:

- **technical areas of cooperation** (Art. II): space, basic high energy physics, environmental protection, health, disaster prevention, agriculture, resource conservation and other areas mutually agreed upon

- **kinds or forms of cooperation** (Art.III): joint projects and programs, meetings where STI exchange is discussed, exchange of information on practices, and legislation and regulations on S&T R&D, visits & exchanges of scientists, other forms as mutually agreed upon.

Funding of projects are covered as follows in Article IV:

With respect to funding, costs shall be borne as mutually agreed on the basis of equitable sharing of costs and benefits.

Executive agents for this agreement are specified in Article VI:

for the U.S.: OSTP Director
for Japan: left blank, but presumably MOFA

Term limit of the STA: specified as five years from signature.

The one major stipulation which later became the focal point of U.S. criticism of the STA was the omission of a management structure, i.e. the creation of a joint committee.

While the USG early on decided that the OSTP would be the U.S. agent, the GOJ had difficult jurisdictional and turf issues to resolve in selecting a GOJ chairman. MOFA, MITI and S&TA could assert their claims and this issue delayed GOJ's response to the USG, and also to PRC and Australia for the same reason.[20]

The three outstanding features of this draft were a frank statement about the principal objective, that funding would be based as agreed on the basis of equitable sharing of costs and benefits and that the Joint Committee was omitted. In light of known bilateral discussions it is difficult to imagine that the GOJ would rush headlong to sign this draft agreement with these three features. Nevertheless, the USG suggested that the signing of the new draft agreement might be accomplished at the end of the next meeting in mid-February 1980 for "symbolic" reasons.[21] Only a few days later, the USG informed the Japanese Embassy that it would prefer to postpone the mid-February 1980 meeting if it meant that the "umbrella" agreement could not then be signed, or there would be no agreement in principle, or specific agreement on joint projects was not immediately forthcoming. The USG appeared overly eager and obviously rushing the GOJ to agree to something at the next meeting.[22]

3. Draft #2, February 1980

When both sides met again in Washington on February 12-13, 1980, they agreed

1. on an initial list of ten projects in space, basic research, health and agriculture.[23] None of these projects can be described as "presidential" in nature, except perhaps the Recombinant DNA research project.

2. a new draft agreement which

a. substantially revised the purposes from the earlier blatant version to a broader statement that the two governments shall develop cooperative activities in S&T R&D in non-energy fields and such related fields as may be mutually agreed for peaceful purposes on the basis of equality and mutual benefit (Art. I).

b. created a joint committee to exchange information and views on the S&T policies of both governments, review cooperative activities and accomplishments under this agreement, and provide advice to both governments on the implementation of this agreement (Art. IV).

c. stipulated that the two governments shall give "due consideration to the equitable distribution of industrial [intellectual was not used] property resulting from the cooperative activities under this agreement and of licenses thereof and to the licensing of other related industrial property necessary for the utilization of the results of such cooperative activities" and shall consult each other as necessary for this purpose (Art. V.2).[24]

d. continued to stipulate that costs for cooperative projects shall be borne as mutually agreed on the basis of equitable sharing of costs and benefits (Art. II.2).

This meeting was described as "an important milestone in the development of a productive partnership for the 1980s [where] this partnership promises the production of greater results than either side could achieve separately in a number of scientific and technological areas which exceed any one nation's economic and manpower resources".[25]

Another USG-GOJ meeting was held in Washington on April 14, only two weeks before the Principals were to sign the agreement. The major revision was in the dilution of the funding provision with the deletion of

on the basis of equitable sharing of costs and benefits.

The USG had retreated from the position of requesting Japan to pay more under certain circumstances to just contributing to the costs of projects "as may be mutually agreed." With this revision, all the seemingly one-sided stipulations in the USG draft had been one by one negotiated out of the agreement and now read as a balanced document without suggestion of one party shouldering more responsibility, whether financial, or in human resources.

Circular 175 Authority. Let us now refer to the Circular 175 process for authority to negotiate this agreement with the GOJ which was begun on November 30, 1979 based on the proposed first draft agreement.[26] It pointed out that even though the USJ S&T programs are "second in importance only to our [USG] program with the Soviet Union in terms of magnitude, breadth, and scientific benefit" there is presently no "all-encompassing umbrella agreement" with Japan. It admitted that during the Press visit to Japan the USG proposed 25 "new and substantial projects" which it significantly said: "Technically, it is possible to carry out this enhanced cooperative program under existing agreements and arrangements." This was echoed several years later in a report prepared as a policy paper for State/OES in assessing the future of USJ S&T relations and even went one step further by raising "The basic question to be faced is whether the political and management burden of the agreement is out of proportion to the value received." [27]

For obvious political reasons the USG felt a Summit level agreement was more desirable at this point and the circular 175 request listed several justifications for this approach:

1. the "imprimatur of and under the execution of the White House" (i.e. the OSTP as executive agent) would emphasize the importance of the new enhanced program and would highlight the program as a positive new note in the frustrating USJ trade relations. S&T relations had now ceased to be a quiet technical relationship as in the past, but had been injected into the political arena to serve the political agenda of the USG vis-a-vis Japan.

2. the "principal objective" of this agreement was to promote joint research and development particularly on large-scale projects. The USG envisioned that this agreement would help the GOJ overcome internal problems concerning "funding authority and coordination on substantial, multi-agency projects." In most documents, three objectives are listed for the 80STA as follows:

> a. To promote Japanese investment on a large scale in basic science projects which US agencies determined to be of interest: a) no projects under $1M were to be proposed and b) a 50-50% USJ contribution US/Japan was foreseen.
>
> b. It was envisioned that only high cost, high risk projects, e.g., space plasma accelerator facility and neutron scattering would be accepted. It should be noted that this agreement was proposed at a time when Japan had a large trade surplus with the US.
>
> c. In addition, of course, such a treaty [sic, should be agreement] was designed to symbolize, and encourage close ties between the two nations, in a safe, relatively nonpolitical arena.[28]

3. the agreement would provide "a home" for conducting high-level S&T policy talks, "an increasingly important activity as we move into the era of 'big science'", i.e. another opportunity to raise with and pressure the GOJ to fund these projects and pursue specific U.S. policy interests at a high level in the GOJ, and an obvious attempt to inject itself into the S/T policy-making process in the GOJ.

4. such a new umbrella agreement "could be of tactical benefit to us [the USG] in helping to gain Japanese acceptance of a maximum number of our [USG] proposal to constitute a meaningful new program and to maintain it at an optimal level".

These rationales for the agreement rather blatantly emphasize the objective of obtaining GOJ funding of not only "big science" but the other 25 [or 23, or 24] proposed "new and substantial" projects. There was no mention or hint of concern with intellectual property rights, access to JSTI and R&D facilities. The drive was for Japanese funding which ultimately was not accepted by the GOJ, at least in the agreement.

The USG pressed for the Principals (the President and the Prime Minister) to sign the agreement since its genesis was a U.S. Presidential initiative, and the subject of several Presidential communications to the Japanese Prime Minister, and a topic at two bilateral summit conferences, and, also, for its political importance in the U.S. and also to balance it with the STA signed by President Carter and PRC's Deng Xiaoping in January 1979.

The 1980 USJ S&T (Non-Energy) Agreement was signed by President Jimmy Carter and Prime Minister Ohira Masayoshi at their Summit Conference on May 1, 1980.

4. The 1980 STA

The press release marking the first USJ STA signed at the Summit level, emphasized that the U.S. and Japan, as "two of the most technologically advanced nations", would, under this new arrangement, conduct jointly R&D programs which "rank high in the national priorities of each country". It asserted that this STA was "a relatively new mechanism [i.e. the 80STA] for developed countries to work together on solving globally important research and development problems" and allows them to "pool resources rather than duplicate efforts in a variety of areas." Japan was invited -- or flattered -- by the U.S. to participate as an equal partner in S&T which meant, in a nutshell that Japan was expected to contribute handsomely to selected U.S. R&D programs.

Some 40 projects were identified for joint endeavor, mostly suggested by the U.S. Almost all the projects were themselves worthwhile and had been earlier described by the USG as "new and substantial" but could hardly be identified as "big science" projects. There were a few exceptions, such as the space projects. Actually nine space projects had already been signed and others close to completion of negotiations. These very projects, which were put forward by the USG as worthy of inclusion in this first Presidential/Prime Ministerial level STA with Japan, were to be targets of severe criticisms in the mid-1980s when the future of this agreement was considered in the USG.

If this agreement, "the new mechanism", was to be successfully executed and live up to the expectations of its U.S. sponsors, then there would have to be a special effort exerted by both the GOJ, and in particular by the sponsoring USG. It was clear from the above description that the GOJ was less than enthusiastic about the agreement, had successfully whittled down to extinction all the special conditions that the USG had hoped the GOJ would accede to, and, in particular, the deletion of any hint that the GOJ would underwrite or contribute financially to any specific U.S. big science project. At a meeting at OSTP on June 26, 1980, soon after the signing of the 80STA, the memo on the meeting stated that "we [USG] need to make another push well before August 15 [1980] [an important date in the GOJ budget cycle] to ensure that our [USG] non-energy project proposals receive [GOJ] funding next year. We were encouraged to do this by our friends in Tokyo." [29] There is no documentary evidence that the USG push was successful.

The U.S. science counselor in the Tokyo Embassy at the time commented in an evaluative report in the mid-1980s that

> the Japanese signed the agreement only under what must be considered duress, even though they avoided being obligated for financial investments.[30]

Under these circumstances, the GOJ would probably not play an active, enthusiastic role in carrying out the agreement. It would play a passive role responding to whatever the USG was prepared to push.

If the USG expectations for this agreement were going to be achieved, a major effort would need to be expended on the U.S. side. But the OES was not a Bureau with substantial prestige in the State Department. In a contract report evaluating the USJ S&T relationship in 1983 as the first step in assessing the future of this agreement, the report commented that

> one of the major problems associated with the S&T Agreement is a lack of central day-to-day management on both sides; coordina-

> tion is left to the individual program or project managers, who are scattered geographically and who change from time to time. The Department of State has been unable to act efficiently as a central coordinator because of turnover of personnel, a lack of technical understanding of the extraordinarily diverse projects, and the inability to assign an officer who could spend an appreciable fraction of his or her time in keeping up with progress or problems and acting as a communicator with the Japanese side.[31]

This situation appeared to persist through 1988 when this writer frequented State OES offices in gathering data and interviewing personnel on the 88STA.

The USG technical agencies were quite satisfied with their existing working S&T relations with the GOJ. They "went along reluctantly with White House [i.e. OSTP] pressure to negotiate UJST [80STA] under the Carter Administration" -- and naturally were "neither surprised nor discouraged that it [80STA] has not been more successful."[32]

While the OSTP, at its insistence, was designated U.S. executive agent, it also did not have sufficient personnel or a political determination to push a substantial enlargement of the bilateral program. It was highly reliant on State/OES to provide staff support for the management of the 80STA. Based on the above, therefore, it appeared that the USG had not taken the necessary steps to augment OES, nor had the OSTP been strengthened with additional personnel to manage the vaunted "new mechanism".

The First Joint Committee was held in Washington on September 24-25, 1981. This was the first, last and only Joint Committee meeting held under the 80STA.[33] At a preparatory meeting with Japanese Embassy officials, the OSTP Deputy Director stressed again "the political importance of the first Joint Committee meeting, noted that the MITI Minister in a recent trip told President Reagan that GOJ will provide more technology to the U.S., and expressed the belief that "non-energy cooperation should be one of the main vehicles for GOJ channeling technology to the U.S." revealing again one of the major U.S. objectives of the 80STA.[34]

Its achievements were indeed meager. The press statement and report on the meeting was a statement of the obvious. The U.S. reaffirmed that its cooperation with Japan in non-energy fields is a "most important pillar of Japanese-U.S. cooperation in science and technology along with their cooperation in energy fields, and that the U.S. has no higher priority in the area of international science and technology cooperation than the expansion and strengthening of cooperative ties with Japan." The Japanese comment was much less superlative: "... they attach special importance to the unique cooperative relationship between Japan and the U.S. ... and expressed their firm desire to facilitate future cooperation." The report notes that "their mutual collaborative efforts constitute the most important single part of its overall international cooperative research and development programs." This described, it asserted, "the increasingly close and mature partnership which exists between the two countries" and they are "determined to continue to strengthen and expand their international cooperative efforts." Forty-two projects under the 80STA are listed with their estimated costs. Many do not even have a small amount listed for either the U.S. or Japanese contributions; most of those listed with amounts are less than a million. Even the Saturn project is only a joint study of the definition phase of the project. So, all in all, these projects may be worthwhile but they are hardly large scale, big science, "presidential" projects. OSTP was unable to persuade the GOJ to contribute to the Superconducting Supercollider (SSC). It was this reluctance and rejection by the GOJ to contribute "big money" to U.S. big science that soured OSTP in its attitude toward the USJ S&T relationship, and this attitude clearly manifested itself later in the interagency renewal discussions concerning the 80STA in mid- and late 1980s; access to JSTI and R&D facilities was not then the focus of special attention.

5. The Legacy

Let us briefly recapitulate, even at the risk of repetition, the background, expectations, experience surrounding the 80STA, especially since it was the target of severe criticism by selected USG officials as the agreement was considered for renewal in the mid-1980s.

The first major S&T program proposed by the GOJ to the USG since 1945 was the so-called "Fukuda Initiative". While it resulted in the USJ Energy R&D Agreement of 1979, President Carter pressed the GOJ to enter into a new agreement covering all other technical areas -- thus the non-energy agreement. The GOVERNING motive for this desire by the USG was to persuade the GOJ to make major financial contributions to "big science" projects, especially the SSC and other USG R&D projects and because the funding of basic research was supposedly slipping in the U.S. and funds from industry for this purpose could no longer be taken for granted. This motive was obviously based on the assumption that since the GOJ had already agreed to make some financial contributions to selected R&D projects, it would be amenable to making additional commitments if sufficient persuasion and pressure was brought to bear on the GOJ. The USG gambled the prestige of the science adviser, and by extension the White House, by sending Dr. Press to Tokyo to open formally the negotiations for this new agreement. The USG's desire for this agreement was so strong that the President sent letters to successive Japanese Prime Ministers and raised the issue at two bilateral Summit meetings. The desire for GOJ funding was blatantly written into Article I of the first draft agreement. The USG quite erroneously read the "tea leaves", so to speak, when it misread Japanese comments and the implications of the tight budgetary situation in Japan even for the GOJ-initiated Fukuda Initiative. The GOJ was obviously unenthusiastic about this agreement. It successfully negotiated out of the agreement all references to possible GOJ commitments to funding joint R&D projects. Since the President had involved himself so completely in pressing the need for this agreement, GOJ officials felt obliged to accommodate the U.S. President by agreeing to have the Prime Minister sign the agreement -- thus it became the first USJ S&T agreement since 1945 to be signed by the Principals. Although it is well known that many summit meetings between the U.S. President and other heads of state are concluded with the signing of an STA for want of something better to sign, the 80STA was "historic" in the USJ context and "unusual" in a broader political context since USJ S&T relations were governed by agreements or MOUs (memorandum of understanding) between departments/ministries or agencies of the two governments.

The Presidential/Prime Ministerial character of this agreement created its own problems: when renewal was considered by the USG it was difficult to consider quietly abandoning the agreement because it was signed by the Principals. It would be embarrassing to both governments if it was not renewed; it would imply, though incorrectly, a major breakdown in S&T relations between the two countries. The fact that the Principals signed the agreement was utilized during the renewal controversy in the USG as a lever to extract "concessions" from the State/OES and from the GOJ and to markedly change the character of the S/T relationship. By having the Principals sign the agreement, it had become a political document the handling of which became much more complicated to manage, particularly in the USG.

Even though the basic tenet or public position was that the U.S. and Japan were now on an equal footing in S&T, or at least in technology, the USG did not wish to invest in GOJ S&T projects since it was more interested -- some might say one-sidedly obsessed, in persuading GOJ to invest in secondary U.S. projects. Could this be described as a relationship on equal footing?

Since the OSTP's major reason for a USJ agreement was to obtain GOJ funding of USG projects, it obviously would be disappointed, if not bitter and frustrated, over Japan's refusal to underwrite USG S&T projects, and its obvious disinclination to accept USG requests/demands embodied in the proposed agreement, a bitterness and frustration that becomes highly evident in the renewal discussions in the USG and in the subsequent bilateral negotiations process. Perhaps because of this preoccupation, the internal USG discussions, and the subsequent bilateral negotiations, there is no evidence that the availability of or access to science and technology information (STI), or access to Japanese R&D facilities was then a concern in the USG nor discussed among USG officials and also not discussed with the GOJ as a major or even a minor issue. If the level of technological competence of Japan was truly believed to be equal -- and not just to flatter it into funding U.S. projects -- it would seem that access or improved or greater access to information and facilities would have been a major concern. The depth of the funding preoccupation to the exclusion of other seemingly legitimate issues was obvious.

As years passed and the future of USJ S&T relations were assessed in the mid-1980s, the USG dissatisfactions with and criticisms of the 80STA of which the USG was itself the architect gathered strength, momentum and diversity. Some were indeed inherent in the agreement itself, others were added as disappointments, better yet as frustrations increased – even though they were not related to the original objectives of the 80STA.[35]

The 80STA was criticized as lacking a strategic plan. Since the financial aspects were deleted from the STA, and since the many smaller projects included in the STA could have been placed in other overall agreements in various technical fields, it could indeed be said that the agreement lacked a strategic plan. But that deficiency stemmed from the fact that the USG had one major and primary objective for this agreement, but which was defeated as the agreement was negotiated. The 80STA is also presented as a failure because of GOJ disinterest which is true but did the USG which pushed the 80STA on a reluctant GOJ show any more interest? Between 1980 and 1985 it surely did not. The agreement was regarded as superfluous and duplicative by USG technical agencies. Just merely judging from the list of projects proposed by the USG to the GOJ, there would be a serious question as to the future of many of the existing S&T agreements covering many technical areas between USG and GOJ. The potential and actual overlap was obvious but was ignored in order to flesh out the 80STA and hopefully give it substance. Even though NASA and the GOJ's Space Activities Commission had essentially concluded negotiations for about 17 space science projects, OSTP insisted that these arrangements be brought under the 80STA -- despite NASA's expectations to the contrary. This was an obvious and blatant attempt to prove that the "Presidential" S&T agreement contained substantial major projects the inclusion of which was one of the primary objectives of the agreement. The USG did recognize internally that budgetary stringencies in both countries resulted in no major joint projects stuitable for inclusion in the 80STA being initiated.

Even though OSTP designated itself as the executive agent of the USG in the 80STA, the first time in USJ S&T relations, it did not, as a staff agency of the White House -- have sufficient personnel to provide the drive for effective day-to-day management of a highly complicated S&T relationship with Japan. Naturally then, this responsibility devolved onto the OES Bureau in the State Department. While OES thus became the staff supporter of the OSTP, it was understaffed, lacked bureaucratic prestige and authority and was a reluctant supporter of the OSTP since State had advocated policy positions contrary to those of OSTP. In light of the above, neither OSTP nor State/OES could muster the bureaucratic staying power to promote various projects with the GOJ. In the eyes of both State and OSTP, pushing the GOJ into accepting a more active relationship, was obviously not high on their list of priorities whether they were personnel or financial.

Even though the National Science Foundation was playing a major role in variety and depths of projects in USJ S&T relations since 1961, it did not propose projects for the new agreement and was not involved in the so-called management of the 80STA.[36]

In contrast to the USG dissatisfactions, Ambassador Endo of MOFA, when queried about problems with the exiting STA from the GOJ perspective, declared there were "no problems" with the 80STA. According to his comments at a House of Councilors's Budget Committee hearing on March 16, 1988, U.S. and Japanese scientists were able to vigorously and freely discuss the issues. From this point of view "good results emanated from the [1980] STA". There were, he said, 49 USJ projects under the 80STA with a few inactive ones. They were mostly information exchange projects among scientists, a few joint research projects. Separately, they were described as providing "meaningful cooperation" (*igi aru kyōryoku)*, modest unelaborate cooperation (*jimi na kyōryoku*). They were operationally not significant as exemplified by the fact that there had been only one Joint

Committee meeting in the life of the 80STA. This assessment was correct, a polite commentary, hardly a ringing enthusiastic endorsement by the GOJ which reluctantly signed the STA in the first place.[37]

Thus the 80STA began its life as a unique and historic milestone in USJ S&T relations, but it soon began to languish for want of political interest let alone will either on the US or Japanese side to make it work. While the GOJ's indifferent and unenthusiastic attitude is understandable, the USG's lack of determination to try to make the agreement work even to its own interest was disappointing, especially since it was the party starting with the U.S. President that pressured the GOJ to sign the agreement in the first place. Although the 80STA was intended to be a bright spot, a new departure, a new mechanism in USJ relations amid the ever frustrating trade negotiations, those anticipations soon evaporated and with it the sense of feeling good quickly waned and again the mutual atmosphere was somewhat poisoned more by U.S. action and deliberate misinterpretation of the content of the 80STA once the financial aspect had been deleted.

In light of the original USG assumptions and expectations about the probable actions of the GOJ, especially their malleability to USG expectations and demands and the subsequent failure to realize these assumptions, these assumptions became the sources of unhappiness, disappointment, frustration, even anger in the minds of some USG agencies, particularly OSTP since it was the one which had pushed for the rejected objectives. These unfortunate circumstances set the stage for what transpired during the tug-of-war surrounding the renewal or non-renewal of the 80STA.

Notes

1. *Expanded US-Japanese Cooperation in Non-Energy Research,* 4/16/1987, p. 1.
2. Bloom, Justin. *The U.S.-Japan Agreement for Cooperation in Research and Development in Science and Technology.* undated but about 11/1985. Appendix I, p. 1.
3. As a new agreement was being proposed, the U.S. and Japan S&T cooperation centered in nine areas of research under varying levels of institutional arrangements as follows: 1) four are based on exchanges of notes or government-to-government agreements (environmental protection, energy, space [in part], and radiation effects); 2) two are based on an agency-to-agency memoranda of understanding (building technology and urban affairs and space [in part]; 3) four are based on references in heads-of-government or cabinet-level joint communiques (natural resource development of UJNR, [basic] scientific cooperation, medical cooperation, and transportation).
4. Bloom, Justin. *The U.S.-Japan agreement for...* 11/85. appendix I, p. 2. The list of 40 projects mentioned by Bloom has not been found. These 40 projects were probably reviewed and only 29 projects presented to the GOJ. Yet even the list of 29 has not been found. As mentioned later a list of 23 projects was found and included in this

study. In USG documents discussing the objectives of the 80STA there was always a deletion. However, in a background document for the October 17-19, 1987 session (Session 1), it was clearly stated that the 80STA "was designed to accomplish the same result in other fields", i.e. GOJ non-energy projects as it had done under the Energy Agreement.

5. Memorandum of Conversation, US-Japan Science and Technology Cooperation, 4/2 /1979. p. 2-4.
6. Expanded US-Japanese Cooperation in non-Energy Research. 4/16/1979. p. 1.
7. Bloom, *The U.S.-Japan agreement...* 11/1985. Appendix I, p. 2.
8. *Scope Paper, Action Program Part II....*
9. *Expanded US-Japanese Cooperation...* p. 4.
10. *Proposal for an Expanded Cooperative Effort in S&T between the U.S. and Japan: Talking Paper* (Paper for presentation to the Japanese, p. 3-4.
11. Bloom, Justin. *The U.S.-Japan Agreement for Cooperation in Research and Development in Science and Technology.* 11/1985 as a reference item during the mid-1980s review of the 80STA. p. 1.
12. Part II Action Program. p. 2.
13. Part II Action Program. p. 4.
14. ibid. p. 8.
15. ibid. p. 8. This same document began a discussion as to whether it was "worthwhile to have a government-to-government agreement." But the reviewer of the document chose to delete this entire section before releasing it.
16. The following discussion is based largely on a "scope paper for the Tokyo visit, September 22-24, 1987, Part II Action Program". Part I was not released under FOI procedures for unclear reasons. Part II was, in places, heavily deleted but the remaining parts appear to provide a sufficiently adequate description of the U.S. view.
17. Tokyo 17020 (9/22/1979).
18. Tokyo 16997 (9/21/1979), Tokyo 16999 (9/21/1979).
19. *Science and Technology (S&T) issues.* undated State briefing paper but probably about 1987 since it was among papers of that period. 1 p.
20. Tokyo 00732 (1/16/1980). Ultimately the GOJ designated the Minister of Foreign Affairs as the GOJ agent.
21. State 8809 (1/12/1980).
22. The U.S. Science counselor in the Tokyo Embassy had warned his colleagues and Dr. Press that the GOJ probably would not make substantial contributions and would not sign an agreement in one session and that these negotiations would probably require at least six months.
23. In space the projects were geodynamics, origin of plasmas in Earth's neighborhood (a feasibility study), Halley flyby/Temple 2 Comet Rendezvous Mission (feasibility study), Saturn Orbiter and Dual Probe Mission (a feasibility study); basic research, neutron scattering with advanced instrumentation; health, center for alcohol, drug abuse, and mental health studies, recombinant DNA research; agriculture, forest tree disease pathology, integrated pest management, biomass (green energy) research.
24. State 48952 (2/23/1980).
25. White House press release, 2/14/1980 and State 042761 (2/16/1980).

26. The signed circular 175 document granting this request was not among the released papers from the State Department. The Circular 175 request was approved in some form as required by USG rules. The evaluative comments in this request are highly revealing of the thinking within the USG about a new agreement with the GOJ.
27. Bloom, Justin. The U.S.-Japan Agreement for Cooperation in Research and Development in Science and Technology. [11/1985] p. 3.
28. In all memos, review documents, talking papers, etc which mention the objectives of the 80STA, three are listed but the first two items are always deleted, leaving the third a fairly innocuous and safe objective when documents were released to me. The one exception was from *Talking points for US-Japan S&T Agreement,* undated but about 1984.
29. Memo of June 27, 1980 on US-Japan Non-energy R&D Cooperation. p. 1.
30. Bloom, Justin. *The U.S.-Japan Agreement for Cooperation in Research and Development in Science and Technology.* [11/1985] p. 3.
31. Bloom, Justin. *An Evaluation of Major Scientific and Technical Agreements in Effect between the United States and Japan.* 8/24/1983. Final contract report to the U.S. Department of State. p. 50.
32. OES memo (6/18/1986) to OES A/S Negroponte, Backgrounder and Talking Points for your meeting with Ambassador Matsuda.
33. OSTP attempted to convene the Joint Committee in the 1980s but without clearing their actions on an interagency basis. Their attempts were unsuccessful.
34. State 216882 (8/14/1981).
35. These "criticisms" in various and sundry forms were garnered from a variety of documents: OES/S Otto Eskin to OES Ambassador [John] Negroponte, Backgrounder and Talking Points for your meeting with Ambassador Matsuda, 6/18/1986, p. 2; State 202834 (6/26/1986); Joint Task Force "Plan of Action for Proposed Renewal of the U.S.-Japan Agreement for Cooperation in Research and Development in Science and Technology", 11/26/1986, p. 6. [The Plan's title page erroneously identified the task force as Joint U.S.-Japan Bilateral Task Force, something that was hoped for but never materialized.]; OES-John D. Negroponte to T-Mr. Derwinski, EPC [Economic Policy Council] meeting on Japan S&T Agreement, about July 10-20, 1987;Memo, OES/SCT to OES Ambassador Negroponte, Background paper and Talkers for your meeting with Senator Rockefeller, 9/22/1987, p. 1; Memo, M. Prochnik to Amb Negroponte, Your interview on the U.S. Japan Science and Technology Agreement with Mr. Ishida of the Asahi Shimbun [in Washington, DC], 10/15/1987, and Tokyo [date not clear], 10/22/1987, Embassy translation of this interview as it appeared in the Asahi Shimbun. Negroponte cited as an example, among others, the 300 Japanese researchers at NIH although he knew per his testimony to the Congress n (described later) that these researchers were chosen on a competitive basis from a worldwide pool of applicants to the NIH.
36. The last comments are described by Justin Bloom in his U.S.-Japan agreement ... 11/1985, appendix I, p. 4.
37. *Sangiin Yosan Iinkai* hearing 3/16/1988, p. 18. Endo was chairman of the GOJ delegation in 1987-88 bilateral negotiations for a new STA. Please note that the number of projects under the 80STA was exceedingly difficult to pinpoint. Endo cited yet another number; it seemed to be a constantly moving target which is only partially true.

IV. The Milieu for the 1988 Science and Technology Agreement

1. As the 1980s began...

U.S.-Japan relationships during the coldwar period became increasingly rocky in the economic and trade area, particularly as Japan's economic power became increasingly evident and even her scientific and especially her technological prowess began to appear to present a challenge -- perhaps perceived as a threat? -- to postwar dominance of the United States. But the politico-military relationship was put on a pedestal and immune to the pertubations of the increasingly difficult and acrimonious trade issues. The S&T aspect of this relationship was yet another dimension. It had developed a quiet, pleasant, mutually beneficial, and, from many scientists' point of view, an exciting life of its own. The S&T relationship was, in a way, a backwater concern, after all, what is there to be concerned about, or fear from Japanese S&T where they are not creative enough to win Nobel Prizes, except in unusual cases. The academic study of USJ S&T relations was usually included, almost as an afterthought, at the end of an economic study.[1] Studies about USJ trade problems, Japanese industrial policy, did not touch upon nor concern themselves with the relationship or impact of Japanese S&T policies to their industrial policies, economic growth, and USJ trade issues. The arguments with Japan had been about textiles, steel, TV, basically manufacturing technology; to this brew was added quotas, tariffs, non-tariff barriers, etc. When automobiles and semiconductors are added to the list then a sense of some anxiety creeps into the U.S. consciousness. Automobiles, well they are an unchallengeable core U.S. industry, undoubtedly the best in the world. Semiconductors, symbolic of U.S. scientific creativity and manufacturing prowess, surely could not be assailed by such as the Japanese if the playing field was indeed level. Something was amiss.

The atmospherics and tone of the S&T relationship began shifting in the latter half of the 1970s -- only 30 years after 1945. There was a burgeoning anxiety that the U.S. was losing its technological edge, not merely to the Japanese, but to the rest of the industrialized world. Under the Carter Administration there was a gentle but growing anxiety, relatively speaking, which did not lead to a pro-active policy positive or negative to the felt impending problem. In contrast, under the Reagan Administration this concern changed into a frenetic, xenophobic reaction and in regard to S&T relations with Japan to outright distrust, suspicion, and deliberate confrontation. By the end of the 1970s, hundreds, if not thousands, of articles, popular and academic, from the shrill to the somber, had been published on every conceivable aspect of U.S.-Japanese affairs. Anxious conferences were held, angry Congressional hearings were convened. We can only cite a few that are felt to have contributed seriously to creating an atmosphere about and an image of Japan. In the shrill category one could cite the article in *Fortune*, The Japanese Spies in Silicon Valley.[2] In the second paragraph of this article it describes Silicon Valley as being 'swept by unease and bitterness, stirred up by the latest thrust of Japan's commercial offensive. ... [Robert Noyce, chairman of Intel is quoted as saying] They are out to slit our throats and we'd better recognize that and do something about it. ... There's a war on and both sides know it." (p.74) Japanese operations are described in strictly military jargon not normal commercial terms. "Most of the Japanese agents ... work out of so-called "liaison" offices set up by such companies as Fujitsu and Hitachi, ... [these offices] are thinly disguised listening post staffed by experts ... who engage in complex intelligence gathering efforts as thoroughly organized as any military operation. (p. 75) ... [the Japanese] lie, steal and cheat. ... The Japanese do not manufacture better [original was in italics] chips, but they do such a thorough testing job that the failure rate ... is significantly lower than it is for U.S.-made parts sold at the same price ... testing that thoroughly would bankrupt them [U.S. chip manufacturers]. (p. 78) ... Silicon Valley executives draw some comfort from the fact that so far the Japanese have been basically skilled copiers and manufacturers, not innovator. But that may be changing, too. Fujitsu promised to be first with a production-quality 64K memory chip, the next step in memory complexity. ... the outcome of the Battle of Silicon Valley is as yet by no means clear." (p. 79) This was obviously a shrill call to action, phrased no less, in military terms.

An article in the *Washington Post* (11/24/1978) by Thomas O'Toole, described the U.S. as seen to be losing its technological edge and world leadership role in some industries notably to Japan in opto-electronics, electron microscopes, and stainless steel, home video tape recorders, TV; even Korea and Taiwan threaten in man-made fibers; in nuclear power

Sweden, France, West Germany and Canada; the supersonic commercial aircraft is British-French. This "decline" was attributed to the thaw in the coldwar, a new skepticism and hostility about the increasingly larger R&D budgets, the value and effectiveness of technology among the American people and Congress, and "U.S. impatience to wait 5-10 years for new products to emerge and profit made from investments." Another example, as an indication of the widespread and diverse sources of this anxiety was a small pamphlet prepared by the Robot Institute of America which painted a dire picture of the U.S. position, e.g. its productivity rate being at the bottom of the industrialized countries. It made nine recommendations to "promote quick and effective transfer of leading edge technologies into new products and manufacturing processes". The robot industry in the U.S. felt its position was in particular danger from the Japanese challenge. Every recommendation was focused on actions for the U.S. to take as a society, industry, government: greater savings, streamlining government-industry relations, greater government support of R&D, tax incentive, BUT no actions calling for Japanese actions or criticisms of Japan. Rather, it said that while Japan and Germany concentrated on national goals, U.S. institutions "became decentralized and divisive, creating an adversarial climate among government, industry and academia, thus rendering it almost impossible to set consensus goals and strategic plans to achieve them." [3] (p.5) Forty years after the end of WWII, the Pulitzer-prize winning writer Theodore White wrote an essay in the *New York Times Magazine* in the summer of 1985 which pushed up the decibel count of the shrill factor several notches. The title in itself, The Danger from Japan ... Japan's trade tactics pose a threat to the U.S.[4] While White praises the Japanese as "technically superb adapters and producers (p.37), ... as brilliant, efficient, aggressive people who prize education as much or more than Americans" (p. 57), his caustic assertions about Japanese objectives often described in military terms must have been chilling, frightening, frustrating with a sense of betrayal that Japan would harbor such designs after we, the U.S. had magnanimously assisted them to build their country after abject defeat. Let us assemble a few of these descriptions scattered throughout his long essay:

> ... the Japanese are on the move again in one of history's most brilliant commercial offensives, as they go about dismantling American industry. (p.22) ... the whole world has become its [the House of Mitsui in the essay but one is left with impression that this also applies to the Japanese government and people] Greater Co-Prosperity Sphere (p.31) [a reference to GOJ's objective to

create this Sphere during the Pacific War] ... The Japanese, as Government policy, are undermining one American industry after another. (p. 38) ... If the present Japanese expansion of production continues, it will, in 20 years, become a greater industrial power than the United States. (p.38) ...The Ministry of Finance provides the launching pad from which MITI directs the guided missiles of the trade offensive. (p. 38) ... The Germans, somehow, evoke little American bitterness because we understand their culture, establish American plants there without hindrance. The Japanese provoke American wrath because they are a locked and closed civilization that reciprocates our hushed fear with veiled contempt. (p. 38) ... The Japanese without mercy, propose to wipe out our supremacy in this [semiconductor] industry, based on our own research and invention. (p. 40) ... [quoting Secretary of Commerce Malcolm Baldridge] Japanese export policy has as its objectives not participation in, but dominance of, world markets. (p. 43) ... [quoting Lane Kirkland, president of the AFL-CIO] To hear the Japanese plead for free trade is like hearing the word "love" on the lips of a harlot. (p. 57)

These are truly fighting words, reminiscent of the language -- perhaps even worse -- than those used in the late 1930s just before the outbreak of the Pacific War. These anxieties echoed well in Congressional hearings and in certain USG agencies as we will find later.[5]

In contrast, Robert B. Reich tried valiantly to point out the dangers of techno-nationalism.[6] He maintained that there was "no way to separate 'our' technological advances from 'theirs'"[Japan's, Germany's or any other friendly nation]. He declared the concerns mentioned above were "misplaced": "The underlying [U.S.] problem is not that the Japanese are exploiting our discoveries but that we can't turn basic inventions into new products nearly as fast or as well as they [the Japanese] can." (p. 63) While he recognized the problems of the U.S. trade balance in high tech products, dependence on one source for selected semiconductors and other electronic parts, he insisted that "Techno-globalism, in sum, has come to be America's central albeit tacit, organizing principle for developing new technologies." The notion of 'American' technology has thus become a meaningless concept ... national boundaries have ... become less and less relevant" (p. 64). He maintained that presuming "the possibility -- indeed the necessity" of trying to make technology developed in the U.S. as "something that can be uniquely American -- developed here, contained within the nation's border, applied in

America by Americans ... as a body of knowledge separate and distinct from that possessed by other nations" (p. 66) as unrealistic and an incorrect policy. There would be three "formidable difficulties" which techno-nationalism cannot overcome: the logistical challenge of confining new knowledge within national borders, how would this principle be applied in reality in light of how technology is now developed by the DOD, U.S. universities and private industry, and finally it is not in "America's interest to bar foreigners from the fruits of our research and development." (p. 67-68) He insisted that the issue was to increase U.S. "investments in America's technological experience" and "technological learning" which have declined. (p. 69) For example, it is more important to give American workers experience in designing and producing the next generation of advanced computer than a U.S. company buying advanced chips from Japan, whether this might be IBM or Fujitsu which had been asked to buy Fairchild Semiconductor.[7]

In contrast to Reich's argument for techno-globalism, an NSC Report of early 1987 described the situation in dire terms singling out Japan specifically: leadership in R&D [in selected fields] seems to already be passing to Japan, ... [other areas] are dominated by the Japanese. The fear was that the Japanese high-tech firms could -- and by implication would -- withhold their advanced chips and related technologies from U.S. firms and thus "impede the ability of the United Sates to compete in any area of manufacturing". The idea that Japanese firms might take such action with the concurrence of the GOJ when the U.S. was, at the very same time, extolling the strength and virtues of the USJ security relationship and our relationship with Japan described as the most important 'bar none' -- a phrase created by U.S. Ambassador to Japan, Mike Mansfield -- is quite remarkable. Based on the actions taken during the 1980s "to protect" U.S. industries, technologies, from foreigners (especially Japanese) Reich's arguments, ideas, and policy recommendations fell on deaf ears and the NSC approach had clearly won the day.[8]

Based on the above, Japan was depicted as a sudden threat that was not looming on the horizon but one that was "in our faces" now so to speak, a Japan that was acting out of the mold into which we had put her. She was directly -- whether consciously or unconsciously, deliberately or just as a result of the flow of human affairs was irrelevant -- challenging the presumed core of our own psyche, technologically and commercially innovative, and, in the process causing us to be uncomfortably dependent upon her for critical military components even though we said to the world and to ourselves that the U.S.-Japan relationship was the most important bar none. Yet, somehow

Japan could not quite be trusted to provide these necessary parts in a time of crisis. The 1980s began with a torrent of bleak, dark, doubting images of Japan which now enveloped the hitherto quiet S&T front and could only bode ill for the politico-military-economic aspects of the USJ S&T relationship as distinct from strictly technological exchange and cooperation.[9]

I would emphasize that the purpose of this chapter is not to conduct yet another detailed study of USJ political, economic and military relationship but to depict the unfolding and evolving milieu in and out of the U.S. government in its rather unusually broad array of manifestations swirling around its S&T relations with Japan. I will start with the semiconductor confrontation, the availability of JSTI, the Wisemen's Group, S&T comparative analyses, USG policy studies on S&T relations, institutional arrangements, high level "private" conferences on these relations, patents, culminating in the so-called Toshiba-Kongsberg Incident and the FSX controversy in more or less chronological order.[10]

2. The Semiconductor Confrontation

The USJ collision over semiconductors in the 1980s was chosen as the symbol of the milieu as we enter the 1980s and in which the STA was discussed, analyzed and negotiated. There are, of course, incidents, disagreements of varying degrees of importance, depth of feeling and anxiety, but the rise of Japan in the manufacture and marketing of semiconductors was regarded as utterly crucial by the political, scientific and technological leadership in both countries to the future economic security and position of their countries in world electronics. They looked upon the evolving dynamics of this equation as symbolic of the potential rise of Japan, the potential decline of the U.S.

When the famous ENIAC computer was created in the U.S. in 1946 just after the end of WWII using 18,000 vacuum tubes, when the transistor was created in the famous Bell Laboratories in 1948, the U.S. reigned supreme in science, technology, industrial might, and military power. There was no serious competition in the electronic world to the U.S. -- unless the image of the struggling former Soviet Union could be stretched for this purpose. In the mid-1950s, the license to use the transistor in Japan was given to one obscure fledgling, oh so fledgling, Japanese businessman,

Morita Akio, who later created the Sony Corporation. While the consequences of this license were profound, at the time, it was, probably more an irritant fly in the ointment of their grand designs from MITI's point of view and of little moment for the inventors of the transistor. Barely ten years -- and this short time should be emphasized -- after WWII ended, when Japan was still not recovered, was still struggling to recoup its living standards to 1935, was hardly taken seriously in anything, let alone electronics, MITI officials -- together, I am sure with Japanese industry -- made a critical decision to make the development of the electronics industry a top priority national endeavor and the Diet passed a law for this purpose in 1957.[11] While this law was a major factor in creating a formidable world class electronics industry in Japan in the future, the perception was that the economic security, well being and a higher standard of living for future Japanese generations lay in electronics and the achievement of this goal would make a major contribution to Japan "catching up with the West", i.e. in the postwar era, the United States. Only a few years later, the GOJ decided to emphasize the computer as its target. It should be remembered that this was a consciously one-sided sense of competition by Japan. The United States government, industry and academia were too sure of themselves to consider this a serious competition - until many years later.[12]

The GOJ thus took steps to protect its fledgling computer industry, by providing tax incentives, "persuading" IBM and Texas Instruments (TI) when they wanted to manufacture in Japan to share patents with Japanese companies, limiting market share, and creating leasing companies to buy Japanese computers as soon as they were made. During the same period, U.S. companies pushed with the creation of the integrated circuit in 1959, the microprocessor in 1971 and the manufacture of successively more powerful semiconductors from 1K, to 4K to 16K RAM (random access memory) chips in 1975. This forced the issue on the Japanese who decided in the mid-1970s to develop the 64K RAM, ahead of the U.S., and organized industry for this purpose. In 1978-79, the Japanese 16K RAM captured 40% of the world market stunning the U.S. industry. The Japanese, threatened with a law suit on dumping by U.S. companies, raised prices somewhat in the U.S. and lowered prices somewhat in Japan. But the damage was done. Then the Japanese introduced samples of the 64K RAM in 1980, produced 9 million in 1981, 66 million in 1982, and by 1986 had 65% of the world market. To deepen the sense of anxiety on the U.S. side, NTT introduced a prototype 256K RAM in 1980.[13]

This resulted in a deep sense of crisis in the U.S. Not only did

Japanese industry appear to be threatening the very core of U.S. technical ingenuity and creativity and manufacturing prowess, but also before long the U.S. military weapons systems would be unhealthily dependent on a foreign source, almost exclusively at that time on Japan. "Something must be done" by whom, for whom, taking what measures, for what purposes was not at all clear on the U.S. side. We were hobbled by U.S. laws, by, what I call, our own unstated, unspoken, "official ideology" about the role of the government and the so-called "free market system". The U.S. wants a more open market in Japan, no dumping; to monitor Japanese actions we called for collection of data on pricing, shipments of semiconductors worldwide, and other measures. Our commercial interests were at odds with our national security relations with Japan and did not coincide with White House priorities. There was also the growing irritation, frustration and anger in the U.S. Congress about Japanese trade practices, the increasing trade imbalance, lack of responsiveness. According to Prestowitz, then an active player in the DOC, the negotiations with the GOJ "never focused on the main issues and ... dealt with symptoms instead of causes. (p. 147) ... Our only tools were bluff and persuasion." (p. 150). Nevertheless, the first of a series of agreements was signed in November 1982; the High Technology Working Group produced a lengthy agreement of nine pages on February 10, 1983 agreeing that monthly data on semiconductor trade will begin immediately, that the U.S. will have improved access to GOJ sponsored projects in supercomputers, fiber optics and advanced semiconductors -- this is a familiar arrangement to be repeated over and over during the 1980s and 1990s. Nothing much happened. Another agreement was signed in November 1983 with the Japanese chairman providing a confidential chairman's note stating that MITI would "give guidance", "encourage" Japanese chip users to buy more U.S. chips. Ironically, to quote Prestowitz again:

> ... like disadvantaged U.S. minorities, we wanted an affirmative-action program that would offset the effect of past discrimination [by the Japanese] by actively working to increase imported chips." (p. 153)

Again, the agreement seemed to work but market conditions first went up for semiconductors and then down and down; Japanese stayed in the market capturing 90% of the worldwide market, U.S. companies withdrew contributing to a massive sense of crisis in the U.S, with a perception of "unfairness" by the Japanese. Section 301 of the 1974 Trade Act gave U.S. companies a recourse to file dumping and unfair trade accusations against

Japanese producers. This they did in mid-1985. But U.S. industry was initially not united in supporting USG action against the Japanese. Eventually, however, the USG also initiated its own dumping case on 256K RAMS, and the other private industry initiated legal actions. In June 1986, the GOJ was informed that the U.S. Cabinet was willing to declare Japan as an unfair trading country. In two months, a third semiconductor agreement was signed on September 2, 1986. In the agreement Japan agreed to assist U.S. companies to sell chips in Japan, to monitor prices and costs of products exported worldwide to prevent dumping; in a side letter, the GOJ agreed to help U.S. companies reach a goal of 20% market share in Japan. Did the GOJ "guarantee" or did it only agree to "try to persuade and encourage" Japanese companies to buy more U.S. chips. The U.S. maintained the former; the GOJ maintained the latter. This bone of contention became so important and unpleasant that the GOJ thereafter would never agree to any numerical target for fear that no matter how they were worded they were would always be "misinterpreted" -- in their eyes -- and twisted to mean a guarantee.[14] Dissatisfied with the progress achieved under the 1986 Agreement, the USG imposed sanctions (tariffs of 300%) on selected Japanese products, but in doing so found that there were far fewer products which did not depend on Japanese semiconductors -- thereby revealing the depth of U.S. dependency on Japanese semiconductors. Two months later at a USJ Summit, since there had been less dumping and the sanctions were partially lifted for Prime Minister Nakasone Yasuhiro's visit. By 1992, the foreign share (mostly U.S.) had reached 16.7%, 1994 22.4%, 1996 27.5%.[15] Seemingly, from the USG point of view, the 1986 Semiconductor agreement had been successful over a ten year period. But the atmospherics, whatever the causes may be, left a heavy pall over USJ relations of distrust and frustration, a determination by the GOJ never to agree again to a numerical target with a somewhat similar inclination by the USG to pursue numerical targets as a way to obtain "concrete results". Numerical targets were not agreed to in the future. Inevitably, this pall hung heavily without direct impact like a dark cloud's persistent, gloomy and dampening effects.

3. The Availability of Japanese Science and Technology Information

a. Congressional Hearings, 1984 and 1985

During the 1970s and 1980s numerous Congressional hearings were held to try to fathom how the Japanese had achieved high productivity and what lessons might be derived for the U.S., high rates of growth, how open (or more accurately how closed) was the Japanese economy, the possibility of revising U.S. anti-trust laws to accommodate joint R&D ventures among U.S. companies to meet this challenge and many other topics. Congressional hearings provided a forum for a steady drumbeat castigating Japan and its practices with little impact on what the U.S. itself should do. There were no hearings specifically on Japanese science and technology and what lessons, if any, might be learned from such hearings. To hold such hearings would appear contradictory: how could one hold serious hearings about Japanese S&T if it is assumed that the Japanese were merely good imitators with occasional creative flourishes. After much persuasion and utilization of connections, the first hearings ever devoted specifically to the availability of JSTI were held in 1984; it was hoped that, if the Japanese S&T challenge was as serious as it was perceived to be, these hearings could be held each year to assess the situation. A second set of hearings were held in 1985 and none since that date except on the STA itself in 1987 and 1988.[16]

The hearings tried to be comprehensive by calling many witnesses; there were many suppliers of JSTI but few users who came to these hearings. The hearings studied Japan's efforts to coordinate STI, U.S. efforts to acquire STI by federal agencies, private sector and university language problems relevant to JSTI.[17] Clearly the role of the Federal government was emphasized in the numerous recommendations to correct the perceived situation: increase funding of language training, translation capabilities, support increased and improved collections of JSTI by U.S. libraries, increased funding of NTIS, increased assignments of technically trained personnel at the Tokyo Embassy, creation of a national information policy, and naturally more studies of the JSTI situation in the U.S. Barriers in the U.S. and Japan were commented upon; in Japan STI is not so systemically organized, thus not so accessible, thus "a barrier"; the copyright systems in Japan and the U.S. were different: Japanese government reports were copyrighted while those in the U.S. were not, thus the dissemination of GOJ reports involved copyright permission, thus "a barrier". On the other hand, the witness opined that "the most critical barrier to acquiring, translating, and disseminating" JSTI in the U.S. was "the lack of awareness of the value of this information." (p. 19). Lack of effective access to JSTI can be attributed "to this negative attitude on the part of Americans rather than any specific attempts by Japan to restrict access to their data" (p. 19); lack of coordination among federal agencies, unwillingness to commit greater resources for the dissemination of JSTI, reluctance by trade associations and private companies to commit resources to establish technical scanning operations to monitor JSTI; the number of American technical specialists who can use the Japanese language in their professional work "is virtually nil" (p. 21). The clear conclusion from these hearings was that the onus was squarely on the U.S. side with a relatively minor problem on the Japanese side, that the Federal government was being called upon to play a far more active role in remedying the situation. Little, if anything, was done about any of these various suggestions. There were no substantive criticisms or attacks on the Japanese system and accessibility -- again, in contrast to the aggressive USG stance only a few years later toward JSTI and its accessibility and the attempt to keep track of Japanese S&T in the U.S. Congress was short-lived, without impact and seemed to reflect in the testimony of the witnesses another disconnect from the attitudes emanating from some federal agencies in the later 1980s.[18]

b. Seminars, Symposia on USJ S&T Relations

It was a milestone that as the 1980s began the first symposium solely devoted to U.S.-Japan S&T relations was held in Washington, DC, in 1981 under the sponsorship of the Japan-America Society of Washington. The keynote speaker, James C. Abegglan, Vice President and Director, Boston Consulting Group in Tokyo, pointed out that the one-way street in the flow of technology from the U.S. to Japan was turning around citing the U.S. steel industry contracts with their Japanese counterparts to provide technical assistance, and the introduction of robot technology into the U.S. from Japan raising the question whether "we can compete with Japan in our ability to use imported technology, but clearly this is a topic whose time has come." (p. 11) Even in the semiconductor field, "Japanese firms are making important contributions to semiconductor know-how. The disquieting aspect of the current situation is continued unwillingness of the Japanese government to completely open its restriction on technology flows. ... especially the know-how generated by the VLSI program" which may lead to intensified friction between U.S. and Japanese semiconductor industries. (p. 71) Nevertheless, the impression persisted that the flow was still to a large extent one way.[19] The second such symposium held five years later in 1986 in Washington struck a distinctly different tone, "Patterns of Interdependence". The technology flow was found in new materials, mechatronics, computers and communications, biotechnology was no longer one-way but approximated more of a two-way street. Gerald

P. Dinneen, Vice President, R&D at Honeywell Corp significantly observed that "Japan is at the point where it feels it has hit a technology ceiling and must create its own technology ... Shifting the emphasis from manufacturing toward basic research could make Japan less competitive in the years to come." assuming that Japan follows in the footsteps of the U.S. in this regards which is a dubious assumption to make. (p. 24-25) The concluding speaker, Justin L. Bloom, tried to paint a more optimistic solution by suggesting 1) an improved access to JSTI, 2) investigation of technological aspects of trade barriers, 3) increase the number of U.S. scientists and engineers in Japan and 4) increase Japanese contributions and role in "big science". However, he could not avoid "somewhat ambiguous conclusions" about creating a new era of enhanced technical cooperation between the U.S. and Japan because of "the abysmal lack of knowledge" -- perhaps "the greatest barrier to a new era" he opined -- in the U.S. about the extent and depth of existing S&T relations. His observations were indeed prescient in what followed in the 1980s (p. 244-147).[20]

Let us turn now from the overview to the more mundane and narrower subject of JSTI which during the next decade will come under more and more, closer and closer effective scrutiny. A relatively small group of academics, independent consultants, concerned USG governmental officials, and others began organizing a variety of meetings, symposia, conferences at government facilities, universities, etc in a vain attempt to arouse the consciousness of government, congress, industry about the need to focus on the availability of JSTI in the U.S. At the time, the stress was on the availability of JSTI, not access to R&D facilities in Japan; there was no concern yet about "symmetrical access" issues. It was innocently assumed -- strictly an unwritten assumption -- that if U.S. S&Es could be apprized of the availability of JSTI they would utilize these opportunities. From 1983 to 1986, an almost yearly conference was held to study the availability of JSTI, how to expedite JSTI availability, how to disseminate JSTI, the need for studying technical Japanese, the need for enhancing library collections of JSTI, etc etc.[21] It attempted to elucidate the systems in the U.S. for collecting JSTI, how it might be acquired through public agencies and private firms, existing library collections for this purpose, and how JSTI might be disseminated. A 1985 meeting became more sophisticated with the introduction of machine translation problems and possibilities, the Japanese were now invited to participate, language teaching technologies and the problems of creating a high tech information company in the U.S. The ONR/NSF 1986 meeting tried to put some specific direction into the entire effort in mastering JSTI.

After several years of attending these conferences, it became clear that the hardy souls were preaching to the converted and that no real progress had been achieved. At least, this writer would be hard pressed to point to any specific successes, new operandi, new actions, new vigor and new galvanizations to exploit JSTI. These conferences were basically not intended to produce recommendations nor actionable proposals. While these kinds of get togethers gradually petered out in the 1980s, another more structured and formal series was begun with an annual JSTI meeting sponsored by the JICST/GOJ and NTIS/DOC alternately on the east and west coasts of the U.S., and later when the Japan Documentation Center (JDC) was created in 1994 in the Library of Congress with the financial assistance of the GOJ through the Center for Global Partnership in the Japan Foundation, it also sponsored annual conferences on JSTI.[22] All these conferences were elaborate affairs with numerous paper presentations on every aspect of JSTI, on how it could be obtained, mined, and utilized.[23] Yet they appear quite repetitious and questionably effective in achieving their goals.

In addition to the U.S., a series of Europe-oriented 2-3 day JSTI conferences were begun in 1987 in England, and were held in 1989 in Berlin, in Nancy France in 1991, in 1996 again in England and the fifth conference in the U.S. in July 1997. These conferences were much broader in scope, more elaborate and more sophisticated in topics covered than any U.S. conference up to that time and were thoroughly international in representation including a strong contingent from the U.S. The topics covered included trends and policies, information sources, analysis and distribution, language and kanji processing and the direct connection, with 45-50 papers. When I attended two conferences (Berlin and Nancy) there was the distinct feeling yet again that while the papers were excellent, thoughtful, fascinating and distinctly more sophisticated in topics chosen and in the analysis and are now years later expanding into other areas (e.g. internet) than JSTI, we were really still talking to the converted, many suppliers of JSTI, collectors of JSTI but few users of JSTI were in attendance. They were more intellectual symposia rather than for the propagation of JSTI -- JSTI was now becoming the vehicle around which a symposium was held. A large Japanese government and private sector contingent always seemingly participated eagerly and were helpful in explaining JSTI systems and availabilities. Consciousness was heightened among the attendees but I am unaware of any breakthrough in the utilization of JSTI by the European and U.S. participants.[24] Like the much smaller U.S. conferences, no proposed actions or recommendations were specifically proposed or acted upon.

It is ironic, yet sad and even pathetic that the U.S. over fifteen years later is still holding conferences of various kinds among ourselves to persuade U.S. researchers that it would be a worthwhile effort to study, mine, and exploit JSTI; almost always Japanese government and industry representatives provided presentations at their own expense to explain "the mystery" of JSTI and what was available. To the best of my knowledge, Japan is the only country that is subjected to such detailed scrutiny on how to obtain its JSTI and the real value, quite aside from the intellectual curiosity and titillation that these conferences arouse, is rather difficult to ascertain.[25]

c. 1988 Ceramics Study

Between 1981 and 1988, advanced ceramics was the only field where as many as four comparative studies were conducted obviously reflecting a high level of concern and anxiety about the implications of Japanese achievements and determination in this field.[26] The 1984 study concluded that the Japanese have a long term commitment to ceramics R&D, management's commitment in this regard is "more solid in Japan than in the U.S." (p. 9), that, despite this commitment, "the Japanese do not appear to have a technical edge over the United States in the field of ceramics in general" (p. 10), that the Japanese could welcome cooperation with the U.S., especially in the development of standards, exchange of non-proprietary information, and cooperative projects between universities and government laboratories. It urged the U.S. ceramics industry, USG and academia to obtain timely access to published research in Japan through this cooperation, develop better communications with Japanese counterparts, capitalize on R&D results from DOD funding, and create a Ceramics Industry Association. Even in the mid-1980s, there were no notes of Japanese reluctance to cooperate, quite the opposite, the report focused on what the U.S. needed to do to keep abreast.

In light of the above emphasis, it is interesting that the only study -- to the best of my knowledge -- on the demand for and utilization of JSTI by U.S. institutions was done in 1988 using the ceramics industry as a case study. This one study from the 1980s stands out in my mind from my experiences in that decade as symbolic of the attitudes prevailing at the time among U.S. scientists, researchers and engineers. The purpose of this survey was to try to find out about the demand for JSTI in advanced ceramics in relation to the availability of JSTI from suppliers. This was a technical field in which the Japanese were a strong competitor to the U.S. It was the first study based on a survey with a quantifiable result; it surveyed large companies, small and medium-sized companies, universities, national laboratories and FFRDCs (Federally Funded R&D Centers), a total of 100 with a response rate of 75 (the latter two groups provided 100% response).

The general conclusion was that while U.S. organizations were aware of developments in Japanese ceramics, a large amount of attention was not given to collecting, analyzing and integrating technical information from Japan and that in contrast to other fields, advanced ceramics groups had few joint ventures or licensing arrangements with the Japanese. The specific causes cited were the lack of long range planning because of short-term pressures in the U.S., lack of language ability, and the cost of collecting, analyzing and translating JSTI was overly burdensome on especially smaller and medium-sized organizations -- the standard litany of "obstacles". Both those organizations with or without JSTI programs felt that the Japanese ceramics industry was equal to or somewhat greater in strength compared to that of the U.S. and that this trend will prevail over the following five to ten years. Ironically those larger and smaller companies without JSTI programs were more optimistic of future U.S. gains than with JSTI programs. This study seems to tell us that there was a consciousness that the Japanese were a strong competitor to the U.S. ceramics industry, greater utilization of JST might be made but the cost of and returns from such an endeavor did not seem to bestir the U.S. industry to action. This study further seemed to imply a distinct contradiction to the clamor for more availability of JSTI either by the government or private sector; the felt need as to which group should provide such a service was evenly divided, notwithstanding the philosophical argument that it should be the role of the private sector to provide this service. We could not draw a general conclusion from this study that there was an overwhelming sense of threat from their Japanese competitors, that there was an urgent need for making available JSTI either by the USG or the private sector, and that there was no felt substantial need to establish JSTI programs. While recognizing that a generalized and extrapolated conclusion must not be drawn from a study of one specialized industry this "laid back" stance stands in contrast to the growing thunder about the Japanese S&T R&D threat spreading throughout the land.[27]

4. 1984 Wisemen's Group Report

In light of the deteriorating status of affairs between the U.S. and Japan centered in economic and trade relations, one of the methods used by both governments was to appoint what was commonly known as a "Wisemen's Group". Both Presidents Carter and Reagan appointed such Groups.[28] The country members of this Group, appointed in 1983 respectively by the U.S. President and the Japanese Prime Minister, were to make recommendations across the board in USJ relations in order to defuse, smooth over, pour oil on troubled waters without igniting them further. It was hoped that it would make genuinely useful and constructive suggestions which could be expected to be carried out by both countries.

The second Wisemen's Group officially was called the United States-Japan Advisory Commission; it's report, Challenges and Opportunities in United States-Japan Relations, was issued in September 1984. As one might expect it began with the usual litany of problems and ended with a statement of the potential: No two countries possess the essential ingredients of a vital bilateral partnership -- economic, political and security ties -- in greater degree than the United States and Japan. There are few areas of basic conflict. "... if we work together, we have an unprecedented opportunity to lead the world into a new century of economic growth and political stability"(p. 107). It pontificated about what each country should do in trade flows, market access, agricultural and forestry trade and in services, investment, industrial policy, energy, defense, communication. As far as science and technology is concerned this report broke new and important ground by including a chapter on science and technology, thus for the first time, S&T had been given equal recognition. By implication this was a recognition by the U.S. of the importance of the Japanese S&T establishment, its achievements and future potential.[29]

In light of future developments in the USG inter-agency process and bilateral negotiations, this report made some rather important observations and recommendations about the USJ S&T relationship.

- The scope of public and private S&T cooperation between the USJ "was unique" (p. xvii)

- "There is now an excellent qualitative balance in the flow of technological ideas between the two countries." (p. xvii) This represents a most egregious disconnect between what such a high ranking panel believes -- or is willing to publicly state -- concerning USJ S&T relations and what the USG says to itself, to the Japanese and to the public.

- It called for a high level review by both governments of the 13 government-to-government S&T agreements for possible "improvements and new directions". This idea was later reflected in an internal USG task force report; but a "joint review" never took place. (p. xvii and p. 98)

- Japan should take a stronger role in generating basic knowledge and the U.S. greater attention to new-product innovation and production technology. (p. xvii) This is somewhat odd in light of the famed American innovative imaginative creativity for new products.

- Encourage joint efforts by USJ high technology firms especially where initial costs of product development were high (p. xviii). In many instances this has happened as a matter of economic necessity.

- Carry out various measures, including Japanese language training of U.S. S&Es so both groups could more effectively work together (p. xviii). Another way to foster such cooperative work would be to place joint projects in Japan (in the past almost all joint projects have been located in the U.S.) The language training of U.S. S&Es was increased; joint projects were not placed in Japan.

- A joint advisory body should examine technical standards for harmonization purposes. This did not occur. (p. xviii)

- An organization similar to the U.S. Academy of Engineering should be created in Japan. This suggestion was carried out.

- Machine translation. In light of the "language problem" -- is a more neutral term than the negative word, "barrier" -- it called on the two governments to work together on MT. The USG invested some monies in MT, conducted a study of MT technology in Japan but a joint effort never took place. Eventually -- 12 years later in 1996 -- the USDOC imported, on a loan basis, a Japanese MT capability in order to expedite the translation of JSTI.

- It also encouraged collaboration on the fifth generation computer which basically did not take place and high energy physics which had a long standing history of collaboration on an equal basis, and in light of the major effort by Japan in life sciences and cancer research it urged a greater US S&E presence in Japan in these fields. This has come to pass in a limited manner.

It is difficult to assess as to whether these recommendations in S&T had a direct impact on later developments. Some suggestions were apparently carried out, e.g. the creation of the Japanese Academy of Engineering is a good example. But the comments on the fundamental relationship, being unique, in an excellent qualitative balance in the technological flow of ideas, were either ignored, disbelieved or deliberately and blatantly distorted in later internal USG discussions and assessments. But the report is an important source describing the evolving nature of the USJ S&T relationship, notwithstanding the whims and prejudices of selected agencies in later years.[30]

5. Comparative Technology Analyses

Faced with the Japanese S&T Challenge, there was a sudden and urgent need to assess where the U.S. stood in various technologies in comparison to Japan, in particular, and to Europe incidentally. It began to dawn on concerned USG officials and others outside the government that the flow of technology to the world, especially to Japan had resulted in perhaps a genuine threat to the predominance of the U.S. in not merely one technical field but other unknown numbers of fields. Basically no serious attention had been given to Japan's persistent and continuous accumulation of knowledge, experience, know-how, and resources, bit by bit, byte by byte. This resulted in U.S. frustrations and anger over its seeming inability to reverse this trend at will -- notwithstanding all the demands made upon it. While recrimination and irritations emanating from the Executive Departments and Members of Congress, there were frequent reports in the press about strategic and tactical alliances, licensing agreements, mergers, between U.S. and Japanese corporations. With rare exceptions, large U.S. corporations were quiet, while the air between the two countries, ever so unfortunately, became more polluted with mistrust and disaffection. It is in this context that comparative analyses began.

Comparative technology studies began to appear in the 1980s and continue to the present (2000). These studies were conducted by both Japan and the U.S.; they were both multi-technology surveys and single technology studies. The GOJ, in general, conducted multi-disciplinary studies covering a broad spectrum of technical fields and appear to have genuinely used them in periodically modifying their S&T policy directions. The USG (and some industrial associations) conducted multi-technology surveys and many comparative technology studies over a long period but it is difficult to assess what actual impact these studies have had on USG S&T policies and even more difficult to comment on what effect, if any, these may have had on U.S. private industries.

Basically two kinds of comparative surveys were conducted by the U.S. and Japanese organizations: comparative analyses of selected technologies and multi-technology surveys. The former were conducted almost completely by the USG, and various kinds of U.S. and Japanese institutions conducted the multi-technology surveys.

a. Bilateral Comparative Analyses

The first known comparative analysis of selected technologies was in 1981, the first year of the Reagan Administration, on steel, electronics and automobiles.[31] But Japanese progress in steel and automobiles was distressingly well known by this time. In the mid-1980s, a new series of bilateral comparative analyses was begun by the U.S. Department of Commerce and came to be known as "JTEC studies"; these U.S.-Japan comparative studies continued in this format for almost a decade until they were gradually widened to include, China, Europe, Russia, and other countries as "WTEC Studies".[32] Between 1984 and 2000 more than 55 such comparative studies covering the principal leading edge technologies had been conducted, more than one half under the JTEC[H] labels.[33] This is an unusually large number of comparative studies to make of two countries who were presumably close allies. Since this was begun during the height of the Cold War - the Soviet Union was described as the Evil Empire by President Reagan -- the existence of a similar series of U.S.-Soviet studies was not known at the time and has not appeared since the end of the Cold War. A similar series of studies was not known to have been conducted comparing European and U.S. technologies. It is indeed ironic that Japan after being derisively caricatured as a copy-cat, imitator, with little creativity, is suddenly the target of a sustained series of comparative studies not of conventional already commercialized technologies but the principal leading edge technologies where presumably much creativity is at work. These comparative studies were all team-led studies (except for a very few) and were funded almost exclusively by some agency of the USG.

One of the early bilateral comparative analyses was on computer science sponsored by the National Research Council in 1982. It concluded that Japan had come from place zero to second only to the U.S. in two decades. In telecommunications Japanese components are now the world's best. Japan will be the most serious competitor of the U.S. in the commercialization of biotechnology. CAD/CIM [computer-aided design, computer integrated manufacturing] in semiconductors, Japanese CAD applications were comparable to the U.S. and in CIM Japan was far ahead. These studies, over the past 16 years constitute a wake-up call to the U.S. and a unique body of knowledge about the technologies of two countries. These studies were conducted in an atmosphere of cooperation, accommodation, goodwill, devoid of acrimony among the scientists and engineers on both sides of the Pacific, in contrast to the fulminations on both sides concerning trade in particular.[34]

b. Multi-Technology Surveys

In contrast to the above description of bilateral comparative analyses of selected technologies, Japan, the European Community and the United States conducted broad brush assessments of technologies and presumed directions in which leading edge technologies were heading in the next 10-20 years, and how each stood in relation to each other and with Europe. The motivations for conducting these surveys were quite different and ultimately, it must emphasized, that these assessments are highly subjective and perhaps -- perhaps even probably -- skewed in certain instances for political purposes, to secure more funding, to raise the anxiety level about the future of S&T in relation to the future economic security and/or military security of the country concerned or in reverse downgrade one's relative position in order not to raise the anxiety level of another country. These broad surveys can be used as input in determining national S&T policies or promoting or emphasizing certain advanced technologies for commercial promotion, or for a dialogue between the government, academia and industry. They could also be used as a factor in deciding upon the future of S&T relations with another country. Despite the differing purposes for which these comparative analyses were made, they identify a highly similar set of technologies which were regarded as most important.[35]

These surveys are included in this study not because they played any known direct role in USJ S&T relations, but rather because the results of these surveys in the 1980s and early 1990s contributed to a sense of anxiety, satisfaction and apprehension, and the felt need to adopt certain policies with a greater sense of urgency; they were one of the many streams of activities that made up the milieu surrounding the evolving USJ S&T relationship in the 1980s and 1990s.

i. Japanese Surveys. The Japanese government began a series of studies in

1971 to assess the direction of technological developments and to forecast the kinds of technologies which would come to fruition about thirty years hence according to the predictions of Japanese experts. These studies have been conducted every five years since then. They were not comparative studies per se between Japan and any particular country; they were probably part of the input in creating GOJ S&T policy.[36] In addition, a Japanese bank, AIST/MITI, a newspaper and a private group reporting directly to MITI's deputy minister created comparative analyses of technology levels between Japan, the U.S. and Europe. In light of Japan's national policy to "catch-up" to the West in S&T it is understandable that they would prepare such studies to assess how far they have come (with a sense of satisfaction) and how far yet they must go with investments, human resources and creativity (perhaps with some sense of dismay). In essence these studies would give the Japanese a sense, a broad brush picture of where they stand. The results, implications of these studies rather than any one study had an impact on the constantly evolving GOJ's national S&T policies. Since these assessments were ultimately subjective, they were based presumably on informed opinion. While all surveys placed the U.S. on the whole ahead of Japan, the latter was closing the S&T gap.

For example, AIST/MITI published the first such comparative survey in 1982 finding that 51 key technologies where Japan was superior and 56 for the U.S.; Japan led Europe in 57 to 29 technologies. The 1984 Industrial Bank survey said Japan shared superior positions with the U.S. and gave Japan the lead in only robotics, sensors and fine ceramics. The MITI Deputy Minister private group gave Japan a lead in five technologies, the U.S. 6 but in ceramics -- in contrast to the above study -- an even position with the U.S. In 1988 MITI published it's first (and last) White Paper on Industrial Technology where it compared technological levels in 1983 with 1988. It concluded that during this period, Japan had equaled the U.S. in a number of technologies, even moved ahead in some but that the U.S. had not equaled or surpassed Japan in any field in which it was lagging in 1983.[37] Are these studies an accurate reflection of reality, were the Japanese deferring to the United States, were they goading themselves to greater concentration and call to further action? Let us assume that they do reflect more or less the reality of the USJ situation in S&T. In that case these studies as a whole must have been alarming to the US government, industry and academia; these studies were not concerned with already commercialized technologies but with leading edge ones of the future, areas where the U.S. presumed it was the natural leader, apparently challenged not by Europeans but by their erstwhile enemy in WWII, and now the U.S.'s most important ally in Asia.

ii. U.S. Surveys. Private Sector Analyses. In contrast to Japan, similar studies in the U.S. were first initiated by the private sector, became the prototype for U.S. critical technology studies and carried on, in the main, by the DOD (and incidentally by the DOC).[38] Coming as they did in the 1980s, and notwithstanding any rationale for guiding national policy, better and more efficient allocation of resources, these studies surely resulted from and are a reflection of the ballooning anxiety in the 1980s and early 1990s about the unseemly -- from the U.S. standpoint -- Japanese S&T challenge.

The Council on Competitiveness, formed in 1986 by major U.S. corporations, issued several reports on the issue of competitiveness beginning in 1988. The formation itself of this Council and the focus of its reports on this subject is symbolic of the concern that existed in U.S. industry about the relative position of the U.S. in relation to Japan and Europe. The findings and recommendations of this industry supported Council in its 1988 report were significant indeed. While recognizing that the private sector has the primary responsibility for U.S.'s competitive position, the Council's report focused on what the U.S. federal government could do, asserting that while "America must respond" "Increased government attention to, and support for, the commercialization of technology is a vital part of that response." Let us cite a few of the findings: the federal government has not focused on the commercial application of technology as a public policy issue which is in stark contrast to the nervousness of the Reagan/Bush Administrations in any involvement in technology assessments and planning, large federal budget deficit plus low savings drive up the cost of capital, USG mechanisms for determining S&T priorities were insufficient, federal investments in education facilities and equipment for S&T are inadequate, lack of consensus on scope, mission and relevance of federal laboratories to 21st century needs. The four broad recommendations with many subparts stressed the need for the USG to improve the macroeconomic environment for better application and development of technology, improve the machinery for S&T policy, increase USG investment in educational S&T infrastructure, and widen the focus of national R&D efforts. While such a report would understandably focus on what the U.S. should do for itself, rather than on what other countries should do, it is nevertheless interesting that, for once, Japan is not specifically singled out at all for actions to be taken. The report said that its concern was "the erosion of that lead [the U.S. lead in S&T]."[39]

This report was updated in 1991 and 1994 and the contrast in assessment between 1988 and 1994 is especially noteworthy. In the 1991 report the role of critical technologies takes complete center stage noting that 1) there is a domestic and international consensus that "critical generic technologies" drive economic growth, 2) that the U.S. position in these technologies is slipping, in some cases been lost altogether, and future trends are not encouraging, 3) foreign governments are pursuing leadership in these technologies, 4) U.S. policy does not adequately support U.S. leadership in critical technologies and U.S. priorities do not sufficiently address these issues, 5) most of the critical technologies are already with us waiting to be transformed into products and services. Again, the report calls upon the President to make technological leadership a national priority, federal and state governments develop policies and implement program to assure that the U.S. S&T has a world class infrastructure, develop networks in industry to enhance its competitive position, set a goal to surpass others, academia should enhance its cooperation with industry.[40]

The 1991 report asserts that its analysis is different from other studies in that it focuses on "the needs of a spectrum of U.S. industries rather than on the needs of particular industries or government agencies" in the following technologies which will be important over the next ten years: materials and associated processing, engineering and production, electronic components, information, power-train and propulsion. The lesson derived from this exercise was that there was an identifiable vital core of technologies for competitiveness, that rather than breakthroughs incremental evolving technologies in process and manufacturing (as the Japanese have eloquently shown) are most important and involve human know-how. The first conclusions drawn from this study was that "the strong U.S. position of a decade ago has deteriorated significantly. U.S. industry has already lost several technologies that are critical to industrial performance and is losing badly in several others."[41]

Only three years later, in 1994, the Council on Competitiveness was able to say that the U.S. had come back in technologies where it was lagging, had maintained its position where it was already strong, and noted additionally that in new critical technologies (multimedia systems, compression technologies, reusable software, rapid prototyping, fuel cells and micromechanical systems) it was competitive. It was an incredible turnaround: from a weak 1991 position to competitive in several fields, from losing badly or lost to weak in four fields. The U.S. "has significantly strengthened its competitive position in critical technologies during the past five years."[42]

Throughout these series of studies, the U.S. was always compared to

Japan and the EC. Can a rather disheartening situation in the 1980s-early 1990s be so quickly turned around from one of despair and malaise to such optimism which also was echoed in the DOD 1995 Critical Technologies Report. Can Europe and Japan have fallen from such earlier eminence, or obversely could the U.S. have genuinely made such unusual recovery in only three years? Did the earlier studies depicting EC and Japan in such glowing terms seriously under-estimate the U.S., over-estimate Japan and the EC? If this was the case, there were unwarranted serious repercussions on how USJ S&T relations were perceived, both in and out of the USG.

Government Analyses. DOC prepared a report, *Emerging Technologies: A survey of technical and economic opportunities* in 1987 and again in 1990. The latter report concluded that "If current trends continue, before the 2000, the United States could lag behind Japan in most emerging technologies and trail the EC in several of them." The DOC studies were a passive endeavor with little impact and soon overtaken by legislation requiring the DOD National Critical Technologies Plan.[43]

The NSC prepared a report in early 1987 comparing U.S.-Japanese levels of technologies which painted a bleak picture indeed: the U.S. is ahead of Japan in six battles, the two countries are dead even in six battles, and the U.S. is behind in 14 battles, the U.S. was losing ground in 21 battles, holding its own in only five. Note the use of military terminology in describing the USJ relationship. This report at the NSC level also came at a crucial moment in the final phase in the development of an evolving USG approach to USJ S&T relations and how and what manner to revise the 80STA. Undoubtedly it must have had an unusual impact on the participants in this process, especially on those who were already inclined to take a strong and adamant stand vis-a-vis Japan in future S&T negotiations.[44]

In accordance with the Defense Authorization Act for 1989, DOD prepared the first Critical Technologies Plan in March 1989; after being criticized for not including a comparison with the U.S. The plan was revised and issued in May 1990 and the next Plan was published in May 1991. It lists 20 technologies and by implication Japan is equal to the U.S. in five fields; if predominately military related technologies are deleted, Japan's position is strengthened. By 1995, the Plan reported that of the 27 chosen technologies, the U.S. now has a substantial lead in 10 areas, a slight lead in 11, and parity in six; Japan is, however, outpacing the U.S. in five areas where the U.S. currently enjoys a substantial lead and in five additional areas where the U.S. has a slight lead. The 1995 report seemed to paint a more rosy picture than previous reports. Nevertheless, the U.S. position is no longer far from assured.[45]

The various comparative technology and policy analyses whether conducted by the private sector or the government (USG, GOJ or EC) for differing purposes tended to select a core of critical technologies common to all. The early U.S. surveys produced "scary" results for the U.S. in the 1980s just as the level of U.S. anxieties particularly about the Japanese Challenge were soaring. While the analyses focused on what the USG in particular and the U.S. private sector should do to remedy the psychologically somewhat unnerving situation, they did not castigate the Japanese (nor the EC) or list demands for "remedial actions" by the Japanese. It must be constantly kept in mind that while these comparative evaluations were the collective wisdom and informed intuition of "the experts", they must remain judgmental in the final sense but, nevertheless, judgments that later had some rather serious consequences in the USG debates with itself and negotiations with the GOJ about the future of USJ S&T relations. The fact that the Japanese were depicted in so many studies as being on par if not ahead of the U.S. in many critical technologies inevitably and unfortunately led to even greater suspicions and distrust which were later reflected in USG interagency discussions and in negotiations with the GOJ. It is suspiciously remarkable that in the short space of only three years -- some half dozen years after the 88STA was signed -- the U.S. was depicted as no longer losing ground but the impression was given that it was rapidly gaining ground. But the number, breadth, scope, purposes of all these studies do provide a vivid mosaic of the almost neurotic concern with the seeming rise of Japan as a technological superpower in concomitance with the seeming decline of technological and creative U.S. The comparative analyses and policy studies were another building block in creating the mosaic of the milieu for the future of the STA.

c. *USG S&T Policy Studies*

In addition to comparative technology analyses and lists of critical technologies, the USG was suddenly awakened to the possibility that its industrial military base might now have become overly dependent on an industry or industries based in a foreign country, stated and unstated this meant Japan in the context of the 1980s. As a result of this rising anxiety, numerous policy studies were conducted for the USG on various aspects of this apparent uncomfortable dependence. They covered such topics as military systems of superconductors, the semiconductor industry and its dependency, defense industrial cooperation with Pacific Rim Nations, DOD's foreign dependence, USG policy options affecting defense trade and the U.S. industrial base, bolstering defense industrial competitiveness, a case study on three weapon systems and their dependence on foreign subcontractors, and a study on industry-to-industry international armaments cooperation with Japan.[46] Except for one, all these studies are on the surface straight forward policy studies but the motivation for conducting them stemmed undoubtedly from the anxiety and fears about the implications of the rise of Japan in the most almost advanced technologies which impinged not only on the economic security and health of the U.S. but also immediately as a perception of a veritable threat to presumed invulnerability of the U.S. military industrial basis.

Let us begin with the 1984 study which focuses on Japan directly. The context of this study is important to understand since it described how the DOD and the USG viewed Japan (only a few examples will be cited from the report): Japan had "now come of age technologically equal or superior to ours (the U.S.) in many areas" (p. 4) (just short of calling Japan a technological superpower), that our relationship with Japan was "one of its [U.S.] most important bilateral relationships, and possibly the most important" (p. 10), that it was assumed that U.S. and Japanese forces "will fight side-by-side in a military emergency involving Japan" (p. 10), that Japan wants to build self-sufficiency in defense systems (p. 11, 37). As a result of its study the report pointed out that

1. a cohesive overall strategy with respect to Japan does not exist, is fragmentary or often conflicting, and is, therefore, urgently needed,
2. Japan has paid a high premium to the U.S. to build up the self-sufficiency of its defense industry and to further its long-term commercial objectives in aerospace,
3. it is now feasible to build a two-way exchange of technology or technological cooperation from the basically USJ asymmetrical situation,
4. much of Japan's dual technologies particularly in processing and manufacturing could contribute to U.S. defense programs and be of immediate interest to the U.S. defense industry,
5. USJ partnership would help retain U.S. participation in and influence on Japan's new courses,
6. Japanese industry will consider exchanging dual-use technologies for defense technology if mutually beneficial, considered case-by-case, their

technology can be protected from misuse,
7. the U.S. knows relatively little of Japanese S&T work and Japan now has the potential to become a competitor in defense technology and products, and
8. most importantly it is necessary to strengthen the U.S. technological base and preserve our technological leadership in order to enable cooperation and competition with Japan and give U.S. industry the ability and confidence to do so.

It recommended that the U.S. should broaden it's technological cooperation with Japan, encourage industry-to-industry initiatives for greater cooperation, define roles and procedures for identifying, initiating, conducting such cooperative projects, provide guidance for U.S. and Japanese industry concerning additional U.S. defense technologies that could be released to Japan, development of two subsystems as joint USJ trial projects, initiate measures for greater understanding of Japanese technologies and finally conduct a high priority comprehensive interagency study on overall trade/defense/economic tradeoffs with Japan for a broader policy context for technological cooperation (p. v-vii). This is truly an astonishing and surprising admission of the lack of USG's preparedness in responding to "the Japanese Challenge"; this was echoed also in the mid-1980s interagency discussions in the State Department. It is significant that this study did not condemn Japan, nor criticize Japan and made no major demands upon Japan. Rather it calls for action by the U.S.: we cannot maintain our lead by conservation and protection alone, we must run faster (p.77).

The other studies, mostly DOD supported, are understandably concerned with the increasing dependence of U.S. military systems on foreign product sources and in almost all the studies Japan is used as the benchmark country. At this time, let us focus on several of these studies:

The *Defense Semiconductor Dependency* report (2/1987) drew some stark, disquieting and also quite unnerving conclusions for the U.S. government, Congress, industry, academia, and the public to hear, digest and to act upon. The report focused on "a critical national problem that at some time in the future may be looked upon in retrospect as a turning point in the history of our nation". "It would be relatively easy to blame these ominous happenings on various inappropriate behavior of foreign competitors (meaning Japan). This, however, would be a gross oversimplification. For a multitude of reasons, the U.S. has not positioned itself to compete effectively in the world semiconductor market."[47] "The relative stature of our technology base in this area [semiconductors] is steadily deteriorating. ... a direct threat to the technological superiority deemed essential to U.S. defense systems exists. ... this is an unacceptable situation."[48] U.S. market share for semiconductors in 1975 was almost 60% then gradually dropped to below 50%; in contrast Japan's share climbed from a mere almost 20% in 1975 to more than the U.S. share in 1986. In DRAM production, the most important commodity product, the U.S. share of the most advanced generation of DRAM has fallen from near 100% to less than five percent.[49] In comparative technology, the U.S. appears to be "behind" Japan in more areas than those in which it is ahead, and not gaining ground in technologies important to the future. "U.S. producers are increasingly becoming incapable of producing the highest technology products with sufficient quality in high volumes and with the timeliness required"[50] The reports most important recommendation was the establishment of a Semiconductor Manufacturing Technology Institute to develop, demonstrate and advance the technology base for efficient, high yield manufacture of advanced semiconductor devices with capitalization of $250 million from the private sector and $200 million per year for five years from DOD. This idea led eventually to the establishment of SEMATECH (Semiconductor Manufacturing Technology) in 1987. The birth of this idea definitely stems from the impact of the technological challenge from Japan and the nervousness which that aroused in the U.S. It is ironic that this recommendation based on the implicit assertion by the report that the challenge requires a coordinated government, industry and academia effort and strategy which had been practiced by Japan in the past and had been denounced by the U.S. When SEMATECH was created it excluded foreign participation (even if the "foreign company" was technically a registered U.S. firm. This also is in contrast to the July 1988 Congressional testimony by USG witnesses that the USG would pursue the GOJ when it created such GOJ-supported R&D to open all facilities under such a project to foreign participation, i.e. U.S. participation (see section IX.10.a.i).[51] This USG/DOD report well exemplifies the almost panic reaction to the foreign (Japanese) challenge that suddenly phoenix-like rose up not on the horizon but at the front door of a core U.S. industry.

In *A report outlining U.S. Government Policy Options affecting Defense Trade and the U.S. Industrial Base* (11/1988), the Japanese are described as "technologically independent" with a competitive, integrated industry base which can quickly identify and absorb the best of others' development. They are a defense trade partner like no other. We must recognize both the challenge and the potential of working more closely with the Japanese and respond constructively to it. The Japanese have substantially greater capacity than we to absorb foreign technology. In many respects they are ideal industrial partners.

The Military Systems Applications of Superconductors, asserted that a surge of Japanese R&D in high temperature superconductivity, "prompted" the Reagan Administration to establish a national program in this field. The present U.S. R&D level is "below critical mass to achieve the desired applications in a timely way. By comparison, the Japanese effort in superconductors is substantially greater than that of the aggregate U.S. commercial and government effort. ... In fact, U.S. industry is already well behind Japanese industry in the development of superconductivity applications." (p. ES-1 - ES-2)

These policy studies were sharply focused on the increasing U.S. dependence on foreign sources for many important military components, citing Japan as the benchmark, the challenge, replete with rather frank discussions where the problems lie in the U.S. and with recommendations on what the U.S. government and industry should do to remedy this situation with no condemnation of Japanese practices, policies, laws and regulations. These studies, to the best of my knowledge, did not have a direct impact on USG interagency debates on the future of USJ S&T relations, but they did constitute part of the milieu and exemplified the rather obsessive nature of the US concentration on Japan in analyzing its own problems.

d. Others

The items gathered in this subsection are basically peripheral and of relative unimportance from this study's point of view. But they are nevertheless examples of the concern exuded by the U.S. in Japanese S&T in the 1980s and 1990s. They were not negative but aimed at enlightening both sides or just non-Japanese about S&T policies, R&D developments, hopes, procedures, etc. through the mass media and through academic presentations.

U.S.-Japan Science Policy Seminars. A series of joint seminars were begun in 1980 under the auspices of the Japan Society for the Promotion of Science and the U.S. National Science Foundation as part of the U.S.-Japan Cooperative Science Program. The six seminars held between 1980 and 1998 in Hawaii were not held in an atmosphere of contention but of mutual enlightenment, discussion and furtherance of understandings of each other's policies and practices which emanated from vastly different cultural backgrounds and imperatives. The seminars began most appropriately with a mutual education about how science policy in both countries was formulated, how research was organized, resources allocated, international cooperation conducted, and on science information for policy. The following seminars concentrated on science policy perspectives (perceptions of the nature of basic and applied science, resource allocation, government practices in support of basic and applied science, cooperation of industry and universities in this regard, and patent policies for government supported research), transforming scientific ideas into innovations, engineering education, S&T to advance goals (supply of S&Es, S&T in regional development, outcomes of S&T), the changing relationships among government, industry and universities.[52] These seminars exemplified the deep and continuing academic dialogue between the two countries on issues that directly impinged on the broader issue of their S&T relations. The apparent openness of these seminars seems to have been several worlds apart from the gathering storm in the USG over S&T relations with Japan and between the two governments over S&T issues. It is regrettable that none -- repeat none -- of this sense of cooperation and cordiality rubbed off on the USG negotiators as the STA was negotiated.

Essays on Japanese S&T in *Science* and *Scientific American*. These essays were not intended to be strictly comparative like those reports in "a" and "b" above but rather an account of the status of Japanese science at any given time. *Scientific American* began an annual series in 1980, "Japanese Technology Today", as an advertising supplement underwritten by Japanese corporations. Despite its sponsorship, the series were, in the beginning, written by an American scholar of USJ business relations and a Japanese science writer and managing editor of the Japanese edition of *Scientific American,* and were highly informative, balanced, and insightful lengthy commentaries on the state of Japanese technology. After half a dozen years, the nature of the series gradually devolved into a series of statements by Japanese corporate leaders and their value correspondingly deteriorated for insight into the status of Japanese R&D. The series ended in 1991, a shadow of its early beginnings but the series does reflect the unending thirst for explanations of the Japanese S&T phenomenon.

Science, the journal of the American Association for the Advancement of Science, began a series on "Science in Japan" in 1986 -- again the timing is interesting -- and was periodically updated in 1992, 1994, and 1996. In contrast to the annual *Scientific American* series, the *Science* reports were, in most cases, much more technically oriented in the information provided. (*Science* also published a series, "Science in Europe" in 1987, 1992 and 1993 and one on China in 1995.)

Reading the comparative analyses would emphasize one's anxieties about the rising challenge of Japan and its implications for continued U.S. predominance in S&T. One would not expect the writings mentioned in this subsection to have an impact one way or the other, and they did not to the best of my knowledge, on the interagency dialogue or the negotiations but they would comprise the small pieces -- small constellations of attempted positive enlightenment -- on the periphery of the USJ S&T jigsaw puzzle.

6. U.S. Legislation and Symmetrical Access

a. Legislation

i. The Japanese Technical Literature Act (JTLA), 1986. This Act was, to the best of my knowledge, the only legislation in direct response to the Japanese S&T challenge. Senators John D. Rockefeller (Rep WVA) and Max Baucus (Dem MT) sponsored the legislation but the Reagan Administration was rather luke warm about the purpose and cost -- initially set at $1 million; in particular the cost would contribute to the ballooning deficit under the Reagan Administration. This disinterest is difficult to understand in light of the obvious anxieties that were prevalent at the time about the implications of the perceived Japanese S&T Challenge, except that perhaps this challenge should, from the ideological point of view, be handled by the U.S. private sector and was not a sphere for governmental interest. Another factor, was perhaps that it was legislation emanating from a Democratic Congress. Notwithstanding this lack of enthusiasm, the bill was passed by Congress with an allocation of $1 million, and signed into law in August 1986.[53] Again, there is no known precedent where a similar act was enacted for German, or European technical literature, even Soviet or Chinese technical literature. It is yet another small indication and reflection of the depth of this anxiety, a frustration about what to do in an adequate manner, and lack of a united political will to tackle this Challenge head-on in a meaningful manner.

This was a "hefty assignment" for one and a half people. Over the next decade the occupants of this Office made valiant attempts to do the best they could to fulfill this broad and daunting mandate. It sponsored publications, conferences, studies, etc but enthusiasm cannot overcome a budgetary handicap and not too enthusiastic Departmental leadership. Under the Bush Administration the Office was downgraded to a mere Program in 1988 and now it has transformed itself appropriately into an Asian technology program under the Under Secretary for Technology. In recent years, it has leveraged its special relationship with the Under Secretary representing the DOC in many international negotiations about S&T matters particularly with Japan. It is a feeble effort as an institution if it was meant to respond to the presumed Japanese S&T Challenge; the occupants of this Office/Program have breathed a political life into it under extremely disadvantageous circumstances.

ii. The Federal Technology Transfer Act, 1986, and Executive Order 12591. The Federal Technology Transfer Act of 1986. This act (Public law 96-480, as amended by public law 99-502 signed 10/20/86) is included here not because it has a direct bearing on the 88STA but because its enactment, at least in part, emanated from the anxieties about the competitive position of the U.S., the possibility that tax funded federal R&D may have been exploited by foreigners (including, of course, the Japanese), and the fact that one small and peripheral aspect of the Act had been invoked in the USJ negotiations as a threat to the GOJ if they did not agree to the USG position.

The concern for better dissemination of R&D results from federal research labs has been a long standing concern of successive Presidents from Kennedy onwards. The anxiety and political will to pass a comprehensive bill coalesced in the mid-1980s resulting in a law that called for a closer link between federal research laboratories and university and industrial research. This federal research had resulted in over 28,000 patents but only about 5% had been licensed. One of the many purposes of this Act was to improve the transfer of commercially useful technologies from federal laboratories to the private sector by allowing these labs to enter into cooperative research with industry, universities and others, by establishing a dual employee award system of royalty sharing and cash awards and establishing the Federal Laboratory Consortium for Technology Transfer. Fifteen percent of royalties received from licensing an invention would be given to the inventor at the lab and the balance distributed among its laboratories. Special consideration would be given to small businesses and consortia of small businesses, and preference would be given to business units located in the U.S. and which would agree to manufacture in the U.S.[54]

Foreign Participation in U.S. Federal Research Laboratories. As a minor part of the 1986 legislation, the directors of U.S. federal research laboratories were given the authority to exclude researchers from countries which did not open its labs to foreign researchers, especially U.S. S&Es. This authority is mentioned here only because the U.S. side threatened the possible use of this authority in the future if the GOJ did not accede to its demands at the negotiating table (see Chapter IX.8). To the best of my knowledge this authority has not been invoked against foreign researchers including those from Japan.

The anxiety about foreign researchers improperly taking -- stealing -- important research results from federal laboratories and exploiting them for their own purposes is well exemplified by a public notice placed in the Federal Register by the DOC asking for information about any foreign researcher who may have "stolen" such information from federal laboratories. There were two responses, thereby implying that the foreign threat was somewhat overblown.[55]

The GAO made a study of U.S. and foreign participation in federal laboratories. It found that 30% of the outside researchers were foreigners, 70% U.S. Of the 30%, four research institutions accounted for 75% of outside U.S. researchers and 80% of outside foreign researchers.[56] Of the outside researchers Japanese researchers constituted 13%, the largest national group, followed by U.K. and PRC each with 8%. Overall "federal laboratories and the United States benefitted more than foreign researchers and their countries through the collaboration on R&D."[57] The GAO report stated that the U.S. federal laboratory managers and administrators did not "perceive a need for additional guidance" and "did not favor formal restrictions on foreign access to federal laboratories."[58]

Executive Order 02591 (4/10/1987). This order was issued in connection with the implementation of the 1986 Technology Transfer Act and two other laws and was intended to set the "tone for the Administration; they convey to agencies and federal employees a sense about how programs and activities should be carried out" and with specific instructions on implementation. Let us turn only to the international R&D aspects of the order: those departments and agencies negotiating cooperative R&D agreements, licensing agreements shall in consultation with the USTR consider whether the foreign companies or governments encourage, permit U.S. S&Es to enter into such agreements, whether these foreign entities have policies for IPR protection, and whether they have adequate safeguards to prevent export of information, equipment, technologies subject to export control under COCOM. The requirements of this executive order were reflected in the draft STA given to the GOJ in August 1987.

iii. Symmetrical Access. The Omnibus Trade and Competitiveness Act of 1988 (Public Law 100-48) included a stipulation that federally supported international science and technology agreements should be negotiated to ensure that a) IPR are properly protected and b) access to R&D opportunities and facilities, and the flow of scientific and technological information, are, to the maximum extent practicable, equitable and reciprocal (sec 5171). Although the text in Part II of this law which included this stipulation did not use "symmetrical access", the Part title itself used this term thereby giving the distinct message as to what was intended by this new addition. Note that the same language was used by the U.S. delegates to the Santa Barbara and Kyoto Conferences in 1985 and 1986 (see subsection 6 below) and in the discussions with the GOJ in 1987-88 where the definition of these terms became the source of considerable heated debate.

b. National Academy of Science, Office of Japan Affairs

The creation of the Office of Japan Affairs in Spring 1988 stemmed directly from the two high level USJ meetings on advanced technology and the international environment in 1985 and l986 (For details about these two meetings see section 7 below). Its primary mission was 1) to provide a resource to the National Academy of Sciences and the broader U.S. science and engineering communities for information on Japanese S&T, 2) to promote better working relationships between the U.S. and Japanese technical communities and 3) address policy issues surrounding a changing USJ S&T technology relationship. The first chairman of the Committee on Japan, the oversight committee, was Harold Brown, former Secretary of Defense and chairman of the U.S. delegations to the aforementioned 1985 and 1986 meetings.

Almost all the studies conducted by this office were organized around workshops from which reports were prepared and published; between 1988 and 1997 it published 19 reports. Each study was funded by outside sources; for example, the U.S.-Japan Foundation in New York funded the first four studies. This foundation's capital stemmed from Japanese funds. Almost all of the studies were conducted in close collaboration with Japanese scholars, industry and various government agencies. To the best of my knowledge there was no equivalent Japanese sponsored endeavor in Japan. The program began with basic studies on the U.S. and Japanese R&D systems, then broadened out to specialized topics and were an analysis of the situation in Japan in comparison to the U.S. Almost every report had a conclusion and recommendations for actions by both Japan and the U.S.; in this sense these reports attempted to be even-handed. There was a tendency for many of the recommendations containing more "should's" for Japan and a constant emphasis on access to JSTI and facilities and how to measure this access -- even though it became apparent that the problem lay more on the U.S. side since there were not enough U.S. applicants to fill the slots available.

The first four were basic benchmark studies on the working environment for research in U.S. and Japanese universities, and three on "Learning the R&D System" on university research, on national laboratories and other non-academic non-industrial organizations and industrial R&D in both countries. Subsequent studies could be divided into 1) selected technologies (biotechnology to which 3 studies were devoted, semiconductors, robotics, aviation), 2) Japanese investment and technology transfer, 3) R&D consortia, IPR, assessing Japan's growing technological capability, and 4) three studies on maximizing U.S. interests and objectives in its S&T relations with Japan. The latter group was a major deviation from its erstwhile comparative analyses. The robotics, opto-electronics and biotechnology studies were used as case studies to assess the "symmetrical access" concept advocated in the summary statement of the 1986 High Level meeting in Kyoto. It found that improved "access" would require actions by the U.S. and Japan and suggested the possibility of creating a binational consortium, new biotechnology liaison institutions, a new information center, and, of course, a mechanism for evaluating access.[59]

In December 1989, an issues paper -- in reality a policy paper -- of the NAS, *Science, Technology, and the Future of the U.S.-Japan Relationship*, was published with the endorsement of the presidents of the NAS and NAE. Granted that the NAS is in theory a private organization, but because of its prestigious position in the U.S. scientific establishment and enterprise, this issues paper -- tantamount to a policy paper -- expresses a high level U.S. commentary and assessment of Japanese science and technology, i.e. that Japan is "an economic and technological superpower". To the best of my knowledge, this is the first acknowledgment of Japan as a technological superpower at such a high level in the U.S. scientific establishment.

> At the core of the significant changes, however, is Japan's emergence as a technological superpower. ... The United States must, therefore, deal with Japan as an economic and technological superpower. ... Japan's emergence as a technological superpower creates significant uncertainties: the U.S. economic and political situation could worsen if effective policy responses are not forthcoming. Inadequate understanding of the changing nature of the U.S.-Japan relationship combined with lack of consensus on how best to proceed, suggest the prospect of serious difficulties ahead.[60]

Measured against its stated mission, the Office of Japan Affairs was successful and made a useful contribution to the body of knowledge about Japanese and U.S. levels of technological sophistication in selected fields, provided a place (*ba* in Japanese) where U.S. and Japanese scholars and S&Es could exchange ideas and enlighten a broader audience about policy issues. But on the crucial issue of influence among USG S&T policy makers it is difficult to discover a positive or negative impact which is disappointing. After only eight plus years of energetic and constructive work the Office of Japan Affairs began winding down its activities and was slated to be absorbed into the International Division of the NAS in 1997. This change resulted obviously from the lessening interest in or obsession with Japan although the U.S. anxiety about Japan's technological prowess has persisted.

c. International Agreements

i. USJ Agreement on Dual Use Technology, November 1983. During the coldwar period since the end of WWII, Japan produced numerous military systems on a co-production basis with the U.S. and paid U.S. firms royalties for the use of their licenses -- basically a one-way street of technology flowing from the U.S. to Japan. While this approach was more expensive for Japan it substantially increased its technological capabilities. It resulted in sales losses to U.S. companies but it served USG strategic interests by contributing to the inter-operability of military systems among the U.S. and its allies, would save on R&D expenses, and presumably readiness would be enhanced.

However, in light of Japan's emerging prowess as a first class technological power, in addition to its economic and commercial importance, the USG began assessing in the late 1970s the desirability of persuading Japan to share its advanced technologies which may have military implications with the USG and U.S. defense contractors. Since the U.S. had provided Japan with necessary advanced technologies to improve and strengthen its defense capabilities, this concept appears on the surface to be eminently fair and reasonable. Yet there were many obstacles on both sides to overcome before it could be transformed into reality.

As a first step, a joint USJ Systems and Technology Forum was created in 1980 late in the Carter Administration to explore greater government-to-government cooperation in the development of defense technologies. Under the new Reagan Administration, and under increasing pressure from the U.S. Congress, "the flow-back" of militarily useful technologies from

Japan to the U.S. which also had important commercial uses became the focus of attention and was regarded as part of a larger concern about the inter-relationship of technologies for commercial and military purposes. After three years, the USJ exchanged diplomatic notes on November 3, 1983 declaring that the GOJ confirms that "the transfer of any defense-related technologies other than military technologies" from Japan to the U.S. "has been and is in principle free from restrictions". To carry out this agreement a Joint Military Technology Commission was created to evaluate and approve requests for transfers by U.S. or Japanese firms. Though seemingly straight forward and obvious, this agreement was regarded as a major breakthrough since Japan made a major exception for the U.S. in its policies concerning its weapons export ban first enunciated in 1967 as the Three Arms Export Principles and broadened in 1976.[61]

Within a few years, based on this agreement Japan had agreed to transfer several "military technologies" relating to 1) a portable surface-to-air missile from Toshiba (September 1986), 2) the construction and remodeling of U.S. naval vessels from IHI (September and December 1986), 3) FSX technologies from MELCO, 4) digital flight control system for the P3C anti-submarine patrol aircraft and 5) ducted rocket engine. In contrast, transfers of dual-use technologies have been practically nil. Implementation is hindered on both sides by the need to define the meaning of reciprocity, the nature of the technologies to be released or exchanged, the role that governments should play, project scale, business and cultural differences. Since it was the USG that was pushing for this technology transfer and/or exchange, it appeared to compromise its position somewhat by appearing to stress this transfer was to strengthen the USG's military industrial base rather than for the mutual benefit as part of the most important, bar none, linch-pin security relationship with Japan and later, under the Clinton Administration, with its initial inordinate emphasis on quantifiable results as sole proof of Japan's sincerity.[62]

On the Japanese side, the firms did not feel sure that their financial and patent interests would be adequately protected -- a tune often heard from U.S. companies about Japanese companies. The GOJ and Japanese firms felt they had paid adequately for licensing U.S. technology in the past; the U.S. side felt otherwise. The animosities that had welled up during the FSX controversy, undoubtedly persuaded the GOJ that it would not wish to get involved in the future in large, politically mandated projects like the FSX.

On the other hand, under the Clinton Administration, dual-use technology was revived as the "technology-for-technology (TFT) initiative". It too had many hurdles to overcome on both sides. The DOD had difficulty deciding on the technology which the USG wished to have transferred. The Japanese companies, particularly the smaller electronic companies, for example, were skeptical of the financial benefits for them through this linkage and the lingering stigma attached to companies involved in defense R&D and production due to the distaste for war-related activities in postwar Japan. Ultimately, the Theater Missile Defense Initiative (TMDI) was chosen by the USG as the candidate for dual-use technology cooperation. For some time, the GOJ rejected the proposed cooperation; the USG presentation to the GOJ apparently raised fears in the GOJ that such cooperation was tied to the TFT initiative. Ultimately, the GOJ did agree in 1997 during the talks to strengthen and broaden USJ military cooperation by re-interpreting the 1960 security treaty.

The USG attempt to persuade the GOJ and industry to share dual--use technology cannot be described as particularly successful due to a number of difficulties and conceptual issues on both sides. The dual-use issue has not been raised in the same manner, to the best of my knowledge with other U.S. Allies. The dual-use issue was briefly described here because it is another example of the changing approach which had been evolving in the USG from the late 1970s onward and particularly during the Reagan Administration and because dual-use as a concept for USJ cooperation and implementation, not any specific technology or initiative (e.g. the TFT idea) did arise as an issue in the subsequent negotiations between the two governments in revising the STA.[63]

ii. Defense Patent Secrecy Agreement. This was a USJ agreement signed in 1956 to protect U.S. patents in Japan which are classified for national security reasons and would be so respected and treated according to Japanese patent law in accordance with the provisions of the USJ security treaty. However, for reasons unknown, an implementation agreement for this agreement was not signed until 1988. The agreement is strictly related to military issues between the two governments and does not from the substantive point of view have any relationship to mainstream USJ S&T relations. It is mentioned here only because it was inserted into the USG inter-agency discussions mainly by DOD about USJ S&T relations and became the source of rather acrimonious arguments among USG Departments as to what extent the GOJ could be trusted in signing an implementation agreement. Because this rather specialized agreement became entangled with the USJ bilateral discussions on the 88STA, this issue is analyzed under Act IV, Bilateral

Negotiations (IX.4.b) and not discussed in this chapter.

7. Kyoto and Santa Barbara S&T Meetings, 1985 and 1986

Dr. Frank Press, President of the National Academy of Sciences pressed two Japanese members of the NAS, Dr. Kobayashi Koji, President, NEC Corporation and Professor Inose Hiroshi, University of Tokyo, about 1982 to hold a bilateral meeting of U.S. and Japanese top level executives of high tech firms, senior faculty at major research universities on high technology and the international environment.[64] At an OECD meeting in Tokyo in 1983, Press again pressed the Japanese to have this meeting. The respective delegations were headed by Dr. Mukaibō Takashi, acting chairman of the Japan Atomic Energy Commission and Dr. Harold Brown, former Secretary of Defense. Two meetings were held: August 15-18, 1985 in Santa Barbara, CA, USA and November 9-11, 1986 in Kyoto, Japan.[65] The first Conference debated basic research, electronics, communications and biotechnology. The "draft summary" of this Conference pointed out that "forces outside of science and technology might damage the climate of scientific and technical cooperation [which] motivated us in organizing this symposium. ... In the long term, political reactions to these tensions will be important and threaten to damage the creation and international applications of new scientific and technological advances" and secondly, Japan must, as a great technological power, "contribute to basic research, participating more fully in the world community which creates the stock of basic knowledge." The draft included two unusual lists called general views of Japanese and U.S. participants each listing their hopes for and dissatisfactions of the other side. This indicates the frankness of the talks and the depth of polite disagreements.[66] The Conference ended with the intention of meeting in Japan about a year later.

The atmospherics of the second conference (11/9-11/1986) must have been quite different and far more confrontational; it produced a press release and a statement with detailed recommendations.[67] The press release urged the two countries to alleviate economic tensions by accepting a new concept -- symmetrical access -- to all elements necessary for the commercialization of new technology. ... 'Symmetrical access' or the availability of equivalently valued knowledge, technology, financing and markets within the two countries, is needed to prevent a cycle of adversarial trade action that would be damaging to both nations. ... [This concept] rather than identical access to a broad range of high technology resources is what has been missing in previous discussion of U.S./Japan trade matters..."[68]

It called for symmetrical access to new knowledge, new technology, capital, markets and symmetrical treatment and operation of foreign subsidiaries. It declared that in joint actions it would conduct studies of six "key research fields" (not specified) with recommendations and programs for facilitating mutual access to "the best institutions" [in the U.S. and Japan], explore the feasibility of accelerating scientific communication in forefront fields via satellite link up, a bilateral study on the common problems faced by the two countries in the Pacific Rim, and urge both governments to take up "the priority of advanced technology trade issues" at the GATT and the Economic Summit, hold workshops on various aspects of access to JSTI and facilities and finally the creation of a Japanese Academy of Engineering.

That there were no follow-up Conferences is perhaps not surprising in light of the tone of the summary statement and recommendations and the footnoted reaction of a most senior Japanese delegate to how he felt he was treated at the Conference. On the other hand, the "symmetrical access" phraseology and concept developed a life of its own. In the Senate passed trade bill, section 3871 established a "Committee on Symmetrical Access to Technological Research" to conduct studies on the degree of symmetrical access by country and recommend executive or legislative actions to be taken. At a House committee hearing on December 10, 1987 on this idea -- a little over a year after the 2nd Conference -- State OES representative raised numerous difficulties about the definition and measurement of "symmetrical access" in his testimony and recommended that this section creating a committee on symmetrical access not be enacted. Ultimately, this section was rejected. Again, it remains ironic that the conclusions and recommendations of the high level bilateral conference were thus given short shrift by State OES which undoubtedly and accurately envisioned itself as the Department to carry out this new mandate and hassle with foreign countries about its interpretation and consequences.[69] On the other hand, the symmetrical access concept appears to have been whole-heartedly adopted by the U.S. delegation (basically OSTP and DOC) in its USJ negotiations on a new STA and became a source of rancorous debate and unpleasant argument between the two delegations. There was an underlying and unstated anxiety, more on the U.S. side, that political retaliatory action could well subvert even the existing openness however unsatisfactory it may be to USG, and political action taken to prevent this access and dissemination through retaliatory political action thereby closing highly restricted doors to galloping Japanese scientific and technological developments, the very opposite of what the USG was hoping from these discussions and negotiations, access to JSTI and research facilities. The follow-on work of these two Conferences was inherited by the creation and work of the Office of Japan Affairs (NRC)(See section IV.6.b).[70]

8. Patents

Patents could be described as a symbol of U.S. creativity thus U.S. patents would, in the public's consciousness, be the preserve of U.S. inventors. There would, of course, always be those few exceptional creative others, whether individuals or corporations. They would be so few, they would be "no problem."[71]

Let us begin with the applications submitted to the U.S. Trademark and Patent Office since this is an initial benchmark of technical creativity. In ten years as shown below, there was a 14% drop in U.S. applications, and 133% increase in application from Japan. That was a formidable and shocking change, eye-catching and disturbing, and would touch on a sensitive nerve in the U.S. psyche.

Year	U.S. Residents	Residents of Japan
1970	72,343	6,093
1975	64,445	8,566
1981	62,404	14,009[72]

In 1970, foreign companies obtained about 18% of the patents granted by the U.S. Trademark and Patent Office. In 1973, the top ten corporations receiving the most U.S. patents were still U.S. corporations except for one German company. Suddenly, in 1983, the U.S. corporations had slipped from nine to five, three Japanese and two German companies. By 1993, the U.S. relative weight had slipped again to four companies and six Japanese companies and no European companies.[73]

In regard to the number of patents granted, in 1975 foreigners in total received 35%, Japanese 9%, 8.5% Germans, French and British combined 7.5% and U.S. 65%. But a decade later the Japanese ratio had jumped to 19% and U.S. 55%. By 1993, Japanese had climbed to 23%, Europeans had gradually slipped backward and U.S. was 54%. Japanese in 1993 owned about double the number of patents in the U.S. than those owned by Britain,

France and Germany combined; Japan thus owned almost 50% of the patents owned by foreigners. If the patents granted to South Koreans and Taiwanese are added Asian countries have 25% of all U.S. patents, over half of the foreign patent holders.[74]

The recognition of this issue among the concerned experts in and out of the government -- and even in the public at large -- was not the niceties of the patent systems in any country, the number of papers cited in obtaining a patent, nor in the technical areas where the patents were granted. Just the fact that Japanese patents alone were commanding such a high percentage of the total patents granted by the U.S., and such a high percentage of all foreign owned patents. After all, we constantly reminded ourselves that the Japanese are not sufficiently creative enough to earn but a few Nobel Prizes. Then in the 1980s we suddenly come face to face with a very different, unanticipated and disturbing development, the possibility that the Japanese are after all -- and not our European brethren -- quite creative and determined. When this "creativity" of some sort is combined with industrial manufacturing prowess, the implications were taken extremely seriously indeed. It was this simple numbers games (numbers of patents) that must be emphasized here as symbolic to the U.S. as to how dramatically the "creativity scene" had changed. These patent numbers added an important but ominous thickening clouds surrounding the future of USJ S&T relations.[75]

9. The Toshiba-Kongsberg Incident and the FSX Controversy

The concluding subsection in this chapter is concerned with two controversies which coincided more or less with the beginning of the bilateral USJ negotiations about the science and technology agreement. The Toshiba-Kongsberg issue involved the reliability of Japan as an ally in the U.S. contest with the Soviet Union and the latter was a problem involving the trust, confidence and reliability between two countries and governments in negotiating an agreement. While these controversies happened over a decade ago, they have cast a long shadow over subsequent bilateral negotiations concerning S&T and other related issues. It is not the purpose here to discuss the merits of these controversies but to describe the kinds of severe psychological reactions on both sides of the Pacific that have not disappeared but have lingered on with serious consequences.

a. The Toshiba-Kongsberg Controversy

The facts of the controversy appear to be quite simple and straight forward. Soviet submarines had hitherto been relatively noisy underwater and thus could be tracked quite easily by the U.S. Navy but in 1986 appeared to be moving about much more quietly to surprise of the U.S. Navy. About a year later in April 1987 it was revealed that Toshiba Machine Co, a subsidiary of the Toshiba parent corporation had sold and delivered to, and installed highly sophisticated milling machines in a Baltic shipyard with a Norwegian company, Kongsberg Vaapenfabrikk (a Norwegian trading company), providing the numerical controllers and software to operate these machines. It was asserted at the time, during the Cold War when Reagan was describing the Soviet Union as the Empire of Evil, that it was these machines and software that provided the wherewithal that made it possible for the Soviets to make their submarines to operate much more quietly underwater and thus much more difficult to track. The "irrefutable evidence" showing this causal relationship was not released. This "treacherous act" by Toshiba and Kongsberg had endangered the safety of U.S. Navy submarines and the very solidarity of the Alliance. This was, furthermore, a most dangerous violation of COCOM (Coordinating Committee for Multilateral Exports), a multilateral organization whose purpose was to coordinate the Free World's attempt to prohibit the export of selected critical technologies and products to Communist nations.[76]

This incident occurred against a background of ever escalating frictions between the U.S. and Japan in trade. In the high technology trade balance, the U.S. had a multi-billion dollar trade deficit with Japan; this balance had until the early 1980s been in the U.S. favor. In addition, there was the argument over alleged Japanese dumping of semiconductors in the U.S. and other parts of the world thereby threatening the very existence of the U.S. semiconductor industry. There were also proposed Japanese takeovers of some high technology firms, e.g. Fairchild Semiconductor Corp. by Fujitsu, Hitachi's industrial espionage against IBM, the potential participation of Japanese industry's participation in SDI (Strategic Defense Initiative) espoused by the Reagan Administration, and several relatively unpublicized COCOM violations by Japanese companies.[77] The feeling in the U.S. at that time was that it was itself slipping seriously in its standing in world science and technology, in productivity, patents, science education, i.e., in its pre-eminent standing in S&T creativity and manufacturing capability, and must have made the U.S. response to the T-K violation even more complicated and severe. Since the U.S. was spending gigantic sums on national security in its desperate contest with the Soviet Union and Japan was spending only a relatively small sum on national defense, this breach of COCOM rules exporting high technology equipment was, so to speak, the last straw.

Immediately, numerous U.S. Congressmen flew into a rage against Toshiba -- and not Kongsberg of Norway.[78] The media showed Congressmen battering Toshiba electronic equipment to smithereens in front of the U.S. capitol. All kinds of drastic legislation was introduced to ban all imports from Toshiba; the State Department was directed by Congress to seek payment of damages from Japan. DOD canceled an order for laptop computers from Toshiba worth $100 million. The U.S. Senate passed a bill stipulating a 2-5 year ban on all Toshiba Corporation products, about $2.5 billion, but, it was estimated, this action would idle 4,000 U.S. workers. While radical measures against Toshiba were popular, there were others in the U.S. who pointed out that while this sale was improper and violated COCOM rules, it did not violate any specific U.S. law; therefore, Japan should take the disciplinary action. An historical perspective should be kept in mind that Japan was far from being the only or even the only most serious violator of COCOM rules; a French government report asserted that 80% of the 2,000 illegal exports to the Soviet Union in violation of COCOM rules, up to about 1987, had originated in the United States.[79]

After some hesitation for which it was roundly criticized, the GOJ began taking some actions: the two top officers in the Toshiba Corporation were forced to resign, the GOJ tightened COCOM and security rules for handling such sales and the Japanese courts fined Toshiba Machine $16,000 and gave two executives suspended sentences. This apparent court leniency further fanned the flames in Congress.[80] The USG eventually decided that imposing severe penalties on Toshiba Corporation would jeopardize USJ relations, that since the two countries had taken steps to tighten their export control systems such punitive measures would be unwarranted. U.S. businesses also weighed in opposing restrictions on their ability to purchase Toshiba components and subsystem for their products. The complications of unilateral U.S. sanctions against a giant company like Toshiba had become apparent and simply taking out one's rage against Toshiba with stringent punitive measures was not as simple as one thought. The potential consequences of such action gradually seeped into the consciousness of many Congressmen.

With the breakup of the Soviet Union occurring only a few years later, the presumed impact of this controversial sale became a moot question, the whole episode took on a somewhat bizarre surreal image and the substantive military technical issues surrounding the incident faded into the background. The "incident" provided a most convenient opportunity for the Reagan Administration and Congressmen to vent, with impunity and self-righteousness, their utter disgust, frustration and anger at Japan as a nation over trade issues, S&T, free loading and now a "sense of betrayal in leaking" presumably vital technical information concerning national security; the GOJ's ineptness in handling the political fall out only added fuel to this firestorm. At the time, no attempt was made by anyone in a public manner to balance the condemnation with the record of other countries, let alone that of the U.S. It is, however, the air of distrust and suspicion engendered by this controversy which only reinforced incipient, latent and existing feelings about Japan, particularly in the U.S. that need to be kept in mind for the purposes of this study.[81]

b. The FSX Controversy

To choose a successor to Japan's F-1 fighter support aircraft -- an indigenously developed and produced aircraft -- for the late 1990s and the 21st century was the task. There was a domestic context for this decision, symbolized in the drive for making Japan less dependent on foreign sources, i.e. U.S. military technology and production for military equipment; this drive was not without its opposition in Japan for various politico-economic and military reasons.[82] Then there was the international context, principally centered in Japan's special relationship with the U.S. The fundamental issue here was the bilateral security relation; in addition there were the bilateral trade issues and the ballooning trade balance, the appreciation of the yen, the wrangling over semiconductors, conflicts over opening of markets in Japan for beef, oranges, public construction, Japanese investments in the U.S., cellular phones, supercomputers, the proposed purchase of Fairchild, a French-owner semiconductor company located in the U.S. All these issues were not directly related to the STA discussion but created "a severe environment" -- translation of a Japanese term, *kibishii* -- for the negotiations surrounding the STA. Thus, the "simple" choice of a successor fighter support aircraft became entangled in an intensively bitter USJ wrangle with long lasting deleterious effects on USJ relations and was yet, at that time, another downward milestone in USJ relations.[83]

The four years, 1985-1989, when the choice of the next fighter support aircraft was being argued over can be delineated into three phases:

up to October 1987 when co-development of a U.S. aircraft was chosen, the negotiations over terms of this co-development in 1988 and the renegotiations of the agreement in 1989.

The Choice of a Fighter Support Aircraft. In September 1985 the Nakasone Cabinet adopted a new 1986-90 defense program which included three options for a successor fighter support aircraft: domestic development and production, conversion and use of the F-4EJ from being the main fighter plane of the Japan Air-Defense Force to the support aircraft role, and purchase of foreign fighter planes. Earlier in May 1985 the JDA (the Technical R&D Institute) had issued a report that Japanese industry had the military technology capability to produce a military aircraft except for the engines and basing the design on an advanced airframe. This report was, of course, a welcome assessment since there had been a growing sentiment that Japan can and should produce its own military aircraft. While there was naturally no connection between the two, there must be lingering feelings of repeating the achievements of the Zero fighter which was an outstanding fighter plane, one of the world's best, when the Pacific War occurred in December 1941 and the dream and pride of again creating such as fighter plane. Concerning option three, three foreign planes were identified by the JDA as candidates: McDonnell-Douglas's F-18, General Dynamic's F-16 and the European Tornado. The emphasis on domestic production of the major military systems is a natural extension of the long standing drive -- over the past century -- for autonomy in military production.[84]

The meaning of domestic development and production for the FSX from the JDA's and DOD's point of view began to clash. The Japanese interpretation was quite fuzzy and subject to misinterpretation by the DOD and even a sense that the Japanese industry and JDA were playing a somewhat duplicitous game. Co-production meant the Japanese would have to give all new technological improvements to the U.S. company free of charge. But co-development would mean that Japan would take the lead in collaborative work and in controlling its technological destiny (Green, p. 14). While it was a somewhat esoteric point it had important implications for the FSX. The JDA was trying to equate co-development with domestic development and production, papering over an inconsistency for domestic political reasons, ultimately with disastrous results for the JDA cause including possible consequences for the USJ security alliance. Even under the JDA concept of domestic development and production (*kokusanka*) the engine would be from a foreign (probably U.S.) manufacturer and the airframe would probably be based on a foreign aircraft.[85]

Initially, the USG/DOD, in light of its separation of trade and security issues, did not ask for discussions on the FSX even at the Secretary level meeting planned for Spring 1986 but the DOD would provide information on the indigenous development of aircraft. But in a lower level meeting in January 1986 the AS/ISA/DOD informally suggested the possibility of co-development. In late 1985, the yen substantially appreciated based on the G-5 group of nations intervention in the monetary markets. The appreciation of the yen, would naturally result in a cheaper purchase price for a U.S. plane. Since Japan had a ballooning trade balance in its favor with the U.S. the demand in the Congress rose in crescendo for Japan to buy U.S. aircraft instead of developing its own version. Such additional purchases by Japan would lower the unit cost and contribute to a DOD budget saving, would enhance USJ inter-operability of military aircraft and would contribute to a lessening of the trade imbalance.

Based on the responses by the USG to a JDA questionnaire about U.S. fighter planes and through its study missions to the U.S. and Europe, the JDA created, ever so reluctantly, a fourth option, joint development; in December 1986, the Japanese Cabinet revised the first option to "development"; this deletion of one word had an ominous implication for Japan's defense industry companies for it now diluted the meaning of domestic development and production. Through U.S. Congressional demands and through GOJ's own actions, the selection of a successor fighter for Japan was now no longer a GOJ decision but one which would embroil many actors in Japan and the U.S. Thus, the dream of domestic production by Japan's aircraft industry evaporated again for the time being but the concept never died; it was always in the background as part of the fundamental clash in expectations, hopes and dreams.

Shortly before Prime Minister Nakasone was to visit Washington in April 1987, the Reagan Administration took retaliatory action against Japan to protect U.S. microchip production. Earlier in July 1986 the LDP in Japan had won a landslide victory thus ironically making the Nakasone Government liable to pressure from the USG to push through stronger defense measures in support of the USG's policy of containing the Soviet Union. The Reagan Administration was thus successful in pressuring the GOJ to increase its FY1987 defense budget by breaching the long-standing GOJ limitation of the defense budget to 1% of GNP. Nakasone (as a former Japanese Imperial Navy officer) was well known for his eagerness to increase Japan's defense budget and contributions and to making Japan "an unsinkable aircraft carrier". The Reagan Administration engaged the Japanese in joint military studies, e.g., the Sea Lanes Defense study stemming from Japanese Prime Minister Ohira's statement that Japan would extend its defense responsibilities out to 1,000 miles from Japan and equipment and tactical inter-operability to meet military emergencies more effectively in East Asia. In the November 1986 elections, the Democrats in the U.S. regained control of both Houses of Congress. Instead of being a hindrance to Reagan Administration initiative, the latter effectively threatened the GOJ with potential/probable actions against Japan by an obviously unfriendly Democratic Congress a tactic often used against Japan by successive U.S. Administrations and strengthened the hands of the Reagan Administration.

On the GOJ's side the task was to avoid buying outright a U.S. aircraft. Through repeated meetings between the U.S. DOD Secretary and the JDA Chief, the establishment of various technical panels to consider the FSX choice, the GOJ was able to buy time to reach an acceptable arrangement on both sides. The chairman of one of the JDA technical teams which visited the U.S. in December 1986 overly bragged about Japanese technical accomplishments and its ability to produce a fighter by itself and thereby irritated -- infuriated -- his U.S. hosts, extremely detrimental to the GOJ cause. This probably led to Senator Danforth of Missouri (in whose state of Missouri McDonnell Douglas (F-18) and General Dynamics (F-16) are both headquartered) sending a letter to President Reagan arguing against a Japanese developed fighter. The President sent a letter to the Japanese Prime Minister. The latter also received missives from Senators Robert Dole (Republican) and Robert Byrd (Democrat) stressing the same points. The U.S. Secretary of Defense now proposed either a purchase of a U.S. fighter aircraft off-the-shelf or a "joint development" based on the F-15, F-16 or F-18. Japanese industry also had accepted the inevitability of such a compromise. The JDA and Japanese industry were also determined to maintain at the least the principle of "Japanese-led" joint development and production of the FSX.

But when the Toshiba Incident discussed earlier exploded on the USJ scene, the GOJ had become highly vulnerable to even greater pressure from the USG; in a way the USG now held a trump card in this process. The urgency and need for the smooth operation of the bilateral security alliance was now brought into play. After still yet more bilateral maneuvering, the JDA and then the Cabinet on October 23, 1987 chose the F-16 from which a remodeled Japanese fighter support aircraft would be created. (Let us keep in mind that the bilateral negotiations on the STA had just begun in this acrimonious bilateral haze).

Through a painful interaction between the domestic politics in the

U.S. and Japan, personal and institutional contacts at various levels, the linkage, for the first time, of defense and trade issues, the U.S. was able, in a most blatant, forceful and direct manner, to inject itself into the GOJ decision-making process for the selection of a new Japanese military aircraft. For the sake of trans-Pacific harmony in the context of the Soviet threat in Northeast Asia and of the close USJ politico-military relationship, Japan bowed to the inevitable but the unabashed U.S. intervention in the Japanese decision-making process, especially through Congressional resolutions, and implicit threats were deeply resented.[86]

Negotiations over The Terms of this Choice. The major task here was to define the real meaning of co-development; this would be an especially important milestone since there were no previous models for this cooperation to follow. It took approximately nine months (from mid-February to end of November 1988) to negotiate an MOU which at the request of the GOJ remained secret at the time but which needed to be submitted to the Congress for approval.[87] The STA bilateral negotiations had reached a new peak of unpleasantness in anticipation of the January 1988 summit meeting between President Reagan and Prime Minister Takeshita Noboru who had taken over from Nakasone in November 1987.

Renegotiation of the MOU. While the DOD under the Reagan Administration had negotiated the MOU presumably taking into consideration USG interests, there were voices of dissatisfaction about the agreement. The JDA naturally assumed that it had negotiated an MOU which reflected the will of the USG without further challenge. Unfortunately, 1988 was a Presidential election year with a bitter and mean campaign. Vice President Bush won but the Democrats retained control of the Congress. The old DOD team was gone; Bush's cabinet appointees were savaged by the Democratic Congress (e.g. the proposed DOD secretary failed to win obtain Congressional support). Because of the frustrations and anger over the FSX, the Democratic Congress revised the law requiring DOD to consult DOC about the commercial implications of such FSX MOUs.[88] Together with these and other factors, a critical mass of bad luck developed which persuaded the new Bush Administration on March 20, 1989 -- a mere two months after assuming power -- to seek "clarification" of the FSX MOU with the GOJ. Assuming it had, in good faith, negotiated an agreement with the USG, the GOJ/-JDA was shocked, frustrated and angered.[89] If the U.S. President was forced into this awkward demand of the GOJ, the Japanese Prime Minister, Takeshita Noboru, was also hobbled by a scandal in his administration which had prompted him to resign by April 24, 1989. As a result, the GOJ/JDA had little negotiating leeway but to accept USG demands: guaranteed 40% work-share, Japanese access to the fire control systems but not the commercially valuable flight control software, the process of identifying improved technologies which were derived from U.S. technologies and were somewhat clarified, the U.S. contractor would pay for technologies developed by Japan and rejection of Japanese insistence on a 50% workshare of the engine.[90] Seemingly, this new agreement would be acceptable to the U.S. Congress but a GAO report stated that the technologies desired by the U.S. contractor were not that unusual and not that important and only reignited Congressional fears of this agreement. Attempts were made to restrict technology transfers to the Japanese prime contractor. The agreement in the final sense, however, squeaked through the Congress on one vote thereby avoiding the limitation on the transfer of technology to Japan for the FSX.[91]

At last the FSX issue was stabilized; MHI and GD signed an agreement on February 21, 1990 and began development work to create a remodeled F-16 to Japanese requirements. MHI became the prime contractor, and GD became the subcontractor Nevertheless, distrust and suspicion on both sides gradually escalated through each stage to new depths. While the basic policy of the GOJ was to make the FSX decision within the context and needs of the USJ security alliance, it needs to be acknowledged that, even among some conservative elements in and out of the government, there was some reluctant questioning of the USJ security alliance, undoubtedly out of frustration and anger over how they felt Japan had been treated in regard to the FSX affair. The FSX has cast a long gray shadow into the future.[92]

The 1980s began with rising anxieties about the Japanese Challenge in the ever enlarging U.S. trade deficit with Japan, the ever continuing confrontation with Japan about trade issues, such as textiles, TVs, automobiles, semiconductors to cite only a few and the suddenly ever haunting rise of Japanese S&T in high technology and presumed U.S. core industries. In the beginning while these contentious issues appeared to involve specific industries and were depriving Americans of their jobs, they were at the technological margin; they did not involve high technology. But as each trade conflict arose, the issues gradually climbed the technological ladder, until they involved core industries (e.g. the auto industry) and the symbol of U.S. creativity, ingenuity, know-how and production symbolized in semiconductors. Another symbol of the changed situation occurred when Japanese patents began to occupy a larger and larger percentage of patents granted by the U.S.

Patent and Trademark Office; suddenly not-to-be-expected Japanese creativity was showing up right in the very heart of the U.S. system. While numerous studies had been made of Japan's economic progress and accomplishments, little attention had been paid to its scientific and technological developments; there was an underlying assumption generally denigrating Japanese endeavors in S&T. However, suddenly, the Japanese S&T Challenge -- many described it as a threat -- loomed not just on the horizon but to the consternation of a nation in the 1980s in America's face. From the U.S. point of view, this was unprecedented, and unexpected especially since the Challenge came from an East Asian country and erstwhile enemy, Japan. We praised them; we put them on a pedestal; we began to fear the Japanese again.

An extremely wide variety of activities were initiated in the U.S. and by the U.S. to try "to understand" the Japan phenomenon, to come to terms with what seemed to be a deteriorating situation in the U.S. and to create a set of policies and atmospheres to change the direction of the U.S. While many heated Congressional hearings had been held on Japan's trade tactics, now some attention was beginning to be given to Japanese S&T (although this interest was not long lasting), the availability of JSTI, and to comparing the status of selected technologies between the U.S. and Japan. These comparative analyses became an almost routine USG preoccupation from the mid-1980s even into the late 1990s. These comparative analyses and other assessments of U.S. S&T levels, appeared, in general, only to emphasize and drive home the growing "challenge" from Japan and raising ever higher consternation among the scientific, military and political leadership of the U.S. in the 1980s; of course, in many instances similar Japanese studies painted the opposite trend also with some anxieties as to their implications for Japan. A small group of enthusiasts in the early to mid-1980s held seminars to try to alert the U.S. and arouse interests in JSTI but they spoke, in the main, only to the converted. By the mid-1980s, S&T, for the first time, had become part of the dialogue on USJ relationships. Hundreds, if not thousands, of articles were written in the media laudatory and condemning of Japanese S&T, trying to unravel "the secret" of this sudden rise and apparently "unstoppable rise" of Japanese S&T. Numerous USG policy studies were conducted to understand and counter or utilize JSTI and a few -- perhaps generously some -- of these recommendations stemming from these studies were implemented. The U.S. Congress under Democratic leadership created a small office in DOC -- against the wishes of the Reagan Administration -- to provide what must be admitted was a small and frail response to this Challenge. An Office of Japan Affairs was created under the NAS under whose sponsorship a substantial amount of research elucidating the nature of Japanese S&T accomplishments and problems and a more accurate portrayal of the Japanese Challenge, provided a useful USJ dialogue on S&T issues. Through DOD studies we became concerned about our dependence on foreign sources (meaning in principle, Japan) for critical electronic components for high-tech weapons systems; this concern from the 1980s continued into the 1990s but with no obvious policies to reverse this trend. Our sense of unease was widespread and deep.

Gradually the tone and decibel count of the anxiety and the response was raised: the USG began to demand that there be symmetrical access to U.S. and Japanese STI and for U.S. researchers to have access to Japanese R&D facilities without having first analyzed the reality of the situation and the probability of its realization on a "symmetrical basis". This access issue became a symbol of U.S. demands on Japan based on the assumption there was much to be discovered and mined in JSTI and that there were ever so many U.S. S&Es lined up and ever so eager to conduct long term research in Japan. Many, in and out of the USG, felt that the rise of Japanese S&T, like its economic achievements, accomplished under the protective U.S. nuclear umbrella over Japan during the tense coldwar years, had resulted in a "free ride" for Japan; thus, Japan now had "an obligation" to remedy the situation, to rectify the presumed imbalance, to lend a helping hand to the U.S. As one among many examples, the USG prevailed upon the GOJ to sign an agreement in late 1983 on dual use technology (i.e. for Japan to provide the USG with desired military technologies) and in 1988 to sign an implementation agreement on the 1956 Defense Patent Secrecy Agreement which neither side had felt sufficiently about for over 30 years to push for implementation; during these three decades Japanese S&T had unexpectedly evolved to the point where the USG was persuaded to pursue the GOJ for an implementation agreement. The tone of these negotiations had slipped from the cordial to difficult, especially when parts of the USG attempted to tie these negotiations to the ongoing STA negotiations in 1987-1988. When the Reagan Administration began to relate USJ military alliance relationships with trade, S&T issues, beginning with the semiconductor agreement, the tone of discussions and negotiations between the USG and the GOJ only deteriorated into new depths of distrust and suspicions in the late 1980s even though State and DOD tried for some time to keep them in separate compartments for the sake of the alliance. The acrimony, distrust and suspicions with long shadows into the late 1990s exploded when the Toshiba-Kongsberg Incident occurred in Fall 1987 and the battle royal over Japan's future support fighter, the FSX, erupted.

During the 1980s and somewhat later, it fretted to itself that indeed U.S. R&D was declining and articles and books only reiterated these anxieties, and simultaneously pronounced Japan as "Number One". With some frustration and anger we humbled ourselves and thereby encouraged Japanese arrogance, and over self-confidence in itself.[93] The U.S. has called for a politically more active Japan, a greater sharing of dual-use military technology, a Japan that takes on a larger role in "burden-sharing", and a greater overall S&T cooperation between the U.S. and Japan in particular and with the world in general.[94]

Simultaneously, the U.S. worried that S&T cooperation would end up helping Japan more than the U.S., e.g. the FSX where we pushed for co-development/production then worried that Japan would get the best of us which quickly leads to distrust and suspicions. The variety of factors discussed in this chapter gradually coalesced into a critical mass, a totality more than its parts, i.e. the proverbial 2+2 was not four but probably six. These factors unfortunately reacted upon each to produce in Fall 1987 the nadir of U.S.-Japanese relations since 1945 as the backdrop for internal USG interagency decision-making and subsequent bilateral USJ negotiations about future S&T relations.

Notes

1. As an example, science and technology was a minor part of a major study of USJ economic relations in *Asia's New Giant: How the Japanese Economy Works.* Hugh Patrick and Henry Rosovsky, eds. The Brookings Institution. 1975. Chapter 8 Technology by Merton J. Peck. *Japan and the U.S.: Economic and Political Adversaries,* Leon Hollerman, ed. Westview Press. Boulder, Colorado. 1979. Chapter 3 Japanese-U.S. Relations in Science and Technology by Takeo Sasagawa.
2. Gene Bylinsky. The Japanese Spies in Silicon Valley, *Fortune* 2/27/1978:74-79.
3. *The Decline of Productivity and the Resultant Loss of U.S. World Economic and Political Leadership*. Robot Institute of America. Dearborn (MI). Policy Document no. 1 (3/1981). 8 p.
4. Theodore H. White. The Danger from Japan, *New York Times Magazine,* 7/28/1985:19-59. He writes with a certain sense of bitterness: "I bristled at the sight of them. I had seen the Japanese blast and flame Chungqing, ... so the luxury of this moment was one I enjoyed. ... No one helped Shigemitsu [the Japanese Foreign Minister who had lost a leg in a Chinese terrorist bombing in Shanghai and was one of the GOJ signers of the Surrender Document] but some American more generous than I brought a chair to sit on as he signed the surrender." p. 20.
5. In Japan there were some books that were titled *Nichi-Bei Gijutsu Sensō* (literally translated The U.S.-Japan technology war but the printed English title was The Japan-U.S.A. conflicts on high technology and a subtitle, strategies for firms to overcome

[USJ S&T] friction by Makino Noboru and Shimura Yukio. Nihon Keizai Shimbunsha. 1984. 278 p.; another book with a similar title was *Nichi-Bei-Ō Gijutsu Kaihatsu Sensō* (The R&D War among the U.S., Europe and Japan), Tokyo. Tōyō Keizai Shimpōsha. 1981. 228 p. These books did not present the U.S. in the same way as White did of Japan but did envision that there would be a long and serious conflict. "War" was probably used more for PR to stimulate sales -- this author purchased copies with relish.

6. Robert B. Reich. The Rise of Techno-Nationalism, *The Atlantic Monthly* 5/1987:63-69. Reich was then professor of political economy and management at the Kennedy School of Government, Harvard University; later, during the first Clinton Administration (1993-1996) was Secretary of Labor.
7. Reich suggested six steps for the U.S. to follow in The Quiet Path to Technological Preeminence, *Scientific American* 261:4 (10/1989):41-46. Those are 1) scan the for new insights, 2) integrate government-funded R&D with commercial production, 3) integrate corporate R&D with commercial production, 4) establish technological standards, 5) invest in technological learning, and 6) provide a good basic education to all citizens.
8. The referenced NSC report was quoted in Reich's article in *The Atlantic Monthly* and in a *Wall Street Journal* article by Bernard Wysocki, 11/14/88 on Japan Assaults the Last Bastion: America's Lead in Innovation. While I was able to garner almost all of the comparative analyses, I did not succeed in obtaining a copy of this NSC report of early 1987. It was prepared by a group of experts from DOC, DOD, NSF, CIA and other agencies for senior governmental consideration. In order to obtain the full flavor of how the NSC report views the rising Japanese Technological Challenge to the point of threatening U.S. military preparedness, economic independence AND threatening the U.S. standard of living, we will quote the NSC Report as referenced on p. 65 of Reich's essay:

   Leadership in research and development in advanced (non-silicon) semiconductor materials and devices seems to already be passing to Japan, especially in the optical electronic fields which may be the basis of the highest performance end products of the future. ... With some significant exceptions computer R&D is still clearly an American strength, but the Japanese are becoming expert in the architecture of mainframes and supercomputers, and obviously in the manufacturing engineering of personal computers. Telecommunication component research is becoming dominated by the Japanese, and while their strengths in downstream system and network industries are more limited, the trend is toward Japanese leadership in some communications systems based on their advantages in semiconductor components. Industrial automation is a field where the Japanese not only already dominate the present market, but also research and development in most sub-disciplines. Clearly the conventional model of U.S. technological leadership in basic research followed by more successful Japanese commercial exploitation is no longer accurate in many of the critical technologies targeted by the Japanese. ...

   By the turn of the century, microelectronics will certainly have major direct effects upon the performance of industries which will directly account for perhaps a quarter of GNP, and which have powerful effects upon military capabilities, economy-wide productivity, and living standards. These include automobiles, industrial automation, computer systems, defense and aerospace products, telecommunications, and many consumer goods. ... If the United States loses competitive advantage in these industries, its productivity, living standards, and growth will suffer severely. Moreover, these industries are dominated by a few nations and firms so that competitive advantage brings significant economic profits and political influence. Thus if the United States becomes a net importer and a technically inferior producer, it would also become a less independent, less influential, and less secure nation.

9. While the above controversies raged, at the same time another controversy was unfolding in the 1980s about the politico-economic analysis of Japan. It was trying to decipher how the Japanese had achieved their up-to-then phenomenal economic success and how to counter their seemingly unstoppable economic machine which, to the consternation of Western economists, was not responding to and acting in accordance with accepted economic principles, i.e. not reacting to standards and principles as established by Western economic theorists and countries. In addition, it was maintained that Japan was a state without a center of power and thus the word of the prime minister was not the word of the government. The protagonists of these analyses advocated various types of get-tough policies and many of their advocacies were indeed tried during President Clinton's first term (1993-1996). These writers created an environment where Japan was regarded as being different and thus must be treated differently and severely to make it conform to Western standards of behavior. This series of highly critical analyses was not discussed in the main text because they were concerned mainly with economics and trade policy with Japan and the economic and industrial policies which the U.S. should follow and not with S&T issues. The principal protagonists and their writings are as follows: Chalmers Johnson, *Japan, Who Governs?* (W.W. Norton & Co., New York, 1995), *MITI and the Japanese Miracle: The growth of Industrial Policy, 1925-1975* (Stanford University Press, 1982, Stanford, CA); Karel van Wolferen, *The Enigma of Japanese Power* (Alfred A. Knopf, New York, 1989), The Japan Problem Revisited, *Foreign Affairs*, Fall 1990; James Fallow, Containing Japan, *The Atlantic Monthly*, 5/1989, Looking at the Sun (Pantheon Books, New York, 1994); Clyde Prestowitz, *Trading Places*, Basic Books, New York, 1988. An excellent summary and assessment of the validity of the assertions and recommendations emanating from this group of scholars was made by Arthur J. Alexander, President of Japan Economic Institute, *JEI Report* 29A(8/1/1997), 15 p.
10. I wrote a trilogy (1989-1991) analyzing the "new amorphous external Challenge" presented by Japanese R&D. The first emphasized the various areas where the U.S. responded with some kind of reaction, e.g. consciousness-raising activities (conference, seminars symposia, congressional hearings on JSTI, educational activities, marketing of JSTI, etc; the second provided a study of the comparative assessments (USJ) of selected technologies; and the third was on one particular approach started in 1983 and which continues to this day (1998), the so-called "JTEC Studies" (later renamed WTEC studies). Cecil H. Uyehara, U.S. Responses to Japanese Science and technology Challenges, Japanese Information in *Science, Technology and Commerce*, IOS Press, Amsterdam/Tokyo. p. 87-111, Proceedings of the 2nd International Conference on JSTI; Appraising Japanese Science and Technology, *Japan's Economic Challenge*, S.Prt 101-121, p. 289-306; JTEC studies -- a common thread, 3rd International Conference on Japanese Information in Science, Technology and Commerce, Preprints, 1991, p. 261-277. I have liberally called upon this trilogy, among many other sources, in preparing this chapter.

11. The Extraordinary Measures Law for the Promotion of the Electronics Industry (Denshi Kōgyō Shinkō Rinji Sochihō) was initially enacted as a law with a time limitation (jigen rippō) but was later revised and its time limitation extended in 1960s and 1970s.
12. Clyde Prestowitz in *Trading Places*. New York. Basic Books, Inc. 1989 presented an excellent description of the Japanese perception: Behind this thinking lay Japan's strong desire to maintain its autonomy and homogeneity. It believed it could not survive as a nation if foreign companies took large shares of its markets. The decision to challenge IBM [the decisive symbol of U.S. prowess in manufacturing computers] was thus seen as a matter of national security. The result was an effort that can be compared in many ways to the U.S. Apollo program. Like it, the Japanese project centered on a few large companies chosen by the government, and led to development of the same consensus in support of the effort and the same sense of national purpose. Americans later complained of a "Buy Japanese" mentality, but just as we would not have used Russian rockets to go to the moon, so the Japanese wanted to use their own computers and parts. Indeed, in 1987, when U.S. rockets and space shuttles were grounded by mishaps, the U.S. government moved to prevent U.S. companies using Russian rockets to launch their satellites. "This was not very different from Japan's actions regarding semiconductors and computers." (p. 130-131)
    This subsection is partially derived from this book, an interview with Clyde Prestowitz, and the Japanese S&T White Paper for 1995 whose theme was 50 years of postwar Japanese S&T. For an insightful description and analysis of the USJ semiconductor imbroglio in the 1980s see chapter 2, Losing the Chips: The Semiconductor Industry, in his book.
13. ibid. p. 135, 137, 144.
14. To quote Prestowitz again: what the USG wanted was "to getting the Japanese government to force its companies to make a profit and even to impose controls to avoid excess production -- in short, a government cartel." (p. 167) ... Over all, we were in a hypocritical position: we kept saying we only wanted market access and not managed trade: yet at the same time, we said that as a result of access we expected the cash registers to ring (p. 170) ... the United States achieved a monitoring system in 1986 that its Justice Department had prevented it from requesting in 1982. (p. 172) ... That one of the pioneers of the U.S. semiconductor industry and one of the main forces behind the complaints that forced the [1986] agreement was now reduced to begging MITI for chips was a measure both of MITI's power and of how far the Americans had fallen. (p. 176) ... Although it has long tried to persuade MITI to intervene in the market in precisely the ways that were earlier described as objectionable. In doing it was enhancing the very MITI power it wishes to see decline. (p.176).
15. *JEI Report*, 31B (8/15/1997):10.
16. These hearings are the following: *The Availability of Japanese Scientific and Technical Information in the United States*. Hearings before the Subcommittee on Science, Research and Technology of the Committee on Science and Technology, U.S. House of Representatives, 98-95. 405 p. March 6,7, 1984; *The Role of Technical Information in U.S. Competitiveness with Japan*. Hearings before the Subcommittee on Science, Research and Technology of the Committee on Science and Technology, U.S. House of Representatives, 99-271, 295 p. The 1984 hearings were held only after I had persuaded two senior Democratic members of this Committee (James Schauer, NY and Norman Mineta, CA) that such hearings were overdue, and who, in turn, persuaded the

subcommittee chairman, Doug Walgren, of the urgent need to hold such hearings. Thus through persistent pressure and persuasion these hearings came into being for a short-lived two years. In January 1981, the U.S. House of Representatives, Science and Technology Committee published a 500-page *Background Readings on Science, Technology and Energy R&D in Japan and China*, prepared by its staff. Concerning Japan, it stated that "Cooperation in science and technology is an important element in our bilateral relations -- in cementing our close ties and creating new knowledge accessible to all nations." (p. 309) and the wide variety of joint R&D projects were continuing "at a steady and productive pace". (p. 310) It also referred to a 1975 binational review of USJ S&T programs by distinguished scientists who "gave a favorable assessment of our activities as a whole, while at the same time pointing out new areas for cooperation and areas for improvement." (ibid.)

17. *The Availability of Japanese Scientific and Technical Information in the United States.* Report prepared by the Congressional Research Service, Library of Congress for the Subcommittee on Science, Research and Technology. 98-2 session. 11/1984. 29 p.
18. Another small example of how seemingly unimportant -- in the mid-1980s -- JSTI was: $750,000 had been authorized for the DOC to increase JSTI availability; this is a super-modest sum for this large task. In the confusion after a Presidential veto, these funds were lost in the parliamentary shuffle.
19. The symposium proceedings were published as *U.S.-Japan Technological Exchange Symposium*. University Press of America, Washington, DC, 1982. 132 p. I was invited by the Japan-America Society of Washington to organize this first symposium and the 2nd one held five years later. VLSI (Very Large Scale Integration) program for semiconductors was sponsored by MITI.
20. The proceedings were published as *U.S.-Japan Science and Technology Exchange: Patterns of Interdependence*. Edited by Cecil H. Uyehara. Westview Press. Boulder, CO. 1988. 279 p.
21. These studies are: 1) *Japanese Scientific and Technical Information in the United States*. NTIS. 1983. 165 p., 2) *Getting America Ready for Japanese Science and Technology,* Asia Program of the Woodrow Wilson Center and the MIT-Japan Science and Technology Program, 1985. 183 p., 3) *U.S. Access to Japanese Technical Literature: Electronics and Electrical Engineering*: Seminar at the National Bureau of Standards (later NIST), NBS Special Publication 710. June 1985. 159 p. and 4) *Monitoring Foreign Science and Technology for Enhanced International Competitiveness: Defining U.S. Needs*, Proceedings of the Office of Naval Research and NSF, 10/1986, 93 p.
22. The JDC actually focuses, in the main, on collecting gray literature on science and technology policy issues in Japan and its materials are supposedly, again in the main, for the use of LC's Congressional Research Service which provides policy research for the U.S. Congress. The JDC's counterpart office in Tokyo which collects the documents is wholly funded by the GOJ through the Center for Global Partnership. Ironically the JDC was abolished by LC in March 2000; the Japan Foundation was no longer providing supporting funds.
23. *How to Locate and Acquire Japanese Scientific and Technical Information*. JICST and DOC/NTIS co-sponsors. 1992 (325 p.), 1993 (474 p.). *Technical Requirements for Accessing Japanese Information: Problems and Solutions*. Japan Documentation Center, LC., Washington, DC, 1994. 161 p.; *Japanese Public Policy: Perspectives and*

Resources. Japan Documentation Center, LC, 1994 327 p.; *CyberJapan: Technology, Policy and Society*. Japan Documentation Center, LC. Washington, DC, 1996. 253 p.

24. Preprints of these conferences: [1st] *International Conference on Japanese Information in Science, Technology and Commerce*, London. British Library Japanese Information Service. 1987. [about 600 p.]; [2nd Conference on] *Japanese Information in Science, Technology and Commerce*. IOS Press. Amsterdam. 1989. 629 p. [Only final report in this series]; 3rd International Conference on Japanese Information in *Science, Technology and Commerce*. Nancy France, Institut de L'Information Scientifique et Technique. 1991. 525 p.; [4th ...; *Fifth International Conference on Japanese Information in Science, Technology and Commerce*. Library of Congress and U.S. Department of Commerce. Washington, DC. 1997. 167 p.
25. The General Accounting Office, investigative arm of the U.S. Congress, also periodically got into the JSTI fray in order to ascertain whether what the USG was doing to monitor and disseminate STI was sufficient and done in a timely fashion. It found that 62 federal civilian and military agency offices (excluding CIA) monitor foreign technology including Japan but that there was little coordination among these offices. United States General Accounting Office. *Foreign Technology: U.S. Monitoring and Dissemination of the Results of Foreign Research*. 3/1990. 36 p.

Even the Military Librarian Division of the Special Libraries Association (SLA) sponsored a Japanese Scientific Information Session at the 78th annual conference of the SLA in June 1987 where three papers were presented: JICST and Access to Information on Japanese Science and Technology in the United States by Edward C. Schroeder, Manager, Business Development, Mitsubishi International Corp, San Francisco; Becoming an Educated Buyer of Japanese Translation Services by Carl Kay, President, Japanese Language Services, Boston, MA; Access to Japanese Sci/Tech Literature -- The User's Perspective by Dawn Talbot, Center for Magnetic Recording Research, UC/San Diego.

In the mid-1980s, three additional JSTI status reports were prepared. 1)The National Science Foundation to the House Committee on S&T, *U.S. Access to Japanese Scientific and Technical Literature*. 4/1985. 16 p. A brief survey without useful conclusions or recommendations. 2) LC Congressional Research Service (CRS). Japanese Science and Technology: Some recent efforts to improve U.S. Monitoring #86-195 SPR, Nancy Miller. 12/23/1986. 26 p. The conclusion was that accessing JSTI "will be expensive" and since there is no agreement on how to improve this access continued analysis and assessment of programs and activities may be necessary. 3) LC CRS. *Japanese Technical Information: Opportunities to Improve U.S. Access* (#87-818 S) by Christopher T. Hill. 10/13/1987. 55 p. Hill's contribution was that he tried to estimate the cost of several options to improve access to JSTI. Outright translation of large numbers of journals would be most expensive ($80 million/year), tax incentives to companies to invest in JSTI access ($50 million in lost revenue), fellowships, language training, travel would cost $25 million. He estimated that this would cost 0.3% of Federal basic research support, or 0.04% of Federal R&D funds. If JSTI made available would result in 0.15% of federal basic research funds, or 0.02% of total US R&D funding, it would, he said, be cost effective for the U.S. But "since there remains substantial uncertainty about the demand for and utilization of STI from abroad, Congress might wish to periodically monitor and reassess whatever commitment it made to access JSTI. Hill's efforts fell on deaf ears; Congress did nothing, nor did the Reagan Administration although JSTI later became the weapon for attacking Japan as will be developed in this saga.

26. The four studies are: *High-Technology Ceramics in Japan.* National Academy Press, Washington, DC. 1984, 64 p.; *Basic Research in Ceramic and Semiconductor Science at Selected Japanese Laboratories.* U.S. Department of Energy. March 1987, 93 p.; *Ceramics and Semiconductor Sciences in Japan.* U.S. Department of Commerce, 1987. 32 p.; *Advanced Ceramics.* National Academy Press, 1988. The DOE and DOC reports were prepared by Robert J. Gottschall of the Office of Basic Energy Sciences, DOE.
27. Survey of Supply/Demand Relationships for Japanese Technical Information in the United States: The *Field of Advanced Ceramics Research and Development.* Report prepared for DOC (Office of Japanese Technical Literature). 1988. 130 p.
28. The first Wisemen's Group was established by President Carter and Prime Minister Ohira pursuant to their joint communique of May 2, 1979. Officially known as the Japan-United States Economic Relations Group, it issued its report in January 1981 (107 p.), an appendix (4/1981, 456 p.) and supplemental report (10/1981, 63 p.). I counted at least 80 suggestions for both countries to follow but there were none pertaining to S&T issues, obviously they were not "hot issues" at that time. It is quite a remarkable document in its unusual selection of admonitions to make to both governments from mutual security to negotiating styles, such as "both governments should work very hard to avoid employing (or seeming to employ) intense, highly visible United States pressure as a catalyst for Japanese policy change" (p. I-101); Japanese negotiators should speak up more countering American criticism as squarely as possible, minimizing misunderstandings, negotiators should beware of attempting to force concessions on specific issues by linking them to the success or failure of a summit meeting (p. I-102) While excellent suggestions they were mostly disregarded in subsequent bilateral negotiations.
29. ibid. p. xvii-xviii, 97-102.
30. In addition, a U.S.-Japanese business group had been meeting annually over 25 years, alternately in Japan and the U.S. to propose solutions to U.S.-Japan trade problems and facilitate business relations. The U.S. group was known as the Advisory Council on Japan-U.S. Economic Relations and the Japanese group as the Japan-U.S. Economic Council. Each year a bilingual report was issued but in 1984, for the first time, a High Technology Position Paper was issued. It was more a policy paper on Japanese telecommunications industry, software protection and GOJ plans for promoting Japanese high technology industries than a commentary on S&T levels in both countries. It contributed to sensitizing and raising the consciousness level of the political leadership of both countries.
31. Conferences were held on comparing USJ industries, e.g. *Comparative Industries: Japan and the United States in the 1980s* by the Japan Society in New York in March 1980 (The Report on the conference was 105 p., Public Affairs Series 12), but they were more concerned with marketing, production and investment, rather than with science and technology issues and thus are not considered in this study.
32. JTEC (Japanese Technology Evaluation Center) or JTECH (Japanese Technology studies). These studies were first sponsored by DOC, then an intelligence agency funding entered the picture and for the past more than ten years has been stabilized under NSF sponsorship with supplementary support from other USG agencies when needed and appropriate. The first contractor was Science Applications International Corporation until 1989 when this responsibility was transferred to Loyola College in Baltimore, MD.

33. A list of the topics of these studies, 1981-1991, is given below with the initials of the sponsoring USG agency, date of publication. The purpose here is to show the breadth of coverage of these studies; providing complete bibliographic information would create an extremely cumbersome footnote so only the date of publication and author or sponsoring organization will be given:
1981: *U.S. Industrial Competitiveness: A Comparison of Steel, Electronics and Automobiles* (OTA),
1982: *International Developments in Computers* (NRC), The Comparative Status of 7 U.S. Industry (NRC),
1983: *Satellite and Rocket Technology* (Bloom),
1984: *Commercial Biotechnology, an International Analysis* (OTA), Computer Science (JTECH), High Technology Ceramics (NRC),
1985: *Mechatronics* (JTECH), Biotechnology (NSF), Biotechnology (JTECH), Opto & Micro-electronics (JTECH), Electro-Optics Millimeter/Microwave Technology (DOD),
1986: *Telecommunications* (JTECH), *Advanced Materials* (JTECH), *Advanced Processing of Electronic Materials* (NRC), *Japanese Opto-Electronics Industry and its relationship to SDI* (H. Glazer),
1987: *Advanced Computing* (JTECH), *Basic Research in Ceramics and Semiconductor Science at Selected Japanese Laboratories* (DOE), *Ceramics and Semiconductor Sciences* (DOC), *Electro-Optics and Millimeter Wave Technology* (DOD), *Japanese Construction* (UK),
1988: *Computer Integrated Manufacturing & Computer-Assisted Design for the Semiconductor Industry in Japan* (JTECH), *Advanced Ceramics* (NAS), *Commercializing High-Temperature Superconductivity* (OTA), *Photonics* (NRC), *ERATO Program* (JTECH), *Defense Industrial & Technology Base* (DOD), *Military Systems Applications of Superconductors* (DOD), *Factory Automation in Japan* (DOD),
1989: *Advanced Sensors* (JTECH), *Superconductivity* (JTEC), *Recombinant DNA in Japan* (DOC), *Basic Research in Superconductor, Ceramic and Semiconductor Sciences at Selected Japanese Labs* (DOC and DOE), *Japanese Manufacturing Technology* (DOD), *Japanese Software* (ADAPSO),
1990: *Space & Transatmospheric Propulsion* (JTEC), *Advanced Computing* (JTEC),
1991: *Space Robotics* (JTEC), *Complex Composite Materials* (JTEC), *Construction Technology* (JTEC), *High Temperature Composites* (TAT/DARPA), *High Definition TV* (JTEC), *X-ray Lithography* (JTEC), *Machine Translation* (JTEC).
From 1991 to the present (1998) almost all such comparative studies were conducted by JTEC/WTEC, a total of about 55.
34. Despite the relatively large investment (funds, personnel and time) in these studies over more than a decade, there has not been any assessment of the value of these studies, how they have or have not been utilized, how they could be utilized either by the USG, or by a Congressional Committee. The selected technologies are not chosen based on a set of national priorities, but by the availability of funds and interests of specific agencies. It is near to impossible to show how any of these studies affected national S&T priorities in government or industry. To the best of my knowledge, no new cooperative R&D programs emanated from any of these studies although one of their aims was to enhance USJ R&D cooperation. Unfortunately, the JTEC[H] studies did not as a principle include specific recommendations for future action.
The popular and business media published numerous articles comparing the

relative standings of U.S. and Japanese technologies. Only a small sampling of articles on bilateral comparison can be provided here merely to indicate how widespread the concern was about Japanese technology: Gene Bylinski, The High Tech Race: Who's Ahead, *Fortune* (10/13/1986):26-57; High Technology: Clash of the Titans, *The Economist* (8/23/1986):3-18; Japan's High-Tech Challenge, *Newsweek* (8/9/1982): 48-59; even *Playboy* gets into the act, The Technology War: Behind Japanese Lines by Peter Ross Range, *Playboy* 28:2(2/1981):88-88, 190-196.

35. p. 38, *Gaining New Ground: Technology Priorities for America's Future.* Council on Competitiveness, 4/1991. 77 p. It also pointed out that of the 23 critical technologies selected by the Council all but two appear on at least one other list (DOD, DOC, MITI, EC lists) and that advanced structural materials, biotechnologies, electronic materials, microelectronics, computers and software, "appear prominently in all of the analyses." p. 38.

    In addition to multi-technology surveys by Japan and U.S., the European Community, the French, German, and British governments each started creating similar studies using various methods of assessments but they were begun much later, all in the early 1990s. See *Critical Technologies in a Global Context: A review of National Reports,* Caroline S. Wagner, Critical Technologies Institute (Rand Corporation), Washington, DC, 5/1997. WP-117, 30 p.
36. These studies, known as *Technology Forecast Surveys* were conducted by STA's National Institute of Science and Technology Policy (NISTEP) with the Institute for Future Technology using the Adelphi method for making these calculations. Between 1971 and 1997 six studies were been conducted and translated into English, each about 300 pages.
37. The details of these comparisons listing the selected technologies are, of course, given in the cited studies but also presented in an integrated fashion on p. 294-296 in my essay, Appraising Japanese Science and Technology, *Japan's Economic Challenge,* Study Papers submitted to the Joint Economic Committee, S.Prt 101-121, 10/90.
38. The Aerospace Industries Associates. *Key Technologies for the 1990s: An Overview.* The Aerospace Industries Association, 1987; Computer Systems Policy Project. *Perspectives: Success Factors in Critical Technologies,* Computer Systems Policy Project. 1990. *Fortune* (10/13/1986:26-44) published an article by Gene Bylinsky, The High Tech Race: Who's Ahead including a scorecard showing that Japan was a close second to the U.S. in computers, new materials and opto-electronics.
39. p. ii, 1-7, Council on Competitiveness. *Picking Up the Pace: The Commercial Challenge to American Innovation* [9/1988]. 55 p., *Competitiveness Index* [1988]. 9 p., Competitiveness Index: Appendices I-IV, Trends, Background, Data and Methodology. [1988]. 1 v.
40. p. 1-5, *Gaining New Ground: Technology Priorities for America's Future.* Council on Competitiveness. [4/1991]. 76 p.
41. ibid. p. 29, 35. In light of the policy implications of the two other significant conclusions they should be noted here:

    - Many areas of U.S. strength have the following characteristics: a) they are close to basic research or are the direct result of basic research without the intervening steps of lengthy technology development (e.g. bio-technologies); b) they do not have heavy capital investment needs (e.g. software); c) they can be initiated largely by individual innovation (e.g. computer-aided engineering); they were strongly supported by USG investment in basic research (e.g. genetic engineering), defense procurement (e.g. rocket propulsion), environmental regulations (e.g. emissions reduction).

- Many of the areas where the U.S is weak or has lost have the following characteristics: a) they have not had sufficient private or public investment in the underlying technology (e.g. display materials); b) there is inadequate risk-sharing among companies in technology development (e.g. electronic packaging); c) they have high capital needs and low capital investment (e.g. electronic packaging); d) they need extensive investment in technology for an extended period of time (e.g. optical information storage); e) they have a significant manufacturing focus (e.g. integrated circuit fabrication equipment); f) they have been targeted by foreign government and industry (e.g. memory chips [by the GOJ and industry]. ibid. p. 35.

If any of these deficiencies were to be acted upon they would be tantamount to adopting industrial policies by the USG, anathema to the Reagan and Bush Administrations and any Republican Congress, yet it was a private sector industry-supported Council that concluded that these deficiencies were significant deterrents to a more competitive U.S.

42. p. 1-2, *Critical Technologies: Update 1994.* Council on Competitiveness. 9/1994. 21 p.
43. *Emerging Technologies: A Survey of Technical and Economic Opportunities.* Technology Administration, DOC. Spring 1990. 55 p.; *The Status of Emerging Technologies: An Economic/Technological Assessment to the Year 2000.* National Bureau of Standards, DOC. 6/1987. 20 p.
44. Reference to this NSC report was made in two articles but I was not able to locate an actual copy to ascertain its actual contents: Robert B. Reich, The Rise of Techno-Nationalism, *The Atlantic Monthly* (5/97):65; Bernard Wysocki, Jr, Japan Assaults the Last Bastion: America's Lead in Innovation, *Wall Street Journal* 11/14/88.
45. A Democratic Congress called for DOD National Critical Technologies Plan, later changed to Report, to be prepared every other year. The Congress also called for a report from the National Critical Technologies Panel under the OSTP, two such reports were issued in 1991 (126 p.) and in 1993 (38 p.). The DOD Plan/Report and the OSTP Panel made the Bush Administration uneasy and was the focus of accusations of fostering a national industrial policy under the guise of such Report/Plan/Panel reports. This concern was reflected in the disagreement between the Congress and the Bush Administration in creating the Critical Technologies Institute (CTI) as mandated by a Democratic Congress. Senator Jeff Bingaman (D-NM) successfully pushed for the creation of the CTI in 1990 through the FY1991 Defense appropriations act notwithstanding opposition from the Bush Administration. The CTI as signed into law by President Bush on December 5, 1991 (PL 102-190). As a compromise it was agreed that the CTI would be a Federally Funded Research and Development Center (FFRDC), administered by NSF with a Council chaired by the Director of OSTP (later changed to the NSF Deputy Director) providing policy analysis for OSTP and FCCSET and with funding provided initially by DOD to NSF; later CTI funding was placed under NSF. This is a convoluted bureaucratic arrangement to avoid the impression that the Bush Administration was indulging in foisting industrial policy on the U.S. (An interesting account on the creation of the CTI is provided by Caroline S. Wagner in a ten-page appendix, Congressional Action creating the Critical Technology Institute, in an internal Rand Corp note, *Advising the President on Science and Technology Policy: A role for the Critical Technologies Institute at RAND*, 1/1996, 30 p.

    DOD issued the following publications: *Critical Technologies Plan* in March 1989 (1 v), May 1990 (1 v), May 1991 (1 v). National Critical Technologies Panel reports were issued in 1991 (126 p.), 1993 (38 p.), 1995 (197 p.), and 1997 (38 p.).

For a general discussion and historical review of the critical technologies concept see Lewis M. Branscomb, Targeting Critical Technologies, *STI Review*, OECD, #14 (1994:33-57); and more generally on USJ S&T relations see Cecil H. Uyehara, Appraising Japanese Science and Technology, *Japan's Economic Challenge*, Joint Economic Committee, Senate print 101-121, p. 289-307.

The difficulty of formulating a reliable methodology for measuring the standing of U.S. Science was well illustrated in *Techniques and Methods for Assessing the International Standing of U.S. Science* (Caroline S. Wagner, Rand report MR-706.0-OSTP (10/1995). A reliable theoretical framework for the assessment of science was not available. Forging a balance, necessarily amorphous between quantitative measures and expert opinion was presented as the key to arriving at S&T recommendations and providing an explicit degree of uncertainty and signposts of change. Based on this analysis, one is tempted to question the apparent sudden change in DOD's critical technology analyses in favor of the U.S.

46. 1984. *Industry-to-Industry International Armaments Cooperation Phase II - Japan.* Report of Defense Science Board Task Force. 6/1984. 142 p. (There was a NATO Phase I study.)

    1986. *Foreign Production of Electronic Components and Army Systems Vulnerabilities.* Report prepared by the Committee on Electronic Components, National Research Council. 1986. 95 p.

    1987. *Defense Semiconductor Dependency.* Report of the Defense Science Board Task Force. 2/1987. 103 p.

    1987. *The Semiconductor Industry.* Report of a Federal Interagency Staff Working Group. 11/1987. 58 p.

    1988. *Bolstering Defense Industrial Competitiveness.* Report to the Secretary of Defense by the Under Secretary of Defense (Acquisition). 7/1988. 65 p.

    1988. *Military System Applications of Superconductors.* Report of the Defense Science Board Task Force. 10/1988. 1 v.

    1988. *The Defense Industrial and Technology Base.* Final Report of the Defense Science Board. 10/1988. 2 v.

    1988. *A Report Outlining U.S. Government Policy Options affecting Defense Trade and the U.S. Industrial Base.* Defense Policy Advisory Committee on Trade. 11/1988. 55 p.

    1989. *Defense Industrial Cooperation with Pacific Rim Nations.* Report of the Defense Science Board. 10/1989. 78 p.

    1989. *A Strategic Industry at Risk.* A Report to the President and the Congress from the National Advisory Committee on Semiconductors. 11/1989. 1 v.

    1989. *Holding the Edge: Maintaining the Defense Technology Base.* Office of Technology Assessment. 1989.

    1991. *Industrial Base: Significance of DOD's Foreign Dependence.* General Accounting Office. 1/1991. 26 p.

    1992. *National Security Assessment of the Domestic and Foreign Subcontractor Base: a study of three U.S. Navy Weapon Systems.* U.S. DOC (Bureau of Export Administration). 3/1992. 121 p.
47. Defense Semiconductor Dependency. Transmittal letters (unnumbered).
48. ibid. p. 1.
49. ibid. p. 5.

50. ibid. p. 8.
51. ibid. p. 11. Four other recommendations made were: establish at eight University Centers of Excellence for Semiconductor Science and Engineering, increase DOD spending for research and development in semiconductor materials, devices and manufacturing infrastructure, establish under DOD a government/industry/university forum for semiconductors, and provide a source of discretionary funds to DOD's semiconductor suppliers. (ibid, p.12-13). It is unusual and significant that this report paid not one iota of attention to the much heralded 1986 USJ agreement on semiconductors which one might say provided a breather to the U.S. industry at a crucial time and opened up the Japanese market over the next decade to U.S. semiconductors.
52. These are as follows:
    1980 (1st). *Proceedings of the First U.S.-Japan Science Policy Seminar*. Tokyo. Japan Society for the Promotion of Science. 91 p.
    1982 (2nd). *Science Policy Perspectives: USA-Japan*. New York. Academic Press. 363 p. Ed: Arthur Gerstenfeld.
    1985 (3rd). *Transforming Scientific Ideas into Innovations: Science Policies in the United States and Japan*. Tokyo. Japan Society for the Promotion of Science. 238 p.
    1986 (4th). *Engineering Education: United States and Japan*. Washington, DC, National Science Foundation. 198 p.
    1989 (5th). *Science and Technology to Advance National Goals: Science Policies in the United States and Japan*. Tokyo. Japan Society for the Promotion of Science. 184 p.
    1993 (6th). *Promoting Basic Scientific Research in Universities: The Changing Relationships with Government and Industry*. [No publisher]. 227 p.
    1998 (7th). *Science Policy in the Twenty-first Century: Bilateral Cooperation in a Multi-lateral Context*. The [U.S.] National Science Foundation and The Japan Society for the Promotion of Science. 136 p.
53. The Act is called the Japanese Technical Literature Act of 1986. Earlier, in May 1983 an attempt was made to make $175,000 available to NTIS to make JSTI available in the U.S. In 1984, the House S&T Committee authorized $750,000 for FY1985 for DOC's Office of Productivity and Innovation, again to increase JSTI availability in the U.S. Even this super-modest sum was lost in the parliamentary confusion after a Presidential veto. This minute but unsuccessful effort probably stemmed from the 1984 hearings on JSTI availability. Since the DOC had to find the money from within its budget, it took the Reagan Administration almost a year to create the Office of Japanese Technical Literature with one full-time person and one part-time consultant with a $300,000 budget to
    - monitor Japanese technical activities and developments,
    - consult with businesses, professional societies and libraries in the U.S. regarding their needs for information on Japanese developments in technology and engineering
    -coordinate with other agencies and departments of the Federal Government to identify significant gaps and avoid duplication in efforts by the Federal government to acquire, translate, index and disseminate Japanese technical information,
    - prepare annual reports regarding important Japanese scientific discoveries and technical innovations in such areas as computers, semiconductors, biotechnology and robotics and manufacturing,
    -encourage professional societies and private businesses in the U.S. to increase their efforts to acquire, screen, translate and disseminate Japanese literature,

-publish an annual directory of resources in the U.S. which translate, abstract and disseminate Japanese technical literature.

54. Federal Technology Transfer Act of 1986. 99th Congress, 2nd session, Senate report 99-283 (4/21/1986), 28 p.; Federal Technology Transfer Act of 1986. 99th Congress, 2nd session, Conference Report (10/2/86), House Report 99-953, 21 p.; Wendy H. Schacht. *Technology Transfer: Utilization of Federally Funded Research and Development.* Congressional Research Service, Library of Congress, 5/11/1989, 11 p.; Paul Blanchard and Frank B. McDonald, Reviving the spirit of enterprise: role of the federal labs, *Physics Today*. 1/96, p. 42-50.; *Implementation Status of the Federal Technology Transfer Act of 1986*. General Accounting Office. (GAO/RCED-89-154). 5/1989. 68 p.; Christopher T. Hill. Federal Technical Information and U.S. Competitiveness: Needs Opportunities and Issues, *Government Information Quarterly* 6:1(31-38).
55. *Federal Register*, 4/1988. [CHU: find additional notes on this issue]
56. General Accounting Office. *Technology Transfer: U.S. and Foreign Participation in R&D at Federal Laboratories.* 8/1988. p. 2.
57. ibid. p. 2.
58. ibid. p. 4.
59. Following is a list of these studies: They were all published by the National Academy Press. Year of publication and number of pages only are listed.

    The Working Environment for Research in U.S. and Japanese Universities: Contrasts and Commonalities, 1989. 51 p.

    Learning the R&D System: 1) University Research in Japan and the United States, 1989. 18 p; 2) National Laboratories and Other Non-Academic, Non-Industrial Organizations in Japan and the United States, 1990. 46 p.; 3) Industrial R&D in the United States and Japan, 1990. 39 p.

    Approaches to Robotics in the United States and Japan, 1989. 21 p.

    Japanese to English Machine Translation, 1990. 36 p.

    Expanding Access to Precompetitive Research in the United States and Japan: Biotechnology and Opto-electronics, 1990. 39 p.

    R&D Consortia and U.S.-Japan Collaboration, 1991. 40 p.

    Intellectual Property Rights and U.S.-Japan Competition in Biotechnology, 1991. 43 p.

    Japanese Investment and Technology Transfer: An Exploration of its Impact, 1991. 46 p.

    U.S.-Japan Strategic Alliances in the U.S.-Japan Technology Linkages in Biotechnology, 1991. 98 p.

    Semiconductor Industry: Technology Transfer, Competition, and Public Policy, 1992. 118 p.

    Japan's Growing Technological Capability: Implications for the U.S. Economy, 1992. 235 p.

    Corporate Approaches to Protecting Intellectual Property: Implications for U.S.-Japan High Technology Competition, 1994, 22 p.

    High Stakes Aviation: U.S.-Japan Technology Linkages in Transport Aircraft, 1994. 144 p.
60. Science, Technology, and the Future of the U.S.-Japan Relationship. 1989, 14 p.

61. The exports of weapons were prohibited under these principles when 1) exports are bound for communist countries subject to the embargo established by Western Allies to deny these nations access to sophisticated weaponry, military technologies or dual-use technologies; 2) exports are bound for countries to which the export of weapons is banned under United Nations resolution; and 3) exports are bound for countries involved in international conflict or countries that might be involved in international conflicts. In 1976 this policy was broadened to include restrictions on exports of military technologies as well as equipment related to arms production.
62. In *The FY 1987 Department of Defense Program for Research and Development* (Statement by the Under Secretary of Defense, Research, and Engineering to the 99th Congress Second Session 1986), 2/18/1986:16 the benefits of increased cooperation, i.e. a flow of Japanese military technology to the U.S. were: 1) a strengthened U.S. industrial base, resulting from the infusion or stimulus of Japanese technology in those areas in which Japan has a lead; 2) more efficient use of our R&D resources, as we avoid duplicating Japanese development efforts and focus our attention on developing and introducing better systems sooner: and 3) improved inter-operability with Japanese fielded systems, thereby strengthening the security force structure in the Northern Pacific. While these benefits are correct, and necessary it is surprising that there is no sense in this statement that politically this flow of military technology to the United States would contribute to the overall security relationship, and how two-way flows would benefit both sides. It is no wonder that Japanese firms are reluctant to eagerly participate when such one-sided objectives are baldly stated without the slightest gesture toward broader goals.
63. The implementation procedures for the November 1983 USJ exchange of diplomatic notes was contained in *Japanese Military Technology: Procedures for Transfers to the United States*. DOD (Office of the Under Secretary of Defense), February 1986. 4 p. and four annexes including the diplomatic notes. The Japan Economic Institute (Washington, DC) funded by the MOFA, almost yearly issued a report on defense technology exchanges: Washington pushes for expanded U.S.-Japan Defense Technology Exchanges, 4//8/1994 *(JEI Report*, 14A):10; Japan's Potential Role in a Military-Technology Revolution, 1/13/1995 *(JEI Report*, 1A):27; Transpacific Efforts to Codevelop Military Systems: Room for Growth, 6/6/1997 *(JEI Report*, 21A):5. The above account was, in the main, derived from these materials.
64. It should be recalled that Frank Press was President Carter's science adviser and was not successful in 1979-1980 in persuading the GOJ to commit substantial sums of money to USG R&D projects in 1979-1980 negotiations. After Carter left the White House, Press became president of the NAS.
65. The Japanese delegation consisted of 22 in 1985 and 29 in 1986 and the U.S. 18 and 21, respectively. The Japanese delegates came from the University of Tokyo, Saitama University, NEC, Mitsubishi Electric Corporation, Nomura Research Institute, Mitsui & Co., Sony, Sumitomo Electric Industries, Nissan, Kyōwa Hakkō Kōgyō, NTT, Nisshō Iwai. The U.S. delegates came from MIT, Texas Instruments, Goldman Sachs, Upjohn Co, Atlantic Richfields Co, Thomas Paine Associates, Hughes Aircraft, Citicorp, Harvard, Rice and Johns Hopkins Universities, former U.S. Ambassador to Japan, Charles S. Draper Lab, Centocor, National Academy of Engineering, AT&T. Both delegations were supported by interpreters, staff, consultants, "resource persons". It was an elaborate high level meeting. On November 11, 1987, the Japanese delegation held a report session to which industry, government and other officials were

invited. The first Conference did not release a communique; only a "draft summary" exists which is included in a long report in Japanese published in *Gakujutsu Geppō* 39:1 (1/86):6-25, organ of the JSPS. This "draft summary" was not available from the records of the National Research Council indicating the slight value given to this report on the first Conference by the U.S. side.

66. I could not find any record of the U.S. delegation holding a similar meeting with industry and government officials about the outcome of this first Conference.
67. Apparently there was no record of proceedings of either Conferences. One senior member of the Japanese delegation in an interview in Tokyo explained that he does not appreciate being shouted at and the table pounded by an American delegate to this Conference in order to persuade the Japanese delegation to its viewpoint. It is counter productive he said.
68. "Symmetrical access" was declared not to be the same as "identical access" but it is instructional to refer to the *Webster's New Collegiate Dictionary* (7th edition) about symmetrical: being such that the terms may be interchanged without altering the value, character, or truth. It gives the distinct impression that the two sides (U.S. and Japan) have the same amount, kind, value, in access to STI and R&D facilities. To differentiate that from identical access is indeed stretching the logic and meaning of these words in the English language. It was an attempt to some how put the pressure on the Japanese to open up access to their information and facilities.
69. The testimony raised three categories of problems: what is meant by access to technological knowledge and research, how would it measured, how would it be valued assuming it could be defined and measured. Under each category, additional questions were posed, all difficult to answer with finality. For example, is it availability of foreign science journals and papers, or abstracts, in a foreign language or in English, is it access to government and/or private R&D facilities, is it the price foreigners request for access, what about the perceptions about the value of S&T in different countries, value of and need for some STI may differ from country to country. Then there is the technical capability and capacity of those assigned to the afore-mentioned Committee which may not be adequate. Could DOD or DOC veto the judgment of NSF all of whom would use different judgement criteria. Interpreting the mandate would "prove enormously complicated, time-consuming and expensive. ... [with] inconclusive results. State 375821 (12/4/87).
70. "If those aspects of the game are perceived as unfair, actions are likely to be taken in retaliation that, while aimed at equalizing the zero-sum competition, may damage the trading and technology transfer system, converting it into a zero-sum or even a negative sum situation." This must be avoided, thus, symmetrical access.
71. The statistics cited in this subsection were derived, in the main, from *Science and Engineering Indicators 1996,* National Science Board, National Science Foundation, p. 6-18 - 6-24; *All Technologies Report*, January 1963 -- June 1993 (A statistical series), U.S. Patent and Trademark Office, Washington, DC, 7/1993.; Who are copycats now?, *Economist* 5/20/89:91-94; Not invented here, *Economist* 10/28/89:74-75.
72. World Intellectual Property Organization, cited in High Tech: Leaving Home-1 by Dan Morgan, *Washington Post*, 5/1/83.
73. The three Japanese companies were Hitachi, Toshiba and Nissan Motor Co.; the six Japanese companies were Toshiba, Canon, Mitsubishi Electric, Hitachi, Matsushita Electric and Fuji Photo Film. In 1973 the ten U.S. companies were GE, AT&T, GM, IBM, Westinghouse Electric, E.I. Dupont de Nemours, Eastman Kodak, U.S. Philips,

Dow Chemical. The 1993 U.S. lineup was IBM, Eastman Kodak, GE, Motorola.

74. In a small *USA Today* box report (9/24/85) entitled Patents: USA Loses grip on new ideas, the high profile of Japanese companies was softened by listing the top 30 companies: only nine are Japanese, five are German and the others were U.S.
75. During the 1980s, one of the U.S. anxieties was centered on the presumed implications of how many patents were held by what companies and individuals from what country. Later analyses were developed that would put provisos on the seemingly high Japanese percentages, to soften the initial impact of anxiety, to re-assure by pointing out citations in support of patents, and other factors to dilute the felt urgency about this radical shift in patent ownership. In study of biotech patents in *Nature* (4/4/1996), 1981-1995, of the 1,175 patents awarded in the U.S., Europe and Japan 75% went to the private sector and of those about half to Japanese companies. But it pointed out that those firms which produced commercial products were, in the main, not Japanese. Adding up the patent numbers was ridiculed as handicapping the National Football League by "simply listing the height and weight of the players" without considering experience, skill, luck, and performance over time. (*Business Week*, 4/22/1996:47) The DOC and the U.S. Council on Competitiveness issued a report in September 1998, *The New Innovators: Global Patenting Trends in Five Sectors*, showing that Japan still leads foreign holders of U.S. by a wide margin. Just to compound the issue, the Council on Competitiveness, a private nonprofit group created in 1986, announced in *The New Challenge to America's Prosperity: Findings from the Innovation Index* (3/1999. 94 p.) that based on an "innovation index" comparing 25 countries, the U.S. was the leader in the 1980s and 1990s but "if current trends continue, by 2005, the U.S. will trail Japan, Finland, Denmark, and Sweden because of inadequate spending on basic research, education and a shrinking percentage of technical workers and the areas affected will be advanced materials and solid state physics and even software (*Business Week* 3/22/1999:6).
76. The Toshiba-Kongsberg case was not the only violation of COCOM rules. In 1985-87 there had been four other violations by Japanese companies for exporting contraband items to Eastern Europe, China and North Korea, e.g. censure of Ishikawajima Harima's export of a floating dock to the Soviet Union.
77. Take, for example, the case of the possible purchase of Fairchild by Fujitsu Ltd. Fairchild was owned by Schlumberger Ltd, a French company, and involved with defense contracts. It's business had been slipping in competition with the Japanese. In early 1987, Fujitsu had proposed to purchase 80% of Fairchild Semiconductor Corp. This proposed purchase immediately was embroiled in various USJ military, trade and technology issues. Secretaries of Commerce and Defense and Director of CIA were opposed for national security reasons, in retaliation for Japan not purchasing U.S. superconductors and against further dependence on foreign sources for important defense material. Those who supported the purchase said this would help finance the U.S. budget deficit, would send the wrong message to investors when the U.S. was trying to persuade them to invest in the U.S., would tie Japan more tightly into the U.S. orbit. Senator Exon (D-Nebraska) wrote President Reagan opposing the sale because it "would have serious implications for our [U.S.] national security". French ownership was less of a problem since they were not that much involved this kind of high techlogy. "While Japan is certainly a friend and ally", the government should act "to preserve at least a minimum and prudent level of its vital defense industry within its borders" The Department of Justice was began it's own investigation. A full fledged cabinet meeting was called for by SecDOC to consider this possible purchase. Faced with these problems, Fujitsu gave up the purchase in early

March 1987; National Semiconductor Corp agreed to purchase Fairchild in September. (*Washington Post* 1/10/1987, 3/17/1987, 3/12/1987, 9/1/1987).

78. The Government of Norway was much more adroit than the GOJ; it defused the incident by admitting its errors and taking quick steps to prevent further such errors in the future. To pursue Norway would not be politically profitable; it was a small European country with whom the U.S. had no trade problems, in contrast to Japan. Thus the "Kongsberg" aspect was quietly dropped.
79. A Norwegian police report, probably made in connection with investigating the Toshiba-Kongsberg case, showed that machine tool companies in numerous West European countries had sold computerized technology to the Soviet Union and China for the past ten years and that France may have been shipping the same type of machine tools to the Soviets at the same time the Toshiba diversion was taking place. Cited in *The United and Japan in 1988*: *A time of transition* [actually covering 1987].- (School of Advanced International Studies/SAIS, The Johns Hopkins University. 1988. Washington, DC., p. 25). This report was part of an annual series begun in 1986 and continuing until today prepared by a group of U.S. and Japanese graduate students at SAIS under the supervision of SAIS professors. They (each about 100 pages) were carefully researched summaries, and balanced analyses and a useful handy reference of USJ relations for each year. One could speculate further on the causes behind this evolving image of the controversy.
80. As a political statement, Paul Freedenberg, DOC Under Secretary for Export Administration which had just been created in October 1987, stated that "the Japanese have come further in the last six months than they have in the last 30 years in terms of changing and improving their overall export control system ... many of the largest Japanese multinationals are developing their own internal export control systems ... clearly this episode helped us convince our allies that something significant needed to be done." *Japan Economic Survey* 4/1988, p. 7 and 9.
81. A. Richard Armitage, an Assistant Secretary of Defense at the time, had written a letter to the U.S. House of Representatives in early March 1988 that Soviet submarines had quieter screws three years before (December 1982?) Toshiba (and Kongsberg) had sold the milling machines and software to the Communists. (*Asahi Shimbun* 3/28/88) This letter would seem to undermine the major implication of the T-K sale that it would and did allow the Soviets to make their submarines quieter but it would not change the fact that this sale did indeed contravene COCOM rules and would improve Soviet capabilities in making submarines.
    B. Japanese reactions to the furious charges by the U.S. were not discussed in the main text since the purpose was to describe the severity of the U.S. reaction as part of the atmospherics which must have invaded USG councils and the subsequent USJ negotiations. Let us focus on several sources of Japanese reactions: legal, political and commentators.
    The legal: In his written judgment against the Toshiba executives, Judge Yonezawa Toshio opined that Toshiba had not considered this machine to be used for military purposes and it is not proven [without benefit of the above Armitage letter] that this sale interfered with the maintenance of international peace and security..
    The political governmental: The GOJ position was exemplified by Prime Minister Nakasone's statement that it is possible that a causal effect could exist between quieter [Soviet] submarine screws and the milling machines but that has not been proven.
    Japanese commentators: As might be expected they had quite different

interpretations.

Japan Echo which is published by a commercial company but is also distributed worldwide by GOJ overseas information offices carried four translated articles in it's 14:4 (1987) issue. One by Karatsu Hajime (Dead Letters on the COCOM, p. 27-30) somewhat ridiculed the propriety of the forbidden technologies under COCOM since many products on the open market incorporate these technologies and noted that the U.S. had not produced any specific proof about the allegations against Toshiba. Another by Murakami Yoshio (Japan's Misguided Response to the Toshiba Case, p. 31-34), the Asahi Shimbun Bureau chief in Washington, criticized Japan's response as misguided in gauging the U.S. response. The third essay by Shima Nobuhiko, a writer, (Japan Treads on the American Tiger's Tail, p. 35-38) analyzed the T-K problem in light of U.S.'s "extreme sensitivity to Japanese and Soviet advances in high technology" (p. 35), that maintaining its high tech superiority was vital to the national economy and security. It cited the 1984 edition of DOD's Soviet Military Power as stressing the importance of making submarines quieter, a 1985 DOD report which identified Japan "as a hotbed of high-tech leaks to the Soviet Union." The fourth essay by Maeda Tetsuo, a free lance journalist, (The Toshiba Flap and America's Global Strategy, p. 39-44) was most interesting and insightful because his analysis placed the Toshiba incident in the context of evolving U.S. strategies to contain the Soviet Union. He described three strategic phases in the U.S. containment of the Soviet Union: 1) through the 1950s was based on land barriers through treaties (NATO, CENTO, SEATO, with Korea, Taiwan and Japan); 2) 1960s-1970s, strategy shifted to a blockade using the seas and ICBMs. By the late 1960s submarines carrying Polaris missiles moved ahead of B-52s and ICBMs as the centerpiece of the new phase strategy. The U.S. was able to keep track of Soviet submarines which "seemed to move through the water to the accompaniment of drums." (p. 41). These two phases, Maeda asserted, the U.S. could rely on its power with the cooperation of its allies. In the 1980s, however, this naval containment strategy began "to unravel, heralding a period of U.S.-Soviet parity" (ibid). The Soviets had developed the SSN-N-8 and 18 missiles which were better than the Polaris and Poseidon missiles and submarines which could travel underwater at 42 knots, dive to depths of 700 meters, and the noise problem was overcome with launching of the Mike- and Sierra-class Soviet attack submarines. Maeda cited the DOD Soviet Military Power reports of 1983, 1984 and 1985 concerning Soviet advances in creating quieter, speedier submarines, creating "a submarine gap" in the eyes of the DOD. A third phase strategy began taking shape in the 1980s under President Reagan combining the MX intercontinental missiles, the Trident II missile for submarines, the B1-B strategic bomber and the SDI. Maeda described this new attempt as "a space blockade". In the new phase, the European allies and Japan were described as "partners" with the U.S. The U.S., said Maeda, "no longer has the money or the technology to carry out a global encircling strategy on its own" (p. 42) Earlier the missile gap crisis and the window of vulnerability were used "to instill a sense of crisis" now, he says obviously different means of intimidation were needed since the cooperation of other countries, Europe and Japan, were necessary. Maeda analyzes that the new longer range Soviet Sea Launched Ballistic Missiles (SLBM) meant that Soviet subs could stay offshore and threaten the U.S. heartland. To counter this threat old ships were re-commissioned and new ones built to create the 600 ship Navy Reagan envisaged including the creation of superior attack submarines. The Soviet began sea trials in 1984 of an even more formidable Akula-class attack submarine. Against this backdrop the Nakasone Cabinet approved the increase of the Japanese defense budget to 1% of GNP, a major achievement in light of past history, and he declared Japan to be an "unsinkable aircraft carrier". Japan was warned of the vulnerability of its sea lanes and the loss of Hokkaido in case of conflict with the Soviets. He baldly asserted that the T-K incident played right into the hands of those concerned about the sub-gap and "unintentionally helped publicize" this gap. The anxiety about the decibel count of submarine screws was "not the real issue". (p. 43) This calls for, he asserts "Japan to be drawn into efforts to close the gap." (p.43) Japan could help in upgrading the P-3C anti-submarine aircraft, battleship improvement, joint operations. It is highly significant that in 1986 Japan had already transferred military technology for battleship construction and improvements and later in 1990 agreed to transfer to the USG digital flight controller technology for the P-3C aircraft. Maeda concluded that "Toshiba was but a pawn in a larger campaign to alter the course of Japanese [defense] policy." (p. 44) The implication here to control and alter Japanese policies is highlighted here because this idea surfaces again during the USJ STA negotiations in 1987-1988. Whether this theory is correct or not, it does provide a seemingly more meaningful contextual explanation for what Maeda called "the virulence of the American reaction" (p. 39) which puzzled many Japanese.

C. Allied Signal sold jet engines for use in military and training aircraft, and in 1994 AT&T sold advanced technology telecom equipment to a company affiliated with the People's Liberation Army and in 1996 Silicon Graphics sold a powerful supercomputer to China's Academy of Science which does subcontract work for the Chinese military on guidance systems. *Newsweek* 5/5/97:46.A House of Representatives Select Committee chaired by Rep. Christopher Cox (R-Calif) accused two U.S. companies, Hughes Space & Communications Co, and Loral Space & Communications, leading U.S. satellite makers of passing sensitive information to the Chinese about their rocket failures. (*Washington Post*, 5/27/1999). Then in November 1999 McDonnell Douglass Corp and a Chinese company, China National Aero-Technology Import and Export Corp, were indicted for conspiring to violate U.S. exports laws in the sale of aerospace equipment that ultimately wound up at a Chinese military plant. (*Washington Post*, 10/20/1999).

82. This drive, now known in Japanese as *kokusanka*, variously translated into English as indigenization, national product-ization, domestic production, is not a postwar phenomenon after Japan became a technological superpower, but goes back to the middle of the 19th century when Japan began its modernization in the face of rampant imperialism in East Asia. This drive has naturally never disappeared. When Japan was defeated and occupied there was nothing to do but wait for the next opportunity. The coldwar between the Free World and the Communist world provided this opportunity. Through co-production and licensing from the U.S., Japan rebuilt its defense industry capabilities; in the aircraft area it built its trainer jet aircraft (the T-1) and other transport jets, and other trainer/ground support fighters (T-2/FST-2)

83. A great deal has been written on the FSX controversy. Only a small sampling which I have consulted will be listed here. John W. Dower, Miri-teku o meguru Nichi-Bei Kankei no kiki, *Chūō Kōron* (5/89:134-149); Michael J. Green, Alliance Politics and Techno-nationalism (Paper given at Conference on High Technology Policy-Making in Japan and the United States, 1993, 46 p, George Washington University); Masaru Kohno, Japanese Defense Policy-Making: The FSX Selection, 1985-1987, *Asian Survey* 19:5(5/89):457-479; Nishii Yasuyuki, Bei San-Gun Fukugōtai ga nerau Gunji

Gijutsu Ummei Kyōdōtai, *Asahi Jānaru* (7/1/88: 20-22); Gregory W. Noble, *Flying Apart?: Japanese-American Negotiations over the FSX Fighter plane*, Berkeley, CA, Institute of International Studies (US-Berkeley), 1992, 66 p.; Clyde V. Prestowitz, jr, Foreign Policy Crisis: The FSX Fighter, *Trading Places*, New York, Basic Books, 1989, p. 5-58; Otsuki Shinji and Honda Masaru, *Nichi-Bei FSX Sensō*, Tokyo, Ronsō-sha, 1991, 365 p.; Jeff Shear, *The Keys to the Kingdom*, Doubleday. 1994, 318 p.; Barbara Wanner, American, Japanese Defense Industries Explore Expanded Cooperation, *JEI Report* 32A(8/22/1997):14; Barbara Wanner, Large-Scale U.S.-Japan Defense Technology Exchanges: The Potential Stall after FS-X, *JEI Report* 40A(10/27/1995):15.

84. The idea of designing an indigenous successor FSX was initiated by industry and the JDA in 1967 even as the F-1 was still in design phase. Subsequently, when the military requirements for the JDA changed with the impending return of Okinawa in 1972, ambitious plans were formulated to make indigenous production of aircraft and weapons a central and explicit goal not only of industrial policy but of Japanese defense policy as well. These plans were formulated by the JDA when none other than Prime Minister Nakasone was Director General of the JDA. These plans ran afoul of domestic opposition in the governing party itself, the LDP, who looked upon such ideas as potentially undermining the basis of the USJ security relationship, and being far too expensive for Japan, and internationally, the developing U.S.-Soviet detente, the oil crises of the 1970s, the reaction of Japan's Asian neighbors combined to force Japan to forego these plans for the time being. Instead, military inter-operability with the U.S. was emphasized and reflected in the 1976 GOJ *National Defense Plan*, the 1978 Guidelines on USJ Defense Cooperation and the 1983 Military Technology Transfer Agreement. Yet, since Japan had attained high technological levels, the GOJ's 1979 Jieitai *Sōbi Nenkan* (1979 Defense Equipment Yearbook) declared that this new higher level of technology should be used "to maintain leverage in negotiations with the U.S. over military technologies" (Green, p. 5).
85. U.S. and Japanese disagreements about the meaning of co-production and co-development swirled around the SH-X, an anti-submarine helicopter in 1985-1986. An MOU was signed in spring 1986 by the USJ clarifying that the helicopter was not a domestic Japanese creation.
86. "The compromise on the FSX was not achieved without friction and a considerable degree of ill will towards the United States..." (Green, p. 24); more Japanese recognize that their security and economic interests will be served by Japan's becoming a co-supporter of the global hegemony that the United States can no longer sustain alone. ... Despite waning American power, Japan is subjected to even stronger influences from the United States" (Kohno, p. 479).
87. It was generally believed that the agreement consisted of the following: 6 prototypes to be built (2 in the U.S., 4 in Japan), the U.S. would receive 35-45% share of the work, the U.S. would have access to new technology (gallium arsenide chips, co-cured composite wings and phased array radar), choice of engines would be postponed for future consideration.
88. Defense Authorization Act of FY1989 (October 1988). Reflecting the rising anxieties in the late 1970s and 1980s, the GAO conducted a study, *U.S. Military Co-production Programs Assist Japan in Developing Its Aircraft Industry* (unclassified version, 3/18/1982, 50 p.) whose title, to begin with, reflects its conclusion. The report correctly points out that Japan's aircraft industry grew and developed largely through U.S.

military co-production programs. This approach, it should be understood, was accepted as proper and correct in light of the contest with the Soviet Union in the cold war. In light of past history of Japan's postwar aircraft accomplishments, the GOJ naturally assisted the aircraft industry in its development. The GAO report recommended that the State Department take the lead in cooperation with the DOC, USTR, Treasury and Labor and other relevant agencies, in assessing military co-production programs, and in developing a clear and more comprehensive military coproduction policy recognizing the trade and economic implications as well as political and military goals to be achieved. More details can be obtained from *Defense Production and Industrial Development: The Case of Japanese Aircraft*, by Richard J. Samuels; and Benjamin C. Whipple, *Politics and Productivity: How Governments Create Advantage in World Markets*. Cambridge, MA, Ballinger Books, 1988; *Japan's Aircraft Industry*, Council Report, #2 (1/11/1980:9). United States Trade Council.

89. Green, p. 28. Noble points out in his article that the Japanese were infuriated "by American demands to renegotiate the terms of an agreement Japan had not wanted in the first place, and over which the two countries had spent more than a year wrangling". The day after Bush's announcement, March 21, 1989, the JDA Director General accurately forecast that "the fate of the FSX project would affect possible future military joint ventures between countries". Later he expanded on his comments: It is indeed a pity that the officials within the U.S. administration and Congress have not ironed out their differences of opinion toward the project. ... I expect the country [the U.S.] to act as a major power" and demanded that the U.S. keep its promise. Some LDP Diet members were even for scrapping the FSX agreement and returning to indigenous development.(p. 33) One of the major criticisms of the GOJ by Karl van Wolferen in his book, *The Enigma of Japanese Power* and his essay, The Japan Problem, in *Foreign Affairs* (Spring/1987:288-303) was that there was no reliable center of power in Tokyo with whom to negotiate; the Prime Minister's concurrence could be undermined by other Ministers and/or the bureaucracy. It seems that the same kind of criticism could now be made legitimately also of the U.S. governmental organization, if the President's representatives do not represent the will of the USG.
90. To avoid yet another confrontation, the GOJ agreed to pay MHI for the R&D and GD would receive the results of this technology free from MHI.
91. After the agreement was reached, Japan began playing the European card, "much to the alarm of American aerospace firms. JDA announced in summer 1989 that it would procure $750 million worth of search and rescue jets from Britain. This would be the first non-U.S. weapons procurement by Japan. This weapons relationship accelerated in 1990-91 in products, marketing, and in joint aerospace projects. MHI even embarked on a strategic relationship with Germany's Daimler-Benz. (Noble p. 27)
92. "In Japan, the political leadership -- and especially the defense establishment -- were exhausted and bitter. After this second clash over the FSX, resentment of the United States and of the alliance ran deep." (Green, p. 29) ... The resentment caused by [the] FSX was internalized by many members [of the LDP] and raised many questions about the commitment to [the] allies that quietly linger (Green, p. 43). Noble was far more dire -- and accurate -- in his observations: "The asymmetry apparent in the FSX negotiations [i.e. the blatant use of U.S. national power] left a legacy of distrust, resentment, and ill will on both sides. ... The real long-term cost, though, is a weakening of defense and diplomatic ties between the United States and Japan. ... To the Japanese, the Americans were incompetent, overbearing and unfair. ... [To the Americans], the

Japanese deceptive and unfair. ... This legacy is likely to impede future efforts at scientific and technological cooperation. ... the Japanese experience with the FSX has clearly set a most unpleasant precedent. ... the FSX set an important and unfortunate precedent as the first major case of U.S.-Japanese co-development, ... [While the tactics of the USG using], the good cop, bad cop routine by the President and the Congress" have succeeded they are "becoming more and more costly. Among the old costs are diminished credibility and the irritation of foreign partners. In this respect the FSX case was not new, but it marked a watershed. New [future] costs include the growing ability of Japan to retaliate and the threat to future cooperative ventures." (Noble p. 51-54) The Japanese commentary is more conspiratorial: That the U.S. hopes to control new Japanese technologies developed through the FSX process by bringing them under defense security purview as defined by the USG/DOD and prevent their utilization in the civilian sector and envelop (*kakoikomu*) Japan into the U.S. strategic and politico-military framework. (Nishii Yasuyuki, Bei San-Gun Fukugōtai ga nerau Gunji Gijutsu ummei Kyōdōtai, *Asahi Jānaru* (7/1/88: 20-22). This writer is unaware of a similar confrontation with the European countries with such vehemence and acrimony when the jointly developed European Tornado fighter was manufactured; it would indeed be interesting to speculate as to the causes for such a stark contrast between the U.S.-European and the U.S.-Japan relationship.

93. "Burden-sharing" was yet another aspect in USJ (and NATO) relations which impinged on these relations but not so heavily as others; for this reason was not discussed separately. Numerous Congressional hearings were held and trips taken to emphasize the need for greater burden sharing by U.S.'s Allies in confronting the Soviet Union. The Report of the Defense Burdensharing Panel of the Committee on Armed Services, House of Representatives, August 1988 (114 p. 100th Congress, 2nd session, print 23) called upon Japan to increase its defense spending, increase host-country support of U.S. forces in Japan, increase its official development assistance to countries of economic need and strategic importance to Japan, e.g. Turkey and the Philippines. In contrast, the JEI Report, Sharing the Defense Burden with Japan: How much is enough? (19A 5/13/88:14), points out that it has become difficult to objectively assess Japanese security requirements, the steps Tokyo has taken over the years to improve and expand its defense capabilities especially since these efforts are viewed "through a lens distorted by anger and frustration over chronic trade frictions." (p. 1)

94. Two essays on R&D are mentioned as an example: Charles H. Ferguson, America's High-Tech Decline, *Foreign Policy* 74(Spring 1989):123-144; Richard R. Nelson and Gavin Wright, The Rise and Fall of American Technological Leadership: The Postwar Era in Historical Perspective, *Journal of Economic Literature* 30(12/1992): 1931-1964. From the politico-military point of view, Professor Paul Kennedy briefly held center stage the mid1980s with his disturbing historical interpretation of world history. Paul Kennedy, The(relative) Decline of America, *The Atlantic Monthly* 8/1989:29-38; Paul Kennedy, *The Rise and Fall of the Great Powers: Economic Change and Military Conflict from 1500 to 2000*, Vintage Books (Random House), New York, 1987, 677 p.; his theory about the "decline" of the U.S. uncomfortably coincided with the "rise" of Japan in the mid-1980s that even a Congressional symposium sponsored by the House Subcommittee on International Economic Policy and Trade was held on May 19-20, 1989 for Paul Kennedy to discuss his theory (*U.S..Power in a Changing World*, 5/1990, 131 p.). A famous book by Ezra F. Vogel, *Japan as Number One: Lessons for America.* Harvard University Press. Cambridge,

MA, 1979. 272 p. was naturally translated into Japanese. A Harvard Professor under the imprimatur of the Harvard University Press probably played a large role in inflating Japanese egos. To be thus praised by such eminences after being semi-ridiculed for so long "as those little brown brothers" who are good copiers this must have been heady stuff for the Japanese. Karatsu Hajime is a good example of a Japanese writer who pronounced in superlative terms Japanese accomplishments and the dire problems besetting the U.S.: American Industry Self-Destructs , *Japan Echo* 18:2(1991):50-55; An Outside Look at America and its decline, *Intersect* 6/1989:9-15. In all fairness we must mention that when Karatsu's book in English, *Tough Words for American Industry* was published in the U.S. in 1988 (it had been published in Japanese in 1985) and his articles in *Business Week* were favorably commented on by none other than the U.S. management guru, Peter Drucker. Karatsu was an NTT engineer, later managing director of Matsushita Communications Industrial Co, director of the Tokai University's Institute of Research and Development. He was trained in the U.S. and regards "America [as] my teacher."

V. The Institutional Arrangements in the USG and GOJ

1. The U.S. Department of State

The statute governing the Department of State establishes the Department's responsibilities in science and technology as follows:

> In order to implement the policy set forth in section 2656b of this title, the Secretary of State (hereafter in this section referred to as the "Secretary") shall have primary responsibility for coordination and oversight with respect to all major science or science and technology agreements and activities between the United States and foreign countries, international organizations, or commissions of which the United States and one or more foreign countries are members (U.S. Code, 1982, v. 9, Title 22 Foreign Relations and Intercourse, p. 508).

So long as the international order after 1945 when the U.S. prevailed as the dominant power was accepted, there was apparently no felt need for a special focus on science and technology as a foreign policy issue.[1] But when this "consensus no longer existed" (Bruce-1990:105), when the lines between domestic and foreign policy became blurred and when foreign developments on the environment, disposition of wastes, depletion of the ozone layer, then the Department of State had to confront the issue of how to contend with these new forces. For almost half a century there have been attempts and intermittent reports to persuade the U.S. Department of State to take seriously the need for professional scientific judgment in top policy considerations as exemplified in the Department's Reorganization Task Force, Number 2 in May 1949.

Based on a State Department report in 1950, a small office was created to consider science and foreign policy. It took more than 15 years and the commitment of an energetic foreign service officer who became its director for another change to take place. Then in 1967, the Secretary of State expanded its staff and responsibilities. In 1969, a new addition, a Science Advisory Committee was created early in the Nixon Administration by Secretary of State William Rogers to supplement the small S&T office. But it was not until 1973 that legislation for FY1974 appropriations for the Department created the Bureau of Oceans, Environment and Scientific Affairs. Thus the structure for giving S&T a role in the higher councils on U.S. foreign policymaking were in place, but was the State Department as an institution and the Foreign Service as a group, temperamentally prepared to accept this "S&T interloper" in these high councils, did it have the political will power, sense of direction, commitment and determination, and willingness to provide substantial budget and financial investment in developing personnel to backup the new structure and challenge? Judging that over the next twenty five plus years at least three reports in 1976, 1992 and 1999 were issued each urging the State Department to take S&T seriously, to make S&T a truly integrated part of the entire foreign policy process, the Department has not been willing to do what was necessary.[2]

The 1976 report called for clarification of the OES mission and extension of its horizons, substantial strengthening of resources (funds and professional staff) and a more thorough utilization of the OES staff. In addition to these broadly stated needs, the report also recommended, among many other details, the recreation of the Science Advisory Committee (SAC) and assistance to the U.S. private sector in facilitating access to foreign technology. It appears many of the ideas of this report were woven into the Foreign Relations Authorization Act for FY 1979 (Public Law 95-426), particularly Title V which, in Smith's words, "should be seen as perhaps the high point at least in what was expected of the Bureau" (Smith 92:140). The 1992 report, among other recommendations, called for the creation of a new high profile position for S&T in the State Department, an S&T Counselor to the Secretary of State. The 1999 report noted: "Yet ironically, as the world becomes more technologically interdependent, the trend at the State Department has been to downplay science and technical expertise. It's time to reverse that trend." It urged that all foreign service officers should achieve basic competence in science, technology and health (STH) matters, 25 technically trained science counselors be assigned to U.S. Embassies, and re-create the Science Advisory Committee and many suggestions. The Department's response to these successive reports was the abolition of the position of Deputy Assistant Secretary of State for STH in 1997, and, in the same year, decided to phase out the "science cone" (an area of specialization, the others are administrative, consular, economic, political and U.S. Information Agency), downgraded Science Counselors at U.S. Embassies and increasingly assigned these responsibilities to FSOs with little or no technical experience in these areas. This trend was sadly described as a "litany of decay of science in the State Department."[3]

a. The Bureau of Oceans, International Environmental and Scientific Affairs (OES)

OES, the only Bureau created by statute, was established in October 1974 with jurisdiction over matters relating to oceans, environment, scientific affair, fisheries, wildlife, and conservation affairs and continues in existence to this day. Under the Carter (1977-1980) and Reagan Administration (1981-1988), OES was heavily involved in a number of significant and, at times, highly controversial international S&T projects: nuclear proliferation, synthetic fuels and alternative energy R&D programs, law of the sea treaty. While several senior foreign service officer served as OES Assistant Secretaries over these years and tried to invigorate the Bureau, that drive always ran into traditional strengths of the regional bureaus which have been most influential and had attracted the most competent staff. The other functional bureaus such as economics and international organizations increased their importance as the issues confronted by the U.S. changed. Furthermore, the subject matter of these Bureaus was not alien in intellectual appreciation, understanding and importantly in emotional appeal and sense of affinity in relation to their responsibilities for the conduct of U.S. foreign policy. Science and technology no matter how packaged, no matter how described as the new important and, at times, critical ingredient in U.S. foreign policy, there appears to have been a chasm which has not been successfully bridged between OES and the other Bureaus even when OES was given important new responsibilities under the Nuclear Non-Proliferation Act of 1978, and to coordinate international agreements for marine science and technology. It was an organizational culture gap, not a culture conflict.

Notwithstanding the apparent assignment of significant responsibilities to the OES, there has remained the ambiguity as to whether the insertion of science and technology into foreign policy meant "policy for science" or "science in policy". This dilemma seemed to stand in the way of a clear mandate for OES, an uncomfortable ambiguity which never seems to have been resolved.

Staffing the OES has been a continuing problem. Foreign Service officers are normally generalists in the social sciences, economics, history, etc. The OES would need technical experts. Indeed in the early years over half the staff had technical degrees. Apparently the interface of generalists and technical experts was unsatisfactory; they did not mesh too well. The generalist lacked the will, desire, and knack of appreciating the technical aspects of an issue. The orientation of the technical experts was a different decision-making process, a different philosophical approach to the solution of technical issues and how they could be woven into foreign policy considerations. OES was regarded as a second-ranking bureau in contrast to, for example, the East Asia Bureau. For an up and coming young foreign service officers, the OES was not a choice assignment. This naturally had an impact on the quality of OES staff. The overall budgetary needs were not as fully met in comparison with other bureaus.

While its continued existence is not challenged, Bruce Smith concluded that

> the State Department has been by virtually any measure the least successful among the agencies [of the U.S. government] in blending scientists and science advisers into policy deliberations ... the department has not allocated time, money, staff or psychic energy to the task; and the scientists have been impatient in learning the nuances and complexities of the agency they have sought to advise and have exaggerated their own importance (Smith, 92:137) ... [the OES was] a second-rank bureau [which] was not taken seriously within the department (except on a few issues, notably the proliferation of nuclear weapons ... Until and unless this basic set of incentives changes, the bureau [OES] will continue to struggle with its identity crisis and its uneasy compromise between the worlds of science and diplomacy (Smith, 92:154).

b. The Science Advisory Committee

The SAC, despite the special efforts and goodwill and determination of some, lived a downward roller coaster and checkered life.[4] In light of the many scientific and technological issues facing all Administrations, such as space, marine sciences, nuclear non-proliferation issues, scientific exchanges, environment, one might expect such a committee could play a most constructive and useful role in the high councils of government. Initially it served useful purposes, then fell into disuse, was abolished, recreated, and gradually faded into the background and non-existence.

A SAC was created in the early Nixon Administration years to advise Secretary of State, William Rogers. It was to supplement the science office and reach out to the scientific community. It was moderately successful: it met with the Secretary several times and worked in groups on several S&T issues. But when Henry Kissinger replaced Rogers as Secretary, the SAC fell into disuse. Kissinger apparently preferred to rely on a science friend from Harvard University where Kissinger had done his research. However, when Dixie Lee Ray, a scientist and former Commissioner of the Atomic Energy Commission, was appointed in 1975 as Assistant Secretary for the newly created OES, she abolished the SAC. Under the aforementioned Title V of the Foreign Relations Authorization Act a new life was temporarily breathed into the SAC concept. According to Smith, this Act "should be seen as perhaps the high point at least in what was expected of the bureau" (Smith 92:140). Under OES Assistant Secretary Thomas R. Pickering, Bruce Smith became his planning deputy and they decided to recreate and rejuvenate the SAC. The very act of attempting to create the SAC ran into bureaucratic problems with the Office of Management and Budget which asserted its authority to review attempts to create an advisory committee -- even though technically that responsibility had been shifted to the General Services Administration. The State Department proposal to create the SAC was rejected not once but twice. Since OMB needed to be "consulted" only on this matter, State decided to create a SAC but then fearing the backlash in its budgetary worries did not provide the SAC with funds for it to be functionally effective.

This organizational failure stems from the conflict of the diplomatic and scientific cultures in an organization dominated by generalist foreign services and scientists who are unable or unwilling to try to see needs of diplomacy and melding these with the scientific approach and culture. With no funds the SAC could not appropriately fund the meetings of the SAC which naturally became infrequent. This infrequency immediately is tied to the ability of SAC members to contribute to the dialogue on the interaction between science and foreign policy needs since these needs are fast moving and cannot wait on the advice of an infrequently meeting SAC. Then there is the issue of security clearances for SAC members for nuclear proliferation issues, a major concern of the OES. It was felt that the SAC as a body could not handle the extremely sensitive day-to-day issues involving heads of state of other countries in nuclear, marine, environmental matters. Later, Assistant Secretary John Negroponte who was, for a while, involved in internal USG deliberations on the future S&T relations with Japan in the mid-1980s and also seemed to be trying to re-invigorate the SAC, but ultimately let it just fade away. State was ambivalent about the SAC itself, its value, its funding, its role, the SAC members themselves did not really know what to do. Bruce Smith concluded that SAC "was born at the wrong time" (92-147).

c. Circular 175 Authority

In order "to facilitate the application of orderly and uniform measures and procedures for the negotiation, signature, publication, and registration of treaties and other international agreements" the Department of State formulated a detailed set of procedures to be followed before negotiations are begun with a foreign government. These procedures have been codified in the *Foreign Affairs Manual*, Chapter 700 on Treaties and Other International Agreements, Section 720 on negotiation and signature. Because this is a codification of Department Circular no. 175, December 13, 1955 this procedure is commonly referred to as "requesting Circular 175 Authority".

It stipulates in copious detail the kinds of treaties and agreements and their constitutional parameters, the considerations for selecting a particular constitutionally authorized procedure, consultations with Congress about this choice, kinds of authorizations that can be obtained, what each circular 175 request will contain including clearances from State's Legal Adviser, and the Office of Management and Budget concerning commitment of funds, goods and services, and the responsibilities of the State Office or Officer conducting the negotiations. Basically, this procedure is to assure that there is a unified position in the USG to negotiate and sign an agreement with a foreign government or entity. The 1980 and 1988 STAs were deemed to be agreements signed pursuant to the constitutional authority of the President, rather than a treaty requiring the advice and consent of the U.S. Senate. All interested parties in the USG are requested to sign off on the request for Circular 175 authority, in the case of the 80STA and 88STA this involved many offices in the Department of State and other departments of the USG. One might compare this procedure to the *ringisei* system in Japan. This request normally includes a justification for negotiating with a particular country about a specific subject, a proposed USG draft if available and other minutes, exchange of notes, and a memorandum of law by State's Legal Adviser.

Normally this request is a rather cut and dried dull uneventful process, a pro forma exercise to assure that all parties in the USG are on the same wave length and going in the same direction. Occasionally one can glean interesting interpretations not mentioned previously and an occasional insight and once in a while, this procedure becomes the catalyst for a bureaucratic explosion even though normally this procedure is embarked upon after agreement has been reached among the various elements of the USG. It was for this latter reason that this brief description has been prepared as part of the institutional arrangements in the USG.

2. The President's Science Adviser and the Office of Science and Technology Policy

a. The Science Adviser and his Support

During the period (1981-1988) leading up to the signing of the 88STA which basically coincided with the Reagan Administration, the President was supported by four science advisers. The first, George A. Keyworth served for 53 months and the fourth adviser, William Graham for about 30 months.[5]

George Keyworth was OSTP Director during the early mid-1980s considerations in the USG of the future S&T relations with the GOJ. He succeeded Frank Press, President Carter's science adviser, and was involved in the initial implementation of the 1980 U.S.-Japan science and technology agreement. As will be discussed later, his visit to Japan in Fall 1981 for this purpose was not positive and appears to have set the tone and direction of the OSTP in these internal USG deliberations. William Graham became director at the time of a critical change in the direction of the USG policy discussions concerning the future of U.S.-Japan S&T relations.

Public Law 94-282 defines the responsibilities of the OSTP Director as follows:

> The primary function of the Director [who is also the President's science adviser] is to provide, within the Executive Office of the President, advice on scientific, engineering, and technological aspects of issues that require attention at the highest levels of government.

The Science Adviser was a special assistant to the President, one notch below the National Security Adviser.[6] According to Graham he interpreted that his responsibilities under the above mandate involved personally briefing the President, identifying issues, proposing and arranging Presidential activities, analyzing policy alternatives, preparing options for Presidential consideration, overseeing Presidential-level agreements in science and technology with foreign countries, and representing the Administration in ministerial-level science and technology activities overseas.[7] This broad and reasonable interpretation laid the groundwork for a potential conflict with the operating agencies, e.g. the Department of State in this instance, depending on the issues involved, the depth of potential differences between them, and the personalities involved in both State and OSTP who had to negotiate their conflicts to a successful consensus, if possible. The OSTP Director and staff could play the role of the conciliator, facilitator, the catalyst for compromise in internal USG policy discussions or lead the charge to assert and impose its viewpoint (using its position as part of the White House staff organizations). The science adviser had, in essence, multiple roles to balance and satisfy, advise the President and the OMB, represent the science community in the councils of government, and become involved in science and foreign policy. At times, these roles cannot be fulfilled and could be at odds with each other and are, in any case, governed by the extent to which the science adviser basks in the confidence of the President and his immediate intimate staff.

The science adviser to the President has led a checkered, precarious, and uncertain life during the last 50 years. The purpose in this summary is to provide a brief historical perspective of the vicissitudes of the role of the science adviser since 1945.

The motivation for creating the position of a "science adviser" in the first instance, it seems, was based on a U.S. response to an external challenge which was believed to involve highly technological issues, i.e. the Soviet politico-military challenge to the U.S., not because of the belief in the need for scientific and technological advice for itself and the welfare of the United States. This approach appears to be reflected in how the first "adviser" was appointed by President Truman in 1950 except that he was located in the Office of Defense Mobilization, not directly reporting to the President.

While President Eisenhower was apparently interested in hearing about and receiving S&T advice, it was the Soviet launch of Sputnik I that prompted the creation of the first presidential science adviser and the President's Science Advisory Committee (PSAC) in October 1957. The adviser, James Killian, former President of MIT, was closely involved in nuclear defense policies, strategic forces planning, air defenses, military use of space not scientific and technology policies as a broad subjects for the social and economic welfare of the nation. The high point in the Presidential S&T advice system, according to Bruce Smith dates from Killian's appointment. (Smith 92:164). With the establishment of the triad strategic policy -- land, sea and air based systems -- of the United States were set in place, somewhat slowly the needs of the nation in S&T education and research began to be raised by PSAC members and other national leaders. But this kind of advocacy came to be tainted as a lobby effort by a special interest group, scientists and engineers (S&Es) which was a handicap.

Under President Kennedy the president's science adviser and PSAC was institutionalized in the executive office of the President when both were established under the government's reorganization plan in 1962. The relationship of the science adviser, James Wiesner, also from MIT, with the President became fragile as he invoked the displeasure of the President when he disclosed some information to the Soviets during the Kennedy-Krushchev meeting in Vienna in April 1961. He found himself out in the cold during the Cuban missile crisis and deliberations on strategic military discussions, an area where the science adviser had played a major role. The gradual emergence of the Vietnam issue in the U.S. government beginning with the Kennedy Administration, also had a gradual impact on the unity of the scientific establishment vis-a-vis such national concerns and consequent strain on the efficacy of the S&T advisory system. The "war" scientists were generally in the national laboratories, and "peace" scientists generally in the academic science group. Nevertheless, Bruce Smith commented that the "golden age of presidential science advising" was from the Eisenhower years to the assassination of President Kennedy in November 1963 (Bruce Smith 92:165).

Under President Johnson, the Vietnam war began to intrude in a major way into the S&T advisory system and the split among the scientists resulted in their exclusion from strategic military and nuclear issues and the types of issues that began to be raised about the environment, toxic waste, pollution, human illnesses, involved scientific advice to the President. At the same time the emergence of these very issues raised some questions about the faith in the continual progress and contributions that science would and could make to society and its ability to control and eradicate these noxious problems. Under President Nixon, while the S&T advisory system was continued, the chasm between this system and the President reached a new depth. Instead of following the rules of anonymous advice to the President, the PSAC members publicly disagreed with Nixon over the Supersonic transport and anti-ballistic missiles. As a result, PSAC was not reappointed in the second Nixon Administration.

Nevertheless because of concern about the lack of scientific advice to the President, the National Academy of Sciences and scientists gathered around Senator Rockefeller, and persuaded President Ford, who became president after Nixon's resignation in 1974, to recreate the Office of Science and Technology Policy including the science adviser and the Federal Coordinating Council of Science, Engineering and Technology in May 1976; but note that the PSAC was not re-appointed. (For details and background see Bruce Smith 92:116-119)

Although President Carter was himself an engineer graduated from Annapolis Naval Academy, his science advisory system was almost allowed to be sacrificed to government streamlining and abolished. But it survived and Frank Press, a noted physicist, became his science adviser, again without the PSAC. According to Bruce Smith, Press focused on five areas: 1) he acquainted OMB officials on the trends in academic science, persuaded OMB to provide greater support of basic research and eased tensions between universities and government, 2) helped to reduce outlays for energy demonstration projects with little immediate prospects, 3) reviewed civil technology to improve the climate for innovation, 4) worked to upgrade the quality of scientific research supporting the regulatory process for occupational health, safety, and the environment and 5) supported increases of defense research expenditure. A number of these actions anticipated the Reagan Administration's emphases in R&D. Science advisers, beginning with Press chose to be the President's adviser not the representative of the science community, to the criticism and chagrin of many in the scientific community (Smith 90:118-119). There is no mention or assessment of Press's negotiations with Japan at the end of the Carter Administration to persuade Japan -- unsuccessfully -- to provide funding of secondary U.S. S&T projects and the conclusion of the first Presidential level USJ S&T agreement in 1980. It was the revision of this agreement that created such a stir in the USG in the mid-1980s.

Although over the years the Reagan Administrations appeared to recognize the need for investment in S&T, particularly defense R&D, and the need for stimulating U.S. innovation, a science adviser was not part of the team at the starting line in January, 1981. George Keyworth, a weapons expert from the Los Alamos Laboratory, came on board in August, 1981. His background in weapons research allowed him as science adviser to become re-involved in defense research, nuclear issues but his main focus was to be a team player for the Reagan Administration's emphasis on increases for defense research funding, methods to re-invigorate America's innovative tradition and improve the reality and perception of U.S.'s position in global competitiveness. Since he enthusiastically embraced the Strategic Development Initiative (SDI), he was well received by other members of the Reagan political family in the White House. While, as mentioned earlier, Keyworth served 53 months to December 31, 1985, his immediate two successors served only months and Graham served for about 30 months. It would seem that the science adviser was not in an attractive position reporting indirectly to the President through the Domestic Policy Adviser (initially this was Edwin Meese, later Attorney General). Graham writing in 1988 as the science adviser stated that he is "currently able to act as an objective unbiased arbiter of agency issues and to provide the President a source of advice unbiased by agency affiliation. This current role should be preserved".[8] This rosy perception of his own role, is in contrast to another assessment of the entire S&T advisory system by Bruce Smith which was far less generous:

> In contrast [to the National Security Council], the Office of Science and Technology Policy has never contributed to a crisis or had an obvious breakdown of function. It has been a minor player in high-level executive politics. The effects of a weak science advisory system are difficult to document; unlike the NSC system, there are no life-and-death foreign policy issues that seize the national attention. But the absence of a central direction in resolving small policy disputes, the failure to set priorities and integrate agency policies, and the erosion of quality control over the government's technical programs do have consequences over time. If the OSTP is weak, agency enthusiasms may not be carefully evaluated, critical breakdowns may go undetected and matters that cut across jurisdictions may be inadequately addressed.[9]

Based on the above, one might prudently conclude that the role, effectiveness and involvement of the science adviser in the governmental processes, the federal budget, national debate about the future of science and technology in the U.S., seems at best murky and unclear, at least at this time.

The office and person of the science adviser and the OSTP has become statutorily institutionalized as part of the President's support staff. The science advisers have become advisers to the President and not representatives, in the first place, of the scientific community. Like all consultants and advisers, the science adviser's usefulness and effectiveness rests with his (so far no woman has held this position) relationship with the President and his immediate staff. He must have their confidence and the President must be interested in S&T, such as was the case with President Eisenhower. He and PSAC must be willing to operate within the rules of White House milieu which did not occur during Nixon's first term, particularly in relation to the publicly stated PSAC's opposition to specific Nixon programs. The science adviser is supported by a small staff, hardly adequate to carry out, for example, its claimed responsibilities based on a broad interpretation of the statutory language in the international arena. It must depend on the goodwill and support of the technical agencies in the USG and the staff support of the OES in the State Department. As we shall see, this is an inherently unstable relationship wherein an active science adviser with an enthusiastic staff with a differing agenda can readily collide with the diplomatic interests of the State Department.

b. The Federal Coordinating Council for Science, Engineering and Technology and the Committee on International Science, Engineering and Technology

When President Ford signed the National Science and Technology Policy, Organization and Priorities Act on May 11, 1986, he also re-created the Office of Science and Technology Policy (OSTP) which President Nixon had earlier abolished, and the Federal Coordinating Committee for Science, Engineering and Technology (FCCSET). "The science and technology policies of the federal departments and agencies have increasing significance to the United States economy and research, competitiveness and foreign policy and national security interests. A single organizational focus for the identification, coordination and evaluation of these policies, however, has been lacking." FCCSET, as a senior level statutory agency, is expected to fill the role of an interagency coordinating mechanism for this purpose.

The Director of OSTP will be the chairman of FCCSET which shall be composed of "policy rank officials" from practically every Department and Agency of the Executive Branch: Departments of State, Defense, Agriculture, Commerce, Health and Human Services, Housing and Urban Affairs, Interior, Energy, Transportation, Veteran's Administration (now Department of Veterans Affairs), NASA, NSF, EPA. FCCSET "shall consider problems and developments in the fields of science, engineering and technology and related activities affecting more than one Federal Agency and shall recommend policies and other measures designed, among other items, to:

4. further international cooperation in science, engineering and technology."

This is a broad mandate which would permit OSTP to inject itself into a myriad of Federal S&T projects, programs, and international agreements. It is also at the same time because of its size, a highly unwieldy organization for coordination purposes. OSTP probably does not have enough personnel to carry out the full mandate of this legislation. If OSTP decided to implement the total responsibilities implicit in the FCCSET mandate in an energetic manner, it would surely turn into a surf battle with powerful political and R&D fiefdoms in several Departments.

Even though international cooperation was mentioned as the last of the functions and responsibilities of FCCSET, the Committee on International Science, Engineering and Technology (CISET) was recreated by FCCSET by a notice in the *Federal Register* on Thursday, April 24, 1986 but had been approved by George A. Keyworth on December 20, 1985, just before he resigned as the science adviser at the end of that year. The chairman of CISET would be the Associate Director of OSTP, an Assistant OSTP Director would be the executive secretary of CISET. CISET should report to FCCSET through the chairman of FCCSET who, in turn, would be the Director of OSTP.

Members of CISET would be sub-cabinet officers or the senior agency official responsible for S&T research programs and international policy issues, and should be subject to the approval of the FCCSET (i.e. the Director OSTP). The members are designated by position such as Deputy Undersecretary, Assistant Secretary, Deputy Director or Administrator of NSF, and NASA, etc, far more specific than for the parent FCCSET membership itself. This is an unusually high level Federal coordinating committee. CISET shall:

1. act to strengthen and complement the international science and technology programs and activities of member agencies

2. serve as coordinator, clearinghouse, and technical evaluator of federal scientific research programs and activities in the international arena to further the domestic research effort, promote foreign policy goals, and protect national security interests

3. coordinate research planning and implementation of international scientific activities between agencies and provide technical oversight and evaluation of major programs

4. advise on the formulation of broad national priorities and policies for international scientific cooperation in both the bilateral and multilateral fora

5. identify emerging scientific issues with international implications and recommend policy options to formulate national strategies

6. identify opportunities for new international scientific initiatives in order to strengthen domestic R&D efforts, complement U.S. foreign policy, and enhance national security.

This extremely broad mandate would definitively allow the FCCSET/CISET to involve itself in any and all international S&T activities in which the USG is involved. In theory, it allows the OSTP through FCCSET/CISET to place any program existing or planned, on the CISET agenda and request studies to be conducted on international R&D issues and question the USG stance, or that of any Department or Agency, in international S&T deliberations. It responds to broader legislation that calls upon the State Department as delegated to the Secretary of State from the President to coordinate these international activities. It permits all manner of FCCSET/CISET involvement in international S&T activities short of actually negotiating an S&T agreement with a foreign government which by law is specifically delegated to the Secretary of State by law. While there appears to be a distinction between the roles of State and FCCSET/CISET, the lines of demarcation are now exceedingly blurred, to the say the least, which gives opportunities to the Director/Chairman of OSTP/FCCSET to play in troubled waters, especially to oppose State, which is so often regarded by many of its opponents as more a client of their regions and countries than sometimes for the U.S. According to one of the OSTP interviewees, no steps were taken to clarify and demarcate the State and OSTP roles, because, at the time and in light of the atmosphere surrounding the role of science advice to the President, it was more important to boost the authority, implied or otherwise, of the OSTP; an attempt at clarification might have resulted in an erosion of OSTP authority.

According to the notice in the *Federal Register*, CISET 1) will form an executive steering committee consisting of the representatives from State and NSF and the Chairman of FCCSET, i.e. the Director of OSTP and 2) form four working groups (with terms of reference specified in the Notice), each chaired by the representative of a Department/Agency to CISET:

1. International Science, Engineering and Technology, chaired by NSF (then Deputor Director, John Moore)

2. Science, Engineering, Technology and International Competitiveness, Commerce (then OPTI, A/S Bruce Merrifield, the Merrifield Working Group)

3. Bilateral and Multinational Activities, State (then OES A/S, John Negroponte)

4. Strategic Science, Engineering and Technology and Technology Transfer, OSTP (then Maurice A. Roesch, Associate Director).

At the inaugural meeting of CISET on March 7, 1986, there were two items of interest to this study: 1) there was recognition that there was obvious overlap between the above four working groups which need careful coordination and nurturing and 2) far more importantly, USTR stressed "the need for early awareness of the issue of protection of intellectual property rights" (IPR) in ALL working groups. This emphasis at the inaugural meeting on IPR may be the genesis and origin of the role IPR played in the USJ bilateral negotiations. It should be noted that not until the first draft of the revised USJ agreement did IPR appear to play a role in the USG deliberations.

Based on available documents (from May to December 1986 when CISET Joint Task Force Report on U.S.-Japan S&T Relations was submitted to the CISET Executive Committee) on the deliberations of two Working Groups, Science, Engineering and Technology and Industrial Competitiveness (DOC) and International Science, Engineering, and Technology Education (NSF) a number of issues were discussed which later appeared in the bilateral negotiations and as a major multilateral project.[10] Since the mission of the working groups was to recommend actions to FCCSET/CISET, it is difficult to trace their specific impact. However, since the breadth of participation in the working groups and FCCSET/CISET itself was so large, the groups' members must have been sensitized to many important issues many of which were of interest to all the department and agencies, e.g. equal access to foreign research facilities and IPR issues. The impact of these working groups could have considerable impact in quite an intangible manner. Only those issues which have a relationship to the evolving dialogue within the USG on the future of U.S.-Japan S&T relations will be noted below.

As an example, the afore-mentioned Merrifield Working Group started with such a far-flung array of topics (flexible automated manufacturing systems which later became first a bilateral USJ issue then a highly structured multilateral project, computer interactive videodisc aided education, world scan technology database, innovation and productivity) to consider that OSTP called on OMB and its own representative on the Group to help this group identify and focus on 2-3 important issues. After meeting several times, this Group prepared a report for CISET consideration in mid-December 1986.

1. It was concerned with guarding through patents and copyrights, the technology of U.S. federal laboratories and the possible leakage of this publicly available information to foreign countries and with disseminating this information to U.S. private industry -- although, of course, it is there for the asking. It asserted that "the free availability of such information without proprietary protection, in effect, subsidizes foreign R&D efforts. ...Without increased cooperation between Federal laboratories and the private sector it is unlikely [even with the new authorities given to federal labs under the Federal Technology Transfer Act of October 20, 1986] that the technology losses from publicly supported R&D can be stanched." There was no proof whatever offered to prove this serious assertion then aimed implicitly at Japan.

2. The USG should formulate policy options for ensuring that all international science and technology agreements concluded or supported with Federal funds adequately address the linked problems of equity, reciprocal access and IPR. For example, were inventions stemming from foreign visiting scientist programs being reported and what rights are being retained for U.S. agencies hosts, are U.S. researchers whether public or private, being given "equal access" to foreign government R&D facilities as granted in the U.S., does the U.S. benefit from federally-funded cooperative R&D project with foreign organizations.

3. Consult with U.S. industry to identify areas of foreign science and technology within the scope of the proposed USJ agreement and have them included in the agreement.

Although there were a few people who were members on both this Working Group and the CISET Joint Task Force on USJ S&T relations, these issues were not a major concern in the late 1986 Joint Task Force Report. It is as though these two exercises were parallel and disconnected despite overlapping personnel. These issues were important to OSTP which, I am sure, assured that they were given more than sufficient attention for they were included in drafting the agreement given to GOJ in August 1987. The issues mentioned were all addressed in the discussions with the GOJ in 1987-1988 negotiations about the future of USJ S&T relations and, indeed, included in some form in the revised agreement. While it is not known how and what specifically the IPR Committee did, if anything, it does show that these issues were center stage in overall discussions in the USG about future S&T agreements in general and the one with Japan of especial interest since it was then under active consideration.

The Working Group on International Science, Engineering, and Technology Education, Infrastructure and Facilities met six times between April 10, 1986 and February 12, 1987 and focused, in the main, on the typology of existing modes of USG participation in international science, the adequacy of existing information about foreign science and engineering graduate students, mobility and barriers thereto for U.S. engineers and scientists to study abroad in Europe and Japan, and about foreign centers of excellence in science and engineering. This Group was provided with a special report by the NSF staff at its February 12, 1987 meeting on "extended term international mobility." This date is meaningful in that it comes just after a visit to Japan by a high-level U.S. delegation to discuss the future of the 80STA, and only six months from the beginning of USJ bilateral negotiations on a revised agreement. The issue of the mobility of U.S. scientists and engineers to Japan was directly related just as a topic for analysis and was also a contentious, emotional and confusing issue in these negotiations.[11]

The above relatively brief commentary on FCCSET and its Working Groups described "another stream of influence, thought and activity" which was concerned with the potential vulnerability of all the research data, inventions, ideas coming from Federal R&D laboratories, their utilization by active foreign corporations, in contrast to the apparent inactive or passive U.S. corporate interest, the IPR of inventions by foreign scientists while at a U.S. research facility, access to foreign R&D facilities. While Japan per se is seldom mentioned in discussing these concerns, the participants thoughts, especially those of some of the more actively involved agencies and departments, must have been of the implications of these concerns as they pondered and acted on the future of USJ S&T relations. For OSTP all of these concerns were specially frought with anxiety, fear of, perhaps, antipathy for Japan. These will later show up in the drafts of the revised USJ STA. Although State OES was represented on the CISET and FCCSET and naturally involved in these discussions, these concerns did not appear to show up in their discussions with GOJ officials, nor in the Joint Task Force or in the apparent dialogue with other USG agencies.

The FCCSET/CISET organizational arrangement was complicated and elaborate; it must have consumed an enormous amount of personnel time to prepare for and attend the working group meetings. Since its function was to make recommendations, it is difficult to judge its efficacy and success. In any case, it is not the purpose of this monograph to make that judgment. The creation of FCCSET/CISET in 1985-1986 blurred the lines of responsibilities for overseeing the international S&T programs of the USG between OSTP and State. There is no known attempt to clarify this ambiguity of responsibilities. Perhaps State did not show enough enthusiasm for such a potential "turf" fight; OSTP did not want to detract or delimit in any way the scope of responsibilities of the new FCCSET/CISET arrangement. The creation of FCCSET/CISET also coincided by accident with the ongoing dialogue in the USG about the renewal/revision of the 80STA. The debate centering around the State Department was not achieving results and merely drifted from month to month and year to year, 1983-1986 with no conclusion. In a way, State had allowed the legal authority and prestige of its office to slip away. With the reinvigorated FCCSET/CISET, OSTP adroitly utilized an opportunity to mobilize the other agencies of the USG concerning the above kinds of issues. While one must refrain from saying "mobilizing the varied interests and anxieties of USG agencies in a confrontation with the State Department" at this time, although the underlying tone, drift and issues were all pointing in this direction.[12]

c. *The White House Science Council*

George Keyworth established this Council in February 1982 to advise him on a broad range of S&T issues affecting the nation as a whole. Although the name of the Council might suggest otherwise, actually it did not report to the President but to the science adviser. The lack of the word "President's" in the title, not just merely "White House" which is vaguer and less specifically identifiable as at the highest level.

The first chairman was Solomon J. Buchsbaum, then Executive Vice President (Customer Systems), AT&T Laboratories.[13] The members were scientists and engineers chosen from universities and industry. Notwithstanding the competence, sincerity of the Council's members, it would be difficult for the small OSTP staff to keep these members up-to-date in the often fast moving world of S&T policy discussions and policy making both domestically and particularly on international issues. This is especially true if the Council met typically only about two days a month. The critics would raise the issue of its effectiveness and need under these circumstances. This is somewhat inevitable since no Administration has been willing to provide enough staff support for this purpose and enough funding for Council members to participate more actively in this process. In contrast, Graham asserted that the Council meet "frequently" with him, and, as required, with the President.[14]

3. Institutional Arrangements in the GOJ

In contrast to the USG as described above, the institutional arrangements in the GOJ for responding to the needs for science and technology issues at least for this agreement appear to be relatively uncomplicated. This is in contrast to the comment [made in the mid-1970s] that "The complexity of the policy-making process in the Foreign Ministry is confusing not only to outsiders, but apparently even to those inside the bureaucracy."[15] In the MOFA, as in other parts of the GOJ, the leadership of the Japanese effort came from division chief or director (kachō) who is given, in contrast to the approximately similar level in the USG, unusual leeway and implicit authority to provide day-to-day leadership in managing the issues confronting the MOFA. Although the division is the lowest distinct organizational unit, it is generally equivalent to the office director in the USG. There are minor sub-divisions in the division with specific persons in charge of these units. Division directors in the MOFA are almost invariably filled by career foreign service officers who have served for 10-15 years to reach this level of authority and responsibility. Undoubtedly their performance as "*kachō*" would have a substantial impact on their future assignments and potential ambassadorial appointments. As Fukui notes, "Real action, even in a controversial or crisis situation, thus tends to take place close to the bottom of the decision-making group structure -- the level of the division head."[16] Fukui correctly and colorfully describes the division and its director as "a mini-foreign minister" and his deputy as the "mini-vice-minister."[17] The division director is highly influential but the seniority rule is strictly applied: he respects the senior leadership, does not attempt end-runs (his career would be jeopardized in such escapades), he would brief the minister or vice-minister with the approval of the bureau chief. It is the test of his diplomatic skills in the bureaucracy, in this case the MOFA, to manoeuver between and among the senior leadership.

Fukui asserts that the Foreign Ministry operates under two important traditions: 1) it "is not really equipped with decision-making machinery operating on a ministry-wide basis" and 2) a belief [among senior ministry officials] in what could be called policy-making by improvisation and a distrust of comprehensive long-term planning and continues to operate with a limited number of career foreign service officers to cover the myriad of foreign policy issues facing the nation.[18] These traditions and personnel constraints have consequences. The "autonomous discretionary power of each bureau" is enhanced and accentuated resulting in a "subsystem dominance in which the individual bureaus and divisions are the basic units of decision-making and action and the individual bureau directors and divisions enjoy a large measure of freedom -- within the limits defined by the rule of seniority -- to map out their course of action on any specific policy issue at hand."[19] This observation could be applied to bureaus and divisions in other ministries. Aware of the role that the Circular 175 authority can play in the USG, this writer inquired about a similar GOJ mechanism and was told in an interview with a Minister at the Japanese Embassy in Washington, that in his career of 25+ years in MOFA he had never heard of, nor seen any materials in MOFA regulations which would seem to be similar to the USG's circular 175 procedure. This lack of a unifying mechanism to assure government-wide uniformity, consistency and legality appears to be overcome, managed by the Japanese unwritten modus operandi that requires a wide range of discussions involving all the necessary parties to an issue and the enforcement of the *ringisei,* where all those with a need to know and participate sign off with their personal seals.

Career officers are stretched thin and have many responsibilities and issues to tackle and resolve. This was evident in the scheduling of meetings for the STA negotiations, e.g. the ambassador chairing the Japanese STA delegation was required to be present at Diet committee sessions to report on --in the vaguest possible terms -- those areas of his responsibilities, the STA, space station, nuclear affairs.

While the Japanese STA Team was chaired by an Ambassador, the actual day-to-day leadership of the Japanese effort came from a Division

Director in the MOFA's Science and Technology Bureau -- the equivalent of the OES in State. However, the Japanese delegation, for example, did not include a representative from the North America Bureau which handles U.S. affairs. Like its U.S. counterpart, the S&T Bureau must negotiate with other Ministries to try to present a united, uniform and consistent front to their opposite numbers in the negotiations. From the GOJ perspective, the principal interested agencies were four, in contrast to the larger number of participating agencies in the USG process and these four directors from four agencies conducted discussions and negotiations among themselves to hammer out a GOJ position.

Since the end result of many MOFA activities is major or minor agreements or treaties of some kind which may or may not require Diet approval, the MOFA's Treaties Bureau is able to become involved is almost everybody's business. It is responsible for insuring that treaties would be in line with Japan's foreign policies and are legally correct. It played an important role in preparing -- together with its more or less counterpart in State, the Office of the Legal Adviser, in the formulation of the final wording of the STA agreement This Bureau's leadership often has to attend Diet committee hearings in cases where their areas of expertise may be needed by the Ministers in responding to the queries and criticism from the political opposition.

The Prime Minister's Science and Technology Council sets grand S/T policies for the nation. Negotiating an STA is the responsibility of the duly appointed government Ministries; nevertheless, selected members of the Council were kept informed of the progress of these negotiations but do not inject themselves into the negotiating process. There was an implicit confidence until recently in the bureaucracies of several Ministries to manage the country's affairs without the kibitzing by other institutions and the Diet. For an additional description of the Teams and their members see Chapter IX on negotiating.

Notes

1. While there are a number of books on science and technology in U.S. foreign policy, I have chosen to rely, in the main, on two books by Bruce L.R. Smith, *American Science Policy since World War II* (1990), *The Advisers: Scientists in the Policy Process* (1992), both published by Brookings Institution, to describe the institutional arrangements in the USG.
2. *Technology and Foreign Affairs,* a report by Dr. T. Keith Glennan to Deputy Secretary of State, Charles W. Robinson, 12/1976. 70 p. and 12 appendices.
   Science and Technology in U.S. International Affairs, A report of the Carnegie

Commission on Science, Technology, and Government. 1992. 125 p.
The Pervasive Role of Science, Technology, and Health in Foreign Policy: Imperatives for the Department of State, Committee on the Science, Technology, and Health Aspects of the Foreign Policy Agenda of the United States. National Research Council. 1999. 124 p.

According to Daniel S. Greenburg (*Washington Post* 11/28/97), a study chaired by Ambassador Robert D. Murphy in 1975 also called for State to "significantly improve it's capabilities to deal with foreign policy aspects of economics ... science [and other fields]. I have not been able to locate this study.

3. Schmitt, Roland W. Science Savvy in Foreign Affairs, *Issues in Science and Technology* Summer 1999:41-44.
4. The following presentation was based, in large part, on Bruce Smith's 1992: 139-154.
5. George A. Keyworth served from August 1981 to December 31, 1985 53 months), his deputy John P. McTague became acting Director from January 1, 1986 to May 23, 1986 (5 months), Richard G. Johnston to October 2, 1986 4 months), William Graham to the spring of 1989 (18 months, the first months of the Bush Administration). Obviously, the middle two advisers were merely caretaker Directors of OSTP.
6. Under President Bush, the science adviser was raised to assistant to the President, co-equal in title to the national security adviser.
7. Graham, William, The role of the President's Science Adviser, in *Science and Technology Advice to the President, Congress, and Judiciary*. Ed. by William T. Golden. Pergamon Press. 1988. p. 152.
8. Graham, William R. The Role of the President's Science Adviser, *Science and Technology Advice to the President, Congress, and Judiciary.* Pergamon Press. 1988. 154 p.
9. Smith 90:127-128. While Bruce Smith in his two books on science advisers describes in some detail the science and technology policies of the Reagan Administration, he does not identify the actual roles of the science adviser and OSTP in these governmental processes, perhaps because they played an anonymous and quiet role behind the scenes or because these policies were initiated and implemented quite independently of the science adviser and his office.
10. These comments were derived from documents on various meetings which were all unclassified and given to this writer privately. In contrast, claiming that the minutes of these meetings came under the exemption clause in the FOIA, NSF, of all agencies, denied their release almost in entirety.
11. Memo, Notes on the Extended-term International Mobility Issue" 2/12/1987, for CISET Working Group on International Infrastructure, Education and Facilities.
12. FCCSET was strengthened under the Bush Administration, but abolished and replaced by the National Council on Science and Technology under the Clinton Administration.
13. Buchsbaum was a veteran adviser to the executive branch: he was a member of the "ill fated" PSAC under Nixon, chairman of the Defense Science Board during President Ford's tenure at the White House, chairman of the Energy Research Advisory Board under President Carter and chaired several ad hoc advisory task forces for Frank Press, Carter's science adviser.
14. Graham, William R. The Role of the President's Science Adviser, *Science and Technology Advice to the President, Congress and Judiciary*, ed. William T. Golden. Pergamon Press. 1988. p. 153.

15. Haruhiro Fukui. Policy-Making in the Japanese Foreign Ministry, *The Foreign Policy of Modern Japan,* ed. Robert A. Scalapino. University of California Press. 1977. p. 8.
16. ibid. p. 11.
17. ibid. p. 11.
18. ibid. p. 15-18. In 1996 the MOFA created a new *Gaikō Seisakukyoku* (Foreign Policy Bureau). Since it was recently created it is, of course, impossible to judge its merits, its successes and/or failures. It's creation probably stems from a recognition that looking toward the 21st century, the MOFA needs to have mechanism which will have a longer term perspective and outlook in planning for Japan's foreign policy future. It is probably similar to the Policy Planning Council in State but it has been given co-equal Bureau status, even though it is obviously a staff function. It remains to be seen whether the Bureau can command the respect and authority of the other traditional Bureaus.
19. ibid. p. 16.

VI. Act I, Mid-1980s Review of the 1980 STA

1. The Bloom Assessment

According to article 9 of the 1980 STA, this agreement could be renewed at the end of five years, May 1985. The basic responsibility in the USG for organizing the necessary review effort for this purpose rested with the Bureau of Oceans, Environment and Scientific Affairs (OES) in the U.S. Department of State. It is well-known that the OES was not a choice spot for ambitious foreign service officers and it perennially suffered excessive personnel turnover and an inadequate assignment of personnel, whether civil or foreign service. Notwithstanding these institutional built-in handicaps, OES worked with the Office of Science and Technology Policy in the Executive Office of the President in laying the groundwork for a systematic review of the 80STA in the US government.

According to an OES memo (11/26/85) on talking points in a discussion with the Japan Desk in the Department of State, "at one time it was felt that a total review of overall USJ S&T cooperation (13 agreements) should be undertaken as a basis for deciding how to approach the UJST [80STA] issue. We have now decided that would be an unwieldy undertaking which is not really necessary." Probably this was the reason that OES contracted with Justin Bloom in 1983 to conduct a substantial study of U.S.-Japan science and technology relations since 1945. This would be the first study to analyze and provide evaluatory comments on the specific S&T programs between the agencies of the two governments. The 66-page single-spaced report submitted to the State Department in August 1983 covered all the S&T projects that then existed between the two countries.[1] To the greatest extent possible, the political value, extent and direction of technology transfer, economic importance, social significance, important achievements, financial and manpower costs to the U.S. where possible and necessary, information on participation by other agencies, collateral programs and the future of the programs were also provided. Data on more than 50 projects and programs from aquaculture to fire research to fast breeder reactors, to food additives, were described. The report did not contain a set of specific recommendations which would govern future U.S.-Japan S&T relations but did contain many suggestions for improvements in various projects and program areas. This evaluation and findings can be summarized as follows.

- it clearly depicted the scope, dimension, durability and depth of the *mutuality* of this relationship that has no rival between the U.S. and any other country and most probably between any two countries and that the relationship was "healthy, productive and growing." [2]

- all projects had been initiated by the U.S. except the 1979 agreement on energy; the Japanese were "reacting to American initiatives".[3] This was remarkable itself since the years covered since 1945 were a period when U.S. S&T was regarded by both Japanese and Americans as being more advanced than Japan. Based on conventional wisdom one would expect Japan to initiate such projects in order to acquire more S&T information from the U.S.

- in general the U.S. project managers appeared well satisfied -- in most cases more than well satisfied -- with the results and felt that mutual benefits had been well worth the effort over the years.

- while the U.S. had not invested its funds in Japanese R&D projects, the Japanese government had invested over $100 million in joint projects located in the U.S. [4]

The report also pointed out the following problem areas:

- the depth, breadth, durability and satisfaction -- and presumably satisfaction on the Japanese side -- was not well understood, appreciated, even by presumably knowledgeable engineers and chief executives. The Bloom study cites William C. Norris, Chairman and CEO of Control Data Corporation calling for the U.S. "to take a firm but constructive stand now to establish an equitable relationship with Japan while there is still time." (*New York Times*, July 24, 1983.) Though in a somewhat different form, this unfortunate perception became a virulent and divisive issue within the U.S. government and in later negotiations with the Japanese government even after evidence to the contrary was abundantly provided to the skeptics within the government.

- the U.S.-Japan S&T relationship had, over the years, grown like Topsy, without following a particular policy or plan. Because of the diffused nature of the USG arrangement, and because of the manner in which the U.S.-Japan S&T relationship evolved, there had been no specific coordination, management or oversight of this "hodge-podge of projects scattered through many [U.S. governmental] agencies". This issue immediately becomes involved with the historical sense of independence of each of the U.S. Departments of government. Bloom suggested the creation of a standing interagency committee of coordinators to keep all agencies and the White House currently informed, avoiding duplication, assisting each other in mutually beneficial projects[5] and assessing what new areas of science and technology could be explored with the Japanese government, e.g. biotechnology and material sciences. In another background paper it was bluntly stated that "we [the U.S.] have not bothered to research potential areas of R&D in which we might benefit from the Japanese". We chauvinistically continued to believe that we are the only country with work in science.[6]

- to increase the reliability of the U.S. as a partner in international science, there should be a greater recognition and action in persuading U.S. scientists and engineers to accept assignments in Japan for extended periods and for counter involvement (through personnel and funds) by the U.S. in Japanese projects.

- "unexpected changes in policies, programs and personnel" created problems not only in the U.S.'s ability to carry out its responsibilities but also in its relations with the Japanese. As will be pointed out later, the lack of appropriate assignments of personnel especially, for example, in the OES Bureau in the State Department, had an unfortunate impact on State's ability to assert itself in its negotiations with other Departments and completely hampered their ability to coordinate the wide-ranging U.S.-Japan S&T relationship.

- dissatisfaction with the manner in which the management mechanism, the high level Joint Committee, had been managed. It met only once in 1981 during the first five years of the 80STA and planned an abortive meeting in the mid-1980s.

- the USG agencies had not exploited the possibilities for U.S. researchers to work in Japan.

- While the GOJ had invested limited amounts in USG R&D projects in the U.S. USG agencies in reciprocity did not, wholly or partially, fund research projects in Japan because of disinterest, inertia, or because of legislative prohibition, or because the GOJ had not invited the U.S. agencies to do so. For example, Japan 1) had agreed to invest $36 million in DOE programs in fast breeder reactor R&D, 2) would contribute $18 million to operations and R&D

associated with the Three Mile Island clean up, 3) would spend about $7-8 million in high energy physics, 4) had agreed to pay $70 million over a five year period to support the Doublet III fusion project in LaJolla, CA, and 5) was building a $30 million plasma accelerator experiment to be carried on the U.S. Space Shuttle. The U.S. had, for example, a mere $2 million invested in an earthquake shaking table at Tsukuba science city near Tokyo.

- Superimposed on these issues -- and not stemming from Bloom's report -- was the concern, particularly in OSTP, that the kind of S&T projects included under the 80STA and the hodge-podge character of the entire S&T relationship was unbecoming of a Presidential agreement. This persuaded OSTP (with passive support from State/OES) to try to bring "order" into this relationship, i.e. attempt to centralize the monitoring of the S&T projects and programs, to create an "umbrella" S&T agreement. The latter point became the source of continuing conflict between OSTP and other executive U.S. agencies. It was also felt from a perusal of the Bloom report that while the many R&D projects between the U.S. and Japan were themselves worthwhile, they did not represent the areas of cooperation "that should transpire between two world leaders in science and technology."[7]

2. Uncleared OSTP Visits to Japan and Presidential Visits

OSTP Assistant Director John Marcum visited Japan in September 1983 and proposed to the GOJ that 1) five areas (high energy physics, solar systems exploration, breeder reactor technology, fusion and remote sensing) should be emphasized in U.S.-Japan S&T cooperation, 2) overall review of the bilateral S&T relations should be conducted and announced during the forthcoming visit of President Reagan to Japan in 1984. The above mentioned five areas were those proposed by the U.S. at the Versailles Summit Meeting in 1982. This proposal to the Japanese had not been cleared within the USG but "certainly [Marcum's] underlying objectives had merit." [8]

Such presidential visits were frequently used as occasions to raise issues with the GOJ and hopefully to persuade them -- to force them, as some would say -- to make a decision or act on an issue which had been in negotiations for sometime. This visit was no exception. On October 28, 1983 a memo from Assistant Secretary James Malone to Under Secretary Schneider saw this Presidential visit as an opportunity to move forward, redirect or put pressure on the Japanese to consider some new initiatives in S&T. It noted that

1. the U.S. was "now dealing with a sophisticated, technological equal in many fields of international scientific and technological interest."

2. in order to assure the continued smooth S&T relationship, especially when U.S.-Japan trade relations in high technology areas were highly strained, it was advantageous to "insulate our [U.S. S&T] cooperative programs from the highly charged and largely negative atmosphere of our trade discussions."

3. the S&T programs were "the one very positive aspect of the U.S.-Japan relationship and the prospects of moving the program in directions which will have greater long-term scientific value for the U.S."

4. there were continuing "bureaucratic battles" between the OSTP and other agencies over the "umbrella" concept and that State "had overcome most of their [other executive agencies'] objections by assuring them that this group would not interfere with any agencies' programmatic or budgetary prerogatives."

5. the Bloom report provided the basis for preparing a policy paper for interagency review.

This memo recommended that the U.S. suggest a bilateral review mechanism be created to conduct an annual review of the U.S.-Japan S&T relationship to enhance bilateral cooperation in S&T. The U.S. team would be co-chaired by the Under Secretary of State and the President's Science Adviser and membership would include all the interested U.S. agencies and would operate in a fashion similar to the U.S.-Japan Economic Sub-cabinet Committee. It is perhaps significant that the memo did not mention the 80STA nor did it invoke article 9 of the 80STA concerning the Joint Committee.[9]

3. USG Proposal to GOJ for Bilateral Review

When the U.S. Under Secretary of State attended the OECD meeting in Tokyo in November 1983, he incidentally also met with senior officials of four Ministries (MOFA, MITI, MOE and STA) to discuss, as a trial balloon, the U.S. idea of creating a joint "policy level" mechanism of the U.S.-Japan S&T relationship "to improve the quality of U.S.-Japan S&T cooperation". He pointed out that the U.S. was prompted to make this suggestion because 1) of the large number of agreements, 2) of the diverse nature of the areas of cooperation, 3) of [U.S.] budgetary limitations and the U.S. desire to identify new areas of cooperation, etc. He assured the GOJ Ministry representatives that the U.S. suggestion "was not prompted by unhappiness with the present situation but more by a desire to make a good situation even better and that this cooperation represents one of our highest priorities and we do not want to do anything to disrupt that relationship."[10]

According to the U.S. Embassy cable reporting on these meetings the Japanese response was clear and unequivocal. 1) The "technical Ministries" (presumably the MOE, MITI and STA) were unanimous in their opposition to this proposal. 2) The U.S. proposal could permit the establishment of a permanent institution which would become involved in the management of the cooperative activities and which would give the Foreign Ministry a stronger role.[11] These points were similar to the problems on the U.S. side where the several functional Departments were highly concerned that the OSTP and State/OES would similarly become involved in the management of "their projects."

4. USG Internal Reviews

While one of the functions of the U.S. review team was ultimately to meet with the Japanese in formulating a future S&T relationship, the other function was to act as an "inter-agency group" to assess the existing research projects and create a vision of a future S&T relationship between the U.S. and Japan.[12] This study clearly envisioned a "bilateral review" of U.S.-Japan S&T cooperation to take place Fall 1984. Due to internal USG problems a series of review meetings were held in April 1984 chaired by Under-Secretary Schneider and another in May 1984 and on August 2, 1984 both chaired by OES/SCT. The AmEmb in Tokyo responded on May 15, 1984 most positively to the proposal for a high-level review of U.S.-Japan S&T relations as "the essential step in helping to assure that the U.S. obtains maximum benefit from the rapid scientific and technological advances which are taking place in Japan."[13] The Embassy's three-page analysis of the situation was detailed and could be summarized as follows:

- the Japanese response in the first instance to the proposal for a review was 1) their concern that it might be used to terminate many ongoing activities and 2) superimpose another layer of bureaucracy on top of the existing coordinating and implementing structures. There was also another telling comment about Japanese reaction "that we [the U.S.] would simply use this exercise to press them up for more money, as we have on other occasions."[14] It should be pointed out that the GOJ was highly pressured by the Carter Administration to launch a trilateral project (with Japan and Germany and the U.S.) for coal liquefaction SRC-II in the late 1970s which was then later abruptly canceled early in the Reagan Administration after substantial sums had been invested by the Japanese (and Germans). This memo then referred to NASA's Jim Begg going to Tokyo to urge the GOJ to contribute millions to the space station project which Japan did, to President's Science Adviser Keyworth's trip in 1984 spring to urge GOJ to invest billions of dollars in the super-conducting super-collider (SSC). Japan did not bite this bait; the SSC was canceled in 1993 as part of the reduction of U.S. budgetary deficit. In the late 1970s, President Carter's Science Adviser Frank Press also went to Japan to persuade the GOJ to invest in S&T projects which were not sufficiently important to be funded in the U.S. budget.[15]

- in the absence of such a review, coverage of Japanese science and technology will proceed piece meal with little, if any, consideration of U.S. national interests and urged that this review look to the future, on "newly established areas of excellence" and not get bogged down with existing programs which are "proceeding very effectively" and do not require a high level review.

- concerning the controversial assessment about the flow of benefits, "today, the flow of benefits under most, if not all, of the agreements is reasonably well-balanced... [and] there has been over the past several years substantial flows of Japanese funding to support R&D activities/facilities in the U.S.

- Japanese support of basic research. Partly from Japan's own volition and partly in response to U.S. pressure, Japan is moving to support more basic research and to open government supported research activities and institutions to foreign participation.

- the review – presumably the internal USG review – should assess Japanese S&T activities and capabilities in relation to U.S. activities, determining those areas in which the U.S. could most benefit from a cooperative relationship and identifying the best mechanisms for establishing the cooperation, ranging from simple information exchange arrangements, to cooperative R&D activities, to joint programs, to participation by each country in the construction and operation of research facilities in the other country.

In contrast to the positive response from the Embassy in Tokyo, the agencies attending the aforementioned internal meetings in May and on August 2, 1984 were, according to a memo of August 30, 1984 to Under-Secretary Schneider, described as "gripped by a strong fear of the unknown, and were reluctant to look into important areas for future cooperation for fear that would jeopardize current activities[16] (a fear shared by their Japanese counterparts as mentioned above). The agencies "insisted" on another survey of current activities -- in addition to the 1983 Bloom report. So far the actual survey report has not been found but according to the aforementioned memo the following conclusions were made from this survey:[17]

1. Existing R&D projects have "gone reasonably well, with the exception of the so-called Non-Energy Agreement [the 80STA]."

2. Most of the agencies have tunnel vision -- or no vision at all -- when it comes to projecting where Japanese strengths in S&T will lie in the future. DOD and Commerce recognize that the USG "has not done a good job of assessing Japanese technological developments."

3. If we continue our current path, "we will be largely confined to Japanese participation in U.S. domestic programs, rather than the U.S. buying into cutting edge Japanese science and technology programs."

According to a report to the Embassy and another document on the history of the 80STA, the afore-mentioned August 2 meeting also concluded as follows:

4. "active projects" carried out under the 80STA were worthwhile – a reiteration of point 1 above.

5. the EPA had no active projects and wished to be removed from participation under the 80STA.

6. the existence of the 80STA had permitted the GOJ technical agencies to acquire funding for projects more readily from the GOJ Ministry of Finance.

7. NIH believes that "its projects under the agreement are perhaps its most productive international arrangement." This is a significant USG/NIH assessment in light of how the NIH projects involving Japanese researchers were

paraded by OSTP at later review meetings and during the negotiations with the GOJ as examples of the worse case of Japanese imbalance of researchers in the U.S. vs those in Japan.

8. the cooperation under the 80STA as "more modest than the original intent of the agreement."

9. the consensus was that the 80STA should be renewed in May 1985 through the exchange of diplomatic notes.

10. future meetings of the Joint Committee under the 80STA should be held at the office director level. This last point is interesting in that the Embassy was asked to sound out the GOJ about renewing the 80STA and its interest in holding a Joint Committee but "should not encourage a meeting." The GOJ agreed with the Office Director level but were not interested in calling for Joint Committee meetings. It was the USG which later criticized 80STA management as inadequate due, among other factors, the lack of Joint Committee meetings. Here the USG itself was discouraging such meetings. And should such meetings be held then they should be limited to the Office Director level. This was later regarded as too low in bureaucratic rank for a "Presidential level agreement", another criticism of the 80STA which was one of its own suggestions.[18]

The August 30, 1984 memo alluded to other U.S. government attempts to assess Japanese R&D trends which it maintained were conducted with "narrow agency views in mind" and declared there was no common base for information on these trends, i.e. "we clearly lack the big picture." This is a serious admission and indictment of the governmental process in tracking S&T developments in a country described as technologically sophisticated and equal to the U.S. in many areas. Therefore, the above memo urged approval of yet another study to provide this common base of information and foundation for future policy studies. The study was to identify the scientific and technological areas in which Japan would be a major player in the 1990s and beyond and create the basis for developing U.S. policies for U.S.-Japan government-to-government cooperative programs.[19] In the background statement for this study, it stated clearly that the two governments "had an active and fruitful cooperative relationship in S&T since the early 1960s. U.S. program managers believe that they have benefitted greatly from these cooperative activities." Apparently the need for this study and the related review is that 1) the U.S. and Japan have emerged as the world leaders in S&T and 2) GOJ programs are "moving rapidly into more fundamental research." There was no hint of dissatisfaction with the S&T relationship, nor with the 80STA. It was the changed S&T status of Japan vis-a-vis the U.S. and the former's new emphasis on basic research. This assessment is quite different from the later much more harsh and shrill attacks on Japan and criticisms of the U.S. designed 80STA.

In a draft letter to the Deputy Under-Secretary of Defense for International Programs and Technology (Talbot Lindstrom) concerning this study and asking for DOD fund sharing of the contract, reference was made to a 1984 policy study by the Defense Science Board Task Force on Industry-to-Industry International Armaments Cooperation which urged that DOD 1) define intergovernmental and government-industry roles and procedures for identifying, initiating, and conducting projects involving technological cooperation, 2) initiate measures for improved understanding, status and momentum of Japanese technologies, and 3) perform a high priority, comprehensive interagency study on overall trade/defense/economic tradeoffs and strategy with respect to Japan to provide a broader policy context for technological cooperation.[20] Note that these concepts were incorporated into the above mentioned study contract. A similar letter to DOC Under Secretary Lionel Olmer starkly admits "one fact [that] has clearly emerged: we do not have a good picture of where the Japanese are going over the next five to ten years."[21]

According to a draft proposal for a sole source contract, the existing S&T arrangements between the U.S. and Japan are described as involving "myriads of agreements developed over the years cover[ing] such a broad spectrum and involv[ing] so many players in the USG that we have lost a sense of central focus and purpose." Interestingly, it also cited another 1984 report, this time by the U.S.-Japan Advisory Committee to the U.S. President and Japanese Prime Minister, which, for the first time, included a section on science and technology: "Considering that many of the government-to-government scientific and technological cooperative agreements have been in operation for some years during which each of our countries has shifted the emphasis of its efforts, the time has come for a high-level review to determine possible improvements and new directions for mutually beneficial cooperation.[22]

5. Circular 175 Authority for 1980 STA Extension

Circular 175 Authority for the extension of the 80STA was begun by the State Department in mid-March 1985. Even though implicitly the 80STA was regarded as an agreement which would encompass numerous projects, the word "umbrella" was not specifically used. Suddenly in the first draft of the request for circular 175 authority it states "there is no umbrella agreement for science and technology cooperation between the US and Japan, no umbrella agreement will apply." In the final approved request, this extension was now described as "the umbrella agreement for science and technology cooperation between the U.S. and Japan for all non-energy matters." Although this issue was seemingly clarified, the real nature, definition and meaning of "umbrella" remained a thorny and contentious issue. When this writer was interviewing people and gathering data it was quite difficult to obtain a definition of what an "umbrella" agreement included. It remained fuzzy through the bilateral negotiations in 1987-1988.

6. Renewal Talks with the GOJ

In mid-November 1984, Under-Secretary Schneider visited Tokyo and informed the GOJ that the USG wished to extend the 1980STA for five years.[23] The Embassy had been intermittently meeting with MOFA officials about the extension and possible meeting of the Joint Committee under the 80STA. After discussing the pros and cons of extending the 80STA for 1, 2, 3 and 5 years from the political and legal points of views, the Japanese preference had been to extend the STA for "a normal five year extension."[24] OSTP dissatisfaction with the working of the agreement climaxed when then science adviser George Keyworth felt that his trip to Japan in Spring 1984 was unsuccessful and decided to oppose renewal of the agreement in 1985. The GOJ continued to decline to buy into U.S. big science projects, i.e. SSC. A last minute compromise between State (and Embassy) which pushed for a full five year renewal, and OSTP resulted in an interim extension of two years from May 1, 1985.[25] This the GOJ reluctantly agreed to. The two-year extension of the 80STA was signed by Assistant Secretary (OES) James Malone and the DCM of the Embassy of Japan on April 26, 1985.[26]

7. Act I, An Assessment

In the mid-1980s preparatory period, the USG began with a substantive review by Justin Bloom in mid-1983 of the existing joint R&D projects between the U.S. and Japan in the public sector. This was followed by surveys in 1984 because the involved agencies still did not have the confidence that the USG "had a good picture of where the Japanese are going over the next five to ten years". So another study, this time an elaborate two-stage study was proposed to make yet another assessment. What began with a more or less straightforward systematic approach ended with procrastination upon procrastination. This pattern of delayed action was played out again in Act II. The GOJ was repeatedly informed that the review was proceeding and that the USG planned to have a joint review of all USJ R&D programs; the Japanese politely replied that they waited with expectations for the results of the USG review. As this phase ended the USG review process was envisioned in four parts: phase I was completed with only a decision to extend the agreement for two years, phase II estimated to take six months to a year to complete, was to survey U.S. government studies of Japanese ongoing anticipated R&D in areas of major importance to the U.S.; phase III would identify U.S. priorities in its S&T relations with Japan, especially those areas in which the USG "perhaps should pressure the Japanese" admitting that getting such concurrence in the USG "may be difficult" and Phase IV would be the actual USJ joint review.[27]

This internal USG assessment was intended first to provide guidance to the USG in developing its policy for S&T relations with Japan but it was also intended to result in a joint, U.S. and Japanese, review of their S&T relations. The Japanese on the other hand, based on available information, agreed to conduct "some form" of review. Since the GOJ was not an enthusiastic partner to the 80STA to begin with and was not seeking high level Joint Committee reviews, this review was undoubtedly perfunctory and the cause of some irritation and mistrust in subsequent discussions. While the Japanese were willing to hold office level (so-called working level) joint meetings, they did not feel the need or urgency for such high level meetings and were skeptical of USG motives and objectives underlying the "proposed joint review."

While the individual projects under the 80STA were generally regarded as worthwhile, their scope of undertaking and funds invested were disappointing in comparison to the original expectations of the 80STA even though these expectations should have been eliminated by the signing of the 80STA. They were not sufficiently substantive and presidential. This OSTP dissatisfaction became a major issue in the USG in subsequent years. Dissatisfaction with failure to produce major new departures in U.S.-Japan S&T cooperation as presumably envisioned at the inception of the 80STA, was led by OSTP. But no definition of these "new departures" was forthcoming and OSTP as executive agency played a passive role during 1980-85 in pushing USG and GOJ in desired directions. There were no known discussions in the USG as to why the 80STA could not be used to fulfill these deficiencies especially by OSTP.

Despite the dissatisfaction felt by some, especially the OSTP, there is no documentary evidence that OSTP, the executive agent for the implementation of the 80STA, ever seriously pushed for a more effective implementation. OSTP was pre-occupied with the continued GOJ refusal to partially fund U.S. big science projects, e.g. the SSC. Reciprocal investments in Japan by the U.S. were never suggested by the USG or requested by the GOJ, as a quid pro quo.

The 80STA was still regarded as an important symbol in the development and maintenance of close, friendly ties between the U.S. and Japan especially as an imperfect balance to the increasingly unpleasant trade frictions. There was a new recognition that Japan was a highly sophisticated science and technology nation, that the U.S. and Japan had emerged since the 1960s as "the world leaders in science and technology." The flow of benefits had been, as has been recognized for many years, one-way for most of the postwar years but that in the mid-1980s this flow was now more or less balanced. As this period drew to a close, the Embassy pointed out that the USJ S&T relationship involved 16 separate agreements with a total of 44 projects involving 17 different U.S. federal agencies, "the largest and most diverse such relationship that we have with any developed country"[28] Nevertheless, the USG in Washington persisted in trying yet again to persuade the GOJ to fund U.S. projects.

That while the Japanese were opening their national universities to foreign professors and laboratories to foreign researchers, there was a lingering anxiety and suspicion that more could be done to obtain greater access to the new areas of excellence in Japanese science and technology enterprise which would benefit U.S. science and technology and industry. Nevertheless, access to Japanese R&D facilities, public or private, and science and technology information had not yet been raised as serious USG concerns during this period.

State Department's OES Bureau had failed to coordinate effectively and forcefully lead the divergent interests of U.S. government agencies to create an effective and systematic S&T policy vis-a-vis Japan. For example, there was confusion in the USG as to whether or not, and in what way, the STA would and could be an umbrella agreement, about the length of extension, and even the kinds of studies to be conducted in order to reach this decision. Ultimately no evidence has been found that this multi-phase USG study was conducted. Furthermore, friction between and among U.S. agencies was becoming more evident, especially between State Department and OSTP.

Despite the chaotic and intermittent consideration of the future of the STA, some ideas surfaced in this period which survived through to 1988, e.g. a high level review committee, selecting leading edge technological/scientific areas to be given special emphasis in the future STA between the U.S. and

Japan.

Notes

1. *An Evaluation of Major Scientific and Technical Agreements in Effect between the United States and Japan* by Justin Bloom. August 1983. Final Contract report to U.S. Department of State.
2. ibid. p. 56.
3. ibid. p. 55.
4. ibid. p. 55.
5. ibid. p. 57.
6. p. 2, *US-Japan High Level Review*, undated background paper probably about January or February 1984 since events to take place in March 1984 are mentioned.
7. ibid. p. 2.
8. ibid. p. 2.
9. OES memo, James L. Malone to T- Willion Schneider, 10/28/1983, Sub-cabinet Group on US-Japan Science and Technology.
10. *U.S.-Japan High Level Review*, p. 3.
11. Tokyo 22624 (11/21/1983). Principal Deputy Assistant Secretary of State Harry R> Marshall and staff visited Tokyo 25-29, 1984, it was reiterated to the GOJ that 1) the USG wanted a bilateral review of "all existing S&T programs" and 2) the USG would present the GOJ with a written proposal soon. The USG anticipated a preparatory meeting summer 1984 with an under-secretary level (deputy minister in MOFA) meeting in Fall 1984. The MOFA suggested to Marshall that "in order to bring the GOJ bureaucracy along" that the USG make it's formal approach soon and this would push the GOJ to make a decision. Marshall called upon several Japanese Ministries. The ST&A was favorably disposed but said other Ministries would be opposed unless the USG made a high level approach to the GOJ. MITI pointed out that AIST's existing programs with USG agencies were quite well and that any review must include all programs and produce some specific results. MOE, on the other hand, said there was no need for this bilateral review it was pleased with existing cooperation and there were arrangements for review under these arrangements. MOE feared apparently loss of some bilateral projects to more glamorous projects. (State 70394, 3/12/1984).
12. State prepared terms of reference for the May 1984 review and were also sent to the Embassy for its review on 4/27/1984 (State 123526). The review groups would be divided into Group 1 which would identify major U.S. national interests in S&T which coincide with areas in which Japan has strong R&D potential, then determine which current GOJ programs support those US interests, then determine which could be pursued through existing USG-GOJ arrangements within currently available resources, and if additional resources were made available. Also it would identify major national interests in applied sciences where there were systemic obstacles in Japan to U.S. participation, where the private sector cannot participate because of the long term high risk nature of the research should the USG become involved. Working Group 2 would identify programs where high level attention would be needed to persuade GOJ to resolve outstanding issues and how the private sector might be more effectively involved in formulating programs.
13. Tokyo 9581 (5/15/1984).
14. *U.S.-Japan High Level Review*, p. 2.

15. An internal memo declared that "Some in Washington believed for years that we should be pressuring the Japanese to fund more U.S. basic research. Their idea was that we needed to encourage the Japanese to pick up 'their fair share' of the burden of funding basic S&T research." The afore-mentioned Bloom report, showed that character of U.S.-Japanese cooperation is that the Japanese are spending millions on basic research being done in the U.S. "US technical agencies tend to view Japan as a source of extra budget funding and have gained significant programmatic benefit from cooperation with Japan." The reason for the USG not investing in Japanese basic research stemmed from the belief that very little was being done. But "that situation is in transition ... and [they] are beginning to devote more national funding to longer-term science and technology." A review by State/INR was called for to ascertain what the Japanese were doing in basic research, whether they were ahead of the U.S. in certain fields for the planned April 1984 internal USG review of the USJ S&T relationship. But this report by INR, if prepared, has not been found (Memo, OES (James L. Malone) to INR (Hugh Montgomery) on *Assessment of Japanese Research in the Basic Sciences*. 2/14/1984).
16. State memo, James L. Malone (OES) to Dr. Schneider (Under-secretary), *U.S.-Japan Science and Technology Review: Next Steps*. 8/30/1984.
17. The questionnaire for this survey was sent to the Embassy as State 143343 (5/15/1984).
18. State 232448 (8/7/1984); *History: Non-Energy Treaty with Japan,* 3/13/1985, p. 2.; Tokyo 20375(10/2/1984).
19. It was envisioned that the study could be conducted by the CIA, through a contract with a major think tank, or through a Team approach (one individual and a government team working together). The latter approach was recommended but, so far, it is not known whether this new study was approved, and, if approved, the final report has not be found. It envisioned a two-stage study: the first, conducted by a single consultant to be completed by December 15, 1984, 1) would identify what the U.S. already knows about Japanese S&T and where the gaps are and 2) develop a work plan for a comprehensive study on U.S.-Japan S&T future relations; the second part (with no target date established), a much larger study to be awarded through the bidding process to develop policies which would be critical to 1) government-to-government cooperation, and 2) those areas more appropriate for industry-to-industry cooperation. State 238650 (8/13/84), State 258762 (8/30/84), State memo cited in fn 16.
20. *Report of the Defense Science Board Task Force on Industry-to-Industry International Armaments Cooperation: Phase II -- Japan. June 1984.* Office of the Under Secretary of Defense for Research and Engineering. p. 142.
21. These letters were included in the afore-mentioned 8-30-1984 memo in fn 16 above.
22. *Challenges and Opportunities in United States-Japan Relations.* A report submitted to the President of the United States and Prime Minister of Japan by the United States-Japan Advisory Commission, September 1984. 109 p. Science and Technology chapter, p. 97-102.
23. Under Secretary Schneider as in his previous trip to Asia, visited Korea and Japan and in Tokyo meet with MOFA Deputy Minister, AIST/MITI Director General and S&TA Vice Minister. At the S&TA they discussed the Space Station project, Japan's H-2 Launch Vehicle. While Schneider was looking for an agreement with the GOJ about the Japan's H-2 but S&TA's response was that the "most important parts would be built in Japan, that the GOJ considers the H-2 technology to be a complete departure

from the H-1 technology [which had stemmed from U.S. technology] and, therefore, not subject to any U.S. controls. ... there were [in MOFA's opinion] certain restrictions in the H-1 program that were undesirable from the GOJ's point of view; thus, the GOJ decision to go ahead with the H-2 on an independent basis. ... the GOJ would like the freedom that could be gained from the development of an indigenous capability. When queried by the AIST Director General, Schneider said the USG would like to renew the 80STA for another five years (Tokyo 23708-11/16/1984).

24. Tokyo 4484 (3/5/1985), State 71694 (3/9/191985). Although the length of extension of the 80STA was seemingly clear and decided, there is a record of a telephone conversation on April 16, 1985 between Under-Secretary Schneider and Science Adviser Keyworth where the former "reconfirmed his decision" that the agreement should be extended for five years and that the latter did not object. Nevertheless, since the U.S. had agreed, according to a brief memo of April 14, 1985 with the GOJ orally and in writing to a two year extension this remained the U.S. position.
25. When Ambassador Matsuda of MOFA visited Washington in spring 1985 and met with OSTP Deputy Director, the OSTP inclination was to terminate the 80STA or let it lapse.
26. Yet another approach was broached by the USG when the U.S.-Japan economic subcabinet meeting took place in fall 1984: to propose to the GOJ that an ad-hoc group on S&T be created under the auspices of the subcabinet group to deal with cross-cutting long-term S&T issues. It was thought this step could help to integrate certain aspects of our S&T relationship and economic issues. The USG was straining to devise some mechanism that could lead to a dialogue with the GOJ to facilitate USG's access to Japanese R&D. It appears nothing came of this effort. State 280372 (9/21/1984).
27. *History: Non-energy treaty with Japan*. 3/13/1985.
28. Tokyo 25558 (12/17/1984), a cable from Ambassador Mansfield to the Secretary of State reporting on the Embassy's role in this relationship and appealing for an additional mid-level officer to assist in managing this far flung program. It also described the creation of an intra-Embassy S&T coordinating group for S&T issues since S&T "impacts on all aspects of our bilateral relationship from the political to the economic to the military". Because of this importance the science counselor reported directly to the Ambassador.

VII. Act II, A Meeting but No Promised Proposals: Mid-1985-February 1987

1. The Initial Atmosphere

The period begins with a thickening of the conflict between the State Department and OSTP, desultory movement within the USG to conduct the necessary reviews of U.S.-Japan S&T relations, some meetings between the American Embassy in Tokyo and the GOJ, preparation of an options paper by an outside consultant and its indecisive consideration, the OSTP decision in June, 1986 to mobilize the formal Federal S&T interagency mechanisms to create a set of recommendations for high level consideration. The intent here was obviously to put an end to the slow-motion review process and create a package which could be approved by the USG itself for presentation to the GOJ in early 1987. After the 1985 two-year extension, in OES's own words, the "OSTP has taken a lead role in internal preparation of our negotiating position on the agreement."[1]

In mid-July 1985, an OES representative went to Tokyo to discuss several science and technology agreements, among them the 80STA, with the STA, MOFA and MITI officials. Reports on these meetings are indeed skimpy and only reveal that

1. the Japanese had not conducted a detailed review of the 80STA except to consult "some researchers" who endorse the agreement,

2. S&TA inquired about the role of the OSTP and State OES in the USG's review process of the U.S.-Japan S&T relationship. They were informed that OSTP wanted a "stronger agreement", but that State/OES was responsible for the review process. This inquiry underscored the image of an uncoordinated review by the USG with no clear agency in charge of the review. S&TA noted their special interest in being notified about the contents of this review,

3. MITI politely inquired about whether USG was planning another meeting

since the March 1985 meeting promised by Under Secretary Schneider during his November 1984 meeting in Tokyo did not take place. No response is recorded.[2]

In September 1986, a proposed methodology was suggested for organizing the discussions in the USG concerning renewal and possible expansion of the 80STA. There was great concern expressed for the sensitivities of the agencies and not surprising them with any of these steps. This is obviously a reflection of the sensitive and delicate relationship between the OES in trying to carry out its mandate and the administrative independence of the U.S. executive departments and agencies. The methodology included:

1. List the areas of technical cooperation which might be added
2. Prioritize these technical areas of potential cooperation
3. Compare them with existing areas covered by the 80STA
4. Collate the areas of priority to determine which should be added and finally
5. Approach the agencies concerned to suggest initiating new projects for collaboration with Japan.[3]

In making this suggestion, the following list of 44 projects under the 80STA agreement was provided as a guide. It should be remembered that many of these projects were the subjects of criticism -- and sometimes derision in private interviews -- as inappropriate for a Presidential agreement.

It should be remembered that a study by Don Ferguson had been initiated earlier in the year, that a separate contract was proposed to conduct a much more elaborate study of U.S.-Japan S&T relations and create a set of recommendations.[4] With the uneasiness that pervades the atmosphere, so to speak, State appeared interested in involving the OSTP more prominently in the next meeting with the various agencies to persuade the latter to formulate proposals for changes, developing additional programs more appropriate to a Presidential agreement with Japan and/or replacing existing programs. The purpose here was to call upon these agencies with OSTP support for ideas for new R&D programs (and funds) more appropriate to a Presidential agreement to be carried out with Japan. If these ideas were not forthcoming from the agencies, it would be somewhat harder for the OSTP to blame the State Department for lack of support.

The OES Advisory Committee. This advisory committee created early in the Nixon Administration by Secretary of State William Rogers was never able to reach beyond the fringes of the mainstream of the State Department. The Committee met twice a year and virtually no activity occurred at these meetings. The Committee reported to a second rank bureau, the OES, where it was not taken seriously just as the OES itself was not taken seriously within the department. The committee continued "more or less comatose largely inactive but showing occasional flickers of life under A/S John Negroponte".

While Negroponte was reported to be in favor of using this Committee, there are no signs that this materialized. Two documents have been found that are ambiguously marked for a committee in September 1985. They were re-writes of earlier documents on the U.S.-Japan STA. No records have been found to indicate what specific aspects of the STA were discussed, what actions were taken, if any. In essence the barest amount of lip service was given to the Committee's existence and function. It played no effective and constructive role in the deliberations on the future of U.S.-Japan S&T relations.[5]

The White House Science Council. This Council was another organ to which the Science Adviser could turn for advice in a domestic or international S&T policy issue. George J. Keyworth, science adviser to President Reagan, established the White House Science Council in February 1982 to provide advice on issues that involve the science agencies of the USG and policies that involve multiple agencies, and appointed Solomon J. Buchsbaum of A&T Bell Laboratories as Council chairman. There were a number of points from 1983 through 1988 where the science council could have become involved in providing advice on the future of U.S.-Japanese S&T relationships. These points of entry, so to speak, would be during the mid-1980s review which ended in the renewal of the 80STA for two years, in late 1986 when the Joint Task Force prepared a report on future USJ S&T relations, as a conduit for a joint review of the various USJ S&T programs. The Council could be an effective institution for conducting an informal dialogue with the GOJ.

The Council, according to then Chairman Buchsbaum, met six times a year in plenary session. The Council worked by panels but it never focused on USJ S&T relations by its own initiative or by request of the science adviser. U.S.- Japanese issues were however discussed at the informal evening dinner sessions. Apparently, the 80STA was not thought of as particularly strengthening USJ S&T relations, at least not in the framework of existing relations. In contrast, in the private sector the relations between the two countries were extremely complicated, intense and wide ranging. The 80STA and the follow-on 88STA was not a major concern to the private sector since it had little or no impact on industry.

While the Council did not formally initiate any dialogue with its Japanese counterpart, the Science and Technology Council chaired by the Prime Minister, the latter did request a meeting with the White House Science Council in 1985. Since an abortive attempt was made to hold another meeting, this meeting was the only one held during the period under consideration. Engineering education was the topic discussed by both Councils, not the 80STA and its future.

Obviously successive science advisers did not feel that it was necessary to obtain the support, help and/or endorsement of the White House Science for any particular item, in this instance of USJ S&T relations, past, present and future. One member of this Committee said he was briefed individually by the OSTP staff on the USJ STA. Perhaps the staff and the science adviser were not sufficiently creative to effectively utilize the Council as a whole; perhaps because the Science Adviser's office was inadequately staffed even to handle day-to-day operations let alone also administering a more active Council. Buchsbaum commented that while the scientific and technical community held the advisory committee system to the President in high esteem, "unfortunately, neither the power structure within the White House nor, indeed, the various departments and agencies of government shared this attitude"; the advisory structure "was largely ineffective."[6] In light of the contentious issues within the USG about future S&T relationships with Japan, it was unfortunate that a prestigious group like the Council could not have been used more creatively by the science adviser and the so-called power structure.

2. The First Options Paper

Again, Justin Bloom was called upon to prepare a paper "taking no more than a week of writing time and utilizing the knowledge already in his possession, to set out the essential background factors and principal options ... on the non-energy question." He was given an outline by the State Department in early November 1985 with the understanding that such a paper would be

Table 7.1 List of Projects under 80STA

Title	Responsible USG	Agency GOJ
Crustal Plate Motion	NASA	MPT
Plasmas in the Earth's Neighborhood	NASA	ISAS
Halleys Comet Studies	NASA	ISAS
Saturn Orbiter	NASA	NASDA
Spacelab	NASA	ISAS
Space Tether Systems	NASA	ISAS
X-Ray Astronomy	NASA	ISAS
Ocean Dynamics	NASA/HQ	Tokyo U
Cloud HT	NASA/HQ	S&TA
Snow Properties	NASA/HQ	S&TA
Evapotranspiration	NASA/HQ	S&TA
COMSAT Data Experimental Exchange	NASA/HQ	MPT
Balloon, Transpacific	NASA/HQ	ISAS
Solar Studies	NASA/HQ	ISAS
Typhoons	NASA/HQ	S&TA
Marine Resources	NASA/HQ	MAFF
MOS-1 Data	NASA/HQ	NASDA
Antarctic Meteorite Data	NASA/HQ	NIPR
Neutron Scattering	DOE	JAERI
Accelerator	DOE	JAERI
CO2 Effects	DOE	MOE
Diesel Particulates	DOE	EA
Liquid Gas Safety and Control	DOE	ANRE
Biological Effects of Electric Field	DOE	MITI
Hazardous Materials Handling	EPA	MITI
Nitrogen Oxide Technology	EPA	MITI
Environmental Disease	EPA	EA
Ocean Dumping	EPA	S&TA
Fish Bioconcentration and Toxicity	EPA	EA
Toxicology	HHS	MHW
Alcohol Related Problems	HHS	MHW
Drug Abuse and Mental Health	HHS	MHW
Immunization and Vaccine Development	HHS	MHW
Laboratory Animal Science, Primate	HHS	MHW
Laboratory Animal Science, Non-Primate	HHS	MOE

DNA Recombinant	HHS	S&TA/ MOE
Cardio-Vascular Diseases	HHS	NHW
Snow/Avalanche/Landslide	USDA	S&TA
Food Health Regulation (Salmonella)	USDA?	
Pest Management	USDA?	
Biomass Conversion	USDA	S&TA
Commodity Quarantine Treatment Management	USDA	MAFF/ S&TA
Cytogenetics: Hybridization Limits	USDA	MAFF
Biotechnology	USDA	MAFF

Total 44 Projects. Compiled by USG, mid-1980s.

---

prepared before the end of November 1985.

Bloom produced a 13-page report with 3 appendices outlining five options for the USG, sketching in the political background of the 80STA, its shortcomings, the factors to be considered in renegotiating the 80STA, and the pro's and con's of each option, and strategies for negotiation with the GOJ. Since the report contains several direct references to the OSTP and its attitudes, it was obviously intended as a paper for State consideration only. It was the first and most thorough report which gave the State Department considerable material and suggestions for preparing a recommended USG position. It could have been the catalyst for a USG decision on how to proceed with its negotiations with the GOJ. It is significant that the second major report/analysis on U.S.-Japan S&T relations and the future were prepared by an outside consultant to the State Department. This procedure must reflect either the lack of expertise or analytical capability in the OES Bureau or a lack of qualified and sufficient number of personnel who could be spared to prepare these studies, especially an options paper.[7] The Bloom paper proposed the following options:

1. Maintain the agreement in status quo condition for the indefinite future, renewing it as necessary, and leaving it to the Japanese to propose changes.

This option and approach was supported by the Tokyo Embassy and the State Department, and would have been welcomed by the GOJ. It would "save face" on both sides, and would have a neutral effect on the increasingly difficult trade relationship. It probably would not be viable since it would be opposed by the OSTP, and does not address the so-called deficiencies of the 80STA either on the policy level or in implementation. Bloom reports that the OSTP had been "recently for abrogation...because of a perceived rebuff by GOJ in OSTP's attempt to obtain funds for the SSC."[8]

2. Permit the agreement to lapse at the end of the current extension period in May 1987.

This would be a "loss of face" for both the U.S. and Japan, especially since it was the first Presidential level S&T agreement signed by the two countries and would be an admission of the failure of the 80STA an agreement proposed by the U.S. This option would probably be acceptable to OSTP but, probably reflecting State Department thinking, Bloom added "this point is tempered by realization that OSTP may change its position at any time."[9] It would indeed send the wrong political signals that perhaps other S&T agreements should now be questioned and strongly imply that a damper should be put on further U.S.-Japanese S&T cooperation.

3. Propose to the Japanese Government that talks begin aimed at modifying the agreement to alter or eliminate provisions that have made it ineffective.

This would be a stopgap measure that assumes that the USG could put together in a timely fashion a package of proposals which would be acceptable to the OSTP in the first place, other executive agencies and to the GOJ which would seriously consider them and negotiate within a reasonable schedule. Bloom felt that this approach would give the GOJ an early warning that the USG was having difficulties with the agreement (but the GOJ knew this already at least superficially) and would give the GOJ an opportunity to make its own proposals which would probably be quite unlikely.

4. Take a stronger, more positive approach than Option 3 by offering Japan a set of new agreements to take the place of the S&T agreement.

Whatever the rationale, this option would result in the disappearance of the 80STA, and would imply that several significant S&T ideas would be or are in the offing which was then not the case. The Space Station agreement had been signed with NASA by the GOJ's Space Activities Commission. The Japanese were reluctant to support the SSC with large contributions and with the difficulty of mixing military and civilian requirements in the SDI R&D. Bloom pointed out – as had the Embassy – that for this option to be warmly received by the GOJ, the new proposals would have to go beyond the traditional approach of the U.S. to Japan, i.e., that the U.S. would be willing to invest in significant Japanese-based projects which has not been feasible nor welcome in light of U.S. attitudes toward Japan and the U.S. budget problems and atmosphere created by the trade climate.[10]

5. Try to significantly revitalize the S&T agreement by reverting to the original OSTP concept of making it the umbrella agreement for all U.S.-Japan bilateral undertakings in science and technology.

Superficially it would appear that the two parties are making a significant move to coordinate all the far-ranging S&T projects but would be fought by both Japanese and U.S. technical agencies which have maintained by and large considerable autonomy in managing their bilateral projects. It would be more palatable to the OSTP since it had called for this agreement to be more all-encompassing.

In the brief section on strategies for negotiation, Bloom cautioned about the GOJ's usually inordinate concern with "abrupt changes in position or policy" by the USG and that these shifts, justified or not, contribute to the impression in Japan (and other countries), that the U.S. is unreliable in its international technical undertakings.[11] He urged that any of the above options (except 1) should be "floated before the Japanese in informal talks at the middle management level (office director level in the State Department and MOFA) so that the Japanese side can develop a consensus upward. He warned that the U.S. side should expect and be prepared for protracted negotiations. He even suggested that an options paper be given informally to the GOJ at this intermediate management level for internal discussion. Lastly, Bloom suggested avoiding any attempt to rush to negotiate an agreement at the upcoming Economic Summit in 1986.

A minor effort was made in late November 1985 by OES and the Japan Desk in the EAP Bureau to consider the Bloom options, find some viable options which would be supported by the Embassy. Again, much flurry, little substance and not even an apparent plan for discussions with Japan seem to have come of this yet another endeavor. But elements of Bloom's analyses did appear later in a piecemeal fashion in some proposed actions[12].

3. USJ Meetings in Spring 1986

a. OES Options Paper

A/S OES was scheduled to visit Japan in April 1986 to participate in a panel discussion of the annual conference of the Japan Atomic Industrial Forum. As occurred in 1985, discussion of the 80STA with the GOJ was, it should be emphasized, a secondary mission. For this purpose, OES slightly revised the Bloom paper discussing the options and rather interestingly U.S. and Japanese attitudes towards STI and was, for a bureaucratic paper, unusually frank:[13]

- "The Japanese -- along with a number of USG agencies -- were never enthusiastic about the agreement, which in many ways is duplicative of other accords"

- the 80STA, because it was signed at the highest level has become an important symbol for the Japanese [and this writer would maintain as much a symbol for the U.S. too]

- the U.S. Science Adviser opposed the renewal in mid-1985 and a last minute compromise between State, which pushed for a full five year renewal, and the White House, resulted in an interim extension of two years from May 1985 to May 1987

- OES is [still] developing a strategy for dealing with this situation of inter-agency conflict.

Concerning the options, three of five suggested by Bloom in November 1985 were described as "probably not viable".

1. to maintain the agreement in status quo condition for the indefinite future (Bloom's option 1)

2. to permit the agreement to lapse (Bloom's option 2) and

3. to try to revitalize significantly the 80STA by reverting to the original concept of it being an umbrella for <u>all</u> U.S.-Japan bilateral undertakings in S&T (Bloom's option 5).

These options were not viable because

1. the White House, i.e. OSTP, opposes maintaining the status quo

2. permitting the agreement to lapse would be a major blow to U.S.-Japan S&T cooperation with possible ripple effects of a broader nature

3. there would be strong resistance to the umbrella approach both from the Japanese and U.S. technical agencies which prefer the relative flexibility which they now enjoy.

The viable intermediate options were 1) to propose to the GOJ to begin talks aimed at modifying the agreement to alter or eliminate provisions in the 80STA that have made it ineffective and 2) to offer the GOJ a set of new agreements to take the place of the S&T agreement.[14] The first option would give the GOJ an opportunity to offer their ideas on how "to salvage" the 80STA -- except the GOJ never regarded it as needing salvaging, indeed GOJ preferred to extend this agreement as is. As a contrary point, OES described it as a delaying action, and possible embarrassment to the Japanese who could look upon this U.S. proposal as a criticism of their lack of enthusiasm. Actually, both sides were decidedly lackadaisical about the 80STA. The second approach was described as based on U.S. recognition of the defects of the existing 80STA and that USJ technical relations had moved to a higher more equitable level since 1980. On the negative side, was the presupposition that a limited number of new agreements were in the offing -- actually none were (The Space Station Agreement had been signed). But it was recognized that the new large projects would "need to go beyond the traditional approach of US to Japan", i.e. they would "include US investment and participation to some extent in Japanese based projects. This does not seem feasible in the current budget and trade climate" -- USJ were equal but not equal enough for U.S. to balance S&T investments in Japan as Japan had or might invest in U.S. S&T projects in the U.S. Whatever option was chosen it urged, as Bloom did, that these options should be floated before the Japanese in informal talks at the middle management level in the MOFA. The position paper urged that the OES A/S emphasize to Embassy staff that in exchange programs the U.S. must keep pace with Japan's rapidly expanding S&T, and that we are looking in the future at specific fields where US S&T might benefit from increased exchanges, that, at this time, State/OES did not have a USG consensus to propose to the Embassy and the GOJ -- [after three years of work].[15]

This also provided some interesting observations on Japanese and U.S. views and attitudes toward STI. In light of later rather rankerous discussions with the GOJ, it is useful to cite OES assessments on exchange and competitiveness. While "the extremely broad and varied S&T relationship" which results in a "constant exchange of information", the "Japanese have used the exchange to better advantage than we." Since there were very few Americans who competently understand Japanese the language barrier was real; it asserted that the Japanese.

- are much more organized and determined to gather information related to S&T in the U.S. than Americans are in Japan.

- closely monitor foreign patents.

- are avid users of U.S. technical information services.

- have gained access to U.S. technology in U.S. companies or entering into marketing or production agreements with them.

- have been very open with their U.S. scientist counterparts in basic research.

- are already working at the frontiers of research with such materials as gallium arsenide ... and carbon fibers; biotechnology firms already have the most advanced fermentation techniques.

- are now doing creative original work in cancer.

The Americans are described as not as determined to gather S&T information in Japan, have not taken advantage of Japanese openness in basic research, believing that basic research still is not where Japanese strength lies.

At the April 1985 meeting with Ambassador Matsuda, Director General of Science and Technology Affairs in the MOFA, he commented (reminded the U.S.) that the 80STA had been extended for two years "because of U.S. concerns", and that we need to discuss the next extension. In response to U.S. concerns, Ambassador Matsuda suggested two possible courses of action:

1. Drop the review level of the agreement to reflect the reality of the cooperative activity under the 80STA

2. Upgrade cooperation under the agreement.

Significantly he did not suggest terminating the agreement but a relatively minor adjustment within the existing framework of the 80STA, not a new agreement. In deference to the U.S. desire for a joint review, he suggested such a review could take place later in 1986. In response, A/S OES promised that he would "get back to Matsuda ... in the next few weeks" after discussing these ideas in Washington.[16]

Between April and June 1986, a flurry of activities took place culminating in yet another phase in the USG attempt to find a negotiating consensus on the upcoming renewal, revision or termination of the 80STA in May 1987.

b. The Embassy's Response to the OES Options Paper

The Embassy reiterated what was already well known: that the concerns about the 80STA were "driven" by OSTP, by Science Adviser George Keyworth. The Embassy felt the problem began with the September 1981 Joint Committee where Keyworth felt that it was not worth his time to review only a small portion of the USJ S&T relationship under the 80STA structure. While this may be quite correct, the 80STA was basically a USG creation but modified through negotiations with the GOJ. The Embassy insisted that the GOJ knew about USG concerns with the 80STA through numerous meetings with U.S. officials but apparently there had been no suggestions to the GOJ about proposed solutions. In choosing an option, the Embassy called for "serious consideration" be given to continuing the 80STA in its present form but change the level of the Joint Committee from the OSTP to OES (Office Director or Deputy Assistant Secretary of State level) without modifying the 80STA. While the Embassy insisted that this would convey to the GOJ that the USG was "unhappy with then level of activity", this approach was what the GOJ was itself advocating. The presumed USG message would surely be lost on the GOJ.[17]

c. The Embassy's New Science Counselor

Richard Getzinger, in preparation for his new assignment in Tokyo as science counselor visited Washington in early May 1986 and called upon -- the memo more frankly stated that he "managed to work in" these calls to -- NBS, USDA, NASA and HHS to discuss the future of the 80STA. Such details are mentioned to point out that the future of the 80STA always seemed to be a secondary consideration, not the major issue at hand and the more cooperative attitude toward discussing issues with the GOJ. Among the talking points prepared for him for these meetings was a reiteration of the GOJ and Embassy positions and the following:

> ...Japanese views on this change [in the 80STA] need to be carefully solicited before making a formal proposal. We will be discussing the options with the Japanese informally in the next few months to determine what impact the proposed change would have on their internal politics or on cooperation with the US under the agreement.[18]

This solicitous approach to including the GOJ in the process was expressed earlier by former science counselors who participated in Washington meetings on U.S.-Japan S&T relations. The following conclusions were drawn from these discussions with the several agencies:

1. all would like the agreement [80STA] to be renewed for five years

2. these agencies would accept a change from OSTP to State/OES as the executive agency, i.e. no objections to the downgrading of the agreement.

Getzinger went to Tokyo sensitive to the need to get a consensus within the GOJ before State/OES goes back to the USG Agencies with a more formal consensus seeking proposal.[19]

d. New Signals to the GOJ

MOFA's Assistant Director General Shigeda visited State/OES on June 10, 1986. In the talking points for this meeting State/OES expressed the belief that

> "there is more support in the USG for the first of these options, i.e. to continue the agreement in its present form but change the level of the joint conference. One possibility would be to change the joint committee function from OSTP to State at the OES

Deputy Assistant Secretary level."[20]

On the way to this meeting with Shigeda something happened which cannot be explained through available documentation.

The same staff that drafted the above memos, now told Shigeda that

1. ...while the downgrading the level of the agreement was a possibility, ... this would not be acceptable to the White House [i.e. OSTP]. OSTP felt that the agreement should either be revitalized and refocused or terminated. Shigeda apparently confirmed OSTP's desire to refocus the 80STA based on his meeting with Wince.[21] Since Wince had met with Shigeda before the meeting with State/OES, perhaps Wince at OSTP had upstaged State/OES and the latter had to readjust its views to maintain a common front vis-a-vis the GOJ.

2. in the overall review of the U.S.-Japan S&T relationship which OSTP was initiating "there will be considerable pressure for a significant upgrading and restructuring of cooperative activity ... [which will] concentrate on high tech areas where the two countries are at the leading edge of development and have implications for the areas of commercial competitiveness" and urged the GOJ to consider what areas it might wish to include in such a list.[22]

e. Matsuda-Negroponte Meeting in June 1986

Director General/Ambassador Matsuda of MOFA (equivalent to Assistant Secretary) met with his counterpart OES A/S Negroponte on June 20, 1986 -- only ten days after the meeting with Shigeda. As usual, OES prepared a five page briefing paper consisting of an analysis and talking points for this meeting. Under the section, The Issue, OES provides a short description of the difficulties OSTP asserts it has with the 80STA. OSTP is "dissatisfied with the UJST [80STA] because of its failure to provide more new departures in US-Japan science and technology as envisioned at its inception" and is "unhappy" with this cooperation "which, they [OSTP] believe does not sufficiently benefit the US side." In contrast to this feeling USG technical agencies "are generally content with the status quo and would prefer not to see much tinkering with UJST or the US-Japan science and technology relationship generally." These very agencies were described as going "along reluctantly with White House pressure" to negotiate the 80STA under the

Carter Administration and are "neither surprised nor dismayed that it has not been more successful."

Although OES noted that OES/EAP/Embassy options paper indicated that "the most acceptable course of action" was to renew the 80STA at the MOFA/State(OES) level instead of at the White House/PM level, it also stated that OSTP maintains that either the agreement should be amended and refocused to produce more meaningful cooperation in leading edge technologies or terminated. Says OES: "OSTP clearly plans to place considerable pressure on all concerned to resolve the issue in its favor." [23]

Those present at this meeting were Ambassador Matsuda and two staffers from the Washington Embassy, the A/S OES, Ambassador for Non-Proliferation Policy and nuclear energy affairs and several staffers from OES Bureau but no one from OSTP. Ambassador Matsuda met separately with OSTP in the afternoon of the same day. This is just another example where the USG presents a split personality to the GOJ. The main purpose of his visit was nuclear affairs and the 80STA was merely another issue. The USG prepared talking points for the Matsuda meeting were somewhat unusual in that the USG was giving not a unified USG position but a State and OSTP position and asking GOJ for their comments and even that comes after OSTP had already set in motion an inter-agency study of U.S.-Japan S&T relations. In these talking points:

- The Japanese were asked if they had any further "amplification" of their earlier proposal to drop the level of the agreement to reflect reality or to upgrade cooperation under the agreement.

- That there was support among U.S. government technical agencies for the first option, i.e. continue the agreement in its present form but change the level of the joint review and transfer the responsibility from OSTP (Presidential and Prime Minister level) to State/OES and MOFA. This was described as the "most acceptable course of action for both sides" based on options paper prepared by OES and the Japan Desk and the recommendations of the Tokyo Embassy. Would the GOJ agree?

- OES A/S was urged to note to Ambassador Matsuda that OSTP believes the agreement should be either amended and refocused to produce more meaningful cooperation in leading edge technologies or terminated. What are GOJ views? GOJ had already given its response in previous meetings and via the Embassy in Tokyo.

- What areas of leading edge technologies would GOJ propose to increase/upgrade cooperation? The USG was thinking of bio-engineering, robotics, artificial intelligence, ceramics, material sciences, telecom-communications and fiber optics.[24]

While it is not clear that these ideas were actually conveyed to the Japanese, Ambassador Matsuda described the 80STA as a "unique agreement" as the only S&T agreement signed by the Prime Minister (and probably by the U.S. President too), agreed that the level of cooperation was not "presidential" (as stressed by OSTP), commented that when he met Science Adviser McTague in Washington a year ago (1985), the latter expressed an interest in terminating the agreement since it had not performed as originally intended and proposed a three step process beginning with a 2-3 year renewal with an exchange of notes, a technical review of high tech cooperation and a Joint Committee meeting in Washington to approve the new directions.

After Ambassador Matsuda's visit, OSTP agreed to renew the 80STA for another two years (1985-87) with the understanding that the GOJ agreed to participate in a joint review process in two stages, first a meeting of technical experts in December 1986 and a Joint Committee meeting in spring 1987 to formalize these decisions. State cautioned about carefully approaching the GOJ in light of the history of this agreement and Japanese sensitivity to it. The potential areas of S&T cooperation were list as data processing, telecommunications, advanced structural materials, manufacturing, machine tools and robotics, aircraft, space technologies, nuclear, biotechnology.[25]

Since the 80STA was renewed in May 1985 for two years, the USG, under the divided and sparring so-called "leadership" between State/OES and OSTP was unable to arrive at a consensus for negotiating with the GOJ, promised the GOJ that it would review GOJ suggestions and return in a few weeks with a response (which never occurred), met with the GOJ representative not at one meeting where all USG representatives would be present but at separate meetings (at State and at OSTP) where undoubtedly different viewpoints and assessments were conveyed to the GOJ. The USG started more than one study prior to the two-year renewal in May 1985 in order to arrive at a unified USG position. The USG had conducted several studies on USJ S&T relations, but "no comprehensive review with recommendations for future action has been completed."[26] State was unable to muster enough bureaucratic strength and will power together with the assistance of other USG technical agencies and the American Embassy in Tokyo to persuade OSTP (symbolized as part of the White House) to adopt their position instead of the more aggressive OSTP position. With the USG unable to arrive at a unified position, the GOJ obviously did not have an incentive to conduct a formal complicated in-depth review like that attempted in the past and contemplated now by the USG under OSTP initiative. OSTP believed that Ambassador Matsuda had agreed in a meeting with OSTP (presumably when he met with Science Adviser McTague in spring 1985) that the GOJ would also conduct a parallel full-scale review and assessment following along the lines of the phantom USG studies. Undoubtedly, he mentioned something like an agreement "to conduct a review" but hardly one like the U.S. There was no documentation that the GOJ was apprized in detail about the contents and methodologies of the USG studies. This relatively minor issue of how and when a government conducts a review became a source of major emotional distrust of the GOJ by OSTP. "I was lied to..." How then can one trust the GOJ in more broader issues in the S&T negotiations. This mistrust became expanded to include other USG agencies particularly the State Department in the coming months of negotiations among U.S. government agencies and with the GOJ. The gap between the State Department and OSTP had widened and the internal USG process became a snarling hydra-headed operation.

4. 1986 Circular 175 Authority

Under normal circumstances, securing Circular 175 approval to proceed with international negotiations for an agreement or its renewal is rather mundane. In the case of the 80STA more than normal controversy surrounds this procedure and requests to renew this agreement. Although the renewal date of the 80STA was May 1, 1987, the first draft was prepared in mid-July 1986 and approved on August 25, 1986 -- nine months from the renewal date.[27]

The final memo authorizing the renewal of the 80STA included the following points:

- The shorter renewal period from 1985 to 1987 stemmed from "U.S. dissatisfaction with the nature of cooperation" under the original 80STA without specifying any details.

- In considering further renewal, State had completed an internal review, and an inter-agency review under CISET [Committee on International Science Engineering and Technology] was in progress.

- Based on "informal discussions" with the Japanese, the U.S. would, "as an initial commitment" agree to renew the agreement for two years in return for which GOJ would be "prepared to participate in a joint review of science and technology cooperation under the agreement."

- This review would take place in two stages: a) after the U.S. has decided upon its position through the CISET Joint Task Force review process there would be a bilateral technical review, b) followed by a Joint Committee meeting to make final decisions on and ratification of cooperative activity.

- This understanding would be formalized through a "double exchange" of notes with the GOJ. The first note would "initiate the joint review process"; the second exchange would take place after the Joint Committee meeting which was now scheduled for April 1987. The agreement needed to be renewed by May 1, 1987. Authority for the extension of the agreement was obtained at this time "to make appropriate assurances" to the GOJ regarding renewal -- because the repeated delays resulted in questionable confidence. The purpose was to give the GOJ a draft of the renewal note for discussion and comment.

- It was clearly stated that the 80STA would be extended "without change" for a two year period.

Based on the above, the plan was to renew the 80STA without change for two years after a joint review and to show USG good faith there would be an exchange of notes on the review and the draft renewal of the agreement would be given to the GOJ at the same time. Interestingly, although the authority which had OSTP clearance discusses the reviews of the U.S.-Japan S&T relationship conducted by State and the on-going CISET Review, there is no mention of any GOJ commitment to or plans to conduct any review, let alone a multi-stage review. OSTP had insisted that Ambassador Matsuda of MOFA had agreed to conduct an internal multi-stage review just like the U.S. review; the apparent lack of any such in-depth review by the GOJ was a symbol to OSTP of GOJ's unreliability, unworthiness and became a highly charged emotional issue and source of frustration and anger to OSTP.

The outgoing cable to the Tokyo Embassy conveying the Circular 175 approval and requesting their comments on the exchange of notes did not include a copy of the draft note to extend the agreement two years based on an OSTP memo (9/29/86) to OES when the former cleared this cable.[28] The acquiescence of the OES to this demand/request was a reversal of its good faith approach to the GOJ. At the request of OSTP, the cable also included a paragraph on the renewal period: it "must provide adequate lead time for new projects to be developed, implemented, and produce tangible research results for evaluation ... [and] from the perspective of R&D management, a two-year renewal period would have major drawbacks."

Although there were at least two occasions when the exchange of notes could have been made with the GOJ, there is no record to this effect. This Circular 175 authority request seems to have been an effort in futility except that it showed that the renewal as of early October 1986 still apparently was going to be a renewal without change of the 80STA.

5. CISET Review

CISET, under FCCSET and chairmanship of OSTP, created an inter-agency group called "CISET Joint Task Force on U.S.-Japan Science and Technology Relations"(JTF). The appointment of this Joint Task Force set in motion yet another review process - the third if we regard the 1983 Bloom study as the first attempt to assess the U.S.-Japan S&T relationship, the second being the phantom 1985 one-man study and contractual study which presumably were carried out but apparently never used in the deliberative process. This new review literally dwarfs the earlier studies; it will be a much larger, more elaborate and a more institutionally oriented study than those in the past. In light of the inter-agency nature of FCCSET/CISET, this review would, by definition, involve literally all major departments and agencies of the USG. This is in contrast to the earlier studies which centered around State/OES operations. This review has a basis in Federal law and thus could be more effective as a persuasive tool than earlier attempts. It appears that OSTP merely informed OES of the initiation of this study which would be "used as the basis for determining the US position on UJST renewal."[29]

As the CISET Joint Task Force review is launched, the context of the situation could be summarized as follows:

1. The formal renewal of the 80STA will be preceded by a technical level

discussion with the Japanese on substantive content of the agreement in late 1986.

2. Although the CISET review is focused on the non-energy bilateral agreement, it is OSTP's view, that operation of that agreement during its next two year phase can set the stage for an overall, long-term reorientation and restructuring of U.S.-Japanese bilateral and technology relations. This implies an attempt to make the so-called "non-energy" 80STA which is to be renewed, a truly umbrella agreement under which all USJ bilateral projects and relations will be concentrated and assessed. This is a slightly different objective or hope which had not been articulated in the documentation so far and not mentioned to the GOJ based on a reading of the documentation.

3. Suggestions had been received from agencies on what scientific and technical fields should be discussed with the GOJ for inclusion in the agreement and the generic issues that needed to be discussed with GOJ concerning future S&T relations.

For the first time, a timetable was prepared to achieve a USG consensus -- after all, there were now less than 12 months to achieve this consensus and negotiate a revised agreement with the GOJ.

August 1, 1986 Working Groups (WG) begin CISET review under coordination of Joint Working Group Task Force

September 22 Working Groups submit proposed recommendations to CISET Executive Committee

September 29 CISET Executive Committee reviews proposals. WG's draft final recommendations and formal initiatives

October 24 CISET Executive Committee reviews WG submissions

November 3 CISET Executive Committee submits recommendations to Chairman of FCCSET

November-December Informal discussions with Japanese officials

December 1986 U.S.-Japan Technical Review in Tokyo
-January 1987

April 1987 Joint Committee Meeting (OSTP-MOFA) in Washington re U.S.-Japan S&T Agreement in Washington

May 1987 Formal review and signing of Agreement

OES prepared a briefing paper on US-J S&T relations in early July 1986. The objective of this additional new review was to seek suggestions on how S&T cooperation with Japan could be restructured and improved to better U.S. goals and interests. In what areas are the Japanese most advanced; where could the U.S. benefit from greater knowledge of Japanese developments. What are the trade-offs? What can we offer Japan to get entrance into the fields we are interested in? And to define a limited number of high priority research areas which the U.S. believes should be given emphasis in cooperative projects -- the so-called upgrading of the level of cooperation between the two countries.

It raised a number of questions to be answered in scientific goals, international competitiveness and foreign policy considerations. It did caution, however, in good State Department style, that the U.S. should be "mindful of the high sensitivity and likely resistance on the Japanese side to recommendations which might imply sudden dramatic change in the existing relationship". We judge it prudent, says the OES paper, to begin on a modest level with the expectation that success in these initial stages will make it possible to broaden our objectives as the dialogue moves forward.

a. The Tokyo Embassy's Response to the Initiation of the CISET Review Process

In informing the Embassy of the CISET review and the first meeting of the JTF, OES reported that:

1. after the visit of Ambassador Matsuda, OSTP agreed to renew the [80STA] agreement for another two years (May 1985-1987), with the stipulation that the Japanese participate with us in a review of S&T cooperation under the agreement with a view toward producing recommendations to make the accord more scientifically meaningful to both sides.

2. the CISET review process was to develop a U.S. position on upgrading S&T cooperation which would then be taken to a meeting of Japanese and U.S. technical experts "to formulate recommendations" which would then be "formalized" at a Joint Committee meeting in spring 1987.

3. the Japanese are particularly sensitive about the 80STA and must be approached carefully, that to obtain a meaningful response from the Japanese we must begin by proposing modest projects from which both sides can benefit and which will create a win-win situation. The two previous science counselors felt the GOJ would be amenable to a well reasoned U.S. proposal aimed at long term positive evolution of the S&T relationship and urged that the U.S. keep in mind the GOJ invitation to take part in the ERATO project (a MITI large national project on basic research open to foreign participation).

4. the only issue posed to the Embassy was what might be the Japanese interest in S&T cooperation with the U.S. in semiconductors, data processing, telecommunications, advanced structural materials, manufacturing, machine tools and robotics, aircraft, space technologies, nuclear, and biotechnology. It was not clear from where this list was derived nor the extent of its support among government agencies. It was the longest list of technologies and a slightly different array than mentioned earlier.[30]

The Embassy responded by:

1. identifying new materials, particularly polymers and ceramics, microelectronics particularly gallium arsenide devices, optoelectronics and telecommunications devices, computers, particularly supercomputers, and the fifth generation project, and biotechnology, particularly bionics, and pharmaceuticals as the areas stressed by S&TA and MITI.

2. choosing a somewhat different array of technologies which might interest the Japanese in cooperating with the U.S. in a "win-win situation" as follows:

a. in space sciences, the Japanese are particularly interested in materials fabrication using space station facilities

b. in computers, the Japanese are particularly interested in standards for operating systems (e.g. unix) and telecommunications (e.g. the protocols proposed by the U.S. Corporation for Open Systems).

c. in biotechnology, the Japanese are particularly interested in cooperation in guidelines on experimentation and application of new materials and mechanisms with altered genetics.[31]

b. Japanese "Preparations" for Renewal of the 1980 STA and Responses to the CISET Review

Ambassador Matsuda had "agreed" to conduct a review process in the GOJ, presumably -- in the minds of some, particularly OSTP -- in the same style with the same doggedness as the USG had done from 1983 to 1986. According to the Embassy's meeting with MOFA's newly appointed Director for Scientific Affairs (S. Hinata), the latter knew of the renewal for two years, the proposed Joint Committee meeting in spring 1987, but not the meeting of technical experts in late 1986. The Embassy reported that "it was clear that [MOFA] had given little thought about how the GOJ would undertake its own internal review of cooperation under the agreement", that the GOJ welcomed any inputs from the U.S. about possible new scientific programs and that these "would be taken by the GOJ as a very positive sign of U.S. interest in and support of the agreement especially in light of the OSTP attitude about possible termination of the 80STA and the successive USG promises to provide U.S. ideas followed by U.S. procrastination in the entire review process.[32]

Since perhaps many of the scientific and technical areas which the USG had been talking about may well fall within the purview of the S&TA, its representative in the Japanese Embassy in Washington, the science counselor, called on OSTP on September 11, 1986 to obtain a further clarification. The science counselor unintentionally, I am sure, reinforced a feeling that the GOJ had no intention of carrying out its promised review similar to the elaborate USG review when he said that "he did not believe the GOJ had initiated a similar review". He was, of course, immediately reminded that Ambassador Matsuda of MOFA, had "promised such a review". OSTP informed him that the internal review process would lead to technical discussions and a Joint Committee meeting next April 1987.[33] Based on interviews particularly with OSTP staff, Japanese reluctance, inertia, not to conduct this "multi-staged" review as the USG described it, caused a sense of the Japanese having "lied" and created a great sense of betrayal, and untrustworthiness. Unfortunately, probably Ambassador Matsuda somewhat casually said "yes, of course we'll review the situation" or more probably yes, I understand (*wakatta*, in Japanese) which would be even more ambiguous. He probably felt that a GOJ review would take place according to their schedule in their manner as they saw fit; it might or might not be the same or similar to the USG approach. Undoubtedly, he should have made much clearer to the OSTP representative who was much lower in rank than he (who was equivalent to Assistant Secretary) and a woman, what he was agreeing to. OSTP, on the other hand, mistook, misinterpreted, misjudged the diplomatic give and take, and undoubtedly made the assumption that the USG approach was thorough, straight forward, and rational; since undoubtedly OSTP thought they had carefully arranged the whole USG approach, an "affirmative answer" was naively taken to mean consent, agreement, acquiescence, and intention to follow suit. This is a good example in this process of how the mental attitudes, predispositions and language of the participants can and do result in most unfortunate interpretations and consequences.

The USG had repeatedly informed the GOJ that it was interested in the renewal of the agreement, both had exchanged ideas about the length of extension, yet the USG launched new studies each one more elaborate than the previous study and insisted that the assessment must be joint. The OSTP meeting with the Japanese science counselor left some unfortunate impressions on the latter. Probably this meeting prompted the GOJ/S&TA to request a meeting on September 29, 1986 with Embassy officials. The S&TA was concerned about the USG's thinking: did it want to make changes in the agreement as a condition for the agreed upon two-year extension. This did not seem to be an unreasonable request nor a particular intrusion into the USG review process, especially in light of past exchanges between the two governments. At the request of the S&TA, the Embassy science counselor and an OES staffer met with S&TA's Director of International Affairs about a schedule of review in the USG followed by discussions with the GOJ and noted that

- ...the tentative U.S. decision to support an extension of the UJST [80STA] *in its present form for two years* was based in large part on the need to have sufficient time to arrive at a JOINT DETERMINATION [emphasis added] about the future of the agreement.

- since the agreement was signed at the presidential-prime minister level, the projects to be carried out under it should be of special distinction.

- the technical level meeting indicated for later this year [1986] was planned to be a discussion of recommendations from the CISET and FCCSET review, *not a formal presentation of proposals for change* [Emphasis added]. The high level meeting next year [1987] might take up some of these recommendations but that will be determined only after working-level talks in which the Japanese will participate.[34]

As the elaborate CISET process got underway, the GOJ was informed that the USG had made a "tentative decision" to a two-year extension (earlier the USG had said it had agreed with the GOJ to such an extension), that the proposals emanating from the CISET review would be just proposals which the USG and GOJ would then jointly mull over at a technical level meeting in the fall to late 1986, and that the GOJ had done little, if any, substantive assessment for the revision of the 80STA. The latter inaction probably stemmed from the GOJ's thinking about the agreement, and the fact that it appeared the Division responsible for this work in various agencies had to be focused on the Antarctic Treaty Conference in October/November 1986, and be ready to respond to Diet deliberations.

When preparations were being made for OES Deputy Assistant Secretary Robert Morris to visit Tokyo for discussion on the U.S.-Japan Space Station agreement, the Tokyo Embassy commented that the above meeting with the GOJ's S&TA was probably prompted by a September 11 meeting between the Japanese Embassy science counselor's visit to OSTP which resulted in "differences" between what the OSTP had told the science counselor and what the Embassy had informed the GOJ/S&TA at the above September 30 meeting. These "differences" were not clearly stated by the Embassy.[35] The difference seemed to center around timing of the completion of the USG review, the kind of GOJ review which was supposed to take place which is mentioned in the talking points for the Morris meeting in Tokyo in October. This was just one more small State-OSTP friction which seemed to be growing. Since the USG consistently said it was preparing a set of proposals to discuss with the GOJ, it was not unusual that the Japanese Embassy science counselor would inquire of OES whether Morris was bringing "a paper" with him to discuss with the GOJ. OES informed him that the USG was not planning to give to or share with the GOJ "a paper" for joint review. OES did re-emphasize that "U.S. agreement to renewal of UJST [80STA] is conditioned on a review by both sides leading to proposals for refocusing and enhancing [U.S.-Japan S&T] cooperation."[36] It was not unusual that since the USG emphasized the need for a "joint review" the GOJ would expect to receive a proposal, or set of recommendations for future cooperation; this notion may perhaps have been further fostered by the Japanese science counselor's earlier visit to OSTP. After all, it was the USG that was pressing for the joint review exercise. The GOJ was evidently not rushing to conduct its own review. While the friction within the USG increased, the potential for misunderstanding between the USG and GOJ was indeed evident.

In preparing for this meeting a background paper was prepared reviewing the status of affairs. We reiterated to ourselves that the "planned route to renewal" was 1) to complete the internal reviews (the JTF meetings had been concluded, a final report was being prepared), 2) through informal discussions with the GOJ "to reaffirm the policy framework for the renewal process" (note here the use of "framework" which later arises as a source of considerable debate and probably some acrimony during the bilateral negotiations), 3) through a bilateral technical meeting to select proposals for enhancing the agreement, and 4) a Joint Committee meeting in April 1987 to ratify the selections. The renewal period -- between two to five years -- would be used for development and implementation of new projects. The paper warned that "The Japanese should be approached carefully about investing in U.S. R&D projects" and urged that "the U.S. should participate more in formal programs open to foreign scientists, including the Human Frontiers Programs, the Japan Trust and ERATO."

Ambassador Matsuda was informed that 1) the USG had obtained internal Circular 175 clearance to renew the 80STA, 2) the OSTP had asked that some members of the CISET Task Force visit Japan in November "to lay the policy groundwork for bilateral technical discussions to follow in January or February 1987 with a Joint Committee meeting in April 1987."[37] The purpose of this meeting was "to sensitize Japanese R&D officials to the thrust and overall rationale of the U.S position on the need to refocus this presidential level activity.[38] He, in turn, informed 1) the USG that the GOJ also had obtained its own internal authority to renew the agreement which, he hoped, would be a two year extension followed by an automatic renewal, 2) the proposed meeting in November "would be useful in providing an opportunity for the GOJ to learn about U.S. thinking on this matter [how to enhance the agreement], 3) the Japanese review had started "just some time ago" and 4) GOJ would be ready for the November meeting.[39] Only a few days later, the US science counselor and an OES representative met with S&TA on October 20 and AIST(MITI) officials on October 24. The AIST officials described how the GOJ was carrying out its review: looking for areas which would make the bilateral S&T cooperation more fruitful, how such areas might overlap with the U.S.-Japan Agreements on Natural Resources and Environment.

It is important to note that, for the first time, the USG gave each of the above three agencies, a "non-paper" on the proposed schedule to be followed following the completion and acceptance of the review by the CISET Executive Committee. A "non-paper" is a written statement which is intended to convey informally one side's ideas about a particular issue; it is strictly a document without status and informal. While a copy of this "non-paper" has not been found in Washington, it was prepared from a Washington cable to the Tokyo Embassy when the latter was informed that circular 175 authority had been approved.[40] Based on a reading of this cable the following scenario as given to the GOJ was constructed. A multi-stage process is envisioned:

1. beginning with internal reviews of the agreement on both sides

2. followed by informal discussions with Japanese R&D officials to set the POLICY FRAMEWORK [emphasis added] for the renewal process

3. a bilateral meeting would be convened to select proposals for enhancing the agreement. The first exchange of notes would formalize this stage

4. a joint committee meeting would be held in April 1987 for formal ratification of the new plan of cooperation during the renewal period, followed by the second exchange of notes extending the existing agreement.

The State cable commented that this renewal period must provide adequate lead time for new projects to be developed, implemented, and produce tangible research results for evaluation. From the perspective of R&D management, two years would be inadequate, thus suggesting that the probable length would be five years. It should also be emphasized that the USG envisioned not a new agreement, nor a major overhaul of the agreement but a revision that could be managed with an exchange of notes.

Within a short time, Washington cabled Tokyo Embassy that "due to unexpected scheduling conflicts OSTP cannot visit Tokyo in November as planned and suggested December 8-10, 1986, with the hope these meetings would initiate "a dialogue on the framework for discussions leading to a technical meeting early in 1987." [41] The visit was to be part of a trip to Beijing. Both GOJ and China expressed difficulties with these dates. On the U.S. side, the new science adviser, Dr. William Graham was just coming on board in October 1986 which also made a November visit to Tokyo improbable. Washington suggested that the visit to Japan be moved to February 2-4 in Tokyo (and February 9-11 in Beijing) and that a written version of the CISET Report would be made available to the GOJ before the February meeting.[42]

Thus, in conclusion, 1) a proposed timetable had been suggested by the USG, 2) after several delays the informal meeting with the GOJ is now set for early February 1987 and 3) the GOJ would be given a written version of the CISET proposal. It has taken approximately three years to reach this point where informal discussions of a USG proposal would take place with the GOJ.

c. The CISET Issues Paper

The CISET Issues Paper on U.S.-Japan Science and Technology relations was first discussed at a July 11, 1986 CISET meeting. After only one formal draft a final version was prepared on August 8. While the formal participants as members of CISET included representatives from all government agencies, the issue paper was described by OSTP as reflecting "the consensus of the key agencies participating in the review -- USTR, State, Commerce, NSF, and OSTP" an interesting omission of DOD and DOE which played important supportive roles of OSTP in its contest with State in 1987.[43] While the issue paper was distributed as an OSTP/CISET document, the authorship of the paper is not stated even in the first distributed draft. Judging from the generally moderate and constructive tone of the paper in contrast to the vituperative dialogue that erupted later in 1987, perhaps this paper was drafted initially by OES.

The Paper consisted of two parts, policy background (including "goals and issues for U.S.-Japan science and technology relations through the year 2000") and a CISET Technical review. This is the first time in the dialogue about the future of the 80STA that the stated thinking has been projected as far out as the first year of the 21st century. In this sense it is an unusual document in this policymaking process.

The policy background part retraces the U.S. foreign policy objective as the rebuilding of war-torn Japan thereby ensuring the country's future status as a prosperous democracy in firm partnership with the Western Alliance. For this purpose, "the access the U.S. and Western Europe afforded postwar Japan to their scientific and industrial base greatly contributed to the unqualified success of this foreign policy goal."[44] The following sentence from the July 23 draft was deleted from the final paper: "Ironically, the very success of the central US-Japan postwar war II foreign policy objective has led to serious strains between the two countries." It refers to President [Reagan's] commitment to a long term policy that will address the equity and "balance of responsibilities" in US-Japan relations via a process that will strengthen, not weaken, the overall relationship. The same growing awareness was also seen in Japan and that both the U.S. and Japan's mutual prosperity and security in the 21st century depends on how "we pursue shared opportunities and maximize our respective advantages."

> "International scientific and technological cooperation with our advanced partners, such as Japan, must be based on mutual interest and shared responsibility to advance that primary goal. Thus, the time has come to assess and redirect our scientific and technological relations with Japan to concentrate our cooperative activities on long-term efforts that will attack problems at the frontiers of knowledge and advance our respective domestic research agendas in both the basic and applied areas. Such an assessment would also provide an opportunity to resolve the issue of reciprocity in scientific exchange between the two countries and increases the participation of US scientists in Japanese research institutions." [45]

As far as the S&T relations since 1945 are concerned, these

> "interactions have provided tangible, shared benefits and demonstrated a cost-effective pooling of resources, facilities and talent, such as in the areas of earthquake engineering, biomedical research, and fusion." [46]
> "However, the U.S. and Japan have never had substantial discussions about the possibility of formal cooperative agreements in several key areas of science and technology, such as robotics and ocean mining, where the Japanese possess considerable expertise and comparative advantage."[47]

Unfortunately without giving specific evidence on a serious accusation, the paper asserts that "A goodly number of new technologies and processes developed in U.S. facilities have been transferred to Japan and commercially exploited by Japanese industry." Indeed again without proof, it states that "a number of Japanese patents now licensed by U.S. firms encompass technologies first developed in U.S. national laboratories." It fails to point out that similar U.S. firms had exactly the same unfettered opportunity to exploit the same fruits from U.S. government research institutions and apparently did not.

It reports that the Japanese are aware of the rising concern in the U.S. S&T community and USG circles over the "equity-access issue", have made an increase of long term governmental support for fundamental research as a basic thrust of the government, have expressed this new orientation by proposing the "globalization of science and technology" as one of the principal theses for the OECD Science Policy Ministerial in 1987.

The paper set four "long term, mutually beneficial goals" for U.S.--Japan S&T relations through the year 2000:

- To increase the world's knowledge base through joint research in areas where pooling of the two countries' resources, facilities, and talent can accelerate the rate of discovery;

- to expand the global economy so as to enhance the mutual security, economic and physical well being of the people of both countries;

- to enhance the domestic R&D capabilities of both countries; and

- to improve reciprocity in the exchange of personnel and the use of research facilities, and to enhance mutual access to emerging trends and research results in both countries.[48]

In order to achieve these goals it raised 7 issues which the JTF must consider in creating a program for future U.S.-Japan S&T relations.

Under part 2, CISET Technical Review, the initial U.S. goal in negotiating the 80STA, was to "obtain commitments from the Japanese for cost sharing in high cost, high risk projects" the underlying premise and rationalization of which was that since Japan had achieved parity in technical capabilities, "further cooperation at the government level in S&T should be based on an equitable sharing of information, expertise and costs." It did not, of course, mention that many of these U.S. projects were those that were not high enough priority to receive federal funding and that no funding of Japanese S&T projects as part of this reciprocity was ever considered (although recommended by the Tokyo Embassy). It pointed out that the projects under the 80STA were not of Presidential stature without specifying what that means.

The so-called bottom-line product, "the immediate end product of the review" should be a set of recommendations aimed at beginning to refocus and upgrade US-Japan S&T cooperation within the context of the next two-year phase of the [1980] Agreement [extending it from 1987 to 1989]. Those recommendations should be cast in terms of a limited number of high priority research programs which would constitute the framework for a Presidential level science and technology initiative. Pursuant to [Reagan] Administration policy, projects recommended for this initiative should have "significant potential for advancing the U.S. domestic research agenda through international sharing of costs, facilities, and talent."[49]

The conclusion made two important comments, one distinctly delimiting the above request and the other setting out a long term goal and expression of U.S. ideological idealism for future international negotiations for S&T agreements:

- The requested set of recommendations described above was to be "aimed at a MODERATE THOROUGH SIGNIFICANT UPGRADING [emphasis added] of the 80STA."

- The "more long term objective is to initiate a process through which the U.S. Government in cooperation with industry and academia, can restructure science and technology relations with our friends and allies throughout the world in order to increase the general store of knowledge and make effective use of that knowledge for enhanced prosperity and mutual security."[50]

d. The CISET Joint Task Force Report

The JTF formally met as a group three times (July 22, August 8, September 9, 1986), received written suggestions from participating U.S. agencies on proposed cooperative research areas and generic problems affecting the bilateral S&T relationship and suggestions (some written) from the one and only meeting with the private sector on September 12, 1986 which was organized by OSTP. The JTF was composed of representatives from the Departments of Agriculture, Commerce, Defense, Energy, and State, the Central Intelligence Agency, Environmental Protection Agency, U.S. Geological Survey, National Aeronautics and Space Administration, National Bureau of Standards (since changed to National Institute of Standards and Technology) and National Oceanic and the Atmospheric Administration (both part of DOC but represented separately), National Science Foundation, Public Health Service (part of Department of Health and Human Services). The two Task Force coordinators were from NSF and the State Department. Note that as the agency creating the JTF, OSTP itself was not represented on the JTF.[51]

The JTF mission or task is mentioned again because there is a slightly different emphasis than that mentioned in the issues paper discussed above. These tasks were to:

- identify specific scientific and technological areas in which enhanced cooperative research would be beneficial to both the U.S. and Japan.

- explore feasible means for involving U.S. industry in setting the requirements for cooperative research under the bilateral agreement and making more effective use of the results obtained from that research.

- identify likely opportunities for resolving generic barriers to enhanced bilateral cooperation between the U.S. and Japan.

The overview pointed out that the CISET Executive Committee felt that the renewal of the 80STA was an opportune occasion "to initiate a dialogue with Japan" aimed at a long term strategy to achieve mutually beneficial goals because of its high profile nature and because the agreement "is INTENDED TO COVER ALL FIELDS OF SCIENCE AND TECHNOLOGY" [emphasis added], i.e. the umbrella concept which has been controversial in the U.S. and in Japan from the earliest discussions about this agreement and which has never been conclusively resolved.

The JTF report suggested five "procedural recommendations as guides to discussions prior to formal renewal of the 80STA" -- not a revision nor a new agreement – in the following areas:

1. create a permanent interagency executive committee at a senior level to plan and execute further cooperation with Japan.

2. all *new* bilateral R&D projects/programs be considered for inclusion under the umbrella of this agreement. This is described by the report as "a fundamental principle" in the management of the renewed STA.

3. propose to the GOJ that a similar interagency executive committee be created by the GOJ.

4. propose the following areas of expanded R&D cooperation: materials science and engineering (particularly electronic materials, life sciences (particularly biotechnology), information science and technology, automation and process control, artificial intelligence (including software development), global geosciences, joint database development, and joint standardization and nomenclature development.

5. each agency responsible for a project in a specific area would establish an industry advisory group for that technical specialty.

Let us now turn to the details of the report by section:

A. Management of the Agreement. In order to place the present assessment in proper perspective, it provided a short but frank commentary on the shortcomings of the implementation of the 80STA (which will not be repeated here since they have been discussed in an earlier chapter). The report looked upon the renewed agreement "to be a potentially useful device for gaining legitimate access to the results of Japanese research and development" which would also contribute to satisfying the demands for making the U.S. more competitive; that the U.S. should "take advantage" of the new GOJ policy to expand Japanese participation in international science and technology.

Generic issues. It listed five generic issues:

- the need for encouraging and facilitating long-term visits to Japanese laboratories by U.S. scientists and engineers. It did not define long term but that probably meant six months or longer. It is noteworthy in light of the atmosphere and how the issues were handled by the USG in negotiations in 1987--1988 with the GOJ that it did not accuse the Japanese of denying U.S. researchers but stressed the need for encouraging and facilitating these visits.

- obtaining "equitable access" to Japanese data and laboratories. This was a complex matter and "needs further clarification." It did comment that "U.S. scientists have had good access under terms of the bilateral agreement managed by NSF. Making available more scientific and technical data and information became a major Japanese governmental effort and a major focus of the renewed STA.

- opportunities for making better use of underutilized facilities, instrumentation in both countries.

- opportunities for closer cooperation through data base development, standardization and exchange.

- resolving differences in laws and practices in the areas of regulation, liability and intellectual property protection.

S&T areas of cooperation. It recommended in explicit terms, citing again the political significance of the STA, that "the substantive thrust of its [the renewed STA] implementation be limited to a few high priority areas of science and technology".

Creation of a coordinating interagency executive committee. A new formal organization at the Deputy Assistant Secretary level to plan and execute further cooperation should be created by OSTP and that the creation of a similar organization be urged upon the GOJ. The membership of the U.S. committee would be basically the same as the JTF. This committee was envisioned as having seven functions: assess current activities to assure that they are consistent with "the Presidential status of the agreement" - a term that is repeated time and again, to the extent that it begins to look like an obsession, to periodically review activities and resolve problems, develop government wide policies, to act as the "official, acknowledged representative of the USG in establishing and maintaining coordination with industry associations having interests in Japanese science and technology whether under this STA or any other S&T bilateral agreement, to provide scientific representation to periodic meetings with Japanese, determine which USG agencies would take the lead when the selected fields cut across agency responsibilities, and ensure that other existing bilateral or multilateral STAs are recognized and protected."

B. Priority Areas of S&T cooperation. The report commented that some agencies responded to requests for suggested areas of cooperation with Japan, some did not, and some "held negative views toward further cooperation in fields judged to be sensitive or at the leading edge in the U.S." The report attached a summary of agencies responses. The chosen fields were the same as those listed earlier and noted that they "coincide[d] well with the priorities for R&D that have been established by the Japanese government." These proposed specialty areas do not come within the purview of any U.S. agency, thus working in these areas would require close interagency coordination. In contrast, the GOJ has a number of programs along the lines of some of these specialties. In some instances the GOJ has invited foreign participation but, says the report, "to which the U.S. has not responded" -- usually because the U.S. feels that its technology is ahead and cooperation would perhaps be disadvantageous. Yet, it should be kept in mind that the U.S. had been pressing the Japanese to take a more international, broader, and more fundamental interests in basic knowledge.

C. Dialogue with the Private Sector. The report recommended that 1) under the renewed agreement, the U.S. government agency given responsibility for coordinating a broad area of technology cooperation with Japan form an industry advisory group, 2) this responsibility could be given to government contractor organizations specifying that information from Japan be disseminated to interested parties in the U.S. It commented that this approach would dovetail well with the Japanese approach and gave several examples of how this approach is presently working in the U.S., and 3) the OSTP establish as an integral component of the USG's management plan for the renewed agreement "a blue ribbon, non-governmental policy oversight panel with representation from both industry and universities whose principal charge would be to develop and assess long-range directions for bilateral relations between the two countries."

One of the two appendices contained a summary of suggestions from U.S. technical agencies. Only some of the comments and suggestions will be cited here.

NBS (later renamed National Institute of Standards and Technology). "Areas such as automation and biotechnology are also quite sensitive and involvement of foreign nationals is discouraged." This appears, on the surface, a strange commentary when one of the major areas of potential technical cooperation was mentioned as automation.

NSF. On the other hand, this agency suggests robots capable of functioning in hostile environments, automated factory and flexible manufacturing, as possible areas of collaborative research. In contrast to NBS, NSF states that "cooperation in many areas [of biotechnology] would be profitable for both the U.S. and Japan." NSF says the U.S. should test the willingness of the

Japanese to open selected laboratories or areas of cooperation "WHERE WE REQUIRE ACCESS" [emphasis added]. This is a rather unusual choice of words, especially for the NSF. In what way, with what justification can the U.S. "require access" to the laboratories of another country. The U.S. would be somewhat irritated if another country said it "required access" to selected U.S. laboratories.

The Public Health Service. In light of the controversy that later developed about the Japanese scientists in NIH, it is significant that NIH states that "There is a STRONG CONSENSUS [emphasis added] that the non-energy agreement provides an important mechanism for involving Japanese scientists in these cooperative efforts." It pointed out while the Japanese can take advantage of several NIH programs to bring foreign biomedical scientists to the Bethesda Laboratory there are no reciprocal Japanese supported programs for foreign biomedical scientists. There was no mention at all that there was a demand or felt need on the U.S. part for such action by the Japanese nor that the Japanese had been asked and had rejected such requests. This is an important omission in light of later developments.

The report recommended an elaborate and complicated vertically tiered institutional arrangement augmented by supplemental horizontal arrangements with industry. There is a certain intellectual fascination with this proposal yet bureaucratically it did not seem a particularly rational approach to handling the broad, intensive and extensive U.S.-Japan S&T relationship. The JTF exercise 1) is perhaps a manifestation of the frustration felt by many agencies, particularly the OSTP with the obvious inability of the OES (even with State Japan Desk and Tokyo embassy support, and that of many technical U.S. agencies also) to mobilize sufficient support, will power, bureaucratic clout to forge or bring forth a decision, 2) was an adroit move by the OSTP to use a newly invigorated FCCSET/CISET operation to "force" an interagency decision on the future of the 80STA, 3) is a manifestation of USG's anxiety about the implication of an S&T "challenge" (stated positively and constructively) or a "threat" (stated negatively and destructively).

It was suggested that the ideas emanating from the JTF report could be used as a model for negotiating future STAs with other countries. The attempt to create such a complicated hierarchical arrangement many times over with numerous countries would create unmanageable institutional arrangements for other countries and, IN PARTICULAR, for the USG, i.e. the OES and OSTP. The OSTP, as a staff organization in the Executive Office of the President, was over-burdened and unable to cope with the requirements of handling the relations with one country (Japan) since 1980. OES in State would be similarly over-burdened with lack of staff, funds, status and hierarchical clout within the State Department.

While the Report responded to its mandate, it is nevertheless significant that 1) intellectual property rights and trade secrets, and 2) the relation of the results of joint R&D projects to national security, were only tangentially discussed.

The JTF report was submitted to CISET's Executive Committee which consisted of the very same agencies which had drafted the report in the first instance. These issues later became hotly debated topics in the USG internally and in bilateral negotiations, yet they were completely ignored at this seemingly crucial stage in creating an inter-agency position. This silence, despite the presence on the JTF of representatives from the Departments of Commerce and Defense, is noteworthy.

Earlier there were reports that a number of US agencies would be quite amenable to shifting the responsibility for the STA in discussions with the GOJ from OSTP to OES; the Report, however, recommended that OSTP continue to play this role, and took note of the fact that OSTP (by implication the Science Adviser heading a delegation) was planning to have an inter-agency team visit Japan in February 1987 to discuss the JTF report's recommendations. This stance was obviously a continuation of the arrangement reached in 1980 when the 80STA was signed that OSTP, at its insistence, would represent the USG in discussions with the GOJ. Nevertheless, State/OES would be responsible for the ACTUAL negotiations of a revised or new agreement in contrast to such preliminary consultations. Such a bifurcated representation invites confusion and conflict in the USG and naturally confusion in foreign governments.

Throughout the discussions in the USG since 1983 about the future of the 80STA, there was never any mention of a new STA, rather the implication was that the existing agreement would be modified to "strengthen" certain articles and clauses in the agreement. All the recommendations of the JTF report could be accommodated within the legal interpretation of the existing STA. The two governments could readily implement the recommendations under the agreement without any revisions. That was the preferred position of the State Department, the Tokyo Embassy and the GOJ. That would not be politically acceptable to OSTP, and to OSTP alone. In the numerous discussions with the GOJ up to this time, the USG repeatedly informed the GOJ that it would like to "strengthen" the agreement, especially to assure that scientific and technical areas or fields which have "Presidential-Prime Ministerial stature" are included in an amended agreement, not a "new" agreement. It would obviate any possible need to take this agreement to the U.S. Congress or to the Japanese Diet for approval, especially since the original 1980 agreement was not so handled. There was also no suggestion in the discussions with the GOJ that the USG was interested in pursuing the intellectual property rights issue, at least not any more than the passing and fleeting mention in article V.2 in the 80STA.

6. A Consultation with the Private Sector

OSTP convened a meeting with industry representatives on September 12, 1986 to discuss the future of U.S.-Japan S&T relations. It is not known who and how many were invited to attend. The writer was given copies of letters from Rockwell International and Allied Bendix Aerospace. The former suggested a "U.S.-Japan Commission on the Application of Science to Human Needs". This commission, composed of CEOs and their chief technology executives, would be charged with studying ways in which science can be most effectively translated into useful goods and services. The establishment of one or more joint centers for this application was also suggested. The latter suggested not "renegotiating" the agreement but creating an Implementation Annex under Article II.a of the 80STA. This annex would have two objectives: to provide for a methodology for the identification of projects which would have mutual benefits and a mechanism for implementation. None of these suggestions were adopted or adapted for inclusion in the review process.

The convening of such a meeting seemed more like a sop to industry to give them a sense of participation in the review process and fulfill an admonition in the 1976 report on *Technology and Foreign Affairs* by T. Keith Glennan to Deputy Secretary of State to "develop closer liaison with U.S. industry to ensure maximum benefits on matters of common interest" -- if, perchance, one was aware of such a report which was issued in December 1976.

Based on available documents and interviews, this was the only meeting with U.S. industry representatives. It seemed to have had no impact or influence on the USG review process or on the subsequent discussions with the GOJ. There is no evidence that a similar meeting was formally or informally convened with academics.

7. Discussions with GOJ on Revewal of the 1980 STA

The CISET JTF report was moderate, constructive and rather frank considering it was an interagency report making recommendations on how to resolve the renewal of the 80STA which had eluded the seemingly perpetual dialogue and sequential studies by State/OES. It should be remembered that the JTF was coordinated by State and NSF representatives whose agencies have a quite different outlook on S&T relations with Japan than OSTP. NSF has had harmonious and mutually beneficial programs and projects with many Japanese agencies; State (in cooperation with the Tokyo Embassy) maintained a "political" approach to bilateral S&T relations. Furthermore, the report was a task force report from a CISET organization to CISET's executive committee chaired by OSTP. From the beginning of the renewal process in the mid-1980s, there was an undercurrent of potential conflict between the outlooks of State and OSTP, with other agencies appearing to be somewhat passive bystanders. These differences have been mentioned at various points in this study. State (and the Tokyo Embassy) and a number of U.S. departments and technical agencies based on their existing and satisfactory cooperation with the Japanese wished to renew the 80STA in its existing form while the OSTP was negative and skeptical about the U.S.-Japan S&T relations even entertaining thoughts of terminating the agreement. As one interviewee who was involved in the early discussions about the possible renewal of the 80STA said, "The conflict between the OSTP and State simmered below the surface and was well managed."

The report's recommendations were directed to OSTP for it to discuss with the GOJ. The report contained a rather bureaucratically unusual but short section "Negotiations with Japan on Renewal of the Agreement". The report noted that the OSTP headed by the Science Adviser together with representatives of other U.S. agencies would be visiting Tokyo to explain the U.S. approach to the renewal. This short section had but one purpose: The Task Force [which did not have OSTP representation] agreed that it should recommend to OSTP that a "positive, constructive attitude should be shown to the Japanese side." This recommendation was justified by the fact that mutual benefit would result from strengthening the S&T bilateral relationship, provided the recommendations of the JTF are incorporated in the renewed agreement. Usage of this kind of language in a report admonishing the very agency to which it was making the report was, to say the least, unusual. It would seem to reflect the depth of the feelings among the many agencies vis-a-vis the OSTP. One could justifiably say that the simmering had now turned into a rumbling which manifested itself in extraordinary ways as the process proceeded to a decisional climax in early 1987.

8. Interagency Coordination for the February 1987 Presentation to the GOJ

The JTF had prepared a report with recommendations which it was assumed would be the basis of informal discussion -- not negotiations -- with the GOJ. As intimated earlier, relations between the OSTP and OES (and State in general) were on a steep downward slope. Although the agenda was not known, there was a meeting between the Secretary of State, OSTP Director, and OES A/S in the first two weeks of November 1986. Following this meeting, there was to be an another meeting between the OES A/S and Director OSTP. In the memo, briefing paper and talking points preparing the OES A/S for this meeting, the depth of these antagonistic emotions and differences in outlook concerning are starkly evident. The memo begins ominously with:

"Following your good meeting with Graham [Director, OSTP] and the Secretary at which Graham sounded sincere in wanting to improve State--OSTP relations. ..." the background paper and talking points:

- noted that "OSTP has indicated its general satisfaction with the [JTF] proposal ... "

- stated that at "OSTP's insistence" the USG proposed to and the GOJ agreed to meet in November; this was then changed to mid-December because of the arrival of a new OSTP Director; this was again postponed by OSTP to February 1987.

- voiced OES and the Japan Desk's concern with these continued delays "which follow a pattern of poor communication on this particular agreement which will UNDERMINE [emphasis added] the credibility which we have worked hard in recent months to reestablish [with the GOJ]. The result could well be a Japanese conclusion that we are no more serious about this now than in the past. This may lead them [GOJ] to engage in delaying tactics rather than serious negotiation."

- reported that "these points [were] forcefully made to [OSTP]." OSTP's response was that the new science adviser's arrival overwhelms OSTP workload. The agreement was that a written proposal would be made available to the Japanese in early December and that "OSTP makes a firm commitment not to cause further delays." If this is impossible, State will "reluctantly agree to postpone the meeting until January 1987."[52]

In the midst of this gathering storm, OES DAS meets with Hinata of MOFA who was in charge of the Division responsible for negotiating the renewal of the 80STA on December 1, 1986. He is informed that the USG will have an approved version of its proposal by December 15, 1986 to give to the GOJ. The "principal elements" are described as 1) the establishment of a formal interagency management structure on the U.S. side to coordinate R&D programs, with the GOJ being asked to create a corresponding structure on its side, 2) the designation of specific high priority technical fields, 3) an umbrella agreement where all new projects would be considered for inclusion under this agreement. Interestingly, the talking points included regrets for having changed this meeting several times due to changes at "the White House" with the recent arrival of a new science adviser, and that "the White House" had agreed to a meeting in Tokyo in February 1987. For the first time "The White House" was used instead of OSTP, another manifestation of the severely strained relations between OSTP and State. The frayed internal relations in the USG must have been obvious to the GOJ.[53]

Another delay came with the postponement of the CISET meeting on December 3 to discuss and endorse the JTF report. An OES memo to OES A/S of December 2, 1986 complained that the OSTP delays are now jeopardizing the December 15 deadline to give a proposal to the GOJ by that date. What can be done to "avoid another fiasco in U.S.-Japan S&T relations." Based on the record, it appears a letter was sent by OES A/S to the Science Adviser urging a decision on this matter. An unsigned letter from the Science Adviser on December 29, 1986, concurring with the recommendations of the CISET report, concerning the creation of an executive committee chaired by OES.

Again, the deadline for producing a proposal to give the GOJ came and went on December 15, 1986. A Japan Desk memo of January 9, 1987 reported that despite OSTP's general satisfaction with the JTF report mentioned above, the science adviser was "still studying" the proposal and "now tells us that he wants to add a set of overall objectives to be obtained in our negotiations with the Japanese." The background paper attached to this memo is a litany of the problems with OSTP going back to the early considerations in renewing the 80STA.

Amidst this confusion and delay, the Japanese State Minister for Science Mitsubayashi met with OES A/S on January 12, 1987 to discuss the future of the 80STA, Human Frontiers Science Program and the Space Station cooperation agreement. Again, the background papers described the Japanese as "anxiously awaiting more detailed information on our proposals which have been discussed with them only in very general terms to date." The talking points with the Minister covered the same points as before (i.e. a new interagency committee on the U.S. side, asks the GOJ to do the same, specific areas of technical cooperation without naming the fields, and inclusion of all projects under the STA), and to express a hope that written proposals will be given to the GOJ in the near future and that "the White House position will be elaborated by Dr. Graham during his meeting with you later today." -- as if there is a separate view.[54] Why did not this meeting be a joint meeting to express a joint, unified U.S. position at one meeting instead of two separate meetings at two U.S. agencies, one with functional and legal responsibility in the S&T area and one a staff agency of the President with responsibilities in coordinating Federal S&T activities in the international arena.[55]

As head of the U.S. delegation, the science adviser was briefed on January 15, 1987 by the Japan Desk in the State Department on political/security issues, trade, and science and technology. The political security situation was described in glowing terms "as harmonious as we have seen" in contrast to later assessments that 1987 was the worst year for U.S.-Japan relationships since 1945. Concerning trade it recognized the seriousness of the trade friction but described "many of our trade problems of our own making, we must deal with budget deficit, labor-management relations, low productivity growth."

As for talking points in the S&T area, Japan was described as having come into her own in technology development in recent years with 2.77% of GNP ($40 Billion) in R&D, 21% funded by the government, 79% by the private sector, 13% in basic research, 25% in applied research and 62% in development. It stressed that Japan appears willing to participate in international S&T work on a cooperative basis [and] what's needed now was to identify practical means for such cooperation, to avoid past mistakes of seeking GOJ funding for grand projects [it was this very desire which the GOJ turned down in 1980 and 1981 that persuaded OSTP to consider terminating the 80STA], to support the CISET paper idea of renewing the Agreement [the 80STA] and setting up a mechanism to identify areas for and encourage cooperation. [This is a request since presumably the OSTP had already agreed to the CISET report].[56]

OES A/S suggested a two-stage process to the science adviser for him to secure Japanese agreement in principle on the objectives to be sought through renewal and then he [OES A/S] would undertake to negotiate the precise terms and substance of the renewal in more intensive discussions ... as they are prepared to begin. "Your goal" said the OES A/S "should be to secure high level Japanese approval" of three objectives:

1. Agreement on a set of principles for the implementation of U.S.-Japan S&T relationships in an equitable and balanced fashion.

2. Identification of mutually acceptable areas of cooperation in leading high technology fields to be given priority emphasis under the agreement. These areas are: material science and engineering (particularly electronic materials), life sciences, particularly biotechnology, information sciences and technology, automation and process control, artificial intelligence (including software development), global geosciences, joint database development, joint standardization and nomenclature development (same as those used in the CISET Report).

3. Improvement of the management and coordination mechanisms associated with the agreement to assure that our mutual science and technology goals are achieved in the most effective and efficient manner.[57]

Parallel to OES endeavors and briefing papers, OSTP also produced its version, scope papers, outlook, proposals, goals and objectives as mentioned earlier.[58] Several new emphases and ideas began creeping into these papers. In these drafts the term "policy framework" begins to appear as the embodiment of the group of ideas the science adviser hoped to persuade the GOJ to accept. In contrast, in what may be an early draft, the OES was still asserting that "the technical scope of the agreement is largely duplicative of the 12 other U.S.-Japan bilateral S&T agreements, some of which have been in effect for as long as 25 years and is consequently seen as superfluous by many U.S. technical agencies."[59] As these talks began there were 25 active projects under the 80STA. They were being managed by NASA, DOE, HHS, and USDA. Although important from a scientific point of view with the exception of several NASA projects none of these are of a high cost, high impact nature and could easily be subsumed under existing agreements these agencies have with counterpart Japanese agencies. The outlines of the policy framework, objectives, definitive statement to be sought from the GOJ as summarized from these OSTP documents included the following seven key objectives:

1. Make the revised STA the definitive statement of government and substantive priorities in this relationship, as the "USJ flagship S&T agreement" in keeping with its political visibility and strong symbolism as the only S&T agreement concluded with Japan at the Presidential level. This is a distinctly new and more forceful interpretation.

2. Reiteration of the need to improve utilization of existing scientific resources, data and facilities, increased access by U.S. scientists and engineers at junior and senior levels to Japanese research and training facilities and centers of excellence.

3. The creation of a "Bi-national Blue Ribbon Advisory Panel" of members from industry, academia and government to facilitate dialogue and examine long-range directions, and propose recommendations to their respective governments for future cooperation. This idea has been fleetingly mentioned in the past but has now entered the dialogue about the revision of the 80STA as a major issue.

4. Improve utilization of S&T information, data and facilities.

5. High priority scientific and technical areas have been mentioned from the beginning but now, for the first time, must be of critical importance to each country's competitiveness and technological prowess, both partners must possess strong, complimentary research capabilities and adequate resource bases and facilities to engage in equitable joint ventures, that this cooperation will significantly accelerate the rate of scientific progress compared to what is achievable by national efforts alone.

6. The senior interagency executive committee would ensure the attainment of national and scientific goals, examine broad goals and specific policy objectives of any proposed bilateral project, and, on the U.S. side, would make recommendations to the Science Adviser before any bilateral negotiations are begun. The GOJ would also be encouraged to create such an organization.

7. For the first time, an annex was proposed to cover intellectual property and patent rights "as a broad policy framework" for specific clauses which will be drawn up as needed to cover particular technical areas of cooperation.

9. The February 1987 Meeting in Tokyo

The U.S. delegation met with GOJ officials February 2-7, 1987. The senior members of the U.S. delegation were: Dr. William Graham, science adviser and chairman of the delegation, OES A/S Ambassador John D. Negroponte, and Commerce A/S Bruce Merrifield, and Dr. John Moore, Deputy Director of NSF. Since NSF had played a less than major role so far in this renewal exercise, this team was now headed by two persons, almost of co-equal rank, from two agencies who are not exactly the best of friends, who have quite conflicting views concerning policy vis-a-vis Japan for S&T purposes. The US Team met with the Ministers of Foreign Affairs, Education, Science and Technology, Director General of AIST/MITI, some members of the Prime Minister's Council on Science and Technology and selected Diet members, the press, the American Chamber of Commerce in Japan, and experts in selected research facilities.

After the February 1987 visit, four reports were prepared about this meeting: the customary Embassy report and three "trip reports" by the senior members of the U.S. Team.[60] In reading these reports, I was reminded of the Japanese movie, *Rashomon*, where several witnesses to a murder in a forest, including the murderer, the husband, the ghost of the victim his wife and a vagabond each related "the truth". Each report will be summarized here because there are subtle differences and emphases among them: one a straight forward report, a "report to the U.S. President" with a specific agenda in mind, a State Department oriented report to the SecState short and sweet and to the point, and finally a more philosophical analysis of the innovation process and its relation to U.S.-Japan S&T relations.

American Embassy Report. Unlike the other reports by the senior

Team members, the Embassy report was on the agreement and on meetings with GOJ Ministries.

- Both sides agreed to start "detailed negotiations" when the USG Team returns to Tokyo in early April 1987.

- In the meantime, the USG will provide papers to GOJ describing USG proposals on IPR, topics for future joint collaboration, equitable balance and reciprocity in technical exchanges and a new advisory structure. The GOJ was also invited to prepare similar papers.

- USG stressed the "need to make fundamental changes in the nature of the projects and in management structure."

- GOJ's future participation in the SSC would be welcomed.

- MITI, MOE, MOFA stressed the need to strengthen 80STA Joint Committee [not a radical restructuring as suggested by USG], Prime Minister's Council on S&T could serve as Japan's "blue ribbon advisory committee"; willingness to support more Japanese language instruction in the U.S. and Europe and that GOJ had not had an opportunity to coordinate an inter-agency response to USG proposals.[61]

Dr. William Graham's report. As science adviser (and head of delegation), his report was addressed to the U.S. President [Reagan]. As though to imply where his emphasis and concern lay, he focused in the second paragraph of the two-page report, on the imbalance in scientific and technical exchange, "for every five researchers that the Japanese send to the USA or Europe, only one researcher from these countries is placed in Japan.[62] The "basic message" which Graham said he delivered to each senior Japanese government official was that:

- both countries "must contribute to the generation of the world's basic scientific knowledge and the creation of new fundamental technologies so critical to our competitiveness and global security in the 21st century."

- only by "providing *balanced access* to our respective R&D systems can we redress the imbalances of our past relationship" in S&T.

- "I strongly encouraged Japanese participation" in the "world class scientific project", the SSC and "received indications that the Japanese may be willing to support unexpectedly large share of that project". That was indeed a poor reading of the situation since the GOJ procrastinated about any final decision. The U.S. Congress canceled the project in 1993. In hindsight the GOJ was accidentally accurate in procrastinating. The State Department had recommended that, in light of past unfortunate experiences, the USG not pursue the GOJ for monetary support of the SSC. This advice was consistently and regularly ignored by OSTP.

In a somewhat incidental manner, Graham reported that he had apprized the GOJ officials that "a fair trade environment for all parties, a keystone of your [President Reagan's] trade policy would color aspects of our relationships in the civil area, including science and technology cooperation." This was the first time that this linkage was formally made (at least according to available documents) in discussions with GOJ officials at the Ministerial level.

Except for mentioning "probable" financial support of the SSC, there was no mention at all of the great concern for "presidential" projects in selected technologies which had figured so prominently in earlier discussions among USG officials as they prepared for meetings with Japanese. It surely cannot be regarded as an accidental omission if it was truly an issue of the first order, as it seemed to be, in the inter-agency discussions, especially by OSTP staff at these meetings. In addition, there was no mention of intellectual property rights and the need or how to include such topics in any revised agreement, or any national security concerns that may emanate from joint R&D projects. Graham concluded that the Japanese officials he met "appeared to understand the need for strong reciprocity and mutual benefit in a forthcoming agreement."

In addition, a 12-page "non-paper" on US-Japan S&T Cooperation was prepared, presumably by OSTP. It was given to this writer as a packet with the report to the President and was printed on the same kind of paper and using the same font style. A "non-paper" is generally regarded as an official paper whose purpose is to be used for information, reference, an unofficial opinion paper without official standing. It is not clear why this "non-paper" was prepared at this time; judging from the content described below it seemed to be a document prepared to brief the President perhaps, more likely his non-technical, non-scientific staff of S&T technology funding, staffing, education, etc in the U.S. and Japan or perhaps as another background interpretation document from which one might deduce a position on USJ S&T relations. The assertions and data are not supported by footnotes, nor is there any mention from where these interpretations and data came. It is fairly obvious that one of the purposes of the "non-paper" was to give the impression that Japanese scientists and engineers were burrowing into every nook and cranny of the U.S. R&D System to the U.S.'s disadvantage no matter how and why these Japanese scientists and engineers came to be doing this research in the U.S. "The American taxpayer is providing a significant subsidy to the higher education of every foreign student in the U.S." (p. 7)

The important points and assertions of this paper are given in some detail here as an indication of how one agency looked upon the USJ S&T relationship, just as the first serious interagency USG Team went to Japan how it interpreted certain facts about Japanese and U.S. education, R&D systems, their participation in each other's R&D system. Whether mistaken or not, accurate or not, the interpretations of the "selected facts" provided, it seems, the rationale and driving force for certain specific ideas which OSTP pushed with determination -- in contrast to the lack of a detectable core philosophy except to maintain the status quo among other government departments and agencies, especially the State Department.

The non-paper briefly described the support of basic research in the U.S. and Japan. It mentioned that Japan had increased its government basic research budget by 12% between 1979 and 1985 but that support for industrial research had increased by 76%. Concerning basic research support, it did cite the *1986 Japanese White Paper on Science and Technology* on basic research: The strengthening of basic research and the development of human-friendly science and technology along with internationalization have become the basic priorities of the government's policy on science and technology. It also quoted from the MITI R&D 1986 budget: In the past, Japan was able to achieve economic prosperity by importing technologies actively from advanced industrial countries. Now that Japan has become a leading member of the international community of nations, it is necessary for it to promote technical development by giving full play to its own creative initiative and resourcefulness.

It then described science and technology education and facilities in the U.S. and Japan. Substantial space was devoted to foreign and U.S. participation in the Japanese R&D system and foreign participation in the U.S. system and ends with the major R&D initiatives of the U.S. and Japan. In the last section, U.S. and Japanese participation in R&D projects of international organizations, and under the framework of the 1981 Versailles Summit, listed

U.S. projects which "involve potential participation of U.S. allies": manned space station, the space telescope, the Strategic Defense Initiative, AIDS research, mapping the human genome, new fusion programs, the Doublett III-D project and new facility at Princeton, Engineering Research Centers, and the SSC and the new Science and Technology Centers. While the projects are understandable, the several kinds of centers which are strictly U.S. domestic educational projects are strange to see on this list. A similar list of Japanese projects also involving "potential participation of Japanese allies" was given: ERATO for exploratory research in advanced technologies, the Japan Key Technology Center, the Large Scale National R&D Program, JAERI's Fusion program and the Tristan High Energy Physics facility. Similarly, it is difficult to understand why the Key Technology Center is mentioned. It is a semi-governmental organization which provides funds for research to Japanese corporations, and only to a limited extent to foreign participation.

"No data currently exists to measure exact numbers across all sectors of Japanese participation in the U.S. R&D system" (p. 7). Much raw data, i.e. large numbers of students, researchers from Japan attending U.S. institutions are given without any explanations why this is occurring. For example, (1) Japanese students studying in the U.S. greatly increased from the mid-1970s, 25% increase to 13,360 in 1985, which is about 8.5 times the number of Japanese students 30 years ago. It ignores the fact that a large percentage of these students are undergraduates whose parents can afford to send their children abroad for training which was not even considered 30 years ago (p. 7). (2) In 1985 and 1986, there were 316 and 323 Japanese researchers, respectively, at NIH wholly funded by the U.S. NIH (p. 8). But it failed to point out that these researchers applied to NIH for these assignments *in competition with researchers from all over the world.* (3) On the other hand, it did point out that in 1986, under NSF's U.S.-Japan Program 404 Japanese scientists and engineers came to the U.S. and 244 U.S. researchers went to Japan. It accused the Japanese of not providing access to foreigners to comparable R&D facilities in Japan (p. 8). Were there comparable facilities to visit in Japan and were there U.S. researchers available, ready and eager to go to Japan for a year or two? The answer to this issue was not provided. (4) "Cultural traditions, linguistic problems, recent legal barriers and well-entrenched structural differences, have made it difficult for American university researchers and graduate students to visit and work in Japan" (p.11). The paper never comments on whether there was a real demand or clamoring by these very foreigners (including those from the U.S.)

to study in Japan. Until recently there was no rush to study science or engineering at Japanese institutions where, according to U.S. evaluation, university research facilities in the sciences are poor in quality, and often hazardous and where little originality, it was asserted, existed according to those who minimize Japanese university achievements. (5) "Until recent changes in Japanese laws and national policies on internationalizing Japan, American researchers interested in working with Japanese colleagues in national laboratories with a high technology orientation have found it difficult to visit facilities and to establish the linkages necessary for developing mature, reciprocal collaboration" (p. 11). Based on the Justin Bloom study of 1983, anecdotal information and many volumes of JTEC studies this is totally contrary to existing broad, deep and extensive R&D relationships in the public sector.

It does mention that since 1983, U.S. researchers in Japanese programs and visits to Japanese facilities have gradually increased. Somewhat over 100 researchers from U.S. facilities have visited Japanese facilities. There are, of course, many researchers from the U.S. private sector studying in Japan.

This document was used in a somewhat minor and unusual fashion about one month later. This non-paper has been described in some detail not because of its accuracy, its insights, but because of its value as a reflection of how OSTP looked upon and analyzed the so-called "inequalities" in the U.S.-Japan S&T relationship which, in turn, would form the basis for its demands in revising, revitalizing and remedying the structure of the bilateral relationship which it claims was implicit in the 80STA. It should be kept in mind that this "culprit" agreement, the 80STA, was conceived, structured, drafted and pressed upon a highly reluctant GOJ in 1980 by none other than the USG, and within the USG the OSTP, the very Office which is now so vehemently critical of and dissatisfied with the agreement.

Report by Ambassador John Negroponte, A/S OES. As designated head of the U.S. delegation to negotiate the renewal/revision of the 80STA, he prepared a one-page plus report to the Secretary of State. Judging by its contents, the principal speaker appears to have been the science adviser as Team leader. This report states that:

- the science adviser stressed to the Japanese that their stature and responsibilities as a world economic power and potential contributor to the expansion of scientific knowledge must now be given more weight in our S&T relations.

- we are now asking Japan to redress this balance by creating an environment in which each side can profit as equal partners by the provision of increased opportunities for U.S. students and researchers to work in leading research institutions in Japan, potential collaboration [meaning financial support] of the recently announced superconducting supercollider [an idea which OES had earlier urged not be broached under the circumstances], intensify cooperation on global environmental problems, and process engineering.

The A/S said the Japanese had agreed to engage in more detailed negotiations and that he will "press the Japanese" to work with us to "make this agreement the symbol of a revitalized and more balanced relationship" which includes "a set of principles and objectives" which were described in the briefing papers. The IPR part was now more fully described as strengthening these rights "for potential discoveries made under joint research activities."[63] The U.S. proposal was gradually evolving with the addition of new ideas in small increments. Based on the available documents, the source of these small but rather important addition is not clear but one can safely surmise that they did not emanate from the State Department which was basically for a simple renewal of the 80STA, but rather from OSTP or perhaps from Bruce Merrifield's office in DOC.

Report by DOC's A/S for Production, Technology and Information, Bruce Merrifield. As described elsewhere, Merrifield played an active role as DOC's representative on FCCSET/CISET, and in one of the CISET working groups mobilized to assess Japanese S&T/R&D in connection with the effort to arrive at a USG consensus for renewing or terminating the 80STA. His five-page trip report (plus two appendices, a total of 17 pages in contrast to the other short reports) provides yet another version and interpretation of this encounter in the U.S.-Japan S&T relationship. This writer met him several times in the late 1980s as one of many interviews for this study. Merrifield was probably one of the more intellectually stimulating personalities in the personae of this drama. He played an active role, as well as, of course, many others did in the FCCSET/CISET process; he chaired one of its working groups, Science, Engineering, Technology and International Competition. In the JTF exercise, he convened a special meeting at Commerce where he invited Clyde Prestowitz (formerly counsel to the Secretary of Commerce Baldridge) because of the knowledge he presumably had in negotiating with the Japanese in his earlier governmental position, and Robert P. Stromberg, a technology transfer officer at the Sandia National Laboratory in New Mexico since he had many foreign visiting scientists to this Laboratory and had frequent contact with foreign facilities. For the first time through Merrifield's report, we learn that the USG was thinking of creating several annexes to the S&T agreement: cooperative research, funding of basic research, accelerated processing of patents and copyrights by Japan, provision of language assistance to U.S. researchers, accelerated development of machine translation, enforcement of intellectual property rights. Let us, at this point, summarize his report in several categories. We will include in the main text his report on the February meeting in Tokyo, his suggestions about the renewal of the 80STA and place in a footnote his analytical and philosophical comments about the innovation process.[64]

He reported that the Japanese "repetitively raised the question of what purpose the U.S. delegation saw for modifying a quite satisfactory agreement". Also, they professed puzzlement about U.S. concerns of "inequity." Both items were repeatedly explained by the science adviser. An element of surprise and perhaps annoyance with this Japanese approach can be detected. It is quite understandable since both State and the GOJ were in favor of renewing the existing 80STA with some revisions. Through broad, liberal or expanded legal interpretation of the 80STA many, if not all, of the proposed additions by the U.S. could be accommodated within the framework of the existing agreement.

Four areas of "cooperative research and development were surfaced in general terms for separate discussion" but most of the discussion centered on science areas for increased collaboration and reciprocity in basic research. Despite State's recommendation against and anxiety about raising *again* the possibility of Japanese contributions to U.S. big science projects, the science adviser invited Japanese participation in the SSC project, i.e. contributing substantial billions even if not mentioned at the meeting. Apparently he even suggested that the signing of an S&T agreement be accompanied by announcing that Japan agreed to collaborate in some such major program. The GOJ officials must have immediately brought to mind their earlier experiences with U.S. pressure to participate in such projects with abrupt terminations.

The GOJ was skeptical that if a new or revised agreement was expected/demanded it could not be worked out in such a short time, i.e. by May 1, 1987. With obvious tongue in check Merrifield injected "Sign now, discuss later".

In renewing the agreement, Merrifield felt the following ideas, among others, should be incorporated into the revised agreement:

- Redressing some of the inequities mentioned in the footnotes with an on-going process of review, evaluation and readjustment as needed.

- The Japanese should substantially increase the funding of the stay in Japan of U.S. scientists and engineers from junior to seniors scientists.

- The Japanese should accelerate the processing of U.S. patents and copyright applications in Japan and increase patent protection for process patents.

- Development of machine translation should be accelerated as one option for cooperative efforts.

- The legal basis for enforcement of intellectual property rights must be established and a commitment made to the diligent prosecution of violations.[65]

- Focus on each of the forms described in the "Pipeline innovation process" as described in the footnote in renewing the agreement, especially since the U.S. leverage is "quite strong because of the unique capability that we have to generate basic research from which other stages of activity stem".

10. Act II, A Meeting but No Promised Proposals

The February 1987, meeting in Tokyo was the culmination of a major effort in the USG to devise a plan, stance, framework, objectives, etc, in revising the 80STA. Again, all written indications seemed to point to a revision of the 80STA by modifying articles and clauses in that agreement and adding a number of annexes, not a "new" agreement. There seems to have been no mention of any national security aspect to this agreement.

There apparently was no mention to the GOJ what kind of "revision" the USG had in mind, the extent of these revisions, whether the USG was or was not preparing such a document at this time other than papers on specifics topics. There was no mention in USG documents on the status of preparing a formal set of revisions to the 80STA, who might be preparing such documents by what deadline. There was a complete blackout on this issue. This is unusual since the USG had promised to provide various proposals and papers to the GOJ and begin negotiations in early April 1987. Given the need and promise to give a copy of these papers and perhaps a proposed revision to the GOJ two weeks ahead of any meeting, this left only 4-5 weeks for preparing this draft proposal.

Despite this major effort to coordinate USG views to create a strategy vis-a-vis Japan, and despite repeated assurances given to the GOJ about providing a written proposal, no written proposal was given to the GOJ. In the several meetings, the GOJ was given only a generalized description with few specifics. The "detailed negotiations" were postponed until early April 1987. Ambassador Negroponte again assured the GOJ, as he had done in the past on a number of occasions, that the GOJ would be given answers and a written proposal "at least two weeks before our arrival" in early April 1987.[66]

Several specific ideas were mentioned in 1) the access area to increase opportunities for U.S. students and researchers to work in leading Japanese research institutions, 2) intensified cooperation on global environmental problems and such applied areas as process engineering, 3) potential collaboration [i.e. with Japanese funding contributions] on the recently announced superconducting supercollider project. State had urged the Team not to mention potential funding by the GOJ of big U.S. projects in light of the unfortunate experiences in the past. The science adviser did not heed this advice. 4) creating a set of principles and objectives which would underpin USJ S&T relations, and 5) inclusion of annexes on IPR.

The next step for the USG, therefore, was to prepare for the April 1987 meeting with a set of the long awaited U.S. proposals to give to the GOJ.

Notes

1. OES Memo A/S Negroponte to Derwinski, EPC [Economic Policy Council] Meeting (July 1987) on Japan S&T Agreement, p. 3.
2. The GOJ posed the inquiry but there was no response. Indeed, the comment section by the OES staff officer reporting on this meeting was deleted as impinging on national security when released to me.
3. OES memo, 9/26/1985, Proposed Methodology for Renewal of US-Japan Science and Technology Agreement.
4. Neither of these reports have been found and the record does not show that whatever results generated were duly and seriously considered by OES, OSTP or any other USG agency.
5. An OES handwritten note admonishing a staffer about the kind of papers on USJ S&T relations to prepare for A/S Negroponte described the latter in rather unflattering terms: You have to remember that he [Negroponte] doesn't know anything so we must educate him enough about the general picture so he doesn't look like a fool ... The printout [of USJ projects shown above] is not really very helpful ... to a novice like him.

6. Cited by Smith in the aforementioned book on p. 127. Smith's source was "On Advising the Federal Government" by Bucksbaum. p. 71-72.
7. Bloom, Justin. *The U.S.-Japan Agreement for Cooperation in Research and Developments in Science and Technology*. [Nov. 1985], unclassified, 13 p. and 3 appendices. State Department retyped Bloom's paper (without the three appendices, added a title "Options on US-Japan S&T Agreement" and classified it as confidential and dated it "Nov. 27" [1985]. At first the entire paper was denied to me in 1990. Upon appeal it was released in entirety in 1994, some seven years after my first FOI request.
8. Bloom 55, p. 5.
9. Bloom 55, p. 6.
10. Bloom 55, p. 9.
11. Bloom, 55 p.10.
12. The OES report on this report was inconclusive; it was suggested that the GOJ "be sounded out" about these options. Memo of conversation. 11/27/85. Strangely, while the options paper was sent to the Embassy, it was ignored until the Embassy was requested specifically for its recommendations.
13. OES memo, 3/26/86. It contained three items : an overview memo, and two issues papers on the renewal of the 80STA and USJ agreement (3/10/1986) on the space station.
14. Although this issues paper was unclassified in 1986, the FOI reviewer several years later deleted the two intermediate positions which were considered merely scenarios. This writer assumed for this discussion that the two intermediate positions were Bloom's option 3 and 4.
15. Briefing paper: US-Japan Science and Technology. 3/10/1986. 6 p.
16. Tokyo 7554 (4/17/1986). Based on available documents, the USG never did "get back to" the MOFA about their proposals.
17. Tokyo 8126 (4/24/1986).
18. OES memo (5/15/1986), UJST discussion at NBS, USDA, NASA and HHS and attached talking points and OES memo (5/30/1986), Renewal of US-Japan Science aand Technology (non-energy) Agreement (UJST): Recent Developments.
19. ibid.
20. OES memo, 5/30/1986 Renewal of U.S.-Japan Science and Technology Agreement (UJST): Recent developments and an OES memo of June 9, '86 OES/Blanchard to Deborah Wince, OSTP.
21. State 188544 (6/14/1986).Wince was Deborah Wince, Assistant Director, OSTP, formerly a program officer in NSF and later under the Bush Administration as Assistant Secretary for Technology Policy in DOC.
22. ibid.
23. OES memo 6/18/1986, Backgrounder and talking points for your [Negroponte] meeting with Ambassador Matsuda. p. 2-3.
24. OES memo 6/18/1986, Backgrounder and talking points for your [Negroponte] meeting with Ambassador Matsuda.
25. State 248113 (8/7/1986). It is truly surprising that the USG list included such technologies as aircraft, machine tools, etc in light of the US's known sensitivities about Japanese intentions and capabilities in these areas.
26. Briefing paper, *US-Japan Science and Technology: CISET Review*, 7/1/1986, p. 1. Except for the Bloom studies and options, the other studies (by Don Ferguson and Henry Pollock) were never mentioned elsewhere, let alone citing recommendations; State's FOI office said the latter two studies could not be found. If they ever really

existed, they have vanished into thin air.

27. OES (Jack W. Blanchard) to OES Actg A/S Richard J. Smith, Request for Circular 175 Authority to Renew the United States - Japan Agreement on Cooperation in Research and Development in Science and Technology, 8/25/1986.
28. State 316022 (7/10/1986).
29. OES memo 6/18/86, Backgrounder and talking points for your [Negroponte] meeting with Ambassador Matsuda. p. 3.
30. State 248113 (8/7/1986).
31. Tokyo 17183 (9/8/1986). It should be noted that important parts of this cable were deleted when released to this writer.
32. Tokyo 15880 (8/18/1986).
33. State 316022 (10/7/1986).
34. Tokyo 18958 (10/6/1986).
35. Tokyo 19334 (10/14/1986).
36. State 324538 (10/16/1986).
37. OES memo, 10/9/1986.
38. State 316022 (10/7/1986).
39. Tokyo 19929 (10/22/1986).
40. State 316033 (10/7/1986).
41. State 351963 (11/8/1986).
42. State 366325 (11/24/1986).
43. OSTP memo, 8/6/1986, CISET Issue paper on U.S.-Japan Science and Technology relations.
44. CISET issue paper on US-Japan S&T relations, 8/6/1986, p. 1.
45. ibid, p.21.
46. ibid, p. 3.
47. ibid, p. 3.
48. ibid, p. 4-5.
49. ibid, p. 7.
50. ibid, p. 8.
51. Plan of action for proposed renewal of the U.S.-Japan Agreement for Cooperation in Research and Development in Science and Technology. Submitted to the Executive Committee of the FCCSET Committee on International Science, Engineering and Technology by the Joint Bilateral Task Force. 11/26/86.
52. OES memo about Nov. 15, 1986, Your [OES A/S] meeting with Dr. William Graham on Tuesday, November 18, 1986.
53. OES/SCT memo, 11/26/1986, to OES/S Dr. Robert Morris, Your meeting with Mr. Hinata, Monday, December 1 [1986].
54. OES memo, 1/9/86, Your meeting with Science Minister Mitsubayashi ...
55. The USG gave GOJ Science Counselor in Washington on 1/26/87 general information and outline not detailed specifics of the USG proposal: the February 1987 meeting was to "establish an agreed policy framework under which these negotiations will be carried out at a later date." The framework would include principles of reciprocity of access to JSTI and R&D facilities, protection of IPR, overcoming institutional differences hindering "mutually beneficial S&T cooperation", identification of areas of advanced technological research, creation of a U.S. executive committee and hopefully a GOJ committee and an advisory binational panel of private sector S&T leaders. State 25433 (1/30/1987).

56. OES memo, 1/14/1987, "Talking points for your meeting meeting with science adviser William Graham, Thursday, January 15, 1987 ..."
57. Unsigned OES letter (1/15/1987) from A/S Negroponte to OSTP Graham.
58. US-Japan S&T Cooperation: Goals and Objectives, 1/14/87, 7 p. was prepared by OSTP. This memo was then converted into a draft cable to the Tokyo Embassy, entitled, US-Japan Presidential S&T (Non-Energy) Agreement: Scope Paper for Refocusing the next phase; it appears never to have been sent. The principal U.S. Objectives are also brought together in yet another document [by OSTP], US-Japan Science and Technology (Non-Energy) Agreement. A number of the ideas in these papers later appear in the draft revision of the 80STA and later in the 88STA.
59. Visit of White House Science Adviser Dr. William Graham to Tokyo, February 2-6, 1987, Scope Paper on renewal of the non-energy agreement, 1/28/1987.
60. Probably John Moore, NSF Deputy Director, also probably wrote such a report but it was not made available to this writer.
61. Tokyo 2437 (2/12/1987).
62. Memo from Graham to the President, Renewal of the presidential level science and technology cooperation agreement with Japan.
63. OES memo (2/25/1987), OES A/S to Secretary on U.S.-Japan Science and Technology Relations.
64. Trip Report (1/30 - 2/5, 191987), U.S.-Japan Science and Technology Agreement, D. Bruce Merrifield, February 20, 1987. Merrifield, to the best of this writer's knowledge, was the only person among those involved in the future handling of the 80STA who tried to understand the "innovation pipeline" or process as it related to the U.S.-Japan S&T relationship in Appendix II of his trip report. A longish summary is provided here not because it became a central think piece in the USG process; based on available documents

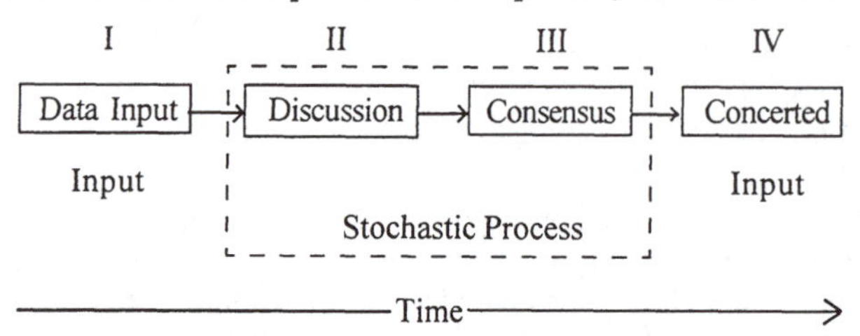

it did not. This summary is provided as a rare example of the thinking on the part of basically one person who, it appears, while harboring some rather strong feelings and judgments had tried to create a more rational basis for a revised, more balanced, and creative S&T relationship between the two countries which were known as the two major players in world S&T at that time. Another reason for this description is that some of its ideas in adapted form did find their way into the final agreement between the two countries. Merrifield's pipeline illustration is reproduced below for the reader's convenience.

In the first of three steps, invention, the U.S. spends each year about $13 Billion (both government and the private sector), according to Merrifield, more than ten times that of Japan or any other nation spends or can spend; furthermore this "advantage has been generally unappreciated and largely wasted" in the U.S. Japanese and other nations have mined this resource freely and developed it through "vertically integrated consortia that exceed the capacity of most individual U.S. companies to compete". He says the

Technology Transfer Act of 1986 will largely dry up this resource for unilateral exploitation, requiring Japan and other nations to pursue joint venture arrangements for access to "next generation" technology.

The second step is where "90% of the cost, the time required and risk" is involved. The Japanese have done exceedingly well in this process; the U.S. companies have been "less competent in doing this and have failed effectively to protect their leading edge advantages."

The final phase of manufacturing, marketing and distribution "will rapidly restructure this phase... creating a new industrial revolution exceeding anything that previously has been known."

Forms of Cooperation in Basic Research. He divides this into "big" and "small" science. In the "big science" category he includes space, fusion, SSC, human genome, sea mining, i.e. those areas where complimentary skills exist, where cooperation could reduce risks, and avoid redundant efforts. (These are the kinds of projects that the U.S. has tried unsuccessfully to persuade Japan to invest in U.S. projects such as the SSC).

In the "small science" area, there are viral protein structures, archeology, volcano boundary faults, bandgap engineering, and molecular synthesis, investigation of superconducting materials. This group has thousands of individual investigators in hundreds of labs, singly or in small teams to further this research. Since, he comments, that about 50% of graduate students, as investigator hands in the U.S., are foreign born, they have early access to discoveries. U.S. labs have not adequately protected new discoveries with patents. There is a need here to have an additional agreement to retain U.S. ownership of these discoveries, perhaps by creating 50/50 joint ventures with U.S. and Japanese companies.

Forms of Cooperation in Technology (The Translation Steps). As in the previous section, there is "big technology" and "small technology". The former are those which "are beyond the risk or resources capacity of individual companies, but ones which, if successful, open up a multiplicity of new commercial opportunities that were not possible before." Those mentioned as candidates in this category were the development of a 16-megabit DRAM semiconductor chip, flexible computer integrated manufacturing, video disc education programs, restructuring, "reskilling" the work forces and the development of "The Orient Express for intercontinental travel".

The suggested effective model for cooperation in this category was to create a new legal entity to acquire ownership of existing technology under license from partners involved; but then contract out further development wherever skills and resources exist, the so-called R&D Limited Partnership Model, a model that is very flexible and provides equitable returns based on contributions made.

The "small technology" form could be creatively adopted by companies working with each other sharing in an equitable distribution of rights and profits.

USJ R&D Relations. He described several "inequities" in the U.S.-Japan S&T relation. The U.S. has built up a fantastic academic infrastructure which is enormous but a "largely wasted U.S. advantage." He believed the Japanese were acutely aware" that the recently passed the Federal Technology Transfer Act of 1986 "could dry up [their] access to this critical source of advanced technology." The Japanese have, therefore, begun investing in basic research in Japan. Since this takes time, in the interim the Japanese are accelerating funding of basic research in the U.S., investing in start-up firms, and creating joint ventures which assure that the manufacturing is reserved for factories in Japan.

He said the U.S. developed a strong S&T relationship with Japan "to ensure development of a prosperous democracy in firm partnership with the western alliance." This strategy was described as successful "but had failed to build a balanced capability in Japan". Underlying this assertion is a dubious, if not wholly incorrect, assumption that somehow the Japanese R&D achievements were the outgrowth of a U.S. plan which had somehow failed also to put in place a basic research capability. He also asserted that the "Japanese companies have been freely exploiting the unique U.S. capabilities [meaning presumably the scientific and technological ideas, systems and data emanating from U.S. Federal laboratories]. The Technology Transfer Act of 1986 now authorizes the directors of these federally funded labs to withhold this kind of information unless comparable access is provided by foreign (Japanese) institutions. (This issue is discussed in section IV.6.a.ii).

65. While all these suggestions were not deleted from the documents released to me through the FOI process, the State Department reviewer put back the black pencil through six other "suggestions" which were conveniently lumped in one group.
66. Letter (2/13/87) Negroponte to Ambassador Tetsuya Endō, new chairman of the Japanese delegation.

VIII. Act III, Rejection to Grudging Consensus: February - August 1987

1. Forward to Rejection, February - March 1987

a. What to present to the GOJ?

After a major but unsuccessful push to muster enough drive and momentum to pull together a USG position and papers to present to the GOJ at the early February 1987 meeting in Tokyo on the renewal of the 80STA, the U.S. Team chaired by OSTP Director William Graham and OES A/S John Negroponte returned to the U.S. and were immediately faced with a looming deadline of presenting to the GOJ an integrated package of proposals including various annexes on IPR, Cooperative R&D, etc within only a few weeks by Mid-March 1987. Again, the USG had set itself an impossible target to meet, especially when so many USG departments and agencies were involved. The GOJ had up until now only been given an oral summary of some generalizations about USG ideas for revising the 80STA without any written material. It is not clear from available documentation what the GOJ specifically was informed, except that later the USG would claim that the GOJ knew of USG ideas, thinking in fairly specific terms. This is difficult to believe when the USG itself was still strenuously pulling together its own position.

This period begins with the launching of another major effort to prepare for the early April 1987 negotiations with the GOJ. When the draft proposal to give to the GOJ was circulated within the USG for final clearance toward the end of March, one of the worst collisions occurred between State and Commerce over a small section of the IPR annex. This led to an even deeper confrontation and mutual distrust between and among Departments of the USG. Every effort was then made through successively higher levels of review to tie the hands of the U.S. negotiating team in its discussions with the GOJ. These tactics reflected not merely utter mutual distrust among U.S. officials but also, of course, and, as a matter of course, of the GOJ. Eventually, quite an extraordinary draft revision from the content and format point of view, was presented to the GOJ in late August 1987 as the basis for negotiating a revitalized 80STA.

Before we describe and analyze the semi-final push to pull together a USG draft for presentation to the GOJ, let us look at a document which reflects the thinking and philosophical approach of the OSTP to the revision of the 80STA, the kinds of demands it was imposing on the USG to persuade the GOJ to accept. It is an undated five-page single spaced document with no agency designation. The circumstantial evidence clearly defines it as an OSTP document. It was probably prepared after the February meeting in Tokyo but before the April 30, 1987 deadline for another extension of the 80STA. The entire tone and specific analysis and demands are hardly what OES supported. The last paragraph is reflective of the frustration and anger in the OSTP position:

> ...Dr. William Graham is determined to take full advantage of the enormous leverage [with the GOJ] that renewal of this agreement [80STA] provides. He is determined that the President should not sign any agreement that does not contain real teeth. The timeliness of expiration on April 30 provides an opportunity to send a very strong signal to the Japanese, that can enhance the climate for the trade crunch that is developing.[1]

Since the entire document naturally cannot be quoted, let us select those more important strains of thought, action demands and interpretations of the existing situation.

- the renewal would serve as a model for other nations.
- place a limit on U.S. agency sub-agreements independently being pursued. This is at the heart of the long simmering OSTP conflict with the various Departments and Agencies with S&T arrangements with foreign countries (not just Japan).
- the renewal should be accompanied by "stringent limitation" on access to U.S. S&T by Japan.
- the renewal should be for a few months pending renegotiations and renewed for only a short period, "2 years."
- create an evaluation and review process to monitor compliance with "measurable objectives to be reached within well defined periods of time."

- the 80STA is described as providing Japan with almost unlimited "access to U.S. advanced technology that Japan does not have the capacity to develop itself." It does not prove how the 80STA allows this happen; it merely makes an assertion which is actually baseless and not related to that agreement. The "inequities" which this agreement presumably "permit[ted]" and for which necessary remedial actions were necessary as follows:

1. Unbalanced number of Japanese researchers in the U.S. as compared to U.S. researchers in Japan; "equal access must now be provided". There is no discussion whatsoever about why this inequality occurs.
2. Japanese must provide English translations or abstracts of their technical journals.
3. Japanese patent processes must be speeded up to two years like the U.S. system and not publish applications as permitted in Japan.
4. Japanese selective project funding in the U.S. does not justify exclusive rights to resulting technology. More equitable arrangements now must be made and monitored for compliance.
5. Access to U.S. "big science projects" must involve shared funding.
6. Where equal access is not granted to foreign R&D facilities, access to USG R&D technology will be restricted in accordance with the enactment of the Technology Transfer Act of 1986. If this authorization is implemented, Japanese access to U.S. basic research in U.S. national laboratories will effectively dry up. This is a warped interpretation of the effectiveness of this law.
7. A U.S. organization which acquires rights for university or government developed technology may not sub-license that technology to a foreign organization, but should form a joint venture.
8. No non-exclusive license should be given to a foreign nation's company.
9. Periodic readjustment must be an essential annex, quantifiable objectives and times for compliance will be developed "as a precondition of agreement."

The existing 80STA does not impede, or specifically encourage, allow or disallow, or prevent the above "inequalities" from occurring as asserted by OSTP. To assert that these "inequalities" were caused by or emanated from the 80STA is an unfortunate and mistaken interpretation, if not a deliberate distorted understanding of the 80STA. Many of the proposed restrictions and remedies will be found in various drafts of the proposed revisions to be given to the GOJ later in 1987. The hostility manifested in this listing of the so-called "inequities" is not proven by legal analysis of the

80STA. They are bald assertions which deny the actual prevailing situation. (see section on training for analysis of opportunities in Japan). It is difficult, if not impossible, to look upon the above assertions and demands as consistent with the ever-repeated statements by the U.S. that the U.S. and Japan as equals in world-class R&D should now cooperate in new and higher levels of R&D cooperation in the leading edge technologies to make major contributions to the world through such endeavors. The assertions and demands seem to point in one direction: to try to do our utmost to hem in, restrict, circumscribe Japanese R&D activities and, to the extent possible, bring them under U.S. control. Assuming that the two countries are more or less equal in their levels of R&D achievements in most technologies -- although that evaluation is not necessarily well accepted by any means -- the U.S. could hardly expect to succeed in controlling, restricting, or circumscribing Japanese activities to the extent implied and demanded. The tone, direction and thrust of this informal paper which must have received rather limited circulation in the USG sets the stage, so to speak, for the direction in which the various aspects of the proposed revision would proceed. The active though highly negative stance toward USJ S&T relations taken by the OSTP stands in stark contrast to the rear-guard passive posture of the OES.

Another deadline set by the USG (with GOJ concurrence) loomed ahead. The USG had promised to provide the GOJ a total package for negotiation "well before" (by March 15, 1987) the April 6-7, 1987 meeting in Tokyo and the extension of the 80STA would expire at the end of April 1987. Thus, the bureaucratic wheels began turning -- churning? -- at a somewhat faster pace and with somewhat greater urgency. Instead of working together as a Team to prepare one draft proposal, it appears State (and its supporters) and OSTP each prepared drafts.[2]

In light of the OSTP delays in the past in holding committee meetings, OSTP was regarded as the most critical factor. Therefore, OES A/S dispatched a friendly letter on February 27, 1987 to Science adviser Graham calling upon his office for the timely provision of important inputs from OSTP in preparing the statement of policy goals, priority areas and initiatives, and the role of the high level advisory panel by March 6 in order to meet the deadline of March 15 to the GOJ. Appended was a handwritten note by Negroponte stressing the importance of the OSTP staff to "help in every way possible to meet the drafting deadlines."

As was recognized at the time, it was going to be well nigh impossible to prepare, in the first place, and then to obtain the agreement of the major Departments and Agencies, to the various parts of the proposed revision package to give to the GOJ within two weeks. As the March 15 deadline approached, OSTP circulated a document on Friday, March 13, 1987 requesting comments by Monday, March 16 in order to despatch them to the GOJ by March 20, 1987. At the same time, a planning session was to be held on March 18, at State presumably to discuss the proposal to give to the GOJ. It was described as a "policy framework for the U.S.-Japan S&T partnership, and priority areas and mechanisms for cooperation." Upon careful scrutiny of the document, it turns out to be quite extraordinary, obviously something thrown together from several sources to try to meet the deadline. After the science adviser returned from his February 1987 meeting in Tokyo, a "non-paper" on U.S.-Japan S&T relations had been prepared. (see the previous section for a description of this document)

The list of addressees included the regular long standing participants, Negroponte of State/OES, Merrifield of DOC, Moore of NSF, Taft of OMB, but this time in addition, Deputy USTR Mike Smith and Frank Gaffney of DOD. Toward the end of 1986 OSTP initiated discussions with Mike Smith of USTR about the revision of the 80STA. This turned into a close working relationship between OSTP and USTR.[3] As Deputy USTR, Mike Smith provided effective political backing and was a strong supporter of the OSTP in its contest with the State Department and others. The sudden appearance of the DOD (especially in the person of Frank Gaffney) will be seen to have an immediate and unsuspecting impact on the proposed package to be presented to the GOJ.[4]

This document circulated for clearance began with Annex I as a "Policy Framework" expressing general principles for the revitalization of the Presidential Science and Technology Agreement between the U.S. and Japan, national S&T policies and R&D goals of the U.S. and Japan. This is then followed by basically a verbatim copy of the non-paper mentioned in the previous section. This was a social science-type analysis of the support of basic research in the U.S. and Japan, science and technology education and facilities in the two countries, participation in the international R&D system, foreign (including Japanese) participation in the U.S. R&D system, foreign (including U.S.) participation in the Japanese R&D System, and major R&D initiatives of the U.S. and Japan. To include such a controversial and detailed analysis of these issues in a "policy framework" to be given to another government, in this case, the GOJ, is difficult to comprehend. An everlasting debate would ensue between the U.S. and Japan over the selected facts, what they meant historically, what they implied for bilateral policy decisions. It is difficult to understand what purpose would be served by including such a social science-type commentary in a "policy framework" for two governments to consider.[5]

Following this "social commentary", article III (Annex I) with ten parameters to measure "balance and reciprocity" in bilateral R&D is inserted. Finally there is Annex II on the areas of technical cooperation and the creation of mechanisms, largely by the Japanese to increase and enhance the number of U.S. scientists and engineers in Japan. The details in this annex will be provided later as the total package of revision takes shape in March 1987. It should be noted that this "package" which was presumably to be given to the GOJ by March 20, 1987, fulfilling a promise from the February 1987 meeting did not include other annexes on the management structure and IPR. There are no indications that this document was ever commented upon, revised or made more complete as a package, let alone fully cleared to give to the GOJ with the purpose of using it seriously as a negotiating instrument. This document is a mere blip in the evolving saga; it is included to show the kind of approach to Japan as manifested in the President's office of the science adviser, the apparent frustration that pervaded the atmosphere, that such a document could be seriously circulated for USG purposes and that competing drafts (States and OSTP) existed instead of one draft prepared by an interagency team.

b. Farcical and Truncated Dialogue with the GOJ

While the USG agencies and departments waged their bureaucratic battles around the future of USJ S&T relations, an insipid and meaningless dialogue -- perhaps more a monologue rather than a dialogue -- took place with the GOJ.

As the USG debated with itself about the revision of the 80STA, the visit to the U.S. of the Japanese Prime Minister, Yasuhiro Nakasone at the end of April 1987 coincided with the potential expiration of the 80STA with no agreed position in the USG, let alone an agreement with the GOJ. During this period, Japanese Foreign Minister Kuranari Tadashi called upon the U.S. Secretary of State on March 6, 1987 and was informed, among a number of other issues, that Ambassador Negroponte would begin negotiations in early April with the GOJ about the revision of the 80STA. The Secretary stressed that the revised STA must be based on "equitable principles ensuring *symmetrical access* to all elements of S&T innovation [note the change in wording from "facilities" which was generally used]. The U.S. would ask Japan to facilitate access for U.S. scientists to Japanese research facilities without proving or asserting that such access is denied. The U.S. would strongly welcome the recent increase in Japanese investment in basic scientific research.[6] These were benign comments to make to the Foreign Minister and did not reflect the true situation nor did it give any indication whatsoever of what the USG was considering at that time -- an elaborate revision with multiple annexes and a complicated management structure.

Because of the upcoming visit of Prime Minister Nakasone to the U.S. at the end of April 1987 and the fairly obvious conclusion, even to the outside observer of the USG process like the GOJ and its Embassy in Washington, that the probability of receiving a proposal from the USG in a timely fashion to negotiate a renewal of the 80STA before the end of April 1987 was basically impossible, the Japanese science counselor at the Embassy of Japan proposed to the State Department on March 16 that the 80STA be renewed in its present form for two years while negotiations continue on a broadened form of the agreement. State responded that it had intentions of providing the GOJ with a proposal by March 20 -- how could that statement be made with a straight face in light of the above developments -- but suggested that a shorter extension might be warranted under the circumstances. The USG commented that it hopes to structure the proposed revisions in such a way as to maintain its status as an executive agreement between the two governments, thereby not requiring Diet or Congressional approval. The USG informed the Japanese science counselor the U.S. side would have an answer to their proposal "in a few days."[7] There is no record that a response was provided to the GOJ in Tokyo or to the Embassy in Washington.

The USJ "dialogue" continued in the Japanese press. On March 22, 1987, *The Tokyo Shimbun* carried an article on the USJ S&T agreement; the story emanated from the Foreign Office and was positively interpreted by the Tokyo Embassy to "help lay the groundwork for quick action by the Japanese cabinet in approving an extension to the 80STA." It announced that bilateral negotiations at the Director General level and A/S level would take place in early April and would discuss 1) the expansion of cooperative activities in advanced technology e.g. the recombinant DNA, 2) the rectification of the unilateral flow in the exchange of scientists, and 3) the establishment of a USJ Advisory Committee on S&T policy issues separate from the Joint Committee mentioned in the 80STA. It was envisioned that these negotiations would proceed smoothly and be completed some time in the summer of 1987. This would require a short extension of the 80STA.

The latter two points indicate the GOJ's willingness to negotiate a renewal of the 80STA based on the meager knowledge possessed on USG thinking and preparations. Furthermore, they thought that, though difficult, these negotiations could be concluded sometime in the summer 1987. This clearly indicates how little the GOJ was informed about USG thinking and probable approach that it was preparing for these talks -- notwithstanding the repeated assurances for answers in a few days, or greater involvement of GOJ in the dialogue as to the future science and technological relations.

As the internal USG discussions reached another climax, a rather obscure inquiry was despatched to the Tokyo Embassy by OES as to what R&D projects were included under the "umbrella" 80STA. When the U.S. delegation visited Tokyo in February 1987, it was given a list of such projects -- which it thought was a Japanese list -- under this agreement. Apparently it had been compiled by the U.S. Embassy. It contained 52 projects: ten were "claimed" by NASA, USDA accepted 7 projects, HHS six projects, DOE 2, 12 projects not claimed by any USG agency and the remainder in varying stages of consideration and limbo. The Embassy responded that the GOJ had a list of 44 projects, all of which were on the Embassy list. A final clarification, though promised, was not found.[8]

c. The March 25, 1987 Draft Agreement

A planning meeting was held in the State Department on March 18, 1987. It was limited to representatives from State, OSTP, NSF, Commerce and USTR. From State both OES and the Japan Desk were represented; USTR by legal counsel who had drafted the IPR annex and the Japan Office; DOC was now represented by three offices, EA (Gerald Underwood), International Trade Administration Japan Desk (Maureen Smith), and TD (Rob Eckelmann) in contrast to earlier instances when only one office, OPTI, represented DOC. While it is not absolutely clear what documents were discussed, this writer has a document marked "For discussion 3/18[/87]" which consists of the opening policy framework and four annexes.

A draft cable to the Tokyo Embassy was prepared on March 20 informing the Embassy that "next week" the Department would send for delivery to the GOJ (and simultaneously to the Japanese Embassy in

Washington) "a draft note on broad principles and objectives" and four annexes on priority areas of cooperation, on mechanisms, initiatives and measures of progress, on management structure and IPR. A cable (3/25/87) was drafted including the diplomatic note and four annexes, presumably reflecting the March 18 meeting. But only five days later (3/30/87) after the State-DOC disagreement described in section "d" below, another almost identical draft cable was prepared adding a new annex on national security obligations. This is an important addition, and will become the source of substantial negotiating problems with the GOJ.[9]

A comparative analysis of the above two drafts reveals that many changes were made -- apparently through discussions at this meeting -- to the March 18 version. There is no hint of disagreement in the documents presently available, no suggestion that a major storm was brewing only a few days hence. The Department of Commerce was well represented at this meeting by three offices with not necessarily similar philosophical views as to the future of USJ S&T relations, one with an intellectual approach and another, based on interviews, extremely harsh, emotional and, one might say, vitriolic viewpoint and deep distrust of Japan. But the focal point for clearance in DOC was changed to ITA/Japan Desk in later drafts.

Since negotiations with Japan in early April were imminent, a Circular 175 authority was being circulated for clearance within the USG.[10] Because of the importance of the March 25 version as reflected in the aforementioned two draft cables in the subsequent "dialogue" in the USG, a fairly detailed outline will be provided below. Although not included in the March 25, 1987, Annex 5 is included here because it was an important addition.

The Policy Framework, is divided into three parts:

1. under the importance of R&D three broad goals were mentioned: a) generate fundamental new knowledge to expand the world's pool of S&T understanding, b) swiftly transfer new technologies and applications to the marketplace, and c) nurture and expand the next generation's talent base.

2. cooperation in international R&D where the U.S. and Japan acknowledge a) the need to create more equitable opportunities for interaction and to work for a balance of contributions and benefits, and b) that at times one country (i.e. Japan, though not specifically mentioned) may need to make greater

commitments in particular components of the R&D process, such as basic research and infrastructure.

3. revitalization of the USJ STA, where it is stated that a) the S&T content of the agreement should match the commitments of its Heads of Government signatories, b) this agreement sets forth the overall policy framework and national goals in the S&T relationship between the U.S. and Japan, c) S&T projects and programs of the highest national priority shall be carried out by the U.S. and Japan, d) the S&T policies, priorities, activities and plans, balance and reciprocity of the overall cooperative relationship between the U.S. and Japan shall be analyzed and identify impediments to balance opportunities, access and benefits with a view to overcoming those impediments.

The above policy framework responded to the perceived inadequacies of the 80STA by establishing, for example, that this agreement would provide the overall policy framework for the USJ S&T relationship, that projects of the highest national priority would be carried out, and that the "balance and reciprocity" of the overall relationship should be examined and impediments to balanced opportunities identified. This would provide the means, justification and opportunity for DOC, DOD, and OSTP to persuade, pressure or attack the GOJ for its "performance" because there was at that time a deep and abiding conviction among these agencies that Japan was not living up to world expectations as arbitrarily established without specification by these USG agencies.

Annex I on Major Initiatives and Priority Areas. It specifies that when the U.S. and Japan collaborate in international R&D and large scale national projects, it should be done within the above framework, and that the investments in infrastructure and state-of-the-art facilities of countries shall bear the costs equitably. This is probably attempting to set the groundwork for large scale GOJ investment in the SSC project which ultimately was canceled by the U.S. Congress in 1993 without Japanese investment.

Again, in an attempt to preclude so-called inconsequential R&D projects from inclusion under this agreement, article II of Annex I specified that a project under this agreement must meet certain criteria: each partner should possess strong complementary or counter balancing research capabilities, adequate resource bases, and appropriate centers of excellence to engage in joint ventures, the project should reflect national R&D priorities and contribute to an equitable distribution of investment and pay-off to each partner's national needs, and have the potential to accelerate the rate of scientific progress compared to what is achievable now and offer tangible contributions to the world's knowledge base.

Lastly, the following general fields were selected as "priority areas of cooperation": materials science and engineering (particularly electronic materials), life sciences (particularly biotechnology), information science and technology, automation and process control, global geosciences and environment, joint database development, joint standardization and nomenclature development.

Annex II on Mechanisms, Initiatives and Areas for Joint Review. In this Annex the language was substantially toned down, beginning with the title, "... Areas for Joint Review" instead of "Measures of Progress." While language to achieve the same objectives was used in the revision, it was implicit that the USG would be able to use this section as a weapon to constantly raise issues of equity, balanced contribution, access, investments, personnel, practically any subject that it might fancy. Another one of the perceived inadequacies of the 80STA was being remedied by requiring Japan to take numerous actions on behalf of U.S. scientists and engineers.

Although this was proposed as an agreement between two technological superpowers, the language originally used in this Annex was extraordinarily one-sided and insulting in the context of a bilateral agreement. The revised wording, revised at the behest of the OES and Japan Desk, is somewhat improved.

1. The earlier wording started off by asserting "given the imbalances that now exist ... each country must undertake special commitments in order to fulfill the promise of a true partnership." This was changed to "Each country undertakes to implement appropriate steps in order to establish a fair and equitable relationship."

2. The USG is merely going to "encourage and support language training programs for technical personnel at U.S. Universities..."; the GOJ, however "will establish at Tsukuba University [at science city north of Tokyo] and other key institutions in the public and private sectors, intensive language programs for visiting American scientists and engineers to facilitate communication in R&D environments as well as in daily life activities in Japan."

3. Several Japanese Ministries would "give priority attention to increasing the number of Americans participating in national R&D programs and collaborating in the priority areas" listed above. USG agencies would merely encourage Americans to participate in these programs, and the GOJ would "provide annually to the U.S. government a comprehensive list with all necessary particulars of current opportunities for American researchers to participate and be employed in the Japanese R&D system."

To increase the number of U.S. researchers in Japan and promote the long term objective of balanced access and development training opportunities the GOJ "will establish and widely advertise a significant number of substantial and prestigious "Japan Fellowships" in science and engineering for American[s] at "Japanese centers of excellence."

Under a proposed Joint Interagency Executive Committee, the U.S. and Japan would create a "working level task force" "to identify, recruit, and monitor American scientists and engineers' access to and participation in Japanese R&D programs," examine barriers and other structural problems on both sides that impede or inhibit increasing the numbers of U.S. researchers in Japan, and set up a system to obtain accurate, yearly statistical data on Japanese S&T researchers in the U.S. and Japanese R&D systems.

The section now euphemistically called "Areas for Joint Review", previously called baldly "specific measures of progress for joint evaluation" was most toned down and drastically rewritten. Originally it began with the U.S. and Japan measuring each government's support of basic research against each government's "political commitment" to support basic research. Thereafter it lists a series of instructions to assess, to evaluate, to compare, to review the flow of scientific and technical information between the two countries, national policies and norms that encourage scientists and engineers to publish their basic and applied research results in internationally referred [sic] -- should be refereed -- scholarly journals, domestic investments to maintain and expand state-of-the-art higher education and training facilities, measure the contributions in establishing world-class facilities at higher education levels, balance achieved by GOJ in opening domestic research programs to foreign researchers, logistical support given to U.S. S&T personnel in Japan including language training and cultural training, financial support given to U.S. researchers in Japan, and Japanese in the U.S., balance achieved in U.S. and Japanese personnel in each other's facilities, commitments of USG and GOJ to support major R&D initiatives to generate new knowledge.

The same ideas were woven into the rewritten version without the provocative words like assess, compare, evaluate. This is extremely unusual detail to be included in a Presidential-Prime Ministerial level agreement. Such implementing details would normally be included in the next lower level understandings. The purpose and determination of agencies like OSTP and DOC was to specify in the greatest detail possible the actions that GOJ would need to take, providing as little room for maneuver as possible, and provide for the provision of certain kinds of information which could then be used against the GOJ, industry and academia as it suited the U.S.[11]

Annex III on the Management Structure. There would be three levels of management: a joint interagency executive committee which would be responsible for day-to-day operations of the agreement. It would be co-chaired by the U.S and Japan at the Deputy Assistant Secretary of State (OES) level on the U.S. side. [Japan side level was left blank in the proposal but it would be MOFA].The Joint Committee would be a high level committee co-chaired by the U.S. Science Adviser and the Minister of Science and Technology. It would focus on S&T policy issues, the specifics of special technical cooperation. There would also be a Joint High Level Advisory Panel, consisting of senior scientific leaders a majority of whom would be from non-governmental institutions in industry and academia. It would review overall advances in S&T in the two countries, identify important areas for cooperation, and recommend new mechanisms.

Annex IV on Intellectual Property Rights. This part caused the sparks to fly, perhaps more accurately to explode in only a few days.[12] It covered dual use technology, confidential information, inventions, copyrights, other forms of intellectual property. Special note needs to be taken of two sections: Protection of Dual Use Technology, and inventions. If technologies are developed under cooperation in this STA, either the US or Japan parties to this cooperation may apply for patent(s) in their respective patent offices and be protected for national security purposes. The injection of national security consideration into these processes at this point was extraordinary in light of the close working relationship between the U.S. and Japan on military issues and the predictable allergic reactions by the GOJ to find such a clause in an S&T agreement. It would complicate the bilateral negotiations, probably poison and enhance the element of distrust several notches higher in future discussions. The clause referred to a 1956 USJ agreement to facilitate interchange on patent rights and technical information for purposes of

Defense. Despite the age of the agreement, the USJ were now (1987) apparently negotiating in earnest about the implementing procedures. The sudden addition of this sensitive national security issue in the USG attempt to mold a package for presentation to the GOJ most probably reflects 1) the change in DOD representation on interagency committees considering the USJ STA and 2) the anxiety generated by the implications of the Toshiba incident, 3) the state of DOD frustration and impatience in obtaining an agreement on the implementing procedures, 4) DOD feeling that attaching to an on-going negotiation with the GOJ implementing a 1956 Defense related agreement might be a useful bargaining lever in obtaining what the USG wanted from both agreements. It, nevertheless, comes as a jolt to see the insertion in a USJ agreement on furthering basic research, particularly in Japan, a clause on the protection of potential defense patents. It would complicate and super-charge the upcoming negotiations with the GOJ about a new STA. In a July 15, 1987 memo OES objected to the phrase "and this agreement shall not otherwise be implemented" as holding the STA "hostage to implementation of the 1956 agreement."[13] Since this issue later became a crisis in the USG (but not with the GOJ)], the inclusion of this seemingly extraneous issue would seem to reflect U.S. anxieties about what may result from a more aggressive bilateral cooperative R&D relationship. Since it was generally felt that the Japanese lacked "creativity", it must be assumed that just the participating U.S. scientists and engineers would be so creative as to produce unusual dual use technologies which could be worthy of national security secrecy. The inclusion of this issue also reflects another attempt by the USG, in this case via the DOD, to try to bind Japan to another potential restriction, another method of control over the use of technology developed in cooperation with Japan. It is another example of the anxiety level within the USG about the implications of S&T cooperation with Japan. Since politically the two countries needed to proceed with the revision, "strengthening" (from the U.S. point of view) of the 80STA, this was but another attempt to build a cordon around these activities.

Under inventions, another attempt was made to spell out various possibilities about how inventions would be handled under given circumstances. Again, these detailed stipulations would normally be executed under an implementing agreement. Again, the attempt here was to try to make them "more binding", more significant, more important, thus making it more difficult, in the eyes of such agencies as the DOD, DOC and OSTP, for Japan to evade them so to speak by including them in a Heads of State agreement, just short of requiring Diet and Congressional approval.

The core of the inventions subsection revolves around two possibilities: If the invention is "made or conceived"

1. as a result of the exchange of information between the Parties, such as by joint meetings, seminars or the exchange of technical reports or papers

a. the Party whose personnel make the invention (the Inventing Party) has the right to obtain all rights and interests in the invention in all countries;

b. in any country where the Inventing Party decides not to obtain such rights and interests, the other Party has the right to do so; and

c. in any country where one Party obtains rights and interests in an invention, the other Party has the right to a non-exclusive, irrevocable royalty-free license to the invention for the Party, *with the right to grant sub-licenses*.

2. by personnel of one Party (the Assigning Party) while assigned to the other Party (the receiving Party) during an exchange of scientific and technical personnel

a. the Receiving Party has the right to obtain all rights and interests in the invention in its country and in third countries, and the Assigning Party has the right to obtain all rights and interests in its country;

b. in any country where either Party decides not to obtain such rights and interests, the other Party has the right to do so; and

c. in any country where one Party obtains rights and interests in an invention, the other Party has the right to a non-exclusive, irrevocable, royalty-free license for the Party and its nationals.

Let us try to transform this legalese into everyday language.

1. In the first case, if someone invents something as a result of an exchange of information (e.g. joint meetings, seminars, exchange of technical papers, etc), then the inventor (let us say he is a U.S. inventor) has the right to obtain all rights and interests in the invention in ALL countries. However, the other Party in this exchange of information may seek rights to this invention in all countries where the inventor does not seek such rights.

2. In the second case, if the inventor, let us say a visiting Japanese scientist to a U.S. facility, creates an invention then the receiving U.S. facility where he is studying has the right to obtain all rights and interest in the invention in "its country", in this example, the U.S. facility and in "third countries", i.e. those countries other than the U.S. and Japan. In this example, the Assigning Party (i.e. the home facility of the Japanese scientist) has the right to obtain all rights and interests in Japan. When neither the U.S. nor the Japanese inventors decide not to seek their rights and interests in the U.S. or third countries (other than Japan) in the case of the U.S. scientist, and if the Japanese visiting scientist does not seek rights and interests in Japan where he is allowed to do so, the other party, i.e. Japan, then the U.S. scientist may seek such rights in Japan, and the Japanese inventor may seek such rights and interests in the U.S. and third countries.

3. There is one more important condition attached to both examples: where, in the above example, the U.S. inventor has all rights (i.e. in all countries in (1) above), and in the U.S. and third countries (except Japan) in (2) above, the other Party (the Japanese inventor) has the right to non-exclusive, irrevocable, royalty-free license to the invention held by the American inventor with a right to grant sub-licenses (when the invention stems from an exchange of information). This means the Japanese (i.e. the other party) could a) commercialize the invention and export it (including to the U.S.) without paying any royalties to the U.S. patent holder where the American has the rights and interests, and b) in the case of (1) above may also grant sub-licenses to other parties. In this example in reverse, the American inventor would have the same right to commercialize the invention and export it to Japan without paying royalties to the Japanese party even when the latter has all rights in Japan.

The right to grant sub-licenses when an invention stemmed from the first case of information exchange would result in "giving the shop away", both parties would then have lost control of and their interests in the patents would be wiped out.

It is not unusual, if not quite normal, in handling patents for the inventor to hold the patent, and "the other party" (when the invention stems from an exchange of personnel) to have the right "to a non-exclusive license to the invention." However, in light of the atmosphere at the time between the U.S. and Japan -- particularly the attitudes harbored by some in the USG, especially the DOC and OSTP -- and the anxiety and the assumption that "the other party" (in my example a Japanese) would be more imaginative, more creative in using this right to develop quickly a useful and popular product for sale around the world and in the U.S. Because there were more Japanese scientists and engineers assigned to U.S. laboratories than Americans to Japanese facilities, it was implicitly presumed that the receiving party (the U.S. side) would therefore benefit. So long as these special conditions hold this could remain a probability. If, however, the GOJ successfully pushed for and funded research which resulted in useful basic knowledge, and many more U.S. scientists and engineers studied for longer periods (1-2 years) in Japan then these special conditions would not hold. Ultimately, the non-exclusive irrevocable royalty free license provision was deleted from the draft given to the GOJ. It could be said this deletion was a lack of belief and faith in the U.S. inventor and U.S. commercial firms to be sufficiently imaginative and nimble enough to bring to market a new product quicker than a potential Japanese firm.

Annex V on national security obligations. This annex undoubtedly reflected the views of the new DOD representative on the interagency committee. The two sections in this Annex were concerned with adequate and appropriate measures according to each countries national laws to safeguard information and prevent the unauthorized transfer or retransfer of all technical information, equipment and data stemming from projects under this agreement. This Annex was added at an interagency meeting on March 30, 1987. The inclusion of this Annex and the previous dual use technology aspect of Annex IV stemmed from the general distrust of Japan stemming from the Toshiba-Kongsberg Vaapenfabrikk case which derived from a commercial transaction not a bilateral R&D project. To the best knowledge of this writer, there never has been any "national security" related information, equipment and data stemming from joint USJ research projects. Annex V is another manifestation of USG's (especially DOD in this case) distrust and hostility and an another attempt to plug all possible loopholes and tighten controls over bilateral R&D projects.

d. March 25, 1987 DOC Declaration

As mentioned previously, based on available documents, many departments and agencies attended the various sessions on the draft proposal to give to the GOJ by March 20, including 2-3 representatives from different parts of the Department of Commerce, e.g. International Trade Administration, Office of Production, Technology and Innovation and others. Since this proposal was going to be submitted to the GOJ, it was necessary to obtain Circular 175 authority to do so. This proposal was in this process, even as the draft was being discussed. Based on interviews, the last clearance for whatever reason -- probably quite by accident -- was DOC. Remember there had been no known objection from any of the DOC representatives (including ITA Japan Desk Chief who was most vehemently critical of Japan) at these meetings. There are no known proposed revisions to the package by DOC. Yet at the 11th hour in the clearance process -- five days after the March 20 deadline to give the GOJ a proposal -- the Deputy General Counsel of DOC sent a terse three paragraph letter to OES A/S stating that

> "I am authorized to advise you that the Department of Commerce strongly objects to this request at this time. Further, we object to any distribution of the proposal to any representatives of the Government of Japan."Our initial review leads us to conclude that the proposal would severely undercut major Administration initiatives to promote U.S. competitiveness by encouraging the commercialization of federally funded research. From a broader perspective, the propriety of extending this agreement at all is, in our view, a matter that requires cabinet-level consideration." [14]

It is not clear who "Our initial review" refers to, but probably the General Counsel's office. Presumably the three DOC offices which had been participating in the interagency discussions had been reviewing carefully the implications of various parts, particularly the patents and copyrights section. It would seem wholly feasible that the ITA representatives who were extremely critical and suspicious of Japan would notice a clause that would undermine and undercut the Reagan Administration's major initiatives to promote U.S. competitiveness. This reference to initiatives presumably refers to the 1986 Technology Transfer Act which, among other things, attempted to control the utilization of research results stemming from Federally funded research. There was an amorphous yet strongly held belief that foreigners, presumably Japanese, were using ideas stemming from Federal research for commercial purposes. This belief was strongly held but it remained amorphous, without proof and substantiation. The R&D results and data are freely available. Therefore, the Japanese must have used this glittering collection of gems from Federal research for their own purposes. That must have happened. The DOC letter is symbolic of the depth of the animosity in DOC toward Japan itself, and any proposed agreement with Japan.[15]

Based on a comparative analysis of the draft renewal agreement and that which was ultimately given to the GOJ in late summer 1987, *the* major difference lies in the invention section under IPR which was briefly analyzed above. The clause giving "the other Party" the right to a nonexclusive, irrevocable, royalty-free license to the invention for the Party and its nationals was deleted from the USG package. This is believed to have been the culprit condition which inspired DOC's strong objection even though DOC representatives had not objected based on available documents to this clause in interagency deliberations.[16]

This terse memo raised the heat many degrees to a high state of emotional animosity between State and DOC, coming as it did at later than the 11th hour. It pushed back the issue of resolving the future of the 80STA and the USJ S&T relationship almost to square one. DOC's sudden rejection of many months of interagency coordination and hours of discussion now placed the USG in a potentially embarrassing position having the only Presidential Agreement on USJ S&T expire during the Reagan-Nakasone summit meeting in Washington.

Based on the severe and sudden rupture in the USG negotiating structure, the USG informed the Tokyo Embassy that "because of unforeseen complications in reaching interagency agreement on proposals ... it will not be possible ... to begin negotiations" as planned in the first week in April. The OES cable further stated that "At this point, it is not possible to determine with certainty when an agreed U.S. position will be achieved". Again, the USG had to cancel a meeting with the GOJ and also be unable to provide a proposal as promised. OES optimistically speculated that "with only a few weeks delay" the talks can be rescheduled. State requested the Tokyo Embassy to cancel the meetings with the GOJ and suggest a six month extension (to October 31, 1987) instead of a two year extension suggested by the GOJ and commented that securing an extension beyond this six month extension does not seem possible.[17]

e. Circular 175 Authority

In light of the impending USJ Summit meeting in April 1987, the first order of business concerning the 80STA was to obtain concurrence in the USG for a six month extension. This was not easily obtained.

During the three weeks from March 25 to April 14, there was a letter from the Secretary of State to the Secretary of Commerce, telephone calls among Assistant Secretaries, from Deputy Secretary of State to DOC Secretary. A meeting was hurriedly setup between OES A/S and DOC Under Secretary to try to rescue the situation. In preparing for this meeting March 26, 1987, OES reported that DOC "unexpectedly informed [OES] that because of improper internal coordination certain interested offices" in DOC were not aware of [State's] proposal and have raised strong objections and that the DOC Secretary was against extending the agreement, "on the basis of a very inadequate briefing at Commerce"[18] The OES talking points for this meeting were:

1. The U.S. negotiating position was to improve access to Japanese technology, create review mechanisms to control and regulate our relationship and a better balance.

2. The Japanese are ready to talk about this issue. The USG agreed with the GOJ on the timing and general substance of the promised discussions during the February 1987 meetings in Tokyo.

3. State is aware of DOC's concern that Japan emphasized development of commercial technology over basic science and we (USG) are seeking to correct this situation. Japan has only recently developed an ability to make real S&T innovations [-- a most condescending attitude towards Japan]. However, this capacity of the Japanese is likely to improve in the future and we may become more dependent on Japan for basic research results than in the past. [If that is the USG belief why push Japan into basic research with such vigor and determination?] Thus, this is not an opportune time to withdraw from an S&T relationship which opens doors otherwise most likely tightly closed.

4. To have the 80STA expire as Japan's Premier Nakasone is meeting with President Reagan would be most embarrassing to the USG.

5. This memo also said that "as best as we can tell, Commerce has not yet clearly defined its substantive concerns" with the USG proposal to Japan.

In order to reinforce the seriousness of the DOC objections at the very last minute, the next day, March 27, the Deputy State of State despatched a letter to the Deputy Secretary of Commerce. In the transmittal memo it is reported that DOC's Smart told OES A/S that he and the Secretary had not been kept adequately informed about the project [i.e., the USJ

S&T agreement renewal] and felt they needed a "breathing spell". Based on advice from the science adviser, OES A/S agreed to postpone transmission of the proposal to the GOJ for a few weeks "to give time to Commerce to put its house in order." The State letter to DOC made the following points:

- the interagency group had worked to accommodate the views of all agencies including DOC and USTR.

- "...the importance of initiating these negotiations while the Japanese appear ready." [This implies a past Japanese reluctance to negotiate. That is a somewhat inaccurate presentation of the Japanese stance. Nowhere do State documents indicate any Japanese objections to these negotiations. They must be waiting curiously and even impatiently at this point after so many USG promises to provide a proposal which was not produced.]

- the renewed agreement "will enhance our access to Japanese technology and begin to redress existing imbalances in the exchange of research and information." [These revisions could have been attained through the 80STA but obviously the USG wanted to create a new atmosphere through a renewal, a drastic revision.]

- State suggested a six month additional extension to the 80STA to create a unified USG position and negotiate with the GOJ. DOC was asked to concur in this extension.[19]

At an interagency meeting on April 3, 1987 a request for a six-month extension (and the addition of Annex V on national security obligations) was presented for concurrence. DOC conditionally concurred: it demanded a review process which would "take into consideration the whole range of trade and competitiveness issues involved"; no draft agreement be provided to the GOJ *"or to the U.S. Embassy in Tokyo"* [emphasis added] until after "an appropriate review of trade and competitiveness issues -- at the Cabinet level if necessary"; provision of an index of all S&T agreements which "will be entered into/amended/extended/renewed in the next six months" and an opportunity to review each item for any action to be taken; and that DOC reserves the right to request higher level interagency coordination [EPC, not NSC] and intends to ask the EPC to task an interagency group to draft policy guidelines for S&T agreements.[20] Apparently DOC insisted it needed more time to study the proposed revisions to the 80STA, and requested a list of all

STAs which were coming up for renewal or for negotiation during the next six months in order to study the trade implications of each BEFORE clearance could be given. State regarded this as "a clear challenge" to its management of S&T agreements.[21] OES and the Japan Desk felt this was a totally unacceptable position which must be turned around. "Allowing Commerce to insert itself into S&T affairs across the board like this would undermine our [State's] Title V responsibilities as coordinator of international S&T relations. A week went by and DOC would not budge. In this process, State had agreed, upon the insistence of DOC and DOD to take the proposed renewal to a Cabinet level review body, the Economic Policy Council.[22] OES A/S had asked the Deputy Secretary of State to phone DOC Secretary Baldridge.[23] This flurry of high level activity was apparently persuasive. Circular 175 authority was granted April 14, 1987.

When the Diplomatic Notes extending the 80STA another six months was first discussed with MOFA, the latter commented that six months was considered "too brief" for GOJ review of complex USG proposals, in particular any involving IPR. MOFA was correct in the time it took to negotiate a revised agreement. When Ambassador Endō, Director General of S&T Bureau in MOFA, representing Japan, signed the extension on April 28, 1987, he suggested that, if the USG had draft position papers on U.S. and Japanese researchers in Japan and the U.S. this could be an opening for developing opportunities for additional access of U.S. researchers in Japan.[24]

This disarray in the USG must have been obvious to the GOJ without knowing all the details and causes of the disagreements and constant postponements of promised bilateral meetings. While the GOJ was initially not at all convinced that a wholesale revision of the 80STA was necessary, it had gradually began to accept the USG's felt need to "strengthen and revitalize the agreement", especially since it was a Head of State agreement.

Twenty-fifth Anniversary of USJ Collaboration in Basic Science. Amidst this discord in the USG, the U.S. Secretary of State sent a congratulatory letter on April 13, 1987 to MOFA Minister Kuranari Tadashi for inclusion in the 25th anniversary report on USJ S&T collaboration mostly in the basic sciences with NSF. While some may discount the laudatory comments as necessary political hyperbole on this special occasion, it is one measure of how one arm of the USG felt about USJ S&T collaboration. Let us quote most of this letter.

> The United States-Japan Committee on Scientific Cooperation has served as an excellent example of bilateral science and technology collaboration for the past twenty-five years. As the oldest international cooperative science program in which the United States participates, the Committee has pioneered cooperation on a bilateral level and significantly advanced the scientific and engineering knowledge of both nations.
>
> In the past twenty-five years the Committee has worked to encourage more exchanges in fields of mutual scientific interest, contributed to the promotion of bilateral and international understanding and helped preserve our environment for future generations. All activities have been performed on a basis of mutuality, avoiding costly duplication of effort and resources.
>
> In sum, the United States-Japan Committee on Scientific Cooperation has merged the strengths of two nations in an effort which surpasses the results that they could have achieved acting independently....[25]

President Reagan also sent a letter stating that the programs under the U.S.-Japan Committee on Scientific Cooperation "have served as a model for many other cooperative programs, not only between Japan and the U.S. but also between our respective countries and other nation around the globe."[26]

USJ S&T in *Science, Technology and American Diplomacy*. As the GOJ is handed this rather unusual document, from the perspectives of the practices of international negotiations and mutual respect, the USG, in sharp contrast to the wrangling dialogue within itself, is describing the state of USJ S&T relations in the 1987 edition of the above publication as follows: This annual series was drafted by the State Department and published annually since 1980 to the mid-1990s as a Congressional Committee Print for the use of the Committee on Foreign Affairs, House of Representatives, U.S. Congress.

> "S&T cooperation has served an overriding U.S. foreign policy interest for more than 30 years with Japan. The original objective was to integrate Japan into the community of advanced Western industrial democracies. That objective has been achieved with *resounding success* [emphasis added], efforts are now being made to refocus cooperation under the U.S.-Japan Agreement on Cooperation on terms appropriate to the two countries' resources

> and priorities today ... Japan has emerged as one of the world leaders in S&T development in an impressive array of advanced technologies. This has introduced a new dimension into the bilateral relationship: the importance of U.S. remaining abreast of Japanese S&T activities.
>
> Formal cooperation in basic science and engineering research commenced in 1961 when President Kennedy and Prime Minister Ikeda established the U.S.-Japan Committee on Scientific Cooperation. ... The Committee is the oldest of its kind and has served as a model for bilateral S&T agreement with many other countries. ... A large number of scientists and engineers of both countries have established close professional relationships in the past 25 years. *Access to advanced public research facilities and information in Japan by U.S. scientists consequently is excellent,* particularly in the field of environmental studies and the management of water and bottom sediment pollution ... Access to private industrial facilities and government-backed public corporations and special projects such as ERATO and ICOT... also is beginning to occur. ... The U.S. proposals, developed by the Committee on International Science and Technology (CISET), aim to establish a firm partnership of peers that assures equity of access and scientific benefit in all cooperative activities, in spite of differing national R&D policies and structures."[27]

2. Re-building a Grudging Consensus, March - August 1987

a. A Policy Framework and Five Annexes

Despite the full participation of the DOC in the earlier clearance process for creating a package for presentation to the GOJ, the level of unspecified complaints by the DOC was suddenly raised to the Secretary of Commerce -- and thus, correspondingly, to the Deputy Secretary and Secretary of State. On May 13, 1987, Negroponte, OES A/S, wrote to Under Secretary of Commerce for International Trade, S. Bruce Smart, reassuring DOC that both State and Commerce share common goals in "enhancing our policies on trade, competitiveness and intellectual property rights" and that State had agreed to a Cabinet level review of the draft Japan agreement as requested by DOC. Although seven weeks had elapsed and the need for creating a unified USG package was doubly urgent, State could still comment:

In the next few days we hope to learn the specific views of the Department of Commerce on the current [rejected] draft.

This delay was the result of a basically de novo analysis, review, assessment of the draft S&T agreement with Japan by DOC and DOD -- after years of joint endeavors. In an internal DOC memo DOC's "key concerns" were listed as follows -- apparently these had not been transmitted to State:

Strategy for Achieving Equity. It noted that while the broad goal of achieving equity was accepted by all, there was "no clear agreement on strategy." DOC called for focusing on "obtaining equal value for continued Japanese access to our [U.S. national labs] research base." It called for access to Japanese applied research which was almost all in Japanese industry. It asserted that the U.S. should "utilize the new authority of the 1986 Technology Transfer Act and 1987 Executive Order #12591] to limit Japanese access to Federal facilities until equity is achieved." DOC disagreed with OSTP's emphasis on encouraging Japan to spend more on basic science and obtain access to Japanese labs.

Applicability to all Areas of U.S.-Japan Cooperation. It maintained that the "high standards of reciprocity" in proposed revisions should be reflected in *all* USJ agreements -- if not, the GOJ would prefer not to agree to projects under the Presidential-Prime Ministerial agreement. It suggested the development of a set of guidelines to apply to all agreements.

Guidance for Other Agreements. "One purpose for sending this agreement to EPC was to seek guidance for S&T agreements with other countries." DOC used the USJ agreement to force consideration of trade and competitiveness implications for S&T agreements in general. It demanded that State consult with DOC, DOD, USTR, and other agencies in coordinating S&T agreements[28] – based on the above evidence State had included all these agencies in the inter-agency discussions and clearance.

This review resulted in numerous major and minor changes in the draft agreement before presentation to the Economic Policy Council in July, 1987. These changes can be characterized as an attempt to tighten by several notches the control mechanism around Japan's S&T operations as they might affect U.S. access to Japanese S&T data and facilities. Although the revised

STA with Japan was conceptually an extension of the 80STA, an agreement which would facilitate unclassified basic and other research between and among the national laboratories of both countries, a fifth annex on the handling of potential national security information was formally added to the draft with an added stipulation that the revised 80STA would not become effective until the full implementation of the 1956 U.S.-Japan defense secrecy agreement. The revised STA suddenly was proposed to become a hostage to another totally unrelated USJ agreement. While the draft agreement with Japan would, as suggested above, attempt to control Japanese activities, the decision of the Economic Policy Council would, it was hoped, circumscribe the State Department's leeway in negotiating with the GOJ.

Before plunging into the details of the DOC review and demands, it would be useful to review how the State Department now viewed the "negotiating strategy" with the GOJ.[29] The "policy framework" continues "an extensive statement of principles and objectives for U.S. Japan S&T cooperation" for a more balanced relationship by encouraging the flow of Japanese scientific information from Japan to the U.S. A fifth annex has been added to the same four annexes. The fifth is a "statement on protection of information and technology transfer obligations for the safeguarding of U.S. interests in national security areas" -- no mention is made or hinted at concerning Japanese national security interests in this area. The U.S., by contrast, relies heavily on the leverage obtained through S&T agreements, such as the one proposed for renewal, to gain equivalent access to the Japanese R&D system. Moreover, "as Japan is now beginning to place more emphasis on basic research, we are likely in the future to want even greater access to the results."

In contrast to this somewhat aggressive stance, the same strategy statement also pointed out that:

> "Many Japanese government labs invite foreign participation, but have difficulty finding foreign researchers willing to go to Japan for any length of time. Language difficulties, lack of sufficient Western style housing at reasonable costs, and other cultural adjustment problems are seen as the primary problems, rather than resistance from the Japanese side." [Even] "he American Chamber of Commerce in Japan (ACCJ) High Tech Committee which met with Dr. Graham and Ambassador Negroponte in February [1987] cited the difficulty in finding U.S. researchers

> with Japanese language ability." [he proposed draft]"requires the Japanese to make efforts to lure US researchers to Japan ... principally at their own expense ... to encourage and assist US scientists to work in Japan", and sets up several bodies to monitor progress.

The above quotation from a State document would seem seriously to undermine the implied U.S. assertions that the Japanese are uncooperative in welcoming foreign researchers, that they somehow are not allowing S&T information to flow out of Japan, whether in Japanese or in English. Why is it necessary for one country to provide so much extra consideration in language training, cultural adjustment, special housing, etc "to lure" foreign scientists to a country, in this case Japan, when apparently by USG's own admission there are a dearth U.S. candidates? The U.S. does not provide such special consideration for foreigners coming to the U.S.

About the content of the USG proposal, the State strategy made this prescient assessment:

> "Given the major new concepts which we intend to table for consideration, many unprecedented, it is probable that much of the ensuing initial discussions may be devoted to explaining the meaning and intent of our proposals."

This is well stated and an accurate portrayal of the situation. Nevertheless, State blithely assumed it would take only "several days" to explain these ideas to the GOJ; they would then need "a month or so for internal discussion and consensus building" and if the USG sticks to the October 31, 1987 deadline (the expiration date of the latest extension) for completion of talks, this "could help significantly in crystallizing the Japanese response." Later, during the negotiations it did indeed take a lot of explaining to the GOJ and the deadline was grudgingly extended again and again. It was a blatant assumption that, under pressure from the USG, the GOJ's internal processes would be "crystallized" to accept the USG draft proposal.

Another Negroponte-Endō meeting. In mid-June 1987, Ambassador Negroponte met with Ambassador Endō Tetsuya, MOFA DG for S&T, about the Antarctic Treaty, space station negotiations, and the 80STA revision. When queried by the GOJ about the delay in providing papers to Japan,

Negroponte merely reported they are "still under review in the USG and that we [USG] are uncertain when the review will be completed", a refrain heard many times in the past. Although the USG claims to itself that the GOJ is "aware of the thrust of our approach" -- whatever "thrust" may mean in this context -- there is no intimation whatsoever that at this meeting the GOJ learned any more about the details of "the thrust" than in the past. It is important to keep in mind this apparent lack of information conveyed to the GOJ over these many months and its impact on the subsequent bilateral negotiations in 1987-1988.

In light of their objections to the earlier interagency draft S&T agreement with Japan, the DOC International Trade Administration (ITA), which included the Office of Japan Affairs, coordinated the DOC review and revision effort. Based on interviews with various personnel in ITA particularly the Japan Office in 1988-1989, the antagonism, more emotional than intellectual, against Japan was particularly vehement. DOC called for numerous changes in the draft agreement almost all of which were accepted by the interagency group. On the other hand, the Japan Desk was quite dissatisfied with the package produced by OSTP, believing that it ignored the interagency process undertaken over several months, that it harks back to positions taken by OSTP staff before that process unfolded and that it went well beyond S&T issues in the US-Japan bilateral relationship.[30] Only the principal changes will be described below.[31]

Coverage and applicability. Clearly DOC called for the application of the broad policy framework to "all agency-to-agency arrangements" between the USG and GOJ. This is intimately related to the long standing and unresolved issue of the nature of an umbrella agreement. DOC maintained that the technical fields listed in the "priority areas for cooperation" were those where "the U.S. has projects of a far greater magnitude than the Japanese." DOC called for their deletion from the draft agreement but were unsuccessful.

S&T Goals. The policy framework or preamble had included three goals in the past. A fourth was added "to protect intellectual property rights so as to preserve the value of innovations derived from joint collaboration".

Investments in Infrastructure. Under cooperation in international science and technology in the policy framework, an important addition was added committing Japan to investing more in R&D infrastructure. The draft proposes that in order for both countries to derive equitable benefits from a balanced collaboration, both governments acknowledge that "at times one country may need to make greater commitments [i.e. funds] in particular components of the R&D process". While this may seem like a balanced statement because it is conceivable one country or the other could have invested more in one field than the other party, thus implying a balanced statement, the implication behind this statement is that Japan is the one to make that greater investment. The DOC revision made this implication explicit by adding the following to the above quotation:

> [components of the R&D process] "and infrastructure commensurate with its scientific and technological strengths. In this regard, the United States believes that the Government of Japan should increase its commitments in such areas as basic research support, graduate and post-graduate education, training at universities and national research institutes, research facilities and the dissemination of information".

Considering the fact that on the surface, at least, the USG proposes an agreement between two technological superpowers, the US and Japan, dealing with each other on presumably equal terms, this is, to say the least, an extraordinary unilateral assessment included in a bilateral agreement. It is difficult to imagine that the GOJ would accept that statement in an agreement. Its very inclusion in a proposal would surely not contribute to amicable discussions. It is in reverse a remarkable and blunt statement about the DOC's attitude toward Japan.

Major Initiatives and Priority Areas of Collaboration (Annex I). In discussing bilateral projects, the DOC attempted to commit Japan to help fund large scale U.S. national projects by mentioning them by name, the space station which was being negotiated at the time, and the proposed Superconducting Supercollider (SSC) which the GOJ was obviously reluctant to co-fund with the USG. This was but another attempt by DOC to try to "commit" Japan, explicitly or implicitly, to contribute substantially to U.S. projects. While the space station project was a U.S. project with international cooperation from the beginning, the SSC was a national -- not an international -- project. When SSC costs spiraled and U.S. funds inadequate then the USG turned to Japan and others for funding support. In a memo for a July 17 meeting, DOC called for deletion of the suggested priority areas of collaboration. It said that this list "concedes the current high level of Japanese access to many key U.S. programs such as NIH biotechnology research." It suggested starting from zero base and equally from both sides. DOC's technical agencies also feared that their USJ programs would be "disrupted" if subjected to OSTP management/oversight.

Mechanisms, Initiatives and Areas for Joint Review (Annex II). It will be recalled that this was the section which specified in what manner the GOJ would assure that more U.S. researchers would go to Japan and through what mechanisms these commitments would be enforced. Not satisfied with the seven clauses detailing how the Japanese effort would be monitored for compliance, DOC added another eighth part. Rather than trying to paraphrase the addition, it is more effective to quote it verbatim:

> "(8) Recognizing that U.S. scientists and engineers face unique problems in gaining access to Japanese scientific and technical information (STI) not encountered in other major STI source countries, an STI committee will be established to examine and develop recommendations on STI access issues. This committee will serve as a forum where STI organizations may raise and resolve concerns. The committee will be comprised of the directors of Japanese and U.S. Government agencies and other organizations which are directly concerned with STI transfer. In addition to dealing with the specific problem of unpublished literature which is produced as a result of Japanese Government-funded research, the Japanese Government will take steps to ensure that all scientific and technical reports produced by GOJ agencies and their contractors which are not published in "open" (refereed) journal literature will be made available to U.S. researchers through a central source such as NTIS."

The same kind of comment and assessment, in a perhaps even more stringent manner, can be made of this new addition by the DOC as was made for the proposed infrastructure commitment by the GOJ.

Oversight and Management (Annex III)

Multiple co-chairs of the Joint Committee. DOC demanded that, in light of its responsibilities under the 1986 Technology Transfer Act, it should co-chair the Joint Interagency Executive Committee together with State and OSTP. In the afore-mentioned Laun-Farren memo of July 17, Commerce called for the addition of USTR as a fourth co-chair because the 1986 Technology Transfer Act gave a significant management role to both USTR and DOC. This means that the U.S. delegation would, in theory, have 4 co-chairs of its delegation. Who is the senior U.S. representative? Obviously, the State Department; but the intent here was to so surround State that it loses any leeway, any maneuverability without the concurrence of DOC, USTR and OSTP neither of which are sympathetic to State's philosophy and style of approach to the GOJ. This is but another example by DOC to create an obstacle to any State initiative and, in the process, create a distinctly unwieldy institutional arrangement.

Joint Committee Plans. The Joint Interagency Executive Committee would develop an annual action plan which the Joint Committee would approve and adopt an action plan which would enumerate the steps which would be taken to assure compliance with the provisions of Annex II discussed above, and to assure that agency-to-agency implementing agreements are fully consistent with the STA. The former is presented neutrally applying to both the U.S. and Japan but DOC's real intent was but another step to tighten the requirements around Japan. Concerning the latter, there seems to be an implicit assumption that perhaps the agency-to-agency arrangements were not necessarily in compliance with the 80STA. Therefore, this possible deviation must be controlled under the revised agreement. There was another unproven but strongly felt assumption in DOC that these agency agreements redounded to Japan's benefit even though all known assessments described earlier in this study maintain that benefits have been mutual.

Intellectual Property Rights (Annex IV). The major change here was the deletion of two clauses in article IV.(1) and IV.(2) under inventions of the version rejected by DOC. These clauses were discussed earlier in this chapter. For example, when an invention is made by a visiting researcher the receiving institution would have the right to the invention. Under these circumstances, the clause to be deleted stipulated that "in any country where one Party obtains rights and interests in an invention, the other Party has the right to a non-exclusive, irrevocable royalty-free license for the Party and its nationals."[32]

There is an operating assumption here that the invention is more likely to be made when a Japanese scientist is visiting a U.S. institution rather than when a U.S. scientist/engineer is visiting a Japanese institution. This is a strange assumption that some how the presumed "uncreative"

Japanese scientist/engineer might create something worthwhile or be a party to this creative act. "Given the current levels of technology flows [in mid-1980s from the U.S. to Japan] we [DOC] believe this is a fair and accurate assumption from which to begin."[33] This interpretation seems to ignore the possibility -- if the U.S. and Japan are both technological superpowers -- that this could easily cut both ways. This was pointed out as not being in the best interests of the U.S. and later a more sophisticated formula was created during discussions with the GOJ.[34]

Article II of this Annex refers to dual use technology. An OES memo commented that in USG-GOJ space station negotiations, the USG was arguing that "peaceful" also incorporates defense, thus suggesting that "dual use technology" was more appropriate.[35] More importantly, this section included a stipulation that the implementation of the UJST agreement would be "contingent on full implementation of the 1956 Patent Secrecy Agreement". State opposed this linkage. The DOC memo maintained that 1) GOJ has not implemented this agreement and that State has had 30+ years to "ask" and request the GOJ to implement it. There was, of course, no discussion of these efforts by the USG (State and or DOD) to push and demand implementation by the GOJ. 2) Japanese industry was receiving U.S. defense and private commercial efforts -- which "often have commercial applications" "free of charge". 3) In light of the "Toshiba export violations", failure to include "strong patent secrecy provisions" would raise questions about how the Bush Administration was handling USJ S&T relations. 4) Inclusion or exclusion of this provision may set a precedent such as in the Space Station agreement and DOC called for inclusion of this contingent provision in all pending USJ S&T agreements.[36]

Security Obligations (Annex V). At the request of DOD in accordance with the President's Executive order of April 10, 1987, a completely rewritten annex on the protection of defense related information and technology transfer was added to the proposed S&T agreement with Japan. This is the second major addition demanded by DOD for inclusion in the revision.

The first part specifies a fairly clear concept that no properly classified defense or foreign relations related information or equipment shall be released under this agreement and that if such information of equipment is developed during collaborative projects shall be "immediately" brought to the attention of both governments.

The second part on technology transfer stems directly from the Toshiba-Kongsberg incident where unauthorized information and equipment on submarine screws was provided to the Soviet Union. The new addition stipulates that the USG and GOJ will take all necessary and appropriate measures in accordance with the international obligations (presumably meaning COCOM), national laws and regulations to prevent the unauthorized transfer or retransfer of unclassified, export-controlled information and equipment provided or produced under this Agreement. Furthermore, detailed provisions for the prevention of such transfers will be "incorporated into all contracts and others arrangements implementing this Agreement."

This is another attempt to tighten the controls surrounding S&T information developed in collaboration with Japanese scientists and engineers. Undoubtedly, OES opposed this revision, knowing that this annex and the dual use technology revision would be extremely difficult, if not impossible, to negotiate with the GOJ, yet was basically powerless to oppose this revision.

This interagency draft became the basis for the agenda for the mid--July Economic Policy Council which set another set of guidelines for the negotiation of a revised STA with Japan.

Secretary of State meeting with Dr. Eric Bloch, NSF Director. Ten days before the Economic Policy Council was to meet on the renewal of the 80STA, the Secretary of State met with Eric Bloch to underscore that State wished to cooperate closely with NSF, especially in its overseas program. State and NSF were natural allies on the USJ STA in opposition to OSTP and DOC. The OES prepared briefing memorandum for this meeting contained a frank commentary on the USJ S&T relationship and the 80STA renewal process:

1. Efforts to renew the 80STA have been complicated by interagency jockeying over basic goals.

2. OSTP seeks to centralize coordination of S&T activities and to increase access to Japanese R&D.

3. DOC wants to redress a perceived imbalance in the flow of commercial benefits.

4. DOD wants Japan to implement commitments it undertook in 1956 to honor and protect classified U.S. patents.

5. Obstacles to increased opportunities for U.S. researchers to visit may be largely linguistic and cultural.

6. The Japanese do not appear to be overly anxious to renew the agreement and filling the hopper with other issues, e.g. such as some aspects of the IPR annex, and dual use technologies, would not be helpful.[37]

b. The Economic Policy Council Decisions

EPC (together with the Domestic Policy Council) was created on April 11, 1985, early in the second Reagan Administration. These two Councils were created to streamline policy and decision-making. Together with the National Security Council they would serve as the primary channels for advising the President on policy matters. EPC would be the "single entity" to advise him on domestic and international economic policy and oversee the coordination and implementation of the Administration's economic policies. The President chairs the EPC but in his absence the Secretary of Treasury, as Chairman Pro Tempore, would serve as Chairman. EPC consists of the Secretaries of State, Treasury, Agriculture, Commerce, Labor, Transportation, USTR, Chairman, Council of Economic Advisers. The intent was to have meetings of the Principals or their Deputies without a large retinue of staffs. These Councils were to meet as needed to seek consensus and the time limit to reach this consensus was set at one hour based on staff papers previously prepared and appropriately staffed.[38]

In preparation for the upcoming EPC meeting on the USJ STA, CISET prepared a ten-page report and met on June 11, and July 17, 1987 to consider this report and formulate a coordinated USG position for the EPC's approval.[39] Quite unlike the earlier CISET report, this report asserted that the revision of the 80STA gave the U.S. "precedential [sic] opportunities" to create a new STA to:

- improve the balance of science and technology contributions in trade, transfer, and investment between the U.S. and Japan

- provide a single framework of U.S. policy objectives, similar to the U.S.-China model [presumably the US-PRC January 31, 1979 STA], for consolidating existing Federal S&T agreement with Japan, as they expire and

- provide a benchmark for future negotiations of U.S. science and technology agreements with other trading partners.[40]

This CISET process provided yet another occasion where differences between the State Department (and NSF) on one side and OSTP, DOC, DOD were clearly manifested. An OES memo noted that while State and other agencies share in the OSTP objective of up-grading the agreement to improve the cooperative relationship, "there is strong disagreement with the confrontational approach being proposed by OSTP to identify problem areas and lay out corrective steps for Japan to follow."[41] The following points seem to be the main bones of contention as preparations for the July 24, 1987 meeting proceeded.

- The second OSTP draft result resulting from the CISET June 11 meeting did not reflect many of the agency concerns raised at that meeting, and according to OES (about July 10-17, 1987) "a consensus had not yet emerged", and that OSTP had scheduled an EPC review of the revised CISET report without a prior CISET meeting.

- State re-emphasized that it could concur with substituting EPC for NSC as the on-going Cabinet level arbiter of S&T relations with Japan. State's acceptance of EPC review in this instance was a one-time exception. It called for a revision of the CISET report which should be "carefully vetted" before going to the EPC.[42] In a briefing paper for the State representative at the EPC meeting, Under Secretary Wallis, OES understood that "EPC will not accept such a role effectively supplanting the NSC". This same briefing paper also revealed that on July 22 representatives of the NSC, EPC, and OSTP and State (significantly without DOC, USTR, and DOD) met and agreed on "a number of significant changes in the approach taken to date by OSTP toward renewal of the USJ agreement. The paper also noted that "no major new projects have been initiated under [the 1980 USJ S&T] agreement since its inception ... and that the U.S. now derives very substantial benefits from its inter-governmental S&T relationships with Japan ... and that State can accept the principles contained in the July 22, 1987 EPC paper prepared by its staff."[43]

- Without a coordinated policy direction at the CISET, the EPC should not consider this issue; it would be premature.

- State categorically stated it is "incorrect that Japan does not perform enough R&D, especially basic research." Citing NSF data it pointed out the following (total R&D expenditures as % of GNP).

Basic Research	U.S.	12.1%
	Japan	12.9
Applied Research	U.S.	22.0
	Japan	25.0
Developmental Research	U.S.	65.9
	Japan	62.2

OES also maintained that it is not clear that new Japanese efforts and expenditures "to redress perceived inequities" would solve the problems, that Japan would agree to make them or that it would be in the U.S. interest for Japan to do so.

- State rejected DOC and OSTP co-chairing with State the Joint Interagency Technical Committee in light of the Secretary of State's responsibilities set by statute.

- On July 21, 1987, three days before the EPC meeting, an executive summary of the CISET paper was distributed by EPC. State/OES responded immediately that 1) there is no interagency cleared CISET paper on the 80STA and 2) S&T agreements are not trade agreements. Although they often have trade implications there are also national security and other policy implications to be considered. Thus, EPC consideration of the S&T agreement is a one-time exception to the normal NSC review process.[44]

EPC considered and approved the EPC objectives and guidelines for renewal of the 80STA on July 24, 1987 with minor modifications. Science Adviser, Dr. William Graham gave the presentation. Under Secretary of State Wallis (together with OES A/S Negroponte) represented State. Others included the designated alternates to the Principals and an array of senior White House staff. While the principal focus of the approved documents was, of course, on the renewal of the USJ S&T agreement, there were some NEW important principles set forth by the EPC for future negotiations of STAs with Japan and other countries. The main points of the EPC decision are as follows:

1. For the first time and at such a high level, "commercial and trade interests" were formally inserted into STA considerations. The revised USJ STA was to include the principles of

balance and cooperation that reflect U.S. science, *commercial, and trade interests* [emphasis added] consistent with national security and foreign policy interests.

2. Although the word "umbrella" agreement was not specifically used by the EPC executive summary, the above principles would apply to "most existing and all future U.S. S&T agreements and MOUs with Japan to achieve consistency in a) U.S. negotiations abroad and b) U.S. domestic planning and policymaking." Instead, for the first time -- based on available documents -- the above would "provide *a model umbrella of principles* [emphasis added] for elevating and coordinating S&T agreements and MOUs with other U.S. trading partners".

3. Under "Principles for Balance and Cooperation," the first among five items is "equitable responsibilities (not necessarily identical) and reciprocal opportunities for participation in each other's S&T enterprise." This implicitly reaffirms the concept mentioned earlier that Japan may need to invest more funds in certain R&D projects than the U.S. The U.S. and Japan share responsibilities for 1) generating new technologies, 2) maintaining an open basic research environment and disseminating S&T information, 3) supporting open academic basic and applied research, 4) nurturing the next generation of scientists and engineers, 5) adequate protection of IPR and disposition of patents and copyrights, trade secrets and knowhow arising from collaborative activities, 6) adequate protection of dual use technologies and classified and unclassified research activities and results subject to the export control regime, 7) sponsoring and financing large-scale R&D projects at the frontiers of knowledge the costs of which should be a proportion of their risks, benefits and management shares.

4. The Defense Patent Secrecy "hostage" article in the USG draft proposal which was supported by DOD, DOC, USTR, and OSTP but opposed by State and NSF and other agencies, was left in the package by the EPC "primarily to be used as leverage for the Patent Secrecy negotiations conducted separately."[45]

5. Guidance for U.S. negotiators. EPC reaffirmed 1) the six technical areas of areas of priority cooperation with Japan agreed upon internally within the USG, left out of the draft given to the GOJ and subject to bilateral negotiations,[46] 2) that Japan must place U.S. researchers at its national laboratories,

provide language and cultural training programs, establish Japan Fellowships, join in a bilateral task force to check reciprocal access, 3) that the IPR annex be included in the agreement, 4) that Japan issue English abstracts on all technical articles not published in open (juried) literature, 5) the Joint Committee review results of cooperation, initiate new actions and provide a comprehensive annual report to both governments, 6) a Joint High Level Advisory Panel be established, a majority of whose members would be from non-governmental institutions, and 7) establish a joint interagency executive committee to support the Joint Committee. On the USJ side, this executive committee would review all USJ S&T agreements to assure that they are consistent with the principles of the Presidential/Prime Ministerial Agreement, the U.S. executive order 12591 on facilitating access and the circular 175 process, assist federal agencies apply the reciprocity provisions, and establishing subgroups to monitor joint bilateral research. If agreement to these principles could not be obtained from the GOJ, "the Council [EPC] may wish to recommend to the President other options as might be available: letting the agreement lapse and taking other unspecified remedial actions for achieving balance and reciprocity in S/T relationship with Japan." That "dire action" was not necessary since an agreement was reached; but to what extent were these principles and guidance fulfilled?

c. Circular 175 Authority

The procedure to obtain authority to negotiate with the GOJ for the revision of the 80STA was begun in early August 1987 by OES. The proposed draft was given to the MOFA on August 28, 1987 and the circular 175 authority was granted on September 9, 1987. The stage was now set for full bilateral negotiations for a complete overhaul of the 80STA, creation of basically a "new" agreement between what presumably are scientifically and technologically equal partners.

3. Act III, At Last a Unified Proposal

Thus ended another Act in the USG saga to try to create a governmental consensus for the extension of the 80STA. The Act can be distinctly divided into two scenes: the first scene one of somewhat high drama at the upper echelons of the USG and the second, a scramble to create under pressure a semblance of a unified USG proposal to transmit to the GOJ.

When the curtain fell on the first scene, the USG interagency process was momentarily in complete disarray. Despite lengthy -- and often acrimonious -- interagency discussions including several DOC representatives, the DOC at the last minute rejected the proposed package to give to the GOJ asserting that it needed time to completely review the package before it could concur, that the USJ S&T relationship was such that the issue of renewing the agreement should be a special issue for the Cabinet and Economic Policy Council to consider. The objective here seemed to be to raise the stakes against the State Department and to the greatest extent possible to give the State negotiator as little leeway as possible in their negotiations with the GOJ. Having forced this review process on State, DOC agreed to a six month extension of the 80STA.

Retrospectively, the initial efforts to assess and coordinate the far--flung agreements and MOUs in S&T between the U.S. and Japan in S&T were begun in a modest manner among the staffs of State and OSTP. The first evaluation (The Bloom study) of these relationships was prepared by a contractor (a former minister counselor/science in the AmEmb in Tokyo). The original concept entertained by State and the Embassy and the GOJ had been to renew the 80STA in its original form. There was dissatisfaction with this approach in OSTP. OSTP perceived that the 80STA had not lived up to their original expectations as a Presidential Agreement and how it might have benefitted the U.S. but did not. OSTP had, at one time, even opposed the renewal of the agreement. State/OES obviously did not have the drive to carry out its approach and the technical agencies in the USG were skeptical of the implications of OES and OSTP attempting to make the Presidential agreement an overall umbrella S&T agreement when, from their point of view, they had evolved highly satisfactory working relations with their Japanese counterparts. As the months and years dragged on with numerous internal discussions in the USG, the under current between State/OES and OSTP became thicker and thicker, and blacker and more venomous. The circle of participants in this dialogue gradually increased to include DOC, DOD, USTR and the bureaucratic alignment of these agencies standing as a phalanx against OES. The entire USJ S&T relationship also came entangled with domestic legislation on the transfer of technology (e.g. the 1986 Technology Transfer Act) and the need to revise the proposed draft agreement with Japan to make the latter responsive to U.S. needs. Anti-Japanese emotions, distrust of and contempt for and, at that time, even fear of Japan, found in DOC, OSTP and DOD only inflamed the disagreements with OES in State. Although a draft proposal had been apparently thoroughly vetted among the interested agencies in late 1986 and early spring 1987, at the last minute (toward the end of March 1987), DOC challenged the propriety of some clauses in the IPR annex and demanded that the draft revision be completely re-reviewed by their legal counsel and others and then considered by the EPC chaired by the President as a condition for agreeing to a six month extension of the 80STA. After having been deeply involved in the recent interagency process to create a draft proposal the DOC move was indeed cynical yet adroit, highly disruptive yet effective, and successful in forcing the consideration and confirmation of some new issues in S&T relations with Japan *and* other countries. The DOC demand was obviously a move to embarrass the State Department -- practically challenging their patriotism -- and limit, to the greatest extent possible, their leeway in negotiating with the GOJ. It blossomed into a full fledged confrontation between State, on the one side, and DOC, OSTP and DOD on the other. When DOC called for the creation of an interagency committee co-chaired by four agencies and to have EPC review each STA, State was finally bestirred to object effectively to this attempted turf infringement. It won its case at the EPC. On the other hand, DOC won its demand to have trade and national security issues considered and to have a set of principles approved for negotiating STAs in the future, including the 80STA.

During this period, the DOD representative had changed from a career civil servant to a political appointee who was (in my view) emphatically critical of Japan, if not quite anti-Japanese in his proposals and actions. Whereas the STA was to broaden, deepen, and provide better access to U.S. and Japanese S&T information and research facilities, suddenly the proposed package contained a national security section and tied the STA execution to the implementation of a 1956 U.S.-Japan agreement on the handling of inventions which may be classified for national security reasons. This new condition would, if shown to the GOJ, seriously complicate the bilateral negotiation process. According to one of the three coordinators of the Joint Task Force report of December 1986, the latest USG draft agreement was "a fundamental discontinuity" with the JTF Action Plan which was to have been the basis of a new S&T relationship with Japan. Since 1987 was a year when U.S.-Japan bilateral relations were at an all time postwar low point, it was stylish to attack Japan, to insert complicated clauses to tie down Japanese actions and commitments, it was bureaucratically much easier to be on the offensive against Japan, rather than try to tone down the many rather one--sided clauses in the draft revision. That role would fall, almost completely on the shoulders of the State Department. Since OES was not one of the more powerful Bureaus in the Department this was basically impossible to accomplish. Any such attempts by State, would create a whole field of opportunity to "bash the Japanese via the State Department". It appeared more patriotic to try to control the Japanese rather than attempt to create a warmer, bilateral, incremental and equal basis to USJ relationship. This period exemplified the depth and breadth of anti-Japanese sentiment as a whole and a strong personal animus against Japanese held by some senior personnel in such agencies as OSTP, DOD and DOC. It would be extremely difficult for a weak OES Bureau and the Japan Desk in State to combat these sentiments and propose actions and ideas from the State Department, without incurring and inciting heaps of scorn and unspoken contempt and vilification upon themselves.[47]

The second scene culminated, after four years since mid-1983 of contentious inter-agency wrangling, in the approval at the highest level of inter-agency coordination in the USG, the EPC, of a set of principles for the 80STA revision and for future STA with other countries and, by implication, of a draft proposal to give to the GOJ. The draft agreement with Japan reflected, therefore, for the first time, a new philosophy that S&T agreements must hereafter include U.S. trade and commercial and even national security interests and stood in unusually sharp contrast to the publicly stated positions of a glowing and constructive S&T relationship.

Notes

1. U.S.-Japan Science and Technology Agreement. p. 5.
2. OES memo, 3/15/1987, Michaud to Negroponte, US-Japan S&T Agreement: working with OSTP draft, 2 p. It refers to "OSTP drafts" and "State and NSF drafts." The State draft is not available, only the OSTP draft. Another OES memo, 2/24/1987, Michaud to Negroponte, Preparations for Negotiations on the U.S.-Japan Non-Energy Agreement describes State's proposal and 3 annexes but with no actual draft. It describes OSTP as the "most critical factor" in this process and suggests OES A/S send a note to OSTP's Graham.
3. The OSTP representative, Deborah Wince, and Mike Smith subsequently married.
4. The involvement of Gaffney and Smith in this process, and the appointment of Bruce Smart as ITA Under Secretary of Commerce in DOC tilted the milieu in the interagency process toward a distinctly anti-Japanese mold.
5. In an OES memo (3/15/1987), US-Japan S&T Agreement: working with OSTP draft, commented that this type of material could be given to the GOJ as a "non-paper." This material would "invite unproductive arguments". OES made several suggestions how to combine and adjust OSTP's draft to State's draft, thus "we could avoid simply rejecting the OSTP while blending its substance with the draft note we [State/OES] have prepared." In private State staffers ridiculed the amateurishness and irresponsibility of the OSTP draft proposal.

6. Memo from EAP (Sigur) to The Secretary on meeting with the Japanese Foreign Minister on March 6, 1987.
7. State 080369 (3/16/1987) to Tokyo on meeting with Japanese science counselor, Ikeda Kaname.
8. State 68752 (3/10/1987), Tokyo 4323 (3/12/1987). This writer found a similar difficulty in 1988 when he tried to obtain an authoritative list of projects under the 80STA from U.S. Department of State (OES). This difficulty exemplifies the ambiguous meaning of "umbrella" agreement. The number of projects on these lists differed, like a constantly moving target.
9. There is no evidence that either of these cables was actually sent to Tokyo. Furthermore, contrary to the existence of these draft cables, in a collection of documents prepared for the October 2-3, 1987 briefing of the GOJ about USG proposals, called collectively, *Diplomatic Note, Talking Points and Annexes 1-5*, Annexes 1-3 are described as growing out of the Joint Task Force Report of November 1986, and annexes 4 and 5 from a series of CISET meetings in late spring and early summer 1987. So much for correct institutional memory.
10. Documents on earlier and subsequent circular 175 authority requests have been available and commented on in this study. Unfortunately, the same documents which were being circulated through the USG in mid to late March 1987 with a draft similar to the March 25 version were not made available.
11. The handwritten notes on a copy of the OSTP transmittal memo of 3/13/1987 asking for clearance over the weekend for its draft was revealing: What does OSTP really want? to get Japan to fund more basic research in U.S., to get more U.S. researchers to Japan, to change U.S. S&T policy, will this data result in more funding needs, how does the USG use the data thus collected, etc, etc.

    From where did the ideas for Annex II come is not clear. I was told that OSTP was the source but this could not be proven. I did obtain, however, an undated draft memo but probably around late 1986 or early 1987 from the DOC Under Secretary for Economic Affairs to the SecCom, commenting on the revision of the 80STA: it would require the GOJ to make available all open S&T reports produced by GOJ labs or contractors through a central source, create an executive level U.S.-Japan committee to examine access to JSTI, modify Japan patent office practices to expedite issuance of patents to U.S. applicants and others concerning NTT and measures to increase Japanese domestic consumption.
12. An IPR Working Group was chaired by a USTR lawyer with representatives from PTO/DOC, and several other offices in DOC Hqs, e.g., Office of Japan Affairs/ITA. I have not been able to find out how this working group and the IPR annex were created – only the results are known.
13. OES (Reifsnyder) to OES (Prochnik) memo, 7/15/1987. Draft IPR annex for Japan-US S&T agreement.
14. Memo, 3/25/1987, Robert H. Bromley, Deputy General Counsel, DOC to John D. Negroponte, State OES A/S with a copy to John C. Whitehead, [Deputy Secretary of State]. According to an undated draft OES memo (after July 10, 1987) an interagency group agreed on a draft text to negotiate with the GOJ in spring 1987. But OSTP insisted on another review BEFORE the draft text could be passed to the GOJ. It was during this second review that the DOC objected to the text, especially the IPR provisions.

15. An anecdotal story told to me in 1988 was difficult to believe possible but was related in all seriousness: OSTP would not clear a State EAP DAS to enter the New Executive Building where OSTP was then located for a meeting. This then resulted in a State Under Secretary refusing to attend this meeting. The impasse was resolved by shunting this confrontation to a higher in State and to the Science Adviser himself; OSTP backed down.
16. Another interpretation based on interviews was that certain elements of DOC were looking for an excuse to effectively stop this agreement right in its tracks in its contest with State and also somehow "punish Japan" for the Toshiba violation of COCOM rules. This presumably would embarrass Japan. I was also told by a ranking member of DOC that he brought the draft to the attention of the Director of the Federal Technology Management Policy Division, in OPTI/DOC. He studied the draft and declared it to be "fatally flawed", an accidental manner of communication. ITA/DOC claimed that it was responsible for management of government patents and it had not been included in the drafting process; indeed OPTI and NOAA had been interestingly the DOC representatives at the A/S level. Yet, another DOC described the U.S. stance as "a hard line that was ill-timed and ten years too late; the U.S. was like a petulant child. If true, this obstruction was effectively achieved for the moment at a substantial political cost within the USG and a serious image problem with GOJ."
17. Cables, Tokyo 5319 (3/26/1987), State 91721 (3/28/1987), Tokyo 5551 (3/31/1987).
18. OES Memo, 3/26/1987, to OES A/S, Your meeting with Commerce Under Secretary Bruce Smart: the U.S.-Japan S&T agreement issue.
19. OES/EAP memo, 3/27/1987, to Dep. Sec. Whitehead, Letter to Commerce Deputy Secretary Brown requesting Commerce withdraw its objections to S&T negotiations with Japan.
20. OES memo, 4/3/1987, to OES A/S Negroponte, US-Japan S&T Negotiations and the Possible Wider Ramifications. This memo stated there were three conditions to DOC's concurrence to extend the 80STA but they were deleted in the State FOI review process. DOC officially described its conditions in its reply to Deputy SecState's letter of 3/30/1987 on 4/17/1988.
21. OES memo, 4/3/1987, to OES A/S Negroponte quoted above.
22. OES memo, 4/16/1987, U.S.-Japan S&T Agreement.
23. OES memo, 4/10/1987, to OES A/S Negroponte, Commerce position on Extension of US-Japan S&T Agreement; OES A/S Negroponte memo to Deputy Secretary Whitehead, 4/10/1987 [Commerce Objections to S&T Agreement with Japan].
24. Cables, Tokyo 7101 (4/22/1987), Tokyo 7437 (4/28/1987), State 121434 (4/23/1987), Tokyo 9727 (4/28/1987), Tokyo 7714 (5/1/1987).
25. Letter (4/13/1987), George P. Schultz to Tadashi Kuranari, MOFA Minister.
26. Minister Kuranari and Prime Minister Yasuhiro Nakasone also sent similar congratulatory letters. It is significant that the SecState letter was cleared by OES, EAP and NSF and not DOC or any other USG Agency. The reason is that the comments were directed to basic science relations; it asked, however, it is highly doubtful that other such agencies as DOC, DOD would have concurred at this time with such laudatory descriptions.
27. *Science, Technology, and American Diplomacy.* 1987. Eighth Annual Report Submitted to the Congress by the President Pursuant to Section 503(b) of Title V of Public Law 95-426. p. 27-28.
28. DOC memo, 7/17/1987, Louis Laun [Assistant Secretary of Commerce for International Economic Policy] to J. Michael Farren, Deputy Under Secretary for International Trade, DOC, for a July 17, 1987 CISET meeting on U.S.-Japan Science and Technol-

ogy Agreement. Although this memo was prepared months after the decisive date of March 25, 1987, it is cited here since it provides some explanation of the motives which prompted the DOC action on March 25, 1987. DOC did provide a proposed list of factors for considering the trade and competitiveness implications of the proposed S&T agreement but this list unfortunately was not available with this memo.

29. This description is based on 1) a truncated document draft by State OES, May 20, 1987 of which only part 8 Negotiating Strategy has been made available, and 2) a June 1, 1987 memo on a meeting between OES A/S and the DCM in Tokyo Embassy.
30. OES memo, 5/27/1987, untitled.
31. Two almost identical draft agreements are available: one as an attachment to an undated memo from J. Derderian, ITS/IEP/EAP/OJ, to the Under Secretary for International Trade in DOC and another attached to a memo dated June 9, 1987 from OES/SCT to 22 members of the interagency group for their consideration. The latter includes changes "suggested by Commerce and DOD as modified by OSTP" and a complete re-write of the IPR section. The description of the principal revisions is based on this version since it is the one circulated for interagency clearance. The undated DOC memo will also be used as background to the DOC position. In addition, the afore-mentioned Louis Laun-Michael Farren memo of 7/17/1987 will also be used in this description.
32. The afore-mentioned Derderian memo states under issue 3, the IPR language was "revised by a sub-group of Commerce IPR specialists to be consistent with U.S. policy as defined in the Technology Transfer Act and the recent executive order signed by the President on April 10, 1987." The controversial "third" paragraph which was believed to be the cause of DOC disagreement (and which was deleted from the draft given to the GOJ) was surprisingly left in the revised DOC draft to the interagency committee.
33. p. 3 of the afore-mentioned Laun-Farren memo of July 17, 1987.
34. OES(Reifsnyder) to OES (Prochnik), 7/15/1987, Draft IPR Annex for Japan S&T Agreement. This memo also opposed holding the 80STA revision as a hostage to the USJ Patent Secrecy Agreement.
35. ibid.
36. p. 2 of the afore-mentioned Laun-Farren memo of July 17, 1987.
37. OES A/S memo to The Secretary, 7/10/1987, Luncheon meeting with Director Eric Bloch, July 13, 12:30 PM.
38. Since senior officials without staffs attended EPC meetings, minutes were a problem and regarded as less than satisfactory. Based on the available record on the USJ S&T issue the minutes were fairly detailed and circulated for clearance.
39. US-Japan Science and Technology Agreement, CISET report for the EPC, circulated by OSTP on June 10 for the June 11, 1987, CISET meeting. Another meeting was held on July 17, one week before the EPC meeting to try to resolve differences. While it is not known who on CISET drafted this ten-page report, it should be remembered that CISET reports to the Director, OSTP, but that CISET does consist of representatives from numerous agencies, including State, DOC, DOD, and many others. It reiterated the litany of presumed ills and deficiencies of the USJ S&T relationship: GOJ should/-must shoulder more responsibilities in rectifying the imbalance in number of students and researchers in the U.S. and Japan, should take steps to publish in English the results of their S&T research, should open up "her premier government sponsored R&D facilities" such as AIST to foreign participation, should increase cost-sharing and sponsorship of large scale R&D projects with foreign participation, present S&T joint projects do not include nor reflect areas in which "the Japanese have attained world-

class expertise or advantage", and the need to have a set of IPR which would be applicable to present and future USG R&D projects. The one released paper -- there was the first draft and the revised draft after two meetings -- was heavily deleted and pages 8-10 were completely denied release and continued to be denied after appeal. The released portion provided enough information to provide the gist of the report. It must be admitted, however, that it is not entirely clear from the one released report whether it was the first or second draft.

40. The US-PRC S&T agreement was signed on January 31, 1979 during the Carter Administration. The management structure was the Joint Commission used also in the USJ 80STA, supported by "executive agents" on both sides which would work closely together and support the Joint Commission. The Commission "shall ordinarily meet once a year" and it may create when necessary permanent or temporary subcommittees and in a revision agreed to accept 500-700 Chinese students and the U.S. would send about 50 to China. Why is the U.S. willing to accept formally such an imbalance when it is so insistent on making this a major issue with Japan -- even after taking into consideration the level of S&T in their respective countries. There were, of course, no IPR nor security clauses, in this STA. After several extensions of the STA, an IPR annex was agreed upon on May 22, 1991, seven years from the expiration of the 1979 agreement in 1984, in comparison with only three years with the GOJ, 1985-88.
41. OES Memo A/S Negroponte to Derwinski, *EPC Meeting on Japan S&T Agreement*. Undated but about mid-July 1987, prepared for the EPC Meeting.
42. OES memo, 7/10/1987, A/S Negroponte to Science Adviser Graham. This memo gave a page-by-page commentary on the second draft CISET report of 10 pages. Unfortunately about 80% or more of the commentary and the CISET report had been deleted when released to me through the FOI process.
43. OES memo, 7/23/1987, to Wallis Under Secretary, "EPC meeting on Renewal of Japan S&T Agreement..."
44. OES memo, 7/22/1987, to OSTP (Eugene J. McAllister) [should have been EPC], Economic Policy Council paper on U.S.-Japan S&T Agreement. The referenced CISET paper was not released in the FOI process. Based on an interview with a senior member of the then EPC staff, the CISET report was too long, so an "executive summary" for EPC consideration was prepared and circulated on July 21, 1987. Furthermore, EPC did not, as a matter of practice, discuss nor act on the actual text of the draft agreement. By implication, therefore, all the details in the draft were indirectly approved, e.g. the IPR and security annexes, all the reviews of various kinds of activities described in Annex that the GOJ must take unilaterally for access purposes.
45. Memo, 2/12/1988, OES to Whitehead, Deputy Secretary of State, Calling Secretary of Commerce regarding U.S.-Japan, S&T cooperation and talking points.
46. The six are: life sciences, information science and technology, manufacturing technologies, automation and process control, global geo-science and environment, and joint database development. The 7th, joint standardization of nomenclature development was deleted at the request of DOC/NBS [now NIST] because it is a private section function.
47. These comments are based on interviews with many officials in these agencies. The actual language used was quite vitriolic, particularly against the State Department which appeared to be more "the enemy", than the Japanese themselves. Based on an interview with the late Dr. Gaston Sigur (11/18/1992) who was with the NSC (1982-86 (Second Reagan Administration), A/S (EAP, 1986-89), the NSC was not involved

at all in this controversy. State tried to soften the generalized feeling Japan had unilaterally gained from the 80STA, held DOC and OSTP. This assessment could not, he said, be validated through case studies. He had delegated the STA issues to his deputy, William Clark. Sigur believed the 80STA was not a bad agreement and that was described as pro-Japan even "agent of Japan" but we [State] felt secure because President Bush, Secretary of State Baker and OMB Director Darman "were behind us."

IX. Act IV, Bilateral Negotiations, etc., August 1987 - June 1988

1. Introduction

"Bilateral Negotiations, etc" was chosen deliberately in an attempt to suggest that this chapter covers more than the USJ negotiations during the period from October 1987 to June 1988. While the bilateral negotiations were the major focus of this period, there were a number of other developments which throw light on the USJ relationship, how both the USG and GOJ looked upon the draft agreement, how the executive departments in each country related to the Congress and the Diet, respectively, internal coordination within the GOJ, two contrasting Congressional hearings in the U.S. before and after the bilateral negotiations, a short post-mortem discussion among several of the Japanese delegates, the role of extensive American Embassy reporting on potential opportunities for foreign access to Japanese R&D facilities, and the impact of the Defense Patent Secrecy Agreement on the STA negotiations.

While the USG contemplated, for a short while, a one-day session -- later changed to a three-day session -- to wrap up the entire agreement, the GOJ thought six months at least would be required to reach an agreement. An attempt was made to try to rush through an agreement in time for a January 1988 Summit meeting but that failed. Using the draft agreement, the USG was invited by the Japanese to give them "a briefing" of the proposed revisions; a senior member of OSTP referred -- obviously with condescension -- to this briefing as a "tutorial" for the Japanese. There were seven formal negotiating sessions and a final high-level session at State to arrange a compromise on the national security issue pushed by the USG. Ultimately, a new agreement -- not a revision -- was signed at a bilateral Summit Meeting at the 7-Nation Summit Meeting in Toronto on June 20, 1988.

2. The Diplomatic Note and Annexes

The draft agreement prepared by the USG was handed to the GOJ/MOFA on August 28, 1987. It was technically a revision of the 1980 agreement but in reality it was a wholly new agreement. It consisted of a diplomatic note and five annexes, a total of 34 pages. The 80STA was not being revised nor being renegotiated, it was being replaced. This is an important point to keep in mind throughout the nine months of negotiations.[1]

The new draft agreement did not consist of just nine articles like its predecessor. It began with a policy framework, articles and five annexes which themselves were complicated and highly controversial for the GOJ. The IPR annex tried to cover as many contingencies as possible, obviously because of the mistrust of the GOJ and others in Japan. The defense security obligations annex would be particularly sensitive in Japan. Unlike its predecessor, this draft was a highly detailed and particularly one-sided document requiring the GOJ to commit itself to a large number of specific actions. The tone of much of the draft was an implied, though not proven, criticism of Japan since it required Japan to take various actions. A good example is the style, tone and kind of actions Japan would be committed to take as exemplified particularly in Annex II on mechanisms, initiatives and areas for joint review in contrast to the actions that the USG would commit itself to take.

Let us now turn to the details of the draft agreement itself. The draft cannot simply be summarized. We must wade through the details to fully understand what is being proposed – even if some parts are repetitious (The passages cited in this section are from the middle column of Appendix 2.) It had been the earlier intention of officials in the USG -- as mentioned in earlier chapters -- to involve the GOJ in putting together a revised version of the 80STA. That cooperative and positive approach unfortunately never materialized. Instead, various parts of the USG met with GOJ officials and "briefed" them on some generalized principles underlying the USG proposal. But it was never clear from USG cables reporting to itself on these meetings what *specifically* the Japanese officials were told about the draft agreement. One might surmise that the GOJ officials were informed that the revisions would include IPR revisions, a new management structure, specification of priority technical areas, and possibly a list of priority areas of technical cooperation. The USG has steadfastly implied that the GOJ was aware of much of the details, except that the available details do not collaborate this assertion. The GOJ claims, in contrast, that it was completely surprised, taken aback with its details and especially with the numerous burdens and responsibilities placed on the GOJ -- especially for an agreement signed at the Presidential/Prime Ministerial level.

In order to implement the mandate in Article I of the 80STA to develop cooperative S&T activities, the two countries would agree on three "principles and objectives": the importance of science and technology, cooperation in international S&T and the revitalization of the 80STA. While the language used in describing these principles and objectives seemingly places Japan on a scientific and technological par with U.S. -- i.e. as a technological superpower -- the admonitions and directives for future action are clearly directed at Japan, not the U.S. The U.S. and Japan "*should* establish national R&D policies that sustain long-term investments in basic research and create dynamic R&D environments that advance the following broad goals to generate fundamental new knowledge, protect IPR derived from joint collaboration, provide for swift transfer of new technologies, and nurture and expand the next generation's talent base." Under cooperation, both countries acknowledge "the need to create more equitable opportunities for interaction and to work for a balance of contributions and benefits" and both countries acknowledge that "at times one country may need to make greater commitments in particular components of the R&D process and infrastructure commensurate with its scientific and technological strengths" and then lists in what areas Japan should increase its commitments (basic research support, graduate and post-graduate education, training at universities and national research institutes, research facilities and the published dissemination of information). Obviously, in light of the discussions leading up to creation of the U.S. proposal, these words of commitment are, of course, directed in theory to both countries but the obvious real intended target was Japan -- an unusual unilateralism in a bilateral agreement at the Summit level. Under the revitalization of the 80STA, it reaffirmed that the technical content of the Agreement "must match the commitments" of the Presidential character of the agreement, agreed that the revised STA sets forth "the overall policy framework" for USJ S&T relations, that projects and programs undertaken under this agreement will be of "the highest national priority", and that the USJ will, "under the policy framework of this agreement" discuss each other's S&T policies, priorities, activities and plans, impediments to balance opportunities, access and benefits "with a view to overcoming those impediments."

Annex I focuses on major bilateral initiatives and priority areas for S&T cooperation. The first part establishes a number of principles or guidelines for cooperation:

1. Bilateral cooperation in major international R&D initiatives and large scale national projects will be pursued under the "policy framework" of this STA but shall exclude those projects that have "separate management mecha-

nisms" from the purview of the Joint Interagency Executive Committee. The umbrella-ness of the agreement, a major concern in OSTP and State from the mid-1980s, was thereby seriously compromised.

2. When the USJ cooperate in such major programs they will share the costs "in proportion to their respective risk, benefit and management shares and considering the balance and reciprocity of their overall S&T relationship". While this is a seemingly balanced statement of co-equal sharing, the criteria for weighing such a delicate balance would be difficult to create and agree upon and could easily become a point of political tension.

3. Although the USG had, as described earlier, a set of proposed priority areas of cooperation, it was not included in the draft and left to negotiation with the GOJ. The criteria for such bilateral cooperation were as follows: 1) each partner should possess strong, complementary or counterbalancing R&D capabilities, adequate resources and appropriate centers of excellence; 2) the chosen areas of cooperation should reflect the national R&D priorities of both countries and contribute to an equitable distribution of investments and pay-off to each partner's national needs and 3) should have the potential to accelerate the rate of scientific progress and contribute to the world's knowledge and technology base.

4. The directive nature of the last item in this Annex suddenly changes from "should" to "will". "Given the structural differences between the R&D systems in the United States and Japan, the Japanese Government WILL FACILITATE (emphasis added) U.S. participation in those national research programs in basic and applied sciences and engineering that meet the criteria described in this article and that are under government sponsorship and receive government funding." This is followed by a list of Japanese government R&D programs which it says "are comparable" to U.S. government R&D programs at U.S. universities and national laboratories.

Annex II on mechanism, initiatives and areas for joint review are divided into two parts: 1) mechanisms and initiatives "to enhance U.S.-Japan interaction in S&T and 2) areas for joint review. The following description will be divided into what the draft says the GOJ "will" do in contrast to what the USG proposes it will do.

The Government of Japan

1. will establish at Tsukuba University and other key institutions in, not just, the public sector but also the private sector, intensive language and cultural programs for visiting U.S. scientists and engineers to facilitate communication in R&D environment as well as in daily life activities in Japan.

2. will maintain and accommodate visiting U.S. scientists and engineers consistent with standard practices in other U.S. S&T agreements. (This would commit Japan to making available any privilege or accommodation another country may grant to the U.S. even though the GOJ had not participated in such discussions.)

3. will actively encourage vigorous recruitment and acceptance of more U.S. scientists and engineers ... at R&D facilities under their respective institutions or supported with their resources.

4. will give priority attention to increasing the number of U.S. researchers participating in national R&D programs and collaborating in the designated priority technical areas of cooperation.

5. will provide annually to the USG a comprehensive list, with all necessary particulars, of current opportunities for U.S. researchers to participate and be employed in the Japanese R&D system.

6. will establish and widely advertise a significant number of substantial and prestigious "Japan Fellowships" in science and engineering for U.S. undergraduates, graduates and post-doctorals at Japanese centers of excellence.

7. will take steps to ensure that scientific and technical reports produced by GOJ agencies and their contractors which are not published in "open" journal literature are made available to U.S. researchers through a central source in the U.S.

The United States Government, on the other hand,

1. will continue to encourage and support language training programs for technical personnel at U.S. universities and other institutions to facilitate the participation of U.S. scientists and engineers in Japanese R&D facilities.

2. will work with universities and other R&D performers to encourage U.S. researchers to take advantage of existing opportunities to work in Japan.

The USG and GOJ will jointly

1. establish a working level task force from government, academia and the private sector (presumably industry) to develop a system to identify, recruit, and monitor U.S. scientists' and engineers' access to and participation in Japanese R&D programs. No mention is made of Japanese participation in U.S. R&D programs.

2. develop recommendations and actions concerning barriers and structural problems on both sides that impede or inhibit increasing the number of U.S. researchers in Japan -- impediments on the U.S. side to Japanese researchers were not mentioned.

3. yet, in the same section it calls for the task force to set up a system to obtain accurate, yearly statistical data on Japanese and U.S. researchers participating in each others R&D system.

4. establish an STI committee to examine and develop recommendations on improving STI access in Japan because "U.S. scientists and engineers face unique problems in gaining access to Japanese STI not encountered in other major STI source countries".

5. review developments and trends in each government's policies for the support of science and technology especially at universities, and government-supported research institutes.

6. review efforts by each country to stimulate and achieve a balanced flow of scientific and technical information between the U.S. and Japan, and to ensure that new STI comes to the notice of the world scientific community through publication.

7. review efforts in each country to maintain and enhance state-of-the-start educational and training facilities at universities and national research institutes, to promote advanced training opportunities for the next generation of scientists and engineers.

8. review government investments in the establishment and enhancement of world-class R&D facilities in each country at university and national laboratories to generate fundamental new knowledge and generic technologies.

9. review the levels of flows of scientists and engineers between the U.S. and Japan to premier educational and research facilities.

10. review Japan's initiatives to open national R&D programs and S&T institution to foreign participation specifically to encourage U.S. scientists to work in R&D institutions in Japan.

11. review actions by each country to encourage increased participation by U.S. researchers in Japanese S&T programs by providing logistical support, especially language training and cultural familiarization.

The number of actions that Japan would be required to take are much greater than the two rather passive and tepid actions that the USG requires itself to take and are in stark contrast to what would be a balanced S&T agreement between two technological superpowers. In light of discussion, debate and dialogue in the USG comparing the "open" nature of the U.S. system and "closed" nature of the Japanese R&D system, one can only deduce, that the underlying intent, if not the result of these reviews despite the fact that the reviews are couched in terms of "each country", were actually aimed at Japan being the party to take the numerous "remedial actions". Based on any standard of equity and balance, the above list of actions which Japan would be committing itself to undertake together with those actions that would stem from joint reviews, cannot be described or analyzed as a level playing field. The implicit, unspoken, unwritten intent of these measures would be to tighten U.S. control and manipulation of Japan's S&T policies and programs. It is not difficult to imagine how the USG would react if a strong and close ally, for example, a European NATO power, chose to propose such a draft agreement to the USG. Or, one might ask whether the USG would have the gall and audacity to propose such a one-sided draft agreement to a senior NATO partner.

Annex III focuses on the management structure of USJ S&T relations. Dissatisfied with the simple structure of the 80STA, the USG proposed an elaborate three tiered management structure beginning with a working level joint executive committee, a high-level Joint Committee which reports to the

President and Prime Minister, and a high level Advisory Panel. The Joint Committee would be headed by the U.S. Science Adviser to the President and the Minister of Foreign Affairs. The Advisory Panel would consist of "senior scientific, engineering and S&T applications leaders representing a cross--section of the science and technology communities of the two countries". This was an attempt to involve the private sectors in the management of public sector S&T relations. The proposal even suggests where these Panelists might be found: The White House Science Council of the U.S. side and the Prime Minister's Council for Science and Technology in Japan.

Annex IV on intellectual property rights includes stipulations on IPR, protection of dual use technology, confidential information, inventions, copyrights, cooperation. Intellectual property provided under or derived from this agreement would be protected by both countries in accordance with their laws and regulations. Only two sections will be commented on here: first, both parties will accept for filing patent applications classified or otherwise held in secrecy for national security purposes under the Defense Patent Secrecy Agreement of 1956 and its implementation procedures. This kind of information shall not be exchanged under this agreement. According to this section, the "revised" 80STA would be held hostage to -- i.e. not put into effect -- unless the GOJ agreed to implementation procedures for the 1956 agreement. This linkage of tying one agreement's implementation with another was insisted upon by the DOD - with OSTP and DOC support -- which was negotiating at that time with the GOJ for a set of implementation procedures. It is another example of the distrust of the GOJ.[2]

The other major item concerns inventions. This was presumably the topic that was the immediate cause for the DOC denunciation of the draft agreement in late March 1987. The principal thrust here is that any invention is made by a person visiting or assigned to a facility in the U.S., the U.S. not the inventor will have the right to obtain all rights and interests in the invention in all countries. This seems to assume that there is a higher probability that Japanese scientists or engineers will create an invention while in the U.S. rather than a U.S. scientist visiting in a Japanese facility -- presumably because there are more Japanese scientists at this time visiting in the U.S. than U.S. equivalents in Japan and it seemed to imply that visiting Japanese S&Es might be quite creative...

Annex V on security obligations has two parts: one stipulates that no information or equipment requiring protection in the interests of national defense or foreign relations of either Government and appropriately classified shall be provided under this agreement; the second stipulates that the transfer or retransfer of information and equipment provided to or produced under this agreement will be subject to export control laws and regulations of each country. Although there were no known improper transfers of information and equipment under the 80STA, the criticism, anger and distrust of Japanese companies and the government was running at a fever pitch. The DOD, eagerly backed by the OSTP and DOC insisted on the inclusion of the stipulations in Annexes IV and V to prevent another occurrence of the Toshiba-Kongsburg Vaapenfabrik incident when these companies allowed certain militarily critical information and equipment to be provided to the former Soviet Union.

Although it is not clear when this draft agreement was actually prepared, presumably it was sometime toward the end of 1986 and by February 1987. One should keep in mind that the report by the Joint Task Force was dated November 26, 1986 and was in tone and recommendations in sharp contrast to the draft given to the GOJ. The management structure and a number of other ideas can be found in both the report and the draft agreement. The draft agreement called for a three tiered management structure. The JTF report had recommended a working level executive committee in the USG and GOJ and a high level and advisory oversight committee composed mainly of representatives from the private sector. The latter was to be created at the departmental or agency level. The tone of the JTF report was constructive and called for a dialogue with Japan about future S&T relations. On the other hand, the draft agreement was provocative and dictatorial and unilateral in actions to be taken almost completely by one country, Japan. Since the interminable discussions in the USG leading to the creation of the USG proposal clearly indicated, the U.S. system was presumed to be open to all, the Japanese system, in contrast, was not, these declared intentions were clearly aimed, as a tactical move, at using this agreement to pressure Japan from all sides to "improve" its R&D system as desired by the USG. As a proposed bilateral agreement these expositions were not subtle but obviously, clearly, unequivocally aimed at Japan, not the U.S. while using seemingly bilateral language.

Originally the plan had been to involve the GOJ in the dialogue to create the revised STA but that did not occur. At various times, the GOJ was "briefed" on USG ideas, intentions, scope of revisions desired by the USG but nowhere is the record unequivocally clear that the USG in these meetings actually discussed the details as outlined above with the GOJ. While the GOJ was aware the USG desired "a strengthening" of the 80STA it clearly did not expect to receive such a radical revision specifying in such detail what Japan "must do" in contrast to what the USG would passively do for its part of the bargain. The considerable detail on what the GOJ, Japanese universities and the private sector must do is symbolic of the depth and breadth of the distrust and suspicion of Japan among certain U.S. departments and agencies at the political level and the obvious attempt to mold and control Japanese S&T development according to U.S. specifications. This, notwithstanding the repeated declarations by the USG about the great and critical need for USJ cooperation in all aspects of S&T. This then is the background to the USJ negotiations which are about to begin.

Superconductivity Conference, SEMATECH and MCC. Symbolic of the meteoric rise of U.S. anxieties about the relative and absolute status of itself in world science and technology in the mid-1980s, was the convening of a high temperature superconductivity (SC) conference on July 28, 1987 in Washington under no less than White House (OSTP) sponsorship which conspicuously excluded foreigners -- ostensibly aimed at the Communists but the underlying fear was Japan. The conference was addressed by President Reagan to an audience of over 2,000 U.S. businessmen, engineers and scientists to discuss the commercialization of superconductivity. The President called for an eleven point initiative of $150 million (five times the amount then being spent on SC by all USG agencies) based in the DOD (thus assuring secrecy which had not been practiced in this field in the past and for the tightening of the FOI law to protect S&T information generated by government laboratories. A report on superconductivity had recently concluded that Japan was forging ahead with the commercialization of superconductivity -- again a direct challenge to U.S. creativity and ingenuity. This exclusionary action was widely, broadly and severely condemned by the U.S. scientific community and also by European and Japanese S&Es. But this Conference and the SC initiative were recommended ironically just about one month before the USG gave the GOJ a draft as described above demanding that the GOJ open not only its government laboratories but also industrial laboratories to foreign participation.[3]

In May 1989, the formation of a new consortium of AT&T, IBM, MIT and U.S. government laboratories (e.g. Lincoln Labs at MIT) was announced to work on the commercialization of superconductivity. A similar kind of consortium for SEmiconductor MAnufacturing TECHnology (known as SEMATECH) was also formed in May 1987 to improve semiconductor manufacturing with an infusion of DOD funds and private industry support. Similarly, in 1983, MCC (Microelectronics & Computer Technology Corp) was also created for computer technology, again with DOD funding support. Japanese companies were excluded from participating. But it was this kind government-industry cooperative activities that we were lecturing the Japanese to open to foreign researchers. The use of consortia for this purpose was an imitation/adaptation of the Japanese approach to pushing the technological envelope for a given purpose. This was, to say the least, a blatant double standard in flaunting techno-nationalism.[4]

Ironically at the same time that the USG appeared to be encouraging an exclusionary R&D policy at the highest level in the USG, U.S. firms were working on a technoglobal basis as described in chapter IV.1 by making agreements with Japanese firms: Motorola and Toshiba in November 1986 for Toshiba to build a new chip making factory in Sendai with Motorola providing the next generation of logic chips. AT&T tied with Fujitsu to challenge the dominant position of IBM in computers. Cray was reported to have agreed to technological cooperation with Hitachi in supercomputers (*Asahi* 7/22/89), Boeing and Japan's three largest aerospace companies agreed to work together to build Boeing's next generation commercial aircraft, 767-X, for the mid-1990s (*Washington Post* 4/14/90), AT&T and NEC agreed to cooperate on basic chip-making technology (*Wall Street Journal* 4/23/91). These are but a tiny fraction of the technoglobal examples of the strategic alliances that were being forged by U.S., Japanese and European corporations on a multilateral and bilateral basis. There would appear to be some disconnect between the USG position vis-a-vis the GOJ and the reality of multiple linkings among high technology firms around the world.[5]

3. U.S. Congressional Hearings-I, October 1987

Only a few days before the first bilateral negotiations were to be held in Washington, DC, the U.S. Senate Subcommittee on Science, Technology and Space of the Committee on Commerce, Science and Transportation held an afternoon hearing October 15, 1987 on the U.S.-Japan S&T Agreement. Senator John D. Rockefeller IV presided over this hearing; he has a long standing interest in U.S.-Japanese affairs. A/S OES had earlier met with

Senator Rockefeller on September 23, 1987 to discuss the revision of the 80STA. According to a 3-page briefing paper prepared for this meeting, concluding the negotiations in one three-day session was mentioned, but the GOJ suggestion of six months, which was much more realistic was not mentioned. It noted that the GOJ "reluctantly entered into it [80STA] at the strong urging of the United States", cited "key criticisms" of the 80STA (See chapter on 80STA), evaluated the projects under the 80STA as "low cost and minor" but most "effective and worthwhile", enumerated the various methods or steps to revitalize the USJ S&T relationship but SIGNIFICANTLY did not mention the one sidedness of the actions to be taken by the GOJ vs the USG. At the Senator's request he was also given a copy of the "preliminary draft agreement".[6] The following description and analysis is based on the record of these hearings. There were five witnesses, three from the concerned executive agencies and two from the private sector.[7]

Since these hearings were held only days before the first formal negotiations between the two governments were to take place, it is obviously understandable that these hearings would be used quite justifiably, to present a united front, a model of mutual politeness, non-confrontation, completely devoid of any clues whatsoever of the deep-seated antagonisms between the various U.S. agencies involved in this process. The hearings gave State, Commerce and OSTP an opportunity to assure the Senate that a new set of principles governing U.S. international S&T relations would undergird the revised and strengthened S&T relationship with Japan and gave two witnesses from the private sector -- for balance purposes -- who are generally regarded as highly critical of Japanese actions, an opportunity to provide another set of views on this subject and to re-emphasize the need to stiffen the backbone of the U.S. negotiating team headed by State. Since negotiations are "secret" as they proceed, no draft agreement was requested and none naturally was proferred. The basic principles endorsed by the Economic Policy Council were reiterated and endorsed but no inkling about the other measures was included in the draft agreement given to the GOJ in late August 1987 which were intended to commit the GOJ to various methods of control and supervision. The Senate subcommittee hearing was undoubtedly intended to convey the message to the GOJ about the unity and determination of the USG and to give informal Senatorial backing to the U.S. negotiators.

The discussions at the hearing resembled a thoughtful but plodding academic seminar where the professor (Senator Rockefeller) politely listened to the presenters and periodically reminded them of other aspects which they had not mentioned and persuaded them to elaborate on and moderate their publicly held views. The professor/senator undoubtedly knew about USG negotiating strategy, through the contacts his professional staff members maintained with the staff of the concerned executive agencies, i.e. State, DOC, or OSTP. The main topics of discussion centered around the "new" principles on U.S. science and technology relations which would be the foundation of U.S. negotiations with other countries in the future beginning with Japan, remedying the imbalance of flow of information and personnel between the U.S. and other countries, and, in particular in this instance, with Japan. An effort will be made to relate the testimony -- "seminar reports" and subsequent Q and A -- of the participants to the on-going disputes among the executive agencies and anticipated negotiations with the GOJ.

Let us start with the DOC's Acting Secretary Smart's and Science Adviser Dr. Graham's discussion of several issues:

Contribution to basic knowledge. Smart comments that the Japanese have "probably reaped more fruits of it [new scientific information] than we have, certainly more than they have contributed in the development of basic information which only now is starting to achieve world class status. One of the shortcomings of the United States' system at the time which needs to be corrected if our competitiveness is to be as full as we would like it, is to become more skilled and more quick [sic] in bringing new basic knowledge to the market place and to take advantage of what we have developed before our competitors beat us to the punch" (p. 11). Dr. Graham echoes this sentiment with his assertion that "the U.S. industry has become accustomed to not being so aggressive. The U.S. industry will have to change its approach to advanced technology, or it will find itself at a long term disadvantage to the Japanese" (p. 17). "There is a need", says Smart, "for Americans, generally speaking to rediscover the work ethic" the process which he felt had begun after we got over the problems of the Vietnam war and was well pleased with the "aggressiveness of American business" (p. 14). He described the Japanese "as hungry for its first Super Bowl as opposed to one [the U.S.] that is defending the championship for the fourth time" (p. 14).

The systemic nature of this imbalance is explained by Graham to reside in 1) how basic and applied research is conducted in the U.S. and Japan: the former done in universities and national laboratories is open, whereas "advanced strategic research" in Japan is "often performed in closed proprietary settings, much like corporate research in the U.S." which is not frequently open to foreign visitors (p. 18). 2) While the U.S. and Japan spend "roughly comparable amounts on R&D as a percent of national product

(12.1% for the U.S., 12.9% for Japan), the USG provides 47% of total R&D expenditures in the U.S. while the GOJ provided only 22%, with U.S. industry providing "more than 67%." (p. 18).

The 1980 and the new revised draft Agreement. In his opening remarks, Senator Rockefeller commented that the 80STA was a "relatively toothless" agreement. The March 25 volcanic eruption at DOC to the USJ STA draft agreement was referred to by Smart that Secretary Baldridge had asked that the agreement be reviewed "so that we could have an agreement that was, in effect, a two-way street" (p. 12). Until the DOC Secretary had "demanded", not just politely asked for, a total review, the DOC itself had been deeply involved in every step of the creation of the proposed draft to give to the GOJ in March 1987. Again, one is left with the impression that the 1980 agreement was somehow a one-way street forgetting conveniently that this very agreement was a U.S. proposal which was whittled down in negotiations with the GOJ and the USG showed little or no interest in trying to make the agreement work either to the U.S.'s advantage or to both the U.S. and Japan's mutual advantage as we are so wont to assert to ourselves and to the Japanese. This assertion was never systematically proven by the USG in an integrated fashion. Only the implied assertions were left standing.

Japanese students and researchers in the U.S. That an "imbalance" existed between the number of Japanese students and researchers coming to the U.S. and U.S. students and researchers going to Japan is not and has never been in dispute. Smart says that in 1985 there were over 13,000 Japanese students presumably both undergraduate and graduate students -- note not researchers -- at U.S. institutions, "most of them in technical fields. There were fewer than 800 American students in Japan, most of them in humanities" (p. 12). Major institutions, such as MIT, host large numbers of Japanese visitors and 50 Japanese companies have industrial liaison programs. We have very few comparatively speaking in Japan. Again, as in the above paragraph, there was no attempt to explain why these numbers may result in some disadvantages to the U.S. According to the Institute of International Education, there were 350,000 foreign students in the U.S. in 1986-87; the Japanese students would then be 3.7% of this total.[8]

There is no comparison with other countries, so we do not know whether the Japanese number is unusually large or small. There was no analysis or assertion that our companies are clamoring to invest in programs in Japan and have been denied such access by the Japanese. The listener at and the reader of this hearing is left with the impression that we the U.S. are impatiently eager and the Japanese so uncooperative in letting Americans into Japanese facilities. Indeed, Senator Rockefeller in his opening remarks commented that "since 1981, most of Japan's universities have, in fact, been open to foreign research visits. Japan has begun to open up its government aided laboratories to American visitors" (p. 2). According to OES in July 1987 just before the EPC Meeting, "we [the U.S.] are already having problems filling dozens of opportunities now open to us [the U.S.] in Japanese R&D Facilities."[9]

Value of USJ S&T Collaboration. In assessing the impact of the USJ S&T collaboration over the past several decades, Graham concluded that "While the U.S.-Japan science and technology cooperation has been mutually beneficial, it has not been balanced in terms of contributions, and costs, *for historical reasons* [an allusion to the lack of interest on the U.S. side to study in Japan for a variety of reasons]. Benefits for the U.S. have been chiefly to our basic research efforts in [U.S.] universities and government-funded institutions, where Japanese graduate students and visiting scientists have provided in this country a supply of well-trained, disciplined manpower to carry out our research programs ... These people make substantial contributions to our [U.S.] research base and we think we can make such contributions to the research base in Japan and other countries" (p.17). This acceptance that Japanese researchers had made a contribution to U.S. research programs and projects is in such stark contrast to internal USG debate at the staff level and in subsequent negotiations with GOJ, it is hard to believe that these words come from representatives of the very same agencies which had exhibited an historical antipathy toward Japan. There was no mention that these Japanese researchers were often chosen in competition with other worldwide candidates. Again, there was no mention as to whether there is a disgruntled line of U.S. researchers chafing to do research in Japan.

Despite its presumed anxiety about foreign (including Japanese) students in U.S. universities, Graham stated "that this year [1987] *for the first time* [emphasis added] we as a nation will be collecting data on the extent of foreign participation in our universities and our national laboratories" (p. 18). Even though Secretary Smart cited some overall numbers about Japanese students in the U.S., he did not have any data showing which Japanese institutions were willing to invite U.S. researchers; under prodding from Senator Rockefeller he concurred such data would be "interesting statistics"(p.13). Even though we were conducting such a survey in 1987 for the first time to obtain accurate data, the issue of Japanese scientists at U.S. institutions was one of the more acrimonious issues during the negotiations. One could not deduce from this hearing the enormity of this issue in the minds of selected U.S. delegates to the upcoming USJ negotiations for a revised STA.

The STA and the Trade imbalance. One of the main assertions is that the time has come to review future S&T agreements beginning with the USJ STA in its relationship to trade and economic security. This itself is a major departure from the way most previous STAs were handled by the USG. Under the careful prodding of "professor" Senator Rockefeller, Secretary Smart says that "to the extent that it gives us access to Japanese technology and helps us to translate that to economic advantage it will clearly help the United States in its competitive position in the world ... I am not as certain as some that it [the USJ STA] will have a major effect on the trade deficit which, in my own feeling, has as its roots in the macro-economic policies of this country and Japan and other countries more than it has in the basic competitiveness of individual nations" (p. 13).

The EPC review of new principles underlying international R&D. Both Secretary Smart and Science Adviser Graham cited the Economic Policy Council's review and adoption of a set of new principles to govern future S&T agreements beginning with the USJ STA. As described earlier, DOC, OSTP demanded that the EPC have an opportunity to review these principles instead of the NSC which would have been the preferred venue by State. These principles seek balance and reciprocity in bilateral scientific interchange in terms of access and opportunities for research and training, the generation and dissemination of new knowledge to the world S&T enterprise, protection of IPR, cost sharing of large international S&T projects, fair compensation for basic research that is used and a recognition of any military security implications that exchange of information may have (p. 12 and 16).

Assistant Secretary Negroponte (State/OES) commented on the following issues.

The extent, complexity and value of the USJ S&T relationship. He brought to the hearing a list of agreements and MOUs (which were not included in the proceedings) ranging from atomic energy, to good laboratory practices in the FDA in the U.S. and its counterpart in Japan to demonstrate the complexity of the USJ S&T relationship and suggested that the many U.S. agencies involved in cooperative work with Japan would not do so, "if they did not see some mutuality of benefit in entering into such arrangements" (p. 22). Based on this reasoning clearly State believed that the relationship was not a one way street.

Japanese researchers at NIH. It is significant that State's Negroponte brings up this issue as one of the major items to explain in the beginning of his testimony. He cites about 330 Japanese researchers at NIH, 260 of them as post-doctoral fellows (visiting fellows; another major group is 50 more senior visiting scientists). He notes that these researchers are chosen for one to three year appointments *through worldwide competition* which is open to U.S. and foreign scientists. Of the 2000 post-doctoral positions, half are filled by foreigners and half by Americans. Those filling these positions must sign an agreement stating that any employee invention becomes the property of NIH. Thus, he asserts, that in the celebrated case of NIH, all benefits did not flow in one direction and reported that there was a "great deal of satisfaction with this arrangement" by NIH (p. 22). Despite this additional explanation about the NIH case, the State delegation chairman was unable to control other U.S. delegates from destructively and aggressively pursuing the "phantom issue" of Japanese researchers at NIH as the prime example of the imbalance between the two countries.

The new negotiating principles. Negroponte enumerated them at the hearing as confirmed by the EPC. It is significant that he deliberately did not mention the EPC as the review agency -- especially since the EPC review was conducted contrary to State's wishes.

This hearing actually began with William C. Norris, chairman emeritus, Control Data Corporation, as the first witness and concluded with Clyde Prestowitz. Norris described half a dozen reasons for the continuing imbalance in the flow of technology and provided some suggestions for remedying this situation. Japan conducted only minimal basic research and had -- together with other nations -- unlimited access to U.S. basic research since this research is openly published. Further, this small amount of Japanese basic research was conducted by government and industry labs in the main and "closed to U.S. companies" (p. 5). U.S. companies do not have access to technologies developed in Japan based on U.S. science "in large part because policies of the Japan Fair Trade Commission inhibit grant-back agreements" (p.5). While the Japanese can acquire technology through small U.S. companies through licensing, equity investment and acquisition of the total company, the U.S. does not have similar opportunities in Japan. Japan's "best graduate students" are sent to the U.S. to obtain PhDs many of whom are provided with "financial as well as intellectual supports" (p.5). While there was a time constraint on the witnesses to provide additional information to prove their serious accusations there was no such attempt by U.S. institutions or evidence that such supplementary data was provided. Norris did comment that "large U.S. corporations have not sought foreign technology as diligently as they should have" (p. 4). In contrast to Secretary Smart, he believed that "technology is the pivotal element in competitiveness. Hence, technology is a key trade issue" (p.4). In order to remedy this "imbalance", Norris suggested the following steps be taken (p. 6-7):

- "The most important and critical needs is the vast expansion in technological cooperation embracing a wide range of science and engineering", especially in large scale cooperation in research in such areas as the superconducting supercollider (SSC) and high energy physics. The USG was unable to persuade the GOJ to invest in the SSC and itself could not find the funds to complete the project it launched. The SSC was canceled in 1993.

- Japan should open up its government supported R&D projects to U.S. participants. The GOJ had begun to open up its laboratories to foreign participation. But again, there was no proof that there was a substantial group of U.S. researchers impatiently waiting for this opportunity.

- Increase U.S. scientists and engineers living in Japan as researchers have done for a long time in European countries. Because there was little probability that this could be meaningfully achieved in the near future, the GOJ "should subsidize any U.S. university where Japanese nationals are studying" to pay for teaching assistants, keeping laboratory facilities up-to-date, junior faculty positions -- not just science and engineering students but would include all Japanese students in ALL fields of endeavor. This would be in addition to paying for out-of-state tuition. There is no mention that this special subsidy should be paid by any other government, only Japan. There is no rationale why Japan is singled out for partially subsidizing U.S. higher education.

- During the transition, the GOJ "could provide an appropriate level of funding for basic research to be administered by the U.S. National Science Foundation."

- The USG should request that the Japan Fair Trade Commission revise its policies so that new or improved technologies developed by Japanese companies which had used U.S. science could be granted to the U.S. firms.

- In order to assure and accelerate the flow of technology from Japan to the U.S. -- if such can be adequately identified -- a large scale program should be established "to transfer Japanese technology to the United States through joint ventures with small U.S. firms and entrepreneurs in the United States."

- Finally, while a system to measure the flow of technology is difficult to develop, that flow can be approximated well enough to determine whether equity is being approached.

Some of these ideas do appear in the draft given to the GOJ in late August 1987, e.g. GOJ defraying some costs of U.S. scientists and engineers in Japan and to create a system to measure the results achieved by the Japanese in opening up their laboratories to foreigners.

Prestowitz emphasized in the beginning of his testimony that the balance in the flow of technology between the U.S. and Japan was "one of the most important things that the United States should be focusing on, not only as part of its trade policy but as part of its overall economic and national security policies" (p. 26). He voiced the same interpretation as did Graham when he noted that "not because people have ill intent but because the natural working of the two systems happens to put the United States at a disadvantage" (p. 32) and called upon both governments to take steps to correct this situation "so that the perception of unfairness which naturally arises from it [the systemic disadvantage] does not poison the relationship" (p. 32). Despite his known critical position vis-avis Japan, he said "I want to emphasize, I do not like the word 'unfair'. It is pejorative. It implies that the Japanese are doing something bad ... even illegal. I do not think it is bad per se. I do not want to judge the Japanese" (p. 32). Prestowitz suggested that "it might be possible for the Japanese to consider expedited procedures, for example, in the case of foreign patents, particularly when their companies are getting very expeditious treatment" in the U.S. This hearing was concluded with the presiding officer, Senator Rockefeller, describing the soon to be negotiated STA with Japan as "a landmark agreement" (p. 35).

The hearing was clearly planned to give the principal executive departments involved, State, DOC and OSTP, an opportunity to formally and publicly explain their positions to the Senate, to show, within the atmosphere of a cordial academic-like seminar without any of the rancor that so permeated the working relationships of the various executive agencies, a united front on the eve of the negotiations with the GOJ. There was no attempt by the Senate Subcommittee to affect the draft agreement, its approach to the GOJ -- especially just prior to the opening of the USJ negotiations. There was no unanimity about the probable impact of the STA on the trade balance between the U.S. and Japan; this connection had been one of the major underpinnings for vigorously pushing for a radically revised agreement. Although the access to S&T information and facilities remained a major concern there was no major concern or commitment made to demand Japanese substantially change their policies on increasing foreign researchers in Japan. (The reality was that the GOJ under persuasive prodding by the U.S. and others was taking steps to try to attract more foreign researchers to Japan.) It was highly significant and symbolic of how important it was for State to publicly explain the worldwide competitive selection process of NIH's researchers and the acceptance of Japanese researchers through this process. Notwithstanding this denunciation of the inaccuracy of using, at least, the Japanese researchers selected to be NIH researchers, the U.S. chairman of delegation was unable to control some members of the U.S. delegation from pursuing the GOJ with vigor, determination, drive and emotion about the large number of Japanese researchers at NIH as symptomatic of the imbalance. (The U.S. chairman was suddenly changed when Negroponte was appointed to the NSC as Deputy National Security Adviser and DASS/OEC Voss was appointed chairman.)

There is no evidence that there was a direct link between the ideas of Norris and Prestowitz and what was finally included in the draft USJ agreement. As mentioned above, some of their concepts were included in the draft. Their appearances surely were nothing more than convenient window dressing which gave the two non-government witnesses another opportunity to make known their observations and their recommendations. The committee staff and Senator Rockefeller had been briefed by the A/S OES; according to an OES memo (9/22/87), Senator Rockefeller requested and was given a copy of the preliminary draft agreement. Since this hearing was informational, not for appropriation or authorizing actions, it was not unusual that there was no attempt to integrate what the subcommittee had heard from government and private sector witnesses. The appearance of these two private sector witnesses reminded one of the earlier pro forma meeting with the private sector in September 1986 ostensibly seeking their guidance, support and counsel in formulating a U.S. negotiating position vis-a-vis Japan. The two sets of testimonies appeared to stand alone and independent of each other without a context and interrelationship. As a totality, however, this hearing could be described as an informal "advice and consent" session of the agreement to be negotiated.

4. Other Internal USG Issues

a. Access to Japanese Science and Technology Information and R&D Facilities

The issue of access to Japanese R&D facilities and S&T information (JSTI) was, as described earlier, a major and important demand in the draft agreement given to the GOJ in August 1987. It was also an emotionally loaded issue with the outspoken members of the U.S. delegation and the cause for considerable tension with the GOJ delegation. Based on available documents, there was no analysis by the USG of this issue for the purposes of negotiating an S&T relationship with Japan before or during the bilateral negotiations. There was no detailed analysis which showed that the GOJ deliberately raised barriers against access, or that USG sponsored researchers -- or for that matter privately sponsored researchers -- were denied access by the Japanese to their facilities and JSTI. There was no analysis as to why there were only a few U.S. S&T researchers in Japan. There is no known documentation that there was a line of U.S. researchers ever so eager to pursue long periods (e.g., one to two years) of residence in Japan for R&D research. But there was agreement by both delegations that there were relatively fewer U.S. researchers in Japan in comparison to Japanese S&T researchers in the U.S.

Annex II of the draft agreement stipulated seven actions the GOJ would take in contrast to two for the USG: the GOJ would establish intensive cultural and languages programs, would maintain and accommodate, would actively encourage, would give priority attention to, would establish and widely advertise, and would take steps ... and would undertake eleven joint kinds of reviews concerning access, policies, programs, recommendations and the creation of various committees and task forces to oversee the entire USJ S&T relationship.

In light of the above, we will review in this section, 1) how the availability of JSTI was regarded, what remedies might be implemented if a problem did exist, and 2) what access opportunities to Japanese R&D facilities existed in Japan for U.S. researchers.

i. Access to Japanese Science and Technology Information. Japanese scientific and technological achievements suddenly "grabbed" U.S. interest in the early 1980s -- coinciding with the first Reagan Administration. This was first an interest of "curiosity" wondering how Japan of all countries could have become a power house of technological accomplishments. This interest gradually turned into an anxiety, a sense of threat, one might even say with justification, a phobia.

In response to this new concern about JSTI, seminars, symposia, meetings and Congressional hearings were organized during the 1980s, analyses and reports were prepared to explain JSTI and to provide recommendations which the U.S. should adopt to respond to this new kind of challenge. The findings and recommendations of these hearings, reports and seminars were to exhort U.S. researchers, academicians and the USG to take Japanese S&T more seriously, become better acquainted with the Japanese R&D system, develop incentive programs and language training opportunities in Japanese for U.S. researchers, a much greater and more active role to be played by the USG in the dissemination of JSTI. The recommendations placed the burden for tracking JSTI on the U.S. side and made basically no substantial demands on the Japanese to make a greater effort to facilitate the availability of JSTI. In the first 1984 Congressional Hearings on JSTI, the former U.S. Science Counselor in the American Embassy in Tokyo, Justin Bloom, suggested that let alone not enough JSTI, the JSTI already available in the English language was "under-utilized". The difficulty for the U.S. to decide on how to respond to this situation was well stated in a 1983 report:

> "The United States has not had to face the imperative of 'catching up' in science and technology since the early years of this century. As a result, it does not have an adequate system that government and industry can use to explore technological developments in other nations. This is especially true with Japan, for none of the technologically advanced nations of the world are

> less well understood than in Japan. Stereotypes of this country as an imitative, non-innovative nation have been changing slowly, even as evidence rapidly mounts to refute such a view. Japan has been a net exporter of new technologies for almost a decade" [i.e. from the mid-1970s].[10]

There is a major chasm between the findings of these hearings, reports and seminars on what should be done about making JSTI more available in the U.S. and what the USG was expecting the GOJ to accept in the draft 1987 STA. Perhaps the above paragraph reflects the frustration, annoyance, disappointment and anger that resulted in the large number of demands on the GOJ concerning the provision of JSTI as exemplified in the August 1987 USG draft STA given to the GOJ.

The issue of JSTI availability is more extensively discussed in Chapter IV on the milieu in and out of the USG about the future of USJ S&T relations and in Chapter X on implementation. The purpose of this section was merely to remind the reader that JSTI had been discussed in many fora before the fall 1987 bilateral negotiations began but that the context, milieu, conclusions of these discussions were vastly different from the implications of the USG demands in the draft USG STA. There appears to have been again a disconnect between these fora discussions and conclusions and the content of the draft agreement.

ii. R&D Research Opportunities in Japan. As mentioned previously, the assertive demands by the U.S. delegation for Japan to assure there would be more research opportunities for foreigners, especially for U.S. researchers, was vividly remembered by Japanese delegates and other senior officials as one of the few issues of contention in the bilateral negotiations. The unstated implications of the U.S. demands in this area were, in the minds and images of the GOJ delegates, that the Japanese were holding back on these opportunities, or worse yet denying foreigners such opportunities. They regarded USG demands/accusations as totally unfair, untrue and unjustified. Another implication behind the USG stance was the assumption that if such opportunities in Japan were truly available, there would be many U.S. researchers eager to fill these positions -- otherwise, of course, why push this issue with such doggedness, persistence even to the point of souring the negotiations. The GOJ agreed from the beginning that there was a large discrepancy in the numbers of researchers in each country but they vehemently disagreed about the causes, remedies and implications of this discrepancy.

In this subsection, we would like to review the kinds of materials and data that appear to have been available to those in the USG involved in the debate over the future of USJ S&T relations and the kinds of topics to take up with the GOJ. While it is impossible to state with confidence what specific materials and data were available to these persons and which ones they consulted, the following groups have been selected and presented in more or less chronological order. The first item is from the U.S. Embassy as input to a publicly available annual report on *Science, Technology and American Diplomacy*, followed by two reports on foreign students and researchers in Japan and the U.S., one by NIH on doctoral recipients and foreign researchers and the other on foreign students in general in Japan and the U.S. The USDA and DOC published two appeals in the *Federal Register* on information concerning leaks and possible foreign exploitation of technological and scientific information from U.S. national labs and another about access by U.S. researchers to foreign R&D facilities. The last three parts are based on a substantial series of cables from the American Embassy in Tokyo on R&D opportunities in Japan, an NSF assessment on access to Japanese labs and other existing and planned GOJ international research projects. This range of reports does provide a variety of information on the "access" issue which was strenuously pushed by the USG at the bilateral negotiations.

The Annual Report, Science, Technology and American Diplomacy

As the USG began its preparations for possibly renewing the 80STA in the mid-1980s, the Tokyo Embassy provided some comments on the USJ S&T relationship for inclusion in this annual report.[11]

Basic science. Concerning the issue of access and Japan's interest in basic science, the Embassy is unequivocal:

> "Access by our [US] scientists to advanced public research facilities in Japan [for basic research] is ... excellent. This background puts us in a particularly good position to participate in and benefit from Japan's new science policy which for the first time places a premium on basic research and the production of new knowledge."

It commented that both countries had recognized by at least the mid-1980s that the costs of research in breeder reactors, fusion reactors, and high energy physics were beyond the capabilities of any one country. Consequently, the USJ were holding discussions about joint endeavors and funding in these fields and in the superconducting supercollider.

Japanese R&D facilities and R&D results. "Most Japanese governmental research institutes are now [mid-1980s] open to foreign research and several programs are underway for actually recruiting and providing financial support for them. The faculty of the Japanese National Universities have been opened to foreigners and several (including at least two Americans) are now professors at Japanese national universities." The MOE was providing new funds to chairs for visiting foreign scholars and a survey showed that "numerous opportunities will be available" for foreign researchers in private research institutes. All this will provide for earlier and more detailed access to the results of Japanese R&D.

The report then summarized USJ cooperation in the manned space station project, nuclear R&D and safety, nuclear non-proliferation, fishing, whaling, the strategic defense initiative (SDI), and access to Japanese technology for military purposes.

While the Embassy input into the annual report might be dismissed as political public relations since it is a publicly available report, there is no hint of dissatisfaction, equivocation. There is no inkling of unhappiness with the rate of Japanese changes to attract more foreigners to Japanese universities and R&D facilities.

Foreign Students and Researchers in Japan and the United States. Much has been said about the "imbalance" of the number of Japanese researchers and students in the U.S. as compared to those in Japan in the press, in the inter-agency discussions, in the negotiations. There was agreement that the absolute numbers were "unbalanced", but there was little, if any, discussion of the percentages of these numbers against the universe of foreign students and researchers. There was such anxiety about Japanese students that proposals -- presumably serious proposals -- to require Japanese students to pay more for their tuition to make up for the real cost, i.e. a surcharge which was ignored and did not come to pass but was indicative of the tone and nature of the discussion. In this subsection, we will cover two somewhat different aspects of this general subject, 1) a statistical presentation of the number of students and researchers in both countries, 2) the first year research experience of U.S. post-doctorates and hindrances in this research.

NIH Survey

The Fogarty International Center in NIH conducted a survey of doctoral recipients in 1987 inquiring where post-doctorates spent their first research experience and the hindrances to doing research in foreign countries.[12]

Of the 62,179 with PhDs from 1942 to June 1987, 1,943 (3%) said they had spent three months or more of research in Japan after obtaining their doctorates. When this data is limited to 1952-1986, there were 928 who spent some time in Japan. When this data is divided by historical periods almost 50% of the 928 spent their research time in Japan during 1952-86: 1952-56:6.5%, 1957-61:5%, 1962-66:14.3%, 1967-71:8%, 1972-81:18.3%, and 1982-86:48.2%. From this data, a rather small percentage of the total number of U.S. post-doctorates who had studied abroad went to Japan but in the last four years of this survey, 1982-1986 (a period highly relevant to this study) a higher number of U.S. scientists/engineers was going to Japan, presumably a recognition of the changing position of Japan's standing in science and technology. Nevertheless, Japan was among the top five developed countries visited by U.S. medical researchers and 6th for biological research, although there was considerable difference in number of researchers going to Japan in contrast to those going to the next more popular countries.

Overall, financial support was identified by almost all groups as a factor in going abroad. The pair of factors, more financial support and greater access to information about foreign research opportunities was most cited as an obstacle to foreign research. Better foreign language training did not appear to be an important factor, except for the older age group (over 55). But the situation concerning Japanese language training was a little different, i.e., language deficiency was one of the problems. Of the 597 (almost two-thirds of those who did go to Japan answered this question, presumably for one-third hindrances were not a problem) who responded to the question about hindrances or constraints, 21% cited language barrier, the largest hindrance. "Lack of opportunity" with 15% was second, but the meaning of this term was not defined and thus remained ambiguous.

In summary, one could merely say that a substantially growing number of U.S. scientists were studying in Japan and that the hindrances were more U.S. oriented, e.g., family disruption, language barrier, job security and the lesser hindrances appeared to be lack of information about opportunities which could be both a Japanese and a U.S. problem. This did not seem to substantiate that even indirectly, the Japanese were creating barriers to block access.

NSF Study on foreigners in U.S. Science and Engineering

An 18-page report (3/5/1987) with 15 tables was prepared by NSF for OSTP for use as background information in developing a USG position on the renewal of the 80STA. While "complete data" on either Japanese or U.S. scientists in the U.S. or Japan were "not available", a special effort was made to garner information from various U.S. laboratories, agencies and organizations and the conclusion was the obvious: that while a large number of Japanese S&Es are coming to the U.S. "relatively few" were going to Japan. Nevertheless, it would be useful to put this data in some perspective. In 1985 only 3% (344,000) of the U.S. enrollment in higher education were from abroad, more than half were undergraduates, and 1.5% were in S&E. While Japanese student enrollment has greatly increased it kept pace with other foreign groups and has remained at 4-5% of the total foreign students.

Between 1981 to 1985 over 50% of PhDs in engineering in the U.S. have been given to foreigners, 40% for mathematics and computer sciences, 23% for physical sciences and Asian students particularly East Asian students account for 70% of the engineering PhDs, 50% for physical sciences. While 57% of PhDs awarded to Japanese in 1985 were in the science fields, 20% were for engineers. Japanese awards were 2% of those given foreign students in the sciences, and 1% in engineering, thus an even smaller percentage when the total number of PhDs are included. In the top ten countries whose students received Science and Engineering PhDs in 1985, Japan ranked 9th (with 93 out 2,425 from these ten countries) with China first with 792, followed by India (453), Korea (316), Iran (225), Canada (147), Nigeria (105), Egypt (100), U.K. (99) and Thailand (93). Interestingly, little or no anxiety was felt about such a large number of Asians obtaining U.S. PhDs.

Concerning foreign students in Japan, NSF obtained data from the MOE. Of the 15,009 foreign students in Japan in 1985, 618 (5.3%) were from the U.S., approximately the same percentage as Japanese students in the U.S.; of these 160 were graduate level. About 16+% of the foreign students were studying with GOJ support (97 from the U.S.).

Concerning U.S. S&Es going abroad, the report said "little information is currently available." According to data garnered from Bell Labs, Argonne and Brookhaven national labs, and the Brain Research Institute, 8% of the foreign visits were to Japan. Two percent (12) of the Senior International Fellowships (509) given by NIH's Fogarty International Center went to Japan, and Japan was the tenth most selected country; UK headed the list followed by Switzerland, Germany, France, Israel, Sweden, Australia, Netherlands, Italy. Japanese visitors to U.S. institutions constituted about 8% of all visitors.[13]

Based on the above, reliable data on Japanese and U.S. students in each other's countries especially in the sciences and engineering were not available, but based on available data, Japanese students and researchers constituted a small portion and proportion of the total and Japan was not a popular country to visit and Japanese receiving U.S. PhDs were a relatively small proportion of the total number of foreigners receiving PhDs and an even smaller percentage when all U.S. PhDs were included. The number of U.S. S&Es receiving Japanese PhDs was not known (none probably). In comparison, Japanese receiving science and engineering PhDs in the U.S. was relatively large (but quite small compared to other foreigners). Since even to visit Japan was not at the top of the list, there is probably no line panting to receive PhDs in Japan. Based on this data there would seem to be no need to be overly concerned with the perceived threat from the numbers of Japanese students in the U.S. and the number of PhDs received by them, unless, of course, one wishes to make the assumption that Japanese receiving PhDs in the U.S. and other researchers are far more effective and far more likely to return home and effectively create, copy, re-engineer, etc than other researchers from other foreign countries.

Two Appeals in the *Federal Register*

United States Department of Agriculture. The USDA published a notice in the *Federal Register* in 1987 requesting information of the utilization and exploitation of R&D information obtained by foreign S&Es when they were at U.S. Federal Laboratories (the interest here was really focused on Japanese S&Es). There were unconfirmed anecdotal stories that such exploitation was occurring. The anticipated information did not flow in from this notice.[14]

United States Department of Commerce. Based on President Reagan's Executive Order 12591 (4/10/87) and The Federal Technology Transfer Act of 1986, the DOC published a notice in the Federal Register of April 21, 1988 on access of U.S. Scientists to Foreign Research Facilities. It sought to obtain from the private sector "specific information" concerning 1) the denial by foreign governments of opportunities to do research in foreign facilities or to enter into formal cooperative relationships and 2) the effect of current policies governing foreign access to federal laboratories and on private sector willingness to enter into cooperative agreements with such laboratories. Responses were requested by June 1, 1988. According to the Office of the Director of Federal Technology Management in DOC on August 2, 1989 (over one year beyond the deadline), it had not received -- although this is hard to believe --any responses except one from the U.S. Electronics Industry Japan Office (5/25/1988).

This response stated that "electronics research conducted at Japanese universities is accessible to American companies" although there are factors (housing, low stipend, etc) that make a stay at a Japanese university unattractive and that "the record shows that Japanese private companies do provide great access to their research to Americans" and list the major Japanese electronics companies which have accepted U.S. engineering and computer science graduate students in their research laboratories. It commented that the GOJ sponsored new Human Frontier Research Project and the International Superconductivity Technology Center "are current attempts by the Japanese government to increase access, in a controllable manner, to Japanese research." It urged that the "U.S. adopt a posture of strict reciprocity of access to government sponsored research" which would spur the GOJ "to issue a binding and comprehensive set of rules applicable" to all government agencies in English and Japanese, definitively stating the principles for open access to and use of Japanese government-sponsored research". In the concluding paragraph it stated that "baseless tirades against the Japanese research establishment may make good press but they make bad trade policy". This 5-page letter did not specifically answer the DOC's two major points but it maintained that both the private sector and university laboratories were open to U.S. scientists but that it urged the USG to push for a uniform set of rules since the openness of government labs was uneven.[15]

Embassy Reports on Research Opportunities in Japan

In support of the inter-agency effort to assess the future of USJ S&T relations and to provide an analysis of the access issue, the Tokyo Embassy sent a series of cables from September 1986 to August 1987 (just before the bilateral negotiations began) reporting on various kinds of R&D opportunities being offered by Japanese government agencies and the private sector to foreign researchers.

In a long cable on September 8, 1986, the Tokyo Embassy began a report with the comment that the GOJ "supports opportunities for U.S. [and other foreign] scientists to do research in a variety of advanced scientific and technical fields" at GOJ expense and described the programs for this purpose at the following government agencies: the Japan Foundation for Cancer Research, the Science and Technology Agency, Ministry of Education, Japan Society for the Promotion of Science, Agency for Industrial Science and Technology (MITI), Japan Key Technology Center (funded by MITI and MPT), Research and Development Corporation, and a new program called Japan Trust for International Research endowed with ¥13 Billion under MITI/MPT supervision. Over 500 foreign scientists, with over 150 from the U.S., had been selected for study in Japan in 1984. It pointed out that one program on Cancer Research had been in place since 1960. While it could not give a definitive answer to the issue as to who might be benefitting from this experience, the Embassy believed that "it seems that there are a number of areas in which U.S. scientists probably benefit more than their Japanese sponsors."[16] It should be noted that there was no comment on any Japanese reluctance to provide these opportunities nor was there any dissatisfaction voiced with regard to these Japanese opportunities -- if such dissatisfactions did exist in-house cables would provide excellent opportunities to do so.

After Science Adviser Graham visited Tokyo in February 1987, the Tokyo Embassy prepared a series of six cables to address the following issues that arose at this meeting and to help Washington in preparing for the OES A/S's visit to Tokyo in early April 1987 which ultimately never took place: reciprocity, particularly Japanese initiatives in funding basic research on an international scale, access to Japanese government, university and private research institutions, IPR, and management of S&T agreements in Japan.[17]

Reciprocity. The Human Frontier Science Program was mentioned as "the most prominent" program in this area but cautioned that there were others such as the Exploratory Research in Advanced Technology (with five U.S. researchers and a joint project with Rockefeller University the planning stage), the Japan Key Technology Center which is open to foreign participation but so far with no such participation, large scale projects under the national R&D program.

Since Japan had lagged far behind the U.S. in supporting basic research, Japan would "have to become considerably more aggressive in opening their research programs to foreigners." A number of indices were suggested to measure progress in reciprocity: the amount of GOJ funding for research programs open to foreign participation, actual number of visiting U.S. researchers being supported by the GOJ or by both countries, number of U.S. researchers in private industry laboratories, and number of translated publications made available for publication. While these may appear to be reasonable criteria, there is no comparison with European countries (including UK). It is even doubtful that such statistics are even available for the U.S. and Europe; an attempt at comparison would probably be regarded as highly invidious, unnecessary and impolitic.

In fairness, the Embassy stated that "the USG may have to undertake significant new initiatives to be in a position to take advantage of these increasing opportunities, some of which already exist." Some of these steps are to create an R&D Fellowship Training Program possibly funded by USG (and Japan) for a six-month language and culture training program, USG support for corporate efforts in Japan by giving tax incentives, access to unclassified USG data, identification of R&D opportunities, language training support by the USG. Since the need for this program for corporations and the U.S., is "a nebulous undertaking" "without a ... commitment from the USG, it will be difficult to convince either the GOJ or the private sector in either country of the seriousness of our interest." It also mentioned that some U.S. companies, citing Kodak and Corning Glass have successfully on their own placed their researchers in Japanese government labs.

While it was recognized that the number of U.S. researchers in Japan had grown slowly, "to a large extent the slow growth is a problem of supply, not demand." Embassy believed, on the basis of preliminary data, that Japanese laboratories have been providing opportunities more rapidly than the U.S. has provided trained scientists. The USG must do more "to actively encourage researchers to focus on Japan if this situation is to change." (Tokyo 03855 concentrated on reciprocity)

Access to R&D facilities. NSF had distributed a directory of 124 Japanese companies which were willing to accept foreign (including U.S.) researchers. This was later refined with a survey asking Japanese corporations about foreign researchers on their staffs. Seventeen companies in construction, pharmacy, chemistry, oil detergent, cosmetics, life sciences, oil detergent, electric, electronic beverages, livestock feed, precision machinery, rubber and textiles reported they had foreign researchers.[18]

The number of foreign researchers in Japan's government labs increased substantially (45%) from 1984 to 1986; because of the drop in the

JSPS program the overall level remained flat. It appeared that the GOJ was broadening the research possibilities among government agencies. The USG also had S&E exchange programs on nuclear energy, fusion research, with the Radiation Effects Research Foundation, NASA and NSF. The NSF cooperative program started from a USJ summit in 1961 and is the oldest cooperative arrangement entered into by either country. In 25 years, 4112 U.S. scientists visited Japan, and 5113 Japanese visited the U.S. A wide range of technical fields were covered by these visits. In addition, NSF has cooperative programs in photosynthesis and photoconversion programs, with Japan's fifth generation project, large scale earthquake engineering programs and a language training matching grant program with the Electronics Education Foundation of the American Electronics Association to send 12 engineers to Japan annually starting in 1986-87.

NSF Assessment on Access to Japanese laboratories

As part of its contribution to the internal USG planning process, NSF's International Division prepared two reports in early March 1987 on access. NSF is the USG agency with the longest continual association with Japanese science and technology. It began its report with the following assessment:

> Access by U.S. researchers to Japanese laboratories depends at least as much on actions by Americans as it does on actions by Japanese. Given the size of our respective scientific establishments, it is unlikely that Japan could absorb as many of our researchers as we could of theirs. There are probably relatively more opportunities for Japanese to significantly advance their research in the U.S. than for Americans to significantly advance theirs in Japan.

The report says that "probably" is used "because we don't really know the extent of first-rate opportunities in Japan -- certainly not as well as they know of first-rate opportunities in the U.S." This stems from a greater effort by the Japanese to inform themselves of our system and learn English, and the differing systems of recognition of achievement. U.S. researchers seek recognition through publication, Japanese by impressing their seniors for career improvement. It also mentioned that such other barriers as losing a place in the queue [for promotion], highly competitive U.S. job market, tenure concerns, access to laboratory space, perceptions that the best work is going on in the U.S. and family concerns (two careers and schools) influenced interest in such possibilities.

There may be a significant body of Japanese basic research which U.S. researchers are unaware of and consequently do not ask for. The report pointed out that Japan has embarked on "two fundamental additions" to their S&T policy: promote basic science and internationalize their R&D system. To this end, since 1983, Japanese national universities could hire foreigners; since 1986, Japanese national laboratories could hire foreign researchers "up to and including laboratory research directors". Could foreigners obtain similar positions in the U.S. or in European countries? The GOJ's Fifth Generation Computer project had agreed to accept three U.S. researchers.

NSF concluded that various indicators "imply that Japan's laboratories are quite open to American researchers. To date, we have done very little to test the degree to which appearance and reality coincide. ... Part of the problem of access to Japanese laboratories is ours. ... It will, however, take more funds than are now being allocated to cooperation with Japan. It will be necessary to overcome the effects of a 45% devaluation of the dollar first. ... we cannot escape the fact that pursuing that access costs." A general point is that Japan has done much to open its facilities to foreign, including U.S. S&E personnel. The modest U.S. access to Japanese S&E has been mainly "a result of lack of preparation, effort, and perceived incentives, rather than closed doors."

NSF, nevertheless, suggested 12 actions that Japan could take to assist the flow of U.S. researchers to Japan, e.g. language training, provision of western-style housing, transportation subsidies, advisory service on opportunities, etc. It suggested almost as many actions for the U.S. side, e.g. developing a catalog of opportunities, language training materials, fellowships for language training, providing bigger stipends, but no comment on how the fundamental issue of tenure, losing one's position in the queue for promotion, perceptions of where the best work is done could be resolved.

NSF suggested that the GOJ be given "a relatively long list of possible or illustrative actions, rather than a relatively short list of specific actions for which we would press hard." The reason given for this approach was significant: "we are probably not sufficiently knowledgeable on Japanese bureaucratic and budgetary considerations to pick the optimal combination of *feasible* and *significant* measures." -- quite an admission from the U.S. technical agency probably best informed about Japan. It recommended that the USG provide "draft general principles and intentions for an annex" "plus illustrative actions which would not be part of the draft annex." NSF warned most reasonably that before we approach the Japanese about actions we want them to take, "we must take into account whether and how we could handle Japanese proposals for similar access to U.S. S&E efforts to which the U.S. Government contributes."

These suggestions were ignored and many detailed actions were included in the USG draft agreement which the GOJ would be committed to take. In the negotiation process, however, they were all deleted from the final agreement and many handled differently.[19]

Other Existing and Planned International Research Projects. In addition to the above, there are other kinds of programs that the GOJ had initiated or planned to initiate which involved international participation.

The International Frontier Research Program. The aim of this program was to discover new knowledge that would serve as the basis of technological innovation in the 21st century especially on the biological background of homeostasis mechanisms of animals and plants and on functional frontier materials, and to introduce new aspects of international exchange and cooperation in the field of basic research in Japan. It was created by RIKEN (Institute of Physical and Chemical Research) in October 1986 "as a means of internationalizing Japanese science and technology." RIKEN was originally founded in 1917 as a non-profit institute supported by the Imperial Household. As a result of WWII it was dormant until 1958 when it was reorganized into a non-profit research organization financed by the government and industry. The president is appointed by the Prime Minister for four years. In April 1987 it had 400 permanent researchers and about 600 visiting from abroad. RIKEN directors commented to an Embassy visitor in April 1987 that while foreign scientists are welcome "they sometimes are reluctant to come to Japan because they may suffer setbacks in their careers at home." Language was not a deterrent in the labs but a handicap in the local community. RIKEN has international agreements with such organizations as Max Planck Institute in Germany, Korean Advanced Institute for S&T, Pasteur Institute of France, DOE, NSF and the Universities of Georgia, Washington and Illinois.

The program consists of several research laboratories, three of which are headed by foreigners (two U.S., and one Frenchman). They make several trips to Japan per year and spend several weeks directing the work of their labs. One of the U.S. directors persuaded the Japanese to accept young U.S. corporate scientists to conduct research at RIKEN. This program is to be differentiated from the "Human Frontier Science Program" (HFSP). The RIKEN program is funded by S&TA in contrast to the inter-agency funding and management of the HFSP. It is conducted in RIKEN's suburban campus near Tokyo while the HFSP and the ERATO programs will be conducted in various labs around the world. The RIKEN program is less oriented to commercial application and does not have industry participation in contrast to the other programs. The budget for this program in 1986 was 1,119 million yen (approx $7.36M at ¥152:$1), ¥1,535M in 1987.[20]

The University of Tokyo School of Engineering. It announced on April 1, 1987 to representatives of EC countries, the creation of a Research Center for Advanced Science and Technology (RCAST) at the Komaba campus. According to the Embassy this was "a deliberate attempt to increase the number of foreign students through more vigorous recruiting; there were then about 800 at the University of Tokyo, 389 at Engineering, 17 from the U.S. with most from Asian countries. The second feature was to invite guest researchers from the U.S. and Europe especially in computers and electronics. The Embassy opined that the Center has a "solid foundation" at the University and has support at the national level.[21]

The Ministry of Education. Although, normally characterized as a bastion of conservatism and resistance to change, it had "a reasonably good record of promoting some international participation" (*Japan Times* editorial of 5/26/1987). The Japanese satellite *Ginga* (Milk Way) launched by the Institute of Space and Astronautical Science (under the MOE) had on board instruments built in part in Britain and the U.S. The National Laboratory for High Energy Physics in Tsukuba science city is the site of "comprehensive foreign cooperation in searching for new types of sub-elementary particles." (ibid.) The ERATO program, another program which had been in existence six years (since the early 1980s), funds basic research projects that include young foreign scientists selected from more than a dozen nations.

NTT. It conducts some joint research projects with foreign companies.

The Japan Key Technology Center. It was created only in 1985 to enhance basic research in technical areas of interest to MITI and MPT and does include a "Japan Trust Fund" to support Foreign Researchers in Japan with travel costs, stipend, allowances for family members and for attending conferences and visiting research institutes. But as of 1987-88 the results of this effort were not clear.

The above is a summary presentation of some programs -- probably not all opportunities -- which had been in existence for some years and others which were newly designed to interest foreigners, particularly U.S. and European scientists to conduct their studies under Japanese auspices in Japan.

Based on the above, the following observations and conclusions could be made about the access to Japanese R&D facilities as negotiations began.

1. There was agreement that there was a discrepancy between the number of Japanese researchers in the U.S. and the comparable number in Japan but there was emphatic disagreement on the causes of this discrepancy; the U.S. side chose to ignore this rather crucial element of the equation.

2. The GOJ had only recently begun to create a greater variety of opportunities for foreigners to study for longer than three months (generally the dividing line between short and long term) research in Japan. Scientific intercourse between Japan and foreign countries was not new, especially with the U.S. This had been going on for many years, but the scale was relatively small, and it could be fairly said that the demand was also decidedly limited; after all there was for a long time the sentiment among Westerners in particular that there was not much to learn from the imitators. The above description would seem to describe a frenzy among the Japanese to prove themselves through internationalization of their science and technology through improved access to their facilities. An editorial in the *Japan Times* (5/26/87) said that considerable effort had been made to open university laboratories and public and private R&D centers to foreigners but characterized these Japanese doors "as yet only ajar, if that" and admonished Japan to open the doors further; it is "Japan's turn" to welcome foreign scientists.

3. Was this an adequate response by the Japanese in the eyes of foreigners, particularly those in the USG concerned with the future of USJ S&T relations? The cited cables were available to all parties in these inter-agency discussions and many were actually tagged for the attention of specific agencies, offices and people. There is no way of knowing if these offices and people read or seriously considered these reports from the Tokyo Embassy. However, judging from the contents of the draft USG agreement and known

emotional tone of the negotiations concerning access to R&D facilities and the known stress on comparable or symmetrical access to these facilities, it would be fair to say that the Embassy's efforts to provide Washington with an educational description of what was taking place were totally ignored in Washington; these reports did not fit into the preconceived notion that was being advocated by selected agencies and departments and that the GOJ was given no credit or recognition whatsoever for its endeavors to entice more foreigners. In light of the atmosphere that seemed to prevail in Washington at the time, these Embassy reports were probably thoroughly discounted and thus ignored as uncritical reports by a pro-GOJ U.S. Tokyo Embassy, were unreliable and that, given the first opportunity the GOJ would back off these various programs opening their facilities to foreigners. The various stringent actions that the GOJ was to take according to the draft USG agreement implied that there was either little or no effort by the GOJ, and that the GOJ could not be trusted no matter what was said; therefore, all kinds of stringent reporting requirements must be imposed on the GOJ.

4. Putting aside whether the described GOJ efforts were adequate or not, the Embassy pointed out the Japanese were providing more spaces for foreign researchers, particularly U.S. researchers than there were U.S. researchers to fill these slots, i.e. the Japanese supply was larger than the apparent U.S. demand.

5. The Tokyo Embassy suggested some indices which might be used to measure the success of the GOJ efforts and the need for both countries to establish a methodology for creating data on foreign researchers in the U.S. and Japan.

b. Defense Patent Secrecy Agreement (DPSA)

During the spring of 1987, two national security or defense-related items were introduced by the DOD into the internal USG efforts to create a united policy toward the renewal of the USJ 80STA. One concerned the use and protection of dual use technology; the other concerned an agreement on the implementation procedures of a 1956 USJ agreement on the protection of patent rights relating to national defense.[22] For reasons unknown, the implementation procedures for this Agreement were never worked out between the two countries from 1956 to the mid-1980s. Negotiations concerning GOJ/Industry participation in the Strategic Defense Initiative during the mid-1980s under the second Reagan Administration, the discussions between the two governments concerning the afore-mentioned implementation procedures and the internal USG debate about the renewal of the 80STA, and personnel changes in DOD representation at the interagency meetings on the USJ S&T agreement, approximately chronologically converged around the spring 1987. DOD proposed that the implementation of the renewed USJ STA would be conditional on a USJ agreement on the implementation procedures of the 1956 Agreement as discussed in section IX.8.e.ii.

When a patent application is submitted to the U.S. Patent and Trademark Office (PTO), the contents of the application is not publicly disclosed until a patent is issued. However, when a patent is issued it is then automatically made public. However, the PTO Commissioner may, within six months of the application, at the request of a defense agency, issue a secrecy order if the publication or disclosure of an invention by granting a patent is considered detrimental to U.S. national security. If this occurs, the secrecy order requires that the granting of a patent is withheld as long as national security requires. For national security reasons, these applications could be thus restricted for many years. When the secrecy order is lifted, the effective date for the application, if a patent is granted, would be that day the application was first submitted. The applicant then has six months within which to submit a patent application in a foreign country if the U.S. has a special arrangement with that country for this purpose.

Under normal circumstances, U.S. inventors (companies or individuals) could submit an application in foreign countries for the same invention in accordance with international agreements concerning patent rights. The U.S. secrecy order puts the application in a special category. In order to assure that foreign countries would respect the special status of these applications, and protect the economic benefits that may flow from these patents, the USG entered into bilateral agreements with fourteen countries in the 1950s mostly COCOM countries for the interchange of secret patent rights.[23]

The DPSA which was signed with Japan on March 22, 1956 was based on the USJ 1954 Mutual Defense Assistance Agreement. Article III(1) of the latter provides that:

> Each Government will take security measures as may be agreed upon between the two Governments in order to prevent the disclosure or compromise of classified articles, services or informa-

tion furnished by the other Government pursuant to the present Agreement.

Article IV of the 1954 Agreement also provides that:

> The two Governments will, upon the request of either of them, make appropriate arrangements providing for the methods and terms of the exchange of industrial property rights and technical information for defense which will expedite such exchange and at the same time protect private interest and maintain security safeguards.

In turn, article III of the DPSA provides that:

> When technical information made available under agreed procedures, by one Government to the other for purposes of defense, discloses an invention which is the subject of a patent application held in secrecy in the country of origin, similar treatment shall be accorded a corresponding patent application filed in the other country.

Article VII of the DPSA provides that:

> Upon request, each Government shall, as far as practicable, supply the other Government all necessary information and other assistance required for the purposes of: (a) affording the owner of an invention or technical information made available for purposes of defense the opportunity to protect and preserve any rights he may have in the invention or technical information; ...

In 1962, the Technical Property Committee under the DPSA was convened for the first time at the request of the DOD. Several more meetings were held until 1966 when the paper trail disappears. DOD files on these meetings and any results were not available and could not be located even for the DOD officials involved in discussions with the GOJ in the mid-1980s. On the Japanese side, the officials appeared to be from the GOJ's Patent Office. The U.S. Department of State and the MOFA were apparently not involved in these discussions in the 1960s. Nevertheless, the DOD representative in the mid-1980s as part of the inter-agency group to formulate a new USG S&T policy vis-a-vis Japan accused the State Department of not negotiating an implementation procedures agreement. Why negotiations were abandoned, terminated, postponed is not known. Based on interviews of the officials involved in the 1960s by the DOD representative in the mid-1980s both sides apparently recall no crucial differences to reaching a consensus. There was probably some deference to Japanese political sensitivity to a "secret" patent system and the Japanese concern about the meaning of the term "for the purposes of defense" in the DPSA. One can also speculate that a change of personnel in the DOD and/or PTO resulted in this issue falling between the cracks. Probably much more importantly, during the 1950s, 1960s and 1970s, there was no felt need on the part of the USG to pursue this issue since the flow of technology was basically from the U.S. to Japan and U.S. corporations apparently did not feel any threat or substantial challenge from Japanese industry in military technology. As part of the coldwar alliance between the U.S. and Japan, the latter received U.S. technology through 1) licensing of U.S. defense systems since the 1950s, 2) data exchange since 1962, and 3) coproduction agreements (70 such agreements were reported between 1976 and 1980 alone). Under these arrangements, numerous weapon systems were manufactured in Japan under coproduction arrangements with Japanese industry.[24] Industrial property rights and technical information *for defense* would be provided to Japanese industry on the basis of proprietary information. While this would control the dissemination of classified information it would not cover the U.S. company's patent rights when the secrecy order was lifted. The granted patent would be effective from the date of application many years previously under the U.S. system and with countries with whom the USG had a DPSA arrangement, but not with Japan since there was no implementation agreement between the two countries for this purpose. There was thus during the 1980s a rising concern, fear and anxiety about the rising challenge Japan represented in science and technology and presumably a growing number of actual cases where U.S. companies whose inventions were protected in all other countries with whom the USG had DPSA arrangements but not in Japan because of the lack of an agreement on implementation procedures under the DPSA. For example, if a company's patent application which had been under a secrecy order limitation, was now regarded by a defense agency as being releaseable and a patent could be granted, that company's patent would not have any standing in Japan, and was thus unprotected in Japan. The date of filing an application in Japan would not be the much earlier date used in the U.S. and other countries but, at the earliest, the date of the declassification of secrecy or the date that company applied for a patent in Japan. In the mid-1980s, the PTO had to repeatedly deny U.S. companies filing privileges in Japan because implementation procedures did not exist.

It appears these implementation procedures did not exist, not because the GOJ refused over almost 30 years, but rather because U.S. companies did not feel any serious threat from the Japanese in these esoteric reaches of military high technology. Furthermore, it was not because, as the DOC asserts in its attacks against the State Department in 1987, the State Department had not negotiated an agreement with the GOJ for this purpose; it had not been involved in the early negotiations in the 1960s; that responsibility lay with the DOD and PTO.

As part of the growing recognition of the increasing role that Japan was playing in the development of commercial AND military technology, and that there was the distinct possibility that the flow of technology would no longer be a one-way street but increasingly a two-way street, a new era was created in December 1983 with the signing of a USJ agreement on the future exchange of military (much of which would be dual use) technology. Implementation procedures for the late 1983 agreement had been negotiated with the GOJ by mid-1985.

PTO had brought its concerns to the DOD's attention in late 1984 or early 1985. It was decided that in light of the prevailing atmosphere surrounding USJ relations, especially over trade issues at the time, reviving the bilateral talks which had petered out in the mid-1960s, and remedying the obvious inequity in the relationship, a small group from the DOD, PTO and State Department decided it would treat this problem quietly as a military or a national security issue since it stems from the 1954 Mutual Defense Assistance Agreement. On the GOJ side, MOFA was responsible for these negotiations.

If this was regarded as an economic/trade issue, the DOC and USTR would have been involved with much theatrics and bombast and several more stirings of the anti-Japanese ferment would have resulted. On the Japanese side there would have been no legal basis for secrecy on the civilian side. It may have required a revision of the Patent Office's legislative foundations and would have resulted in a highly charged and emotional Diet debate on the Japanese side, especially since Prime Minister Nakasone himself was opposed as a matter of government policy to the enactment of a National Secret Protection Law (*Kokka Kimitsu Hogohō*). Military secrets were maintained by Japan based on the USJ mutual security assistance agreement of 1954 as approved by the Diet.[25]

The negotiating solution for this issue was through the 1956 DPSA between the USG and GOJ, not negotiating a new agreement which may be subject to Diet or U.S. Congressional approval. Fortunately, the GOJ regarded the 1956 DPSA as a "treaty" which had required Diet approval and would take precedence over internal laws. The USG, on the other hand, regarded the same agreement as an executive agreement not requiring the approval of the U.S. Congress. Article 26 of the Japanese Patent Law states that:

> Where there are specific provisions relating to patents in a treaty, such provisions shall prevail.

Discussions with the GOJ (MOFA and JDA) began in Fall 1985. On the GOJ side a completely new generation of officials were initially skeptical as to why suddenly after almost twenty years since the last discussions about an implementation agreement, the USG was suddenly interested in concluding an agreement on implementation procedures for the 1956 DPSA. It would be useful to digress momentarily to the hearings of the House Foreign Affairs Committee of the Diet on May 16 and 19, 1956 when the DPSA was considered. The GOJ representative explained that there were two kinds of secrets, national defense secrets under the National Secrets Protection Law (stemming from the 1954 MSA Agreement) and secrets protecting the private rights of the owners of certain knowledge and inventions. It recognized that certain information could involve both types of secrets as envisioned in the 1956 DPSA. Japan could not expect to receive such advanced military technology unless it enacted the 1956 DPSA. The fundamental purpose and philosophy of the 1956 DPSA was throughly appreciated by the GOJ in 1956.[26] While, for tactical internal USG purposes, the three agency team of State/DOD/PTO officials described "defense patent secrecy" as a national security issue, in their discussions with the GOJ officials the stress was placed on promoting economic equity, to protect individual and institutional economic activities and to facilitate cooperation between U.S. and Japanese companies in R&D. Even though the merits of and need for the DPSA implementation procedures had to be explained and defended again, the discussions with the GOJ proceeded without difficulties in an amicable atmosphere. These discussions over more than a year resulted in an understanding with the GOJ about the equities involved in the DPSA.[27] Before the DOD representative left the DOD in January 1987, he briefed the DOD official handling the negotiations with the GOJ based on the latter's possible participation in the Strategic Defense Initiative (SDI) and the senior staff person in the OSTP about the background and present state of discussions with the GOJ about DPSA implementation procedures.

In contrast to the past, highly specialized technical information would need to be exchanged if and when Japan, as a government or through Japanese companies participated in SDI R&D and the information provided would undoubtedly require legal protection under the DPSA. The joint US-Japan development of the next generation fighter support aircraft, the FSX, would also require this kind of patent protection for both Japanese and U.S. companies. The DPSA implementation procedures were, in a way, the missing or final link in the arrangements for potential and probable joint sharing of military technology based on the 1983 USJ agreement on sharing military technology. Whereas in the past neither side had invoked the DPSA, there were now in 1986 about 75-80 applications of U.S. companies under PTO secrecy orders which had been lifted and the six months remaining to file for a patent in Japan were expiring for about 3-4 patents. The holders of these patents needed to have the protection afforded by the implementation procedures now being negotiated by USG and GOJ. From the macro policy viewpoint and micro-company interests, therefore, there was an urgent need for both governments to reach an agreement on implementation procedures for the DPSA. Since the early part of 1987, the DOD official for SDI negotiations with the GOJ now also represented DOD on the interagency committee on the renewal of the 80STA. The DPSA implementation procedures immediately became entangled in the internal USG discourse -- more accurately stated as bitter wrangling among USG agencies -- and soon, principally at the behest of the DOD, the implementation of the renewal of the 80STA was made conditional on conclusion of the agreement on implementation procedures for the 1956 DPSA. State Department objected most strenuously against this approach since they were two separate and discrete subjects, but it was overwhelmed by the opposing forces in the USG. It must be remembered that this atmosphere of tension was exacerbated by the Toshiba-Kongsburg Vaapenfabrik Incident in 1987 and that the state of USJ relations in 1987 were at their lowest level since the end of WWII.[28]

The negotiations over this agreement, like those over the renewal of the 80STA, had been conducted without public fanfare. Toward the end of January 1988, however, a joint inter-agency letter (DOC, OSTP and USTR) was sent to SecState urging that the USG decide not to sign a new STA with Japan unless the GOJ signed an agreement on implementation procedures of the 1956 agreement. In order to calm the USG, the Secretary of State had to provide his personal written assurances in order to decouple these agreements in further discussions with the GOJ concerning the renewal of the 80STA and assuring them (such as the DOC, DOD and OSTP) that the GOJ would sign such an agreement.[29] The GOJ had requested postponement of the signing of this agreement until after the 1988 winter Diet session and after April 1, 1988, the beginning of the Japanese fiscal year. The GOJ feared that if the USG agencies (DOD, USTR, or DOC) went public in their attempt to tie these two agreements together, a political furor would erupt in Diet deliberations just as the national budget was to be passed and make the adoption of the DPSA implementation agreement extremely difficult. Fortunately, this did not occur and the implementation agreement was signed on April 12, 1988. Because of the legal and political problems involved, the Japan Defense Agency (JDA) became the depository of the protected inventions or knowledge.[30]

The U.S. media gave scant attention to this agreement except as a routine small news item. In contrast, the Japanese press provided somewhat broader coverage and the implementation agreement was a main topic in the editorials of the three major Tokyo newspapers (*Asahi*, 4/14/88, *Mainichi* 4/14/88, and the *Yomiuri* 4/18/88). While the editorials gave grudging support to this agreement, they were concerned that the old pre-WWII secret patent system in Japan was being revived indirectly through this agreement, that a crack had now appeared in the postwar principle of openness for all patents, that this is but another example of U.S. anxiety about Japanese S&T developments and their implications for the future. While the U.S. worried that without this agreement Japanese industries which had access to U.S. military technology restricted by secret orders of the PTO could develop, using this technology, more advanced technologies and ignore U.S. patents, the Japanese editorials were distinctly concerned with the possibility that Japanese institutions would, ignorant of the restricted military technology, spend large amounts of R&D funds on a "new" idea or technology only to find that a U.S. institution had already accomplished a similar objective and the Japanese efforts would be for naught. This was indicative of the potential chasm that existed in how each party viewed the other and the extent to which mutual trust was difficult to achieve. (See also the section on media coverage of the bilateral negotiations over the USJ S&T agreement.)

One of the few memorable issues for Japanese officials from these negotiations was the USG attempt to link the broader issue of national security with an S&T agreement, and not the more narrow issue of an implementation agreement for the DPSA. The chairman of the Japanese delegation summarily dismissed the implementation agreement as an issue in USJ negotiations over the renewal of the 80STA.

The strident confrontation over linking the 80STA renewal and the DPSA implementation agreement in the USG is symbolic of the depth of the distrust among selected USG agencies (State vs DOD and DOC and OSTP, in this case, in a secondary role) concerning proposed policies toward Japan over certain narrowly defined aspects of national security/economic issues, and the utter distrust of Japan. This issue stands in stark contrast to the oft repeated public statements by the USG about Japan as our most important ally, bar none.

c. The "NIH Issue"

During the October 1987 Congressional hearing on USJ S&T relations, A/S Negroponte raised the issue of the Japanese presence at NIH and explained that these scientists/engineers had been selected by NIH not by the Japanese. Notwithstanding this explanation, and according to my interviews, the so-called "NIH Issue" was vigorously raised with -- others might depict this in more graphic terms such as a continuous emotional barrage and harassment of -- the GOJ delegation. While the GOJ recognized there was an imbalance in number of Japanese scientists and engineers studying and visiting with various U.S. institutions, the inordinate emphasis on the number of Japanese scientists at NIH as symbolic of this imbalance particularly soured the entire negotiating atmosphere. They felt that to stress the "NIH Issue" was particularly offensive in light of how these scientists were selected by NIH. In addition, they felt that it was doubly frustrating because the U.S. side must have known their emphasis was contrary to and a deliberate distortion of known facts. The purpose of this section is, in light of the unfortunate role played by this rather specialized issue, to clarify the background and reality of the NIH international research programs.[31]

NIH sponsored a series of international research programs to invite selected scientists at all levels of their careers to receive further experience and to conduct collaborative research in their biomedical specialties: the visiting program and international research fellows. The Visiting Program consisted of visiting fellow awards (for junior scientists), visiting associate (mid-level) and visiting scientist appointments (mostly senior scientists). These appointments were for 1 or 2 years and could be renewed for up to 4 years. Those foreign scientists interested in participating in these programs sent their resumes to an NIH scientist in their field. A senior NIH scientist investigator in one of NIH's laboratories must be the applicant's sponsor or supervisor during the entire period of award and appointment. A selected scientist is notified by the Director of the Fogarty International Center (FIC), NIH. These scientists were not allowed to engage in outside work while in the U.S., were given a stipend and other benefits. They were not selected by a foreign government or institution but by NIH, an agency of the U.S. government. The selection process was open to any scientist outside the U.S. who believed he was qualified. Presumably they were chosen because of their individual qualifications and merit and value of the work they proposed to conduct while at NIH.

There is also another NIH program called "International Research Fellowship Programs." Applicants must first be evaluated by their own national nominating committee (Japan had such a committee). These candidates must have sponsors in a U.S. institution who are recognized in their fields and who must be willing to assist the applicant in developing the research plan, act as preceptors for the envisaged research and have the resources to support the research. The applicant must have earned a PhD within 10 years of the application date, speak and write English plus some other conditions. Countries may nominate up to 6 candidates; about 100 were chosen each year. Most of the research was conducted in university labs.[32]

A third category was Guest Researchers where the home institutions, foundations and other sources provide all support for a "guest researcher". There is also a fourth category of research contracts placed by NIH.

Another category is the Scholars-in-Residence. Its purpose was to select outstanding scientists and scholars to interact with the scientific community at the NIH campus to engage in free ranging study of subjects of their choosing. Since its establishment in 1969 to 1987 a total of 156 scientists participated from Western Europe, North America, Japan or Australia. Eleven (7%) came from Japan.

According to the FYs 1986 and 1987 *National Institutes of Health Annual Reports of International Activities*, the following awards had been given to Japanese scientists.

Table IX.1 Awards of Japanese Scientists

Category	1986	1987
Visiting Program	311 (1391)	334 (1465)
International Research Fellows	5 (101)	5 (103)
Research Grants	2 (63)	1 (55)
Guest Researchers	72 (551)	68 (538)

Total	390 (2106)	408 (2161)

Note: Total number of Awards in parentheses. According to data provided by NIH, the number of Japanese in the Visiting Program, i.e. those selected through worldwide competition increased substantially in total numbers in the 1980s from 171 (21.2% of 804) in 1978 to 311 in 1986 (22.35% of 1391) but the percentage remained about the same. In the other categories, Japanese participation had substantially declined from 1972 to 1986.

As the USJ STA negotiations began in October 1987, the Director of the Fogarty International Center, Dr. Craig K. Wallace, offered the following statistics of awards to Japan "in context":

1. Approximately 3,673 foreign scientists have participated in the NIH Visiting Program since 1950; 1,338 (33%) of these have been Japanese [due to a higher earlier percentage and a lower percentage in later years].

2. The NIH Guest Research Program, which provides space, facilities and technical support for foreign scientists, has hosted 2,235 scientists since it inception; of these 341 (15.25%) have been Japanese.

3. Japanese scientists have been recipients of 121 FIC International Fellowship Awards since the program was established in 1958. This is about 10% of total awards.[33]

NIH did "not have a defined international policy or agenda, per se, although fully appreciating that modern biomedical research is international in every respect. NIH had six agreements with Japan, the largest number among all the countries with which it had cooperation agreements. These agreements involved the National Eye Institute (1976), National Library of Medicine (1974), National Cancer Institute (1973), National Institute on Aging (1984) and Health (1965). The programs with Japan were "the longest standing bilateral relationship ... in existence since 1965". NIH listed two "basic reasons" for the continuing success of its USJ programs: 1) committed funding from its inception and, 2) a truly two-way scientific quid pro quo with dedicated scientists on both sides.[34]

Although there were opportunities for U.S. researchers to study in Japan, according to the NIH, they "were narrowly focused, highly competitive or limited in number". Obviously, there was "a rather major imbalance".

FIC's Director Wallace called for "periodic overtures [be made] to the Japanese to expand their support for U.S. biomedical scientists.[35]

In conclusion, that there was a substantial disparity in the number of Japanese scientists coming to the U.S. in comparison to U.S. researchers going to Japan was not in dispute. Nevertheless, NIH regarded their long association with Japanese bio-medicine as a substantial success based on the largest number of mutual agreements. While NIH suggested "periodic overtures" be made to the Japanese to increase their support, they also recognized that Japanese were increasing openings and opportunities for foreign scientists. There was no mention in NIH materials showing that Japanese laboratories and institutions had denied NIH requests for acceptance of U.S. scientists, nor was there any mention of special scientific reasons for the need for U.S. scientists to have access to certain Japanese labs because those labs were involved in some scientific breakthrough. No documents were made available to indicate that Japanese scientists had made major contributions to any specific scientific breakthrough, nor in reverse, were there any indications that Japanese scientists had made breakthroughs in Japan based on research done while at NIH. Furthermore, Japanese participation in the other NIH programs were not as high a percentage as in the Visiting Programs. Even in the latter case, the absolute numbers of scientists and percentage to the entire category had dropped in 1988 to 265 (20%). In light of the above, the use of "the NIH issue" as a battering ram in the 1987-1988 USJ S&T negotiations did not reflect NIH's measured approach, must have been a deliberate distortion of reality because a precise description of the situation could have been easily obtained -- through a single phone call, for example -- and because the initial chairman of the U.S. delegation and OES A/S Negroponte had already volunteered to explain the circumstances surrounding "the NIH issue". That this issue was so used by certain elements of the U.S. delegation to the point of souring the atmosphere of these negotiations, is an unfortunate indication of the weakness in the ability of the State Department (OES and the Japan Desk) to lead and control the U.S. delegation.[36]

5. GOJ's Initial Comments, Context, Attitudes and Approach

On the eve of USJ negotiations, let us try to summarize the GOJ's initial comments from a GOJ document given to the USG just before the briefing session, and, based on interviews of GOJ delegates, the context of the negotiations from the Japanese point of view, what the GOJ thought to be USG's "objectives" or targets evidenced by the USG proposal and the GOJ delegates "private" assessments of the USG proposal as distinct from those mentioned in GOJ's Initial Comments. The points mentioned here provide an important backdrop to the forthcoming USJ negotiations.[37]

a. GOJ's Initial Comments on the USG Proposal

About ten days before the "briefing session" on October 2-3, 1987, the GOJ gave the USG a so-called non-paper, an informal paper expressing the GOJ's ideas on the USG proposal on an informal basis.[38] The ideas in this paper "represent a consensus of GOJ agencies and would form the basis for Japanese positions in the upcoming negotiations.[39] The GOJ comments are divided into general and specific.

The thrust of the first item in these comments was that the S&T arrangements under the 80STA had been carried out as well as in any other USJ arrangement. The USJ should first try to promote cooperative R&D activities under the existing 80STA, researchers should be given full rein to conduct cooperative activities and keep government intervention to the minimum. If these endeavors fail, then the GOJ would "not preclude discussions on such amendment". There is an oblique comment that the U.S. text includes two types of issues, one within the scope of the present agreement, presumably S&T relationships, and another "which goes beyond the scope of the present Agreement" which presumably is a reference to national security obligations under annexes IV and V. It does not preclude "taking necessary measures", if, after both government have exchanged views and such actions are deemed to be appropriate. The GOJ concluded that "there is basically no need for amending the present [80STA]" and that its position "remains unchanged."

At the head of the list of comments on the actual text was the GOJ concern that the existing broad range of USJ S&T activities might be affected by the revision of the 80STA, i.e. "amendment should be designed as not to affect the autonomy of such other agreements and arrangements. This was obviously a major concern of the GOJ, particularly its technical agencies (and also those in the USG).

Another major point of divergence (Annex I) was that the GOJ urged that the possible areas and types of cooperation or the candidates of the projects to be implemented within the 80STA should be identified first and then "a suitable framework" created for these activities and that the private sector should not be included in such government-to-government agreements. This approach later became an important sticking point in the negotiations.

The GOJ flatly rejected any idea of imposing "legal obligations" and implied targets concerning access to R&D facilities in Japan and Japanese STI. These will be pursued voluntarily by Japan as "one of its important policy directions" (Annex II).

The GOJ saw the proposed management structure in Annex III as "too restrictive and complicated". If the USG planned to go beyond the scope of the present agreement [the 80STA] to exchange information and views on S&T policies etc, then this would "cause concern from various quarters in Japan". The first order of business was to seek "revitalization" of the Joint Committee under the 80STA.

The detailed IPR provisions in Annex IV were deemed not necessary for a Presidential level agreement and IPR issues should be dealt with on a case-by-case basis in each implementing agreement as was the present practice. The proposed controls on dual use technology and national security obligations in Annexes IV and V were "a serious matter which needs careful examination" is probably the English equivalent of the original Japanese text which is an oblique way of saying these issues are going to be VERY DIFFICULT to negotiate. It concluded with a simple comment that issues concerning the 1956 agreement on defense patent secrecy were "not appropriate subjects for discussion in these negotiations."

These "initial comments" should have been a clear signal to the USG delegation that a three day "one session" would not wrap up the STA negotiations, that indeed rough seas lay ahead for the USG's approach to creating a new arrangement to "revitalize" the Agreement. In light of the general approach taken by the USG as described earlier and because of its commitment to a far-reaching revision of the 80STA approved at the highest level councils in the USG, it appears the USG gave scant, if any, attention to these GOJ ideas.

b. The Milieu in the Japanese Context

Some of the points that will be mentioned here have been mentioned elsewhere. Nevertheless, it is important to repeat them as part of the Japanese context in responding to numerous USG requests. This writer believes this type of thinking and interpretation is at the core of the attitudes of the members of the Japanese delegation. Even if it may be overly simplistic, not necessarily even historically accurate depending upon one's interpretation of history, they are presented here as important examples of the potential, probable and possible perception gap between the two countries and the two delegations.[40]

- U.S.-Japan auto friction had a "revolutionary impact" on U.S. and Japanese business. The U.S. had regarded Japanese as producing steel, TVs, appliances, refinements of conventional industrial methods, i.e. copy cat perception. The auto challenge proved Japan could create an edge against the U.S. through the automobile.

- The "chip war" involving a high degree of sophisticated technology was predictable and this development "really changed the image of Japanese technology in the U.S."

- The U.S. was not concerned so long as Japan stayed a "child", a copy cat. "That was OK." There was no concern. But when the situation changed then the issue of "level playing field" was raised. Obviously, Japan had achieved its new position by tilting the field, by deviousness in manufacturing, and R&D manipulations, e.g. through the number of Japanese sent to foreign countries, especially the U.S. (Another secondary point was that the "core" trade issues had been "resolved" and three peripheral issues remained -- construction, lawyers and the S&T agreement.)

- The radical changes in the rate of progress and developments in Japanese science and technology since 1945 have prompted the U.S., which has continuing economic and trade problems with Japan, to recognize there was now an urgent need for Japan to restructure (*sai-kōchiku*) its science and technology in a manner more appropriate (*fusawashii*) and commensurate with its S&T strength. In this connection, the U.S. said it wished to create a more equal relationship with Japan in S&T, redress (*zesei*) the imbalance (or free-ride) in basic research and investment and the creation of a symmetrical access to Japanese R&D facilities and information.

- Since the USJ S&T relationship was being considered during the cold war period between the U.S. and the Soviet Union when the latter was described by President Reagan as the Evil Empire, the COCOM violation by Toshiba-Kongsberg Vaapenfabrik brought home the increasingly intertwined nature of high technology and national security and deepened the U.S.'s sense of anxiety and danger (*kikikan*). In addition, dual use technology -- a source of Japanese S&T strength -- began to appear more important than military technology in competition with civilian technology.

- While the USG was determined (iki-gomi) to "rationalize" IPR relations and strengthen IPR protection between the governments, it did not target the much broader subject of IPR in general then being negotiated among the European and Asian countries and the U.S. That IPR now required 10 pages in the present draft in comparison to the few lines in the 80STA is a reflection of the USG's determination.

- The proposed draft attempts, with the reality of Japan's R&D strength in the background, to combine the economic, S&T, national security in one agreement. It is not an agreement showing linear progression from the earlier agreement but rather, it represents a discontinuity from the past.

- The Japanese were somewhat reluctant partners to the 80STA which in their eyes had not created problems. They felt no need for a cumbersome new agreement as pursued by the USG. They did not feel it necessary to "control" R&D but to "guide it" -- this distinction is often difficult to discern, influence yes but not control. They felt no urgency for a "binding agreement" and felt that their approach supported the individual researchers. The GOJ approach was decidedly passive, waiting for the USG proposal.

c. GOJ Delegates' Attitudes toward the USG Draft Agreement

Based on available evidence, the USG informed the GOJ that it wanted 1) to strengthen the 80STA, 2) to strengthen and clarify the IPR aspects, 3) to include national security concerns and 4) make the agreement more "Presidential", but there was no evidence that the GOJ had been given any materials nor had there been any inkling about the specifics -- in contrast to generalized ideas -- of the contents of the USG revision. Even the senior U.S. delegate said that "we are asking the GOJ to think about new ground which was not commonly held in the past", that the GOJ needs "to be educated" about the U.S. position, and that through this revised STA the USG was trying to establish a new level of S&T relationship. *The Far Eastern Economic Review* (3/31/1988:60), for example, reported that the GOJ thought of the STA "as merely a piece of diplomatic *omiyage* (gift), the sort of small gift one head of state offers another when he arrives on a visit. Thus, they were quite unprepared for the agreement's sudden appearance late last year bang in the middle of the bilateral agenda." While the journal's dates were slightly incorrect because the draft STA had been given to the GOJ in August 1987, it was true that the GOJ was very surprised. Based on my knowledge, however, Japanese officials either did not correctly analyze the tone of USG officials and what they were proposing even without specific details or did not wish to face the reality of what the USG was really demanding.

The USG draft agreement was described by various Japanese delegates as shocking, legally clumsy, poorly organized, confused, harsh, one sided, a patchwork, provocative, too many "shoulds" for Japan in contrast to only a few for the U.S. The draft proposal contained many "must do's" for Japan which they felt they could not and would not carry out in any case. The less imposing "should" or "may" were not used. It harked back to the austere days of August 1945 as a diktat to Japan. The draft was described as "without logic" (*rikutsu no nai teian*). They were concerned that any numbers which might be allowed to creep into the agreement, would be used against Japan, used like those in the 1986 USJ Semiconductor Agreement and that they would leave a long shadow on the negotiations (*nagai o o hiku*). The draft STA was intended to be an overarching "umbrella agreement" to embrace all USJ S&T relations. The USG draft was deemed not self-explanatory, especially since, in many cases according to a Japanese delegate, the USG delegates would confer among themselves in order to arrive at a unified interpretation of the draft. The Japanese chairman was surprised with what the U.S. earlier described as the thrust of their revision but was "very surprised" in diplomatic terms when the GOJ received the August 1987 draft agreement. One of the U.S. delegates obviously unhappy with the USG draft bitterly referred to it as "oozing with Judeo-Christian ideology".[41]

A Japanese science attache suggested that the world's engineers and scientists would seek out the institutions where they should conduct research, e.g. U.S. scientists once upon a time studied German and studied in Germany and now scientists flock to the U.S. In time, perhaps Japan would become a favorite country to study science and technology. This cannot be artificially created. Because of the heavy involvement of the DOD in U.S. R&D, it tends to be, he continued, more secretive than in Japan. In Japan, the emphasis was commercial where Japan watches, studies and then chooses an approach -- hopefully the right choice -- for achieving a given technological target. The various USJ R&D agreements and MOUs since 1945 focused on specific technical areas without the encumbrances and anxieties of national security considerations. In addition there was the generalized feeling, idealism, and conviction that the sciences, at least in Japan were, in the main, except for industrial proprietary and military information, open and unrestricted. This sense of openness whether correct or not, whether reflecting reality or not, was probably deeply ingrained in the minds of those from "the technical operating agencies" (i.e. S&TA, MOE and MITI national laboratories) and those participating in the negotiations. They would come to these negotiations with great skepticism about the need to seemingly compromise, perhaps feeling that they would have to deny or reject the long-standing perceived openness of postwar Japanese scientific and technology endeavors jointly with the USG. Why was it necessary, they would ask, that Japan should accept such complicating restrictions as national security and on the flow of open S&T information and findings stemming from these projects on all S&T bilateral cooperative projects which have all been open and mutually successful for the past several decades. These bilateral projects would continue, they would assert, even if Japan did not accept the new USG draft agreement. This conceptual approach to the bilateral negotiations would create tensions in, not only the Japanese delegation, but also lead to skepticism and distrust of what the USG was trying to achieve through its draft agreement. In addition there was a feeling among Japanese delegates that the USG over-rated *(kadai hyōka)* Japanese S&T achievements and standing and for the Japanese delegates to under-rate (*kashō-hyōka*) Japan's own achievements in S&T. These various strains of thought or approaches came into focus and clashed in these negotiations. They felt, in any case, in reality and in fact, Japanese basic science was not highly regarded by the U.S. notwithstanding comments by some to the contrary.

Prior to the arrival of the U.S. delegation in Tokyo in early October, several articles appeared in the Japanese press about USJ S&T relations and the upcoming discussions. While it is not feasible to deduce that these accounts accurately reflect the positions, attitudes, judgments of GOJ officials, they probably reflect in general the feelings of the involved officials and under score the above points. According to the *Tokyo Shimbun* of August 12, 1987 (before the USG draft was given to the GOJ), the USG is proposing a draft which "aims at gaining superiority over Japan in high technology where competition with Japan is becoming fiercer."[42] Another article appeared in the *Nihon Keizai Shimbun*, 9/10/1987, which is somewhat more specific in assessing the intentions of the USG draft: if the GOJ accepts the concept of "sharing" with the USG and the private sector, "the state of R&D within the country [i.e. Japan] will become completely known to the U.S. side and that it may result in placing Japan's science and technology under its [U.S.]

control and, in addition, the U.S. retains the sanctuary of classifying military technology".[43]

Another point that was stressed and was repeatedly heard in other interviews with Japanese officials was the U.S. request/demand for "symmetrical access" to Japanese R&D facilities. The driving force behind this request, according to the GOJ delegates, was believed to stem from the Santa Barbara (1985) and Kyoto (1986) meetings where Japanese and U.S. scientists and engineers had met to discuss USJ S&T relations under private (or perhaps more accurately, semi-official) auspices and on-going trade negotiations. The USG had "absorbed" the views of the U.S. science community which had grown in political impact and strength after the auto and chip issues had been raised. Although the USG did press this point with great determination at the negotiations, the USG record does not indicate at all that the Santa Barbara and Kyoto discussions or recommendations were available, read, or considered in drafting the USG proposal. Some USG officials undoubtedly were aware of these meetings and were influenced by them but this kind of interaction on a horizontal basis is difficult to ascertain. USG's aims as perceived by the GOJ could be summarized in four points:

- The U.S. access to and, if possible, control of Japanese civilian technologies in the most direct and cheapest way possible.
- Strengthen/Improve (kyōka) U.S. industrial competitiveness through protection of IPR.
- Maintenance (*iji*) of superiority over the former Soviet Union through protection of secret information (*kimitsu jōhō*).
- Creation/establishment of an arrangement which can be used to make demands *(chūmon)* on Japanese R&D policies, programs and projects.

6. Other GOJ Internal Developments

a. The Kokusai Mondai Kondankai (International Issues Consultation Group)

The *Kondankai*, was created in August 1987 just as the USG was giving the GOJ the draft agreement for a new S&T relationship, under the *Seisaku Iinkai* (Policy Committee) of the Science and Technology Council nominally chaired by the Prime Minister. While the staff support for the Council was provided by the Science and Technology Agency, one of the non-governmental members of the S&T Council chaired the Policy Committee and the Consultation Group. During this period the chairman was Dr. Okamoto Michio, a physician; as such he was the de facto chief operating officer of the S&T Council, responsible for the day-to-day management of the Council. The latter Group had about 20-23 members selected from the Council and elsewhere. In contrast to the Policy Committee which held closed meetings, the Group with its expanded membership was more open.

Periodically, the chairman of the Japanese delegation for negotiating a new STA, Ambassador Endō Tetsuya, would meet individually with a few members of the *Kondankai* to brief them, i.e. to keep them abreast about the various issues in these negotiations and how they were being resolved. They discussed symmetrical access, intellectual property rights, the new situation facing Japan in science and technology especially as it concerns basic research. Generally speaking, the Group's members were silent and approved of Ambassador Endo's report. But there was only one 2-hour "formal" meeting with the *Kondankai*. The purpose of these meetings was to inform the Committee members of the status of affairs, NOT to seek or solicit opinions on specific issues from this exalted membership. This writer was informed that the draft agreement per se was not shown to the Group members. This process was, nevertheless, an important part of the Japanese style of assuring that "harmony" or consensus among the various interest groups and institutions prevailed; it is an example of *nemawashi* (root binding) in the Japanese context. In this instance, the MOFA representative was taking the necessary steps to insure that the scientific, technical and engineering membership of the PM's Council for Science and Technology were thoroughly informed and supportive of the proposed STA. Whatever opinions were voiced, it was felt that the professional civil and foreign service corps had accurately picked up or had perceived (*satchi-suru*) these issues. Through these meetings and other contacts, Ambassador Endō was described as "knowing the feelings of Japanese scientists and engineers." Another member of the *Kondankai* mentioned that it also discussed relations with the European Community, Southeast Asia and the developing countries, although it was in general terms without specifics. Another example of liaison or *nemawashi* was for a Division Director from the Science and Technology Agency to individually brief members of the S&T Council and the *Kondankai* on the USG draft and report on the probable reactions of the OECD to these changes proposed by the USG, especially in light of the isolated position in which the U.S. found itself in the OECD as interpreted by the GOJ. But this member pointed out that he was personally briefed because of his prior position as an adviser in the MOFA on OECD. The professional corps are quite independent and do not feel compelled to accept the opinions of academics and politicians.

The *Kondankai* did not produce a final report on the 88STA, had no known meeting with the Prime Minister about the new STA with the U.S., but definitely fulfilled its function as an informal channel for intra-governmental coordination. The *Kondankai* was disbanded after the draft STA was signed in June 1988.

b. Gakujutsu Kaigi

The *Gakujutsu Kaigi* (The Japan Science Council) was originally created in early postwar years (1949) as a forum for scientists and engineers to voice their opinions about S&T policies. Generally the composition of the Council membership had a leftist anti-governmental attitude. The membership consisted of delegates chosen by technical disciplines and was thoroughly controlled by the extreme left and well known for its highly critical stance toward the successive conservative governments in postwar Japan. The method for selecting delegates to the annual meeting of the Science Council was revamped in 1985 beginning with the 13th term. Each academic society was duly registered in order to participate in member selection. These societies nominated candidates for membership to the Science Council. 210 persons were nominated for Council membership and were then appointed by the Prime Minister to three-year terms. The Council President, a part time position, was chosen from among the members of the Council. Dr. Kondō Jirō became President in 1985 under the new arrangement. The Council is attached to the Prime Minister's Office but functions independently. The Council represents the opinions of 915 societies (1993) with 560,000 members, from economics, philosophy, psychology, law, political science to medicine (including dentistry), agriculture, engineering, pure science.

The Council President, therefore, provides another channel of direct advice to the PM (in addition to the Council for Science and Technology chaired by the PM) on science, technology and engineering. The Council President noted that because of its long standing reputation for its leftist orientation since 1949 had not dissipated with the new election method, and that he had been newly elected only in 1985, the MOFA probably did not have a full appreciation of these changes and did not discuss the STA draft with the Council. The Council President had an opportunity, however, to discuss the proposed STA as a member of the Consultation Group (the

Kondankai). When the STA was discussed, the possibility of Japanese cooperation with USG's SDI (Strategic Defense Initiative) had been raised. He pointed out that he expressed grave concern and anxiety about the impact of the SDI on the free flow of STI and that a "military net" could be thrown over R&D in Japan. This concern, he said, did not intrude upon the STA negotiations, only a matter of concern in the *Kondankai*. The Council per se does not develop an institutional position, and does not have the authority to speak to industry; the Council President, however, can present his "personal opinions" to the government at the highest level. These opinions, however, are carefully developed after discussing the issue(s) at hand with the top leadership of the Council. He noted that the GOJ did not sign an SDI agreement but Japanese industry did cooperate separately.

The *Nihon Keizai Shimbun* (2/9/1988) reported that President Kondō was strongly opposed to any security clause in the new STA and would, as President of the Council issue a statement of opposition emphasizing that such a clause clearly contravenes Japan's postwar policy of openness in S&T without military secrets. The *Yomiuri Shimbun* (4/2/1988) reported that Kondō called the security clause "extremely dangerous". Japanese scientists, he said, were not used to secrets and we would not want to be hauled before the courts as a criminal even before one knew what had transpired. The "court" referred to here was not a Japanese court but a U.S. court when a Japanese scientist visited the U.S.

Again, as in the case of the *Kondankai*, there was no report, no explicit decision by the Science Council on the proposed STA. But one might say that "all bases had been adequately but informally covered" to assure an internal GOJ consensus.

c. Japanese Diet Interpolations, February-March 1988

The Japanese Diet held three interpolation sessions on, among other issues, the status of the negotiations of the U.S.-Japan S&T Agreement and the implementation procedures for the 1956 Defense Patent Secrecy Agreement on February 23 and March 1, 1988 in the Budget Committee of the House of Representatives and on March 16, 1988 in the Budget Committee of the House of Councillors. These hearings were attended by the Prime Minister, Cabinet Ministers and a large array of government witnesses consisting of Bureau Chiefs (equivalent of Assistant Secretaries in the U.S. system) or their equivalents from the various Ministries and other governmental agencies.[44] On both days, the interpolations in the House of Representatives were conducted by the political opposition, a member (Inoue Kazunari) of the Japan Socialist Party on February 23, 1988 and by a member (Takeuchi Kazuhiko) of the Kōmeitō (Clean Government Party) on March 1, 1988. The first concern was why the S&T agreement was not submitted for Diet approval. Three reasons were given: there were no legal issues (*hōritsu jikō*) requiring Diet approval, no financial commitments requiring additional appropriations for Diet approval and the agreement did not involve any changes in the basic relationship between the U.S. and Japan. The second thrust was an attempt to persuade the government to discuss the contents of the on-going negotiations on this agreement between the two governments. Specifically, the interpolator suggested that, as reported in the media, there were three major issues in these negotiations: a better balance in the exchange of researchers between the two countries, strengthening the protection of intellectual property rights, and the prevention of the leakage of defense-related information to third countries. The government witness, Ambassador Endō, chairman of the GOJ delegation, begged off answering this question since it was now being negotiated. Prime Minister Takeshita also commented at these hearings on this issue by emphasizing that negotiations would be secret or confidential, but the final results would, to the greatest extent possible, be made public and also added that when he met President Reagan at the January 1988 Summit Meeting they had agreed that the pending S&T agreement negotiations were important and should be concluded as soon as possible. The polite interpolative sparring with all the necessary honorifics ended in revealing no additional information about the S&T agreement and its status by the government witnesses.

The interpolations on the implementation procedures of the 1956 Defense Patent Secrecy Agreement was only slightly more revealing. The interpolations were based not on a Japanese government announcement or revelation about the pending agreement on implementation procedures but based on an article in *Defense News*.[45] Although this topic was a major issue and source of fiery emotional confrontation in the USG, it did not become an important issue in the USJ STA negotiations because the U.S. Secretary of State assured the skeptical agencies in the USG that the GOJ would sign the agreement concerning procedures. The interpolator tried unsuccessfully to tie this issue to the national security aspects of the S&T agreement.

The thrust of the questioning was whether through these implementation procedures the government was introducing the prewar patent secrecy system of the GOJ. Director General (DG) Ogawa Kunio of the Japan Trademark and Patent Office (JTPO) unequivocally denied that it was the intention of the GOJ -- through the back door so to speak -- to re-introduce the prewar patent secrecy system which had been abolished in 1948. Furthermore, introduction of this prewar system, he said, would be contrary to the basic principle of openness of the postwar Japanese patent system and, in any case, would require the passage of new legislation by the Diet. He explained that no additional legislation was necessary to implement these procedures when they are agreed upon so long as they are negotiated within the confines of the 1956 Defense Patent Secrecy Agreement which already had been approved by the Diet and thus superceded other domestic laws. Japan would treat these secret applications "in a manner similar to that practiced" in the U.S. by not making a public announcement of these applications.

The DG of the JTPO also explained in response to a hypothetical case that if a Japanese firm had independently created a new device through its own research and had applied for a patent, it would be handled in the normal manner as another patent application, even if it were the same as the secret application in the U.S.

The opposition interpolator chided the government for allowing the USG to make the decision as to what was to be classified in national security and that Japan's brainpower was gradually being drawn into "America's secret boxes" and controlled by the U.S. net of national security classification and doubted whether this kind of control by an outside power was in Japan's national interest.

The interpolations in the House of Councillors were conducted in the main, by Fushimi Yasuharu (Clean Government Party). Fushimi who was a past-President of the Japan Science Council was treated with slightly greater deference but there was no yielding on commenting on, or disclosing the problems and the direction of the bilateral negotiations. He tried to persuade the government representative (Ambassador Endō, chairman of the Japanese delegation), though unsuccessfully, to disclose some information by posing similar questions in several different formats.

Thus throughout the interpolations in both Houses, the government representatives would not discuss, nor confirm or even deny the difficult bilateral negotiations. The interpolator's attempts to obtain confirmation, denial or expansion on Japanese newspaper articles on the confrontational status of the bilateral negotiations, or even to obtain a list of points of discussion between the two delegations were firmly, politely, and consistently and insistently declined by Ambassador Endō -- even though the very contents of the newspaper articles were undoubtedly based on deliberately released information by his own staff, probably by one of the Four Musketeers. One could not deduce from these interpolations in committee that there were any disagreements between the two governments on the details of these negotiations. Everyone knew that the negotiations were far from harmonious on numerous issues. The Diet obviously in open committee at least played a non-role in international science and technology policy discussions. It was a charade which everyone understood and accepted.

7. The October 1987 Briefing[46]

Even before the USG draft was given to the GOJ, the latter had been suggesting that, to begin with, it would be necessary to have a preparatory meeting "on the director's level" (Deputy Assistant Secretary level in the USG). Such a meeting was described "as a vital step" to subsequent negotiations at the Ambassadorial level.[47] This visit was described by the GOJ as "an important and timely input in the development of GOJ positions for the negotiations.[48] The USG agreed to send a small team led by DASS/OES Ambassador de Vos in early October to brief and explain to the GOJ the draft STA. This meeting would be a briefing not a negotiating session. This assessment was an initial optimistic outlook by the GOJ.

DAS de Vos, led a small delegation of middle level staffers from OSTP, DOC (General Counsel and ITA) and State (OES and the Japan Desk) to Tokyo to meet with GOJ officials from MOFA, MOE, MITI, S&TA, MOT and the Environmental Agency on October 2-3, 1987 "to explain" the proposed USG draft. According to the OSTP, the delegation went to Tokyo at the GOJ's request to give them "a tutorial" on the proposed agreement. The USG Team assembled an array of documents for use at this briefing and subsequent negotiating sessions. Among these papers was the resurrection of the OSTP non-paper described earlier, the Diplomatic Note and Annexes with commentary and historical data and talking points to use at these meetings. Most of the information in these papers have been described previously but there are occasional snip-its which throw some light on how the USG Team looked upon several subjects.

The umbrella-ness of the STA. The USJ space station arrangement is "only technically under the agreement" [80STA]. When the space agreement was up for renewal both sides agreed to "park" the Space Station negotiations "under the agreement to avoid delays". But the fusion study and the SDI agreements were outside the 80STA as a multilateral exercise.[49] The USG equivocated yet again among some talking points as follows: "While it is not our intention to place all S&T activities under this Agreement, we believe that all such activities and the agreements under which they are carried out should be subject to the overall policy direction which only this Agreement can provide."[50]

Objectives of the renewed Agreement. They were "to promote new large-scale projects worthy of two nations at the frontiers of knowledge" and the funding would be jointly arranged, "considering the balance and reciprocity of the overall S&T relationship", i.e. the ownership of a facility may be owned by one country (the U.S.) and operation and maintenance would be funded by another country (e.g. Japan).[51]

U.S. researchers in Japanese R&D facilities. For the first time recognition is given in papers for the negotiations that U.S. researchers are already at or will be doing work in the following Japanese facilities: Japan Key Technology Center, ICOT, ERATO, AIST, Japan Foundation for Cancer Research, S&TA Radiation Effects Foundation. The issue is to significantly expand and restore balance which had never existed to begin with.

Participation of the private sector. The USG position was "to break barriers between public and private elements of the R&D enterprise".[52] The NSF, however, wanted to draw a distinct line between the public and private sectors.[53] This position is the same as the GOJ but NSF was not an official member of the USG delegation.

Basic Research Definition. For the first time, basic research was defined under talking points as follows, since "it is important here, to be precise" by referring to OECD's Frascati Manual which defines basic research as " ... experimental or theoretical work undertaken primarily to acquire new knowledge of the underlying foundations of phenomena and observable facts, without any particular application of use in view".[54]

The GOJ commented to U.S. embassy officials that the draft looked as though some parts of the U.S. proposal would require action by the Japanese Diet, a fundamental change in the nature of the non-energy agreement. If this were the case, the draft agreement might be regarded as "non-negotiable" by the GOJ.[55] The USG tried to assure the GOJ that there would be no need to change any part of the existing legal arrangements in each country, i.e. no laws would need to be revised requiring Congressional or Diet approval and also explained that the IPR annex was needed in order to satisfy the political needs of the troika (DOC, USTR, and OSTP) in Washington.

The USG felt the negotiating leverage favored the U.S. It felt it could demand agreement from the GOJ or threaten action if agreement was not forthcoming. A senior member of the Japanese delegation said that the U.S. appeared not prepared to discuss the details of the IPR annex and questioned why it was necessary to have such a detailed IPR annex in a Presidential agreement, that the GOJ was not prepared to discuss the proposed draft agreement unless considerable clarification was provided by the USG. The Japanese also felt that the State Japan Desk appeared to be too passive, and needed to be more involved and obviously had difficulty in coordinating the USG position. The Japanese acknowledged that since the GOJ was in a weaker negotiating position than the U.S. side, it emphasized OECD findings where it claimed the USG's positions were not embraced and finally that Japanese "bureaucrats" negotiate with the political blessing and confidence of their superiors and the Diet. GOJ prepared a draft agreement, a modification of the 80STA, in October 1987 as a pro forma counter proposal to the USG draft. It has not been found and apparently was never seriously discussed. The GOJ had not conducted studies of USJ S&T relations as the USG had done. Based on interviews with the Japanese delegates, their side for the most part quietly listened to the expositions of the USG group and raised various issues. This description does not reveal either the tone of the discussion, nor how these ideas were presented nor how the USG group responded. The Japanese delegation felt that the USG group did not address the issues raised by the GOJ, answered a few questions and returned to Washington, in an attempt to create a sense of a rushing need to conclude an agreement immediately. They felt the USG group did not recognize the GOJ issues as worthy of a response. There was basically no discussion, only an explanation. Indeed, the USG had not sought ideas and suggestions from the GOJ as it had once said it would do; on the other hand, the GOJ had not sought such an opportunity with the USG. It reported that Hinata of MOFA had said that the briefing session had been successful in identifying major issues including GOJ concerns. But according to interviews with Japanese delegates, the USG briefing session had resulted in a serious misunderstanding. Thus we begin the negotiations with two delegations whose understandings of the problems, or differences in the consciousness of the basic issues involved are almost diametrically opposed to each other.

- The USG was dissatisfied with the status quo, with the intention to create or impose some kind of control (*kisei*) on S&T cooperation with Japan which will redound to the U.S. advantage. The U.S. desires a new deal in S&T relations, thus the linkage of S&T with IPR and national security.
- Japan is satisfied with the status quo and wishes to respect the unencumbered (*jiyūna*) initiatives of researchers.
- the GOJ felt the draft was excessively detailed as a legal document, was without clear definitions and inappropriate (*fu-tekisetsu na*) wording, i.e. those implying unfair attribution of fault to Japan.

The negotiations for a new STA were starting out on the wrong foot and within a rather charged atmosphere headed for heavy seas and difficult and trying times.

8. Bilateral Negotiations, etc.

The USG had begun its review exercise sometime around the summer 1983 and gave the GOJ a new draft agreement on August 28, 1987. Even though it took the USG four years to pull together its opinions on the future of its S&T relations with Japan, it now asked the GOJ to accept this radically new document in ONE negotiating session of three days during the week of October 12, 1987.[56] Shortly before the U.S. Team was to leave for Japan in early October for the aforementioned briefing session, an OES-OSTP meeting took place on September 23, 1987 to review the situation. Based on the preparatory meeting notes to A/S OES, the OSTP was willing to let the 80STA lapse "if we [USG] are unable to conclude negotiations at the single [three-day] October negotiating session"[57]. Perhaps this expectation might be dismissed as a tactical move in the negotiating strategy. The underlying assumption behind such a tactic must be based on the feeling that the GOJ could be rushed, might be so intimidated by the need to sign an agreement at a Bilateral Summit, or be bowled over by the details of the new agreement and unrelenting U.S. pressure, or feel so guilty that agreement would be readily forthcoming. Perhaps this "one negotiating session" idea reflects a USG contempt for the Japanese and their ultimate willingness always to accept U.S. proposals.

The GOJ maintained that the translation itself of the long USG draft would require a fair amount of time, followed by GOJ interagency deliberations and analysis of the draft. It expected that contrary to the USG estimate of a one-session negotiation strategy, multiple meetings lasting about six months would be needed to conclude a new agreement.[58] In contrast to the one session consisting of three days as contemplated by the USG, the negotiations consumed seven sessions each consisting of several days to a week. The sessions were held alternately in Tokyo and Washington, beginning with the latter. The delegations met on October 17-19, December 2-5, December 17-22, 1987, January 5-7, January 21-23, March 3-8, March 21-30, 1988. As might be expected, when the negotiations were held in Washington the U.S. had a larger delegation and vice versa when it was held in Tokyo.

In the meantime, the 80STA would have to be extended -- especially reluctantly by the USG -- on a piece meal short-term basis as a pressure tactic against the GOJ. This tactic could be described as working its influence on the GOJ only to a limited degree. It was indeed extended another three times, three months from November 1987 to January 31, 1988, for two months from January 31, to March 31, 1988 and finally another three months to June 30, 1988.

a. The Teams

To say that the interaction among the delegates of one team and the interaction between the delegates of the two delegations play at times critical roles in the process, the hostile or friendly milieu and final outcome is to state the obvious. Yet, the dynamics of these two delegations by these interactions, conflicts and, I must add, cooperation, did have a direct impact on the outcome. Normally these interrelationships and dynamics cannot be detected through the dry reporting evident in cable traffic between Washington and Tokyo. Some inkling of these interactions and their impact on the negotiations became evident through the many interviews conducted for this study in Washington and Tokyo. While this writer will endeavor to describe these flowing interactions, one must keep in mind the legal ramifications of such descriptions.

The U.S. delegation. It was initially headed by OES Assistant Secretary Ambassador John D. Negroponte; the other members came from the State Department (5), U.S. Trade Representative (2), Department of Commerce (5) and one each from OSTP, Economic Policy Council, Departments of Energy, Treasury and Defense, and one observer from the National Science Foundation. That the NSF was not a full fledged member is an anomaly even though it played an active role in the USG process as described earlier and because it managed the oldest and largest USJ bilateral S&T program. One of the U.S. delegates described the U.S. delegation in rather stringent terms: it was haphazardly organized, none were "Japan experts" [except that one or two had had some negotiating experience with the Japanese], participation was episodic, no Japanese language capability except one who was a Japanese-American and who could speak and read Japanese. The delegates tended to repeat their questions, and only a few participated in the dialogue with the Japanese, and they tended to be those most critical of Japan.[59] When each of the involved USG agencies were interviewed, only the DOC staffers provided an unadulterated emotional condemnation of Japan, lauded the OSTP (mentioning Deborah Wince by name) as the savior and effective spearhead of the USG position and adamantly refused to discuss this issue with me, nor provide any documents or suggestions -- even though this strategy would be apparent by reading the documents I had already garnered and by discussing this issue with other USG and GOJ delegates. The conduct, style and use of language, etc used by two U.S. delegates, in particular, left an "indelibly negative impact" on the GOJ delegation and the negotiating milieu. This impression was conveyed to me by senior Japanese officials who had not even attended the negotiating sessions.

Based on available documents, only brief summaries were sent by cable to Tokyo or to Washington by the USG describing the negotiating sessions. The U.S. delegates were described as lacking preparation for the meetings, were not prepared in a systematic manner to answer the barrage of questions put forward by the GOJ, and publicly disagreed among themselves at meetings with the GOJ.[60] There were a number of lawyers, generalists, patent specialists, but only one scientist that this writer knows of. She had a PhD in chemistry and was an AAAS Science Engineering and Diplomacy Fellow assigned to OES in State. Based on my observations and interviews of OES personnel, she became by default the staffer who provided a great deal of the staff work for these negotiating sessions. She was young, energetic, dedicated and knowledgeable but not a specialist on STAs nor on Japan. When one looks at the personnel situation at the OES office handling these negotiations, it was patently weak, if not in disarray. People who had worked on the STA in the mid-1980s had left, resigned or transferred to other posts. Ambassador Negroponte was replaced by the senior Deputy Assistant Secretary in OES, Peter Jon de Vos in November when Negroponte was assigned as deputy to General Colin Powell, national security adviser to President Reagan. He was a Latin-America specialist where he was ambassador to Honduras and later to Mexico after his stint at the NSC. He was regarded as a listener and energetic doer who inspired a convenient mix of confidence and mild awe.[61] The GOJ regretted his assignment to the NSC and the "downgrading" of the head of delegation to a Deputy Assistant Secretary who was not knowledgeable about USJ S&T affairs.

The role of the U.S. Embassy in Tokyo can hardly be described as active and playing an important role. While the STA was regarded by the Embassy as symbolic of USJ S&T relations, if not our relations in a much broader sense, the Embassy's top priority concerns were, however, the bilateral agreement on nuclear cooperation in spring 1987. The U.S. nuclear negotiating team was small but was successful in reaching an agreement with the GOJ without acrimony ... and the space station Freedom. The U.S. space station team was headed by a woman and it also accomplished an important agreement in an amicable manner. The Embassy was also concerned about leakage of information to the Communists. In contrast, the U.S. S&T Team was always rather large. There was a high decibel count of acrimony in the US Team. The Embassy's efforts were wasted. The USG Team was accusatory of the GOJ and took advantage of Japanese restraint where the acrimony was not returned to the US side. The GOJ Team seldom reacted aggressively to U.S. tactics; this caused the USG team to react, stronger and push harder and thus ensuring a repetition of the vicious cycle.[62]

Notwithstanding the senior political standing of the U.S. Ambassador himself, Mike Mansfield, Washington believed that the Embassy suffered from a case of "clientitis", thus the Embassy's reports, opinions were discounted by Washington. The USG team was not interested in the Embassy's ideas. Washington did not trust nor delegate sensitive issues to the Embassy; all final decisions had to be arranged in Washington. The Embassy's lengthy reports on the opportunities for researchers in Japan were ignored in the negotiations. This was much more significantly underscored by the apparent indifference in Washington to the Ambassador's recommendation directly to the SecState for the USG not to pursue a heavy handed approach to the GOJ (discussed in Chapter XI).

Although understandably only the most important S&T policy issues came to the attention of the Ambassador, he was apparently greatly interested in S&T issues. In an Ambassador Mansfield to SecState cable, the Ambassador noted that

> Here in Tokyo, perhaps more so than in almost any other embassy, science and technology impacts on essentially all aspects of our bilateral relationship from the political to the economic to

> the military ... Our formal S&T cooperative relationship with Japan is the largest and most diverse such relationship that we have with any developed country.[63]

For this reason, the science counselor reported directly to the Ambassador and chaired an intra-Embassy S&T Coordinating Group of all those officers involved with S&T related issues. Notwithstanding this attempt to coordinate S&T issues with other aspects of our relations with Japan, the Embassy's role must have been thankless, frustrating and disheartening.

The Japanese delegation. It consisted of representatives from the MOFA, MITI, MOE, and S&TA. The chairman was Ambassador Endō Tetsuya, a career foreign service officer, who was also then negotiating the space and nuclear treaties with the U.S. His staff support consisted of four division directors (*kachō*) from the above agencies; they called themselves the "four musketeers". As *kachō* they provided the drive, strategy, content, and substance of the GOJ approach and direction. Three of the four were in their mid-forties; the S&TA representative was slightly younger at 37, thus they had grown up in the postwar period and could be regarded as a new generation of officials. Three were from Tokyo and Kyoto Universities and the MOFA *kachō* was interestingly from Keio University, one of the few leading private universities (MOFA for generations had been the preserve of the University of Tokyo); they had majored in law (2, MOFA and MITI), engineering (S&TA) and education (MOE).[64]

Some of the GOJ representatives came to the negotiations with a strong feeling that science and technology cooperation whether bilateral or multilateral, was, by its nature, a positive good for all concerned. Why should we, they would say, have "to suffer through this hard slogging when it is supposed to be a positive good"; why the need for concessions by Japan again; R&D projects, as mentioned earlier, would continue even without the problematic proposed new STA. The attitude of having no STA is preferable to a "lousy" STA was somewhat similar in philosophy to how some elements in the USG felt about the process -- though for entirely different reasons.

There was also a feeling among a number of the Japanese delegates that some senior USG delegates were unacquainted (bad enough) and worse yet were indifferent as to how the Japanese government was structured, how it conducted its business. Some believed that some of the senior USG delegates were singularly interested in achieving what was perceived to be the U.S. President's policy (although they had played a dominant role in creating that very policy and objective which was ratified at the highest USG level and thus enhancing their own interests and ambitions). This sense of frustration -- one might even characterize it as a great disappointment combined with annoyance and anger -- was, I was told, a highly personal position, not a reflection of the concerned Ministries. Nevertheless, since these same officials provided the guidance, momentum and strategy for the GOJ delegation, their views naturally played a critical but awkward role -- especially for the MOFA which, like its counterpart in the State Department, was greatly interested in persuading all parties concerned to accept a new science and technology agreement. The GOJ delegation was described by a USG delegate as comparatively more systematically organized, relying more on their Embassy in Washington for support (the science counselor himself and two others attended the meetings in Washington), seemed to work together as a team more effectively, thus the self-characterization as "four musketeers". I was told that the Japanese team taped each meeting and prepared summaries of each meeting. Translating was done sequentially; this was a disadvantage to the USG since the GOJ staff understood English sufficiently well enough to develop the next response.

The Japanese Embassy played a more active role in the entire process beginning with the science counselor and in providing on two occasions the chairman of the Japanese delegation. The first and last sessions in Washington were led on the Japanese side, by Minister Hyōdō Nagao (General Affairs) and Minister Nomura Issei (DCM), respectively, in the Japanese Embassy in Washington. Minister Nomura Issei played a particularly important role together with the U.S. Chair, Ambassador de Vos, in wrapping up the remaining issues in the last session. According to one U.S. delegate, Japanese negotiators were "on the whole, extraordinarily effective ... better prepared, organized, disciplined, focused and [more] goal-oriented than many American negotiating teams."[65]

On the other hand, the DOD delegate while respecting the GOJ's negotiating strategy of persistence, characterized the GOJ Team "as not always having their act together" and with the large number of GOJ participants from many agencies when the sessions were held in Tokyo, their questioning of U.S. delegates was scatter-shot and seemingly uncoordinated. Some of the more caustic USG delegates described the Japanese as less than forthcoming even deceptive, especially about the existence or non-existence of classification laws and regulations in light of the Toshiba-Kongsburg Vaapenfabrik incident.

Both U.S. and Japanese delegates were caustic of their team leaders as too weak, particularly the U.S. chairman. This trait in both chairmen had a serious deleterious effect on the negotiations especially in how they were conducted, how disagreements within and between delegations could be resolved and how the more boisterous, vocal and insistent members of the delegations could be controlled, especially in the U.S. delegation. Both sides regretted that chairman Negroponte was suddenly replaced just as the negotiations began. Both chairmen were out-voted by their more vociferous delegates. The U.S. Chairman, for example, was essentially subject to a veto by DOC, OSTP and DOD representatives; he had to placate them partly because of his own political weakness, lest there be turmoil upon returning to the U.S. from a session in Tokyo. The Japanese Chairman, also was subject to similar pressures. One Japanese delegate described three "crises" (*kiki*) which they faced during the negotiations. Two were precipitated by internal GOJ confrontations concerning a possible compromise with the USG on the national security or security obligations article in the draft agreement and the third was when the USG rejected, just before the negotiation deadline of March 30, 1988, a GOJ compromise and, in addition, demanded/requested the addition of another principle under article I, i.e. security obligations. These are discussed later in the appropriate sections.

b. At last, the Negotiations Begin, October 17-19, 1987

Just a few days before the first formal negotiations were to begin, the *Nihon Keizai Shimbun* published on October 13, 1987 an article on USJ S&T relations. The Embassy commented that the article closely paralleled information given the Embassy by GOJ officials but the translation of the article was not sent to Washington until October 19, the last day of the first negotiating session. The article noted that 1) technology friction was becoming more serious between the U.S. and Japan, 2) the U.S. demand for "symmetrical access" to facilities and information has been "smoldering in U.S. industrial and academic circles" for some time, and 3) that the demand for reciprocity is applying a term which, up till now, had been used only for trade issues. It pointed out that a bilateral conference would take place between NSF and the Japan Society for the Promotion of Science to discuss joint research and acceptance of U.S. researchers in Japan -- which, it took pains to point out, had been happening for the past 25 years -- but that the reciprocity issue would be injected into these discussions. It noted that the "first formal debut" of the term symmetrical access was at the 2nd

Japan-U.S. Conference on Advanced Technology and International Environment in Kyoto in November 1986 (for discussion of this Conference see IV.7). It recognized that Japan should receive more foreign researchers -- again, if they wish to visit Japan. Concerning U.S. participation in GOJ research projects it noted that MITI and S&TA were working on how such participation could be established but countered that in the U.S. many similar USG projects are closed to foreigners because of the claim to U.S. national security and suggested that both sides should confer on how the legal systems could accommodate such foreign participation, citing the case of SDI research restrictions. Concerning access to information, it noted that there were over 8,000 passwords in Japan to access Japan Online Information System (JOIS) but overseas passwords to the JOIS, the "number is too small and so we are ashamed to count it." It explained that the Ministry of Welfare had programs for 80 foreign researchers to study at the National Cancer Center from 1984 to 1986. The same Ministry will have a program for foreign researchers to participate in AIDS and senile dementia studies. The Ministry of Education was beginning a program in FY1988 to invite 100 foreign researchers to Japan. Professor Emeritus Takashi Mukaibō of Tokyo University asserted that Japan was now quite open, that more money is disbursed by the USG for R&D but that much of it is marked secret and that private enterprises cannot be required to accept foreign researchers even though in Japan these enterprises invest much more than the GOJ in R&D -- contrary to the U.S. pattern. He questioned whether funds should be appropriated by the GOJ just for introducing foreign manufactured systems especially since such expenditures are "Public Investment."[66]

In light of what has transpired so far, the USG Team naturally finds itself barraged by seemingly repetitive and interminable questions -- thus, highly irritating, frustrating to those who wished to wrap up the negotiations in "one session of three days" -- from the GOJ Team about the meaning of a framework, its legal meaning, its political implications. GOJ curiosity is further stimulated when the USG huddles to create a unified statement. The GOJ would huddle to come up with a unified Japanese understanding of what the USG meant. This spiraling activity engenders only suspicions about the motives of the questioner and answerer. Are they stalling? Are they telling the truth? Thus, are there hidden meanings in the very idea of a framework? Is this a controlling statement like an article on IPR would be? Is it a statement of good wishes and intentions? What do many of the words and phrases, sentences really mean? What is their cumulative value as an overall statement. What do words like "comparable access", "symmetrical access", "balanced access" really mean intrinsically as words, then as words in the context of an executive agreement between two S&T superpowers? In a way, it became a word game. The USG Team pressed -- according to some, pressed hard for GOJ acceptance of the "framework"; the GOJ pressed equally hard for their version of a revised 80STA. The USG wanted GOJ agreement on a broad set of principles and objectives on S&T and cooperation in international S&T; the GOJ wanted agreement on the kinds of specific projects and areas of possible cooperation around which the necessary framework would be woven. How would the substantial existing projects fit into the new arrangement? How was the presently patently lopsided nature of Japanese students in the U.S. in contrast to U.S. students in Japan going to be resolved? The role of the private sector in a government agreement was demanded by the U.S. (but not supported by the NSF) and rejected by the GOJ; how would these contradictions be resolved? The gaping gap between the different philosophical approaches to the agreement were now formally and clearly evident. At the end of the first session, each had rested their cases without seemingly having made progress at all. In contrast to the interview data the generalized feeling of frustration, disappointment probably even anger, an internal OES memo of early November 1987 gives a markedly different assessment of the situation,

> "Both sides agree[d] that it is important to address issues of balance, reciprocity and science policy, however the Japanese are reluctant to address some aspects of these issues within the framework of this agreement, to improve the management of the agreement. We are most likely to move forward on the IPR and management mechanisms issues during the next round."[67]

While the rather blithe OES assessment may appear in retrospective to have been more accurate, the DEEP CHASM that did exist between the two Teams was as much philosophical as well as emotional and tactics of negotiations.

- Each of the Japanese agencies coming to the negotiations arrived with preconceived notions about S&T cooperation. The USG position was generously, charitably, or accurately portrayed -- depending on one's personal inclination and bias -- as "conceiving S&T in a total context" while some of the Japanese delegates strongly felt that S&T should be considered by itself, for its own sake. Connecting or relating S&T cooperation with economics and trade issues was barely tolerable and acceptable, attempting to make a link with national security was rejected out of hand (*atama kara kyozetsu hannō*). An STA could not conceivably be negotiated without the concurrence of the Japanese science community. To persuade these skeptics and dissenters, part of the four musketeers, would consume much time, effort and patience.

- That the USG Team did not contain representatives from such agencies as NASA or DOE where actual projects were carried out and that the GOJ did have representation from S&TA, AIST and MOE which were the executing agencies of R&D projects, caused some "spinning of the wheels" in the negotiations. The Japanese felt quite strongly that this difference -- of scientists/engineers being unable to discuss, to communicate with their U.S. counterparts in the technical agencies -- was a major cause of the slippage and misunderstanding in the discourse between the Teams.[68]

- The USG Team members, in most instances, did not have sufficient understanding in order to respond to the needs of the Japanese legal system, practices and public opinion. In certain cases, their understanding (*ninshiki*) was erroneous. This reference was definitely related to the debate on classification of information in Japan for national security purposes. This point was a source of bitter dispute.

- In the bilateral discussions, the proceedings moved along the lines of the U.S. demands and a passive Japanese stance, created among some members of the GOJ delegation, a sense of victimization (*higai ishiki*). This feeling, in turn, made it quite difficult for the GOJ team to step forward positively, energetically (*sekkyoku-teki ni*) with its own initiatives.

- The USG Team (especially OSTP) pushed certain issues, e.g. the unfairness of personnel exchange ratios, and the need for classification of national security information stemming from joint R&D activities, with bull dog tenacity and would insist on pushing even when, in Japanese eyes, they were factually incorrect and inappropriate, made a lasting impression on the consciousness of all Japanese delegates -- and many others in more senior positions than the negotiators -- and a continuing impact on the entire negotiations process. The impact of how the issue of Japanese students in the U.S. should be balanced against the small number of U.S. students in Japan and addressed in the future had an inordinate, deep and abiding impact on delegates and the negotiations. This issue will be discussed later.

These then were the major factors impinging on the atmosphere of the bilateral negotiations as they opened and continued to impact on the following six sessions. They cannot be forgotten as we thread our way through these sessions. Since no agreement was even on the horizon at the end of the first session, the 80STA would have to be renewed: the GOJ wanted six months, the USG three months. By insisting on only renewing the agreement for three months, the USG hoped to put pressure on the GOJ to come to an agreement in time for the up-coming Summit in January 1988.

The Japanese delegation chairman, Ambassador Endō, met with the U.S. Embassy DCM and science counselor on November 25, 1987 to discuss the STA negotiations, one of three on-going topics (nuclear cooperation and space station agreement). He explained that the GOJ counter-proposal was "well-advanced ... nearly complete" and that MOFA would like to complete the negotiations in time for the Reagan-Takeshita Summit meeting in January 1988 and before the expiration of the STA on January 31, 1988. And the Japanese delegation would be prepared to negotiate an extra round "to apply the finishing touches" in December or early January 1988. It was tantalizing to assume that the USG pressure tactics were bearing fruit.

Concerning the contents of the GOJ counter-proposal he said the GOJ would prefer that the "necessary changes" be included in one new Presidential-Prime Ministerial level agreement. The GOJ proposal was described as follows: for the management structure, the USG ideas would be "weighted heavily"; IPR provisions will rely on the working group's inputs; instead of including measures concerning steps to change the balance in exchange of personnel in the agreement, the GOJ would include an appropriate statement in accompanying documentation; JSTI availability would be handled by creating a task force to study ways and means to improve the situation under the new Joint Committee; dual technology and national security information would be "difficult areas to resolve". This is hardly the approach to the agreement that should raise the hopes of the USG delegation for an agreement in the immediate future.[69]

Hardly by coincidence, on November 26, 1987, the *Nihon Keizai Shimbun* published a report on the STA negotiations reflecting the official GOJ position conveyed to the USG. This article which used descriptive metaphor about the USJ negotiations reflected the depth, breadth and extent -- without accompanying adjectives -- of the disagreement during this first session between the USG and GOJ. The general thrust of the report was that most USG proposals were being accepted in order to avoid technological friction. It said it was impossible to "tide over the situation with the attitude which it has so far taken of expanding cooperation within the framework of the agreement in force. However, it [GOJ] shows repulsion [sic] that there are many parts which cannot be accepted ..." This would suggest that there are yet still serious disagreements to overcome before a new agreement could be signed. The USG is reported to have told the GOJ that it absolutely would not respond to an extension beyond January 31, 1988. The GOJ decided on this policy "because if the US side's dissatisfaction is to be left unheeded, the technological friction may catch fire, all at once."[70]

Perhaps the GOJ motivation expressed in this article for an early agreement with the USG, gave the USG Team a sense that, notwithstanding Ambassador Endo's offer described above, the GOJ could be pushed to accept a totally USG approach to a new agreement. Can such a conclusion be drawn from this style of "reading the tea leaves?"

c. Session 2, December 2-5, 1987

The USG preparations for the second session were quite extensive, thorough and the Embassy provided a detailed assessment of various working alternatives in various fields.[71] On access, the GOJ felt that it had received the message from the USG and it had begun taking steps to remedy the situation as outlined in one of the preparatory papers mentioned in the footnote. Consequently, the GOJ felt that this issue was not a subject that needed discussion in the agreement. "There may be some room for negotiation" concerning the choice of forum for discussing personnel access. Concerning the management structure, the GOJ was trying to create its own alternative but a "key issue" would be the terms of reference for the three tiered management structure. In the choice of projects, some were waiting for the USG to propose (S&TA's Miyabayashi), another felt the specific areas should not be included (MOE's Kusahara) and MITI's Inaba suggested the addition of materials research. There was no consensus for technical areas. In regard to the participation of the private sector, this was thought to be a "difficult area" for negotiation. It was thought that it would be difficult, if not impossible, to require private sector involvement in a government-to-government agreement. If the GOJ funded a project 100% "would it be possible to agree to share results of the project with the U.S.? MITI suggested that this issue be taken up in a small meeting to see how this could be done outside the scope of the new agreement. The GOJ wishes to handle IPR on a case-by-case basis under a separate arrangement and the IPR working group should first determine whether any reference to IPR is really needed in this agreement and then find the proper language for an IPR annex. The Embassy's conclusion was that "only in matters which the GOJ sees as infringing on its own policy making independence" (i.e. especially with respect to written commitments to take steps to improve access to its S&T), does it appear that the Japanese side will balk at accommodating U.S. views.[72]

The U.S. Chairman, now changed to Ambassador de Vos, delivered some opening remarks expressing the hope that during this second session we could make significant progress with "substantive responses" from the GOJ to USG proposals. While it was a general review of USG ideas, he also commented on the broad scope of the umbrella-ness of this agreement: "we can think of few, if any, issues in the S&T area that are not within the scope" of this STA. Also, rather importantly, he noted that both sides had agreed on a temporary basis that since negotiations are continuing elsewhere, "to set aside our own negotiations on this subject [patent secrecy]."[73]

In light of the above, session 2 was devoted initially to a discussion of the Japanese counter-proposal in the form of a new agreement. The GOJ proposal was described by the USG as having "little improvement in substance" where the GOJ proposed a management structure somewhat like the USG, and called for other aspects of the USG proposal be put in agreed minutes, record of discussion or GOJ policy statements. Based on the GOJ position which had been mentioned to the USG Team several times, this presentation was probably a reasonable alternative. The USG Team considered the GOJ proposal and produced its own alternative for presentation on December 3 which was consumed by GOJ questions about this revised draft. The next day, December 4 was spent discussing a new revised GOJ draft and the USG revisions.[74]

According to my interviews, the chairmen instead of driving hard for agreed language in a systematic manner toward a new agreement, tended to set aside difficult passages for future consideration, in this way piling up unfinished disagreeable business. This postponed the day of reckoning and aggravated an already charged atmosphere. It gave an even greater sense of lack of progress.

The IPR working group meeting on December 5, 1987 "produced no progress", "a non-starter and the Japanese proposal presented at the end of the session represented, in the eyes of the USG team," "a step backward" from the 80STA because it left out "equitable" in the distribution of IPR benefits, and one of "extreme disappointment" with GOJ's "unwillingness to engage in constructive discussion based on the U.S. proposal". The USG unhappiness was based on Japan's continuing insistence that the agreement contain a general statement almost word for word from the same as that in the 80STA, i.e. the GOJ stated again its preference for the application of IPR on a case-by-case basis. The USG IPR Team headed by the DOC insisted that there could be no STA without the proposed USG IPR provisions included in the new agreement, that it was incumbent on the GOJ to respond to the USG "in a constructive manner" -- obviously that did not include re-stating the GOJ's position on IPR from the beginning as described above. Rather than impinge on GATT negotiations, the proposed USG IPR package could be used as a model of IPR protection at these multilateral discussions. Based on discussions about the USG proposal, GOJ had serious concern with "secret" in trade secrets (this word was later changed) and with lack of any consideration for the visiting scientist in allocating IPR benefits from collaborative work. The USG Team also emphasized that its IPR package must be adopted because the EPC had instructed the USG Team to obtain such provisions in a future S&T agreement. By making this demand/threat the USG Team was beginning to blatantly attempt to impose unilaterally its decisions made at the highest level of the USG. This assertion also totally ignored what the GOJ may wish to have as an IPR package. By invoking the EPC at the second session seemed to imply that the USG Team was beginning to feel the pressure for producing results after having asserted that merely one session was sufficient to wrap-up the negotiations.[75]

Thus, the USG could report that only "some progress" was made; it was apparently agreed that there would be periodic USJ cabinet-level meetings to discuss S&T policies of mutual interest -- with such difficult issues as IPR, dual use technology, reciprocal access and even whether the new agreement should be considered "special" because of its high level and basis of a policy framework or simply a general umbrella agreement remaining on the platter. On the other hand, the USG must appear to the GOJ as quite intransigent and uncooperative on insisting that its totally new package on IPR for example must be adopted when there had been no actual but perceived problems with the existing 80STA. Language used, type of voice projected, insistent discussions of certain topics, e.g. access to facilities and S&T information, and many other tangible and intangible issues were leading to higher levels of tension.

It was agreed that the USG would prepare another draft for session 3 which would be held in Washington, DC just before Christmas -- but, of course, breakup in time for Christmas. Momentarily the U.S. held a slightly upper hand when the GOJ said it would like to try to arrange for a new S&T agreement to be signed at the January 1988 bilateral Summit. This slight advantage was perhaps somewhat down-graded, compromised a little by appealing to EPC decisions to try to force the GOJ to accept the USG IPR package.

d. Onward to the January 1988 Bilateral USJ Summit

i. Session 3, December 17-23, 1987. Since the Reagan-Takeshita summit meeting for mid-January 1988 loomed in front of the two delegations, the December 17-23, 1987 and January 5-7, 1988 sessions have been combined in this section. It will be recalled that Ambassador Endō proposed working toward obtaining an agreement in time for the Summit. In light of the outstanding issues this would seem more like an empty gesture for he must have known, understood and appreciated that the issues between the U.S. and Japan could not be readily bridged UNLESS one side or the other suddenly gave in or substantially compromised. Having made the suggestion, the USG naturally tried to pressure the GOJ to accept its version for the sake of the Summit. But that also was unwarranted optimism that the USG could suddenly pressure the GOJ into accepting an agreement of which it was highly critical from the beginning; at one point it even suggested that if the USG draft was so unacceptable that it was perhaps non-negotiable. While there was some little progress, the trend line in the negotiating atmosphere was definitely down.

In accordance with its promise the USG produced a revised version of the draft agreement and gave it to the GOJ on December 11, 1987, about a week before the third session began. It was basically the same USG draft except that, in the eyes of the GOJ, it was "more polished", the more blatantly one-sided responsibilities placed on Japan had been toned down, were less provocative, less one-sided. The USG draft still contained all five annexes including the defense patent agreement of 1956 and this STA would not come into effect until implementation procedures were agreed upon (Article V.3), detailed provisions about the release of information which might arise from open joint R&D project under security obligations (Article VI). This STA would continue to remain hostage to the USJ negotiations concerning the implementation procedures covering the afore-mentioned 1956 agreement. The USG draft revision compromised on the "umbrella-ness" of the agreement by stating that while the agreement would provide "the policy framework" for large scale projects and major R&D

initiatives, such projects "may have separate management mechanisms" and thus "would not fall under the technical and management review of the Joint Working Level Committee (Article 1.3)." It also stipulated that "universities [both public and private since there was no delimiting description], national laboratories and private sector research centers" may participate in joint projects. After the tedious and repetitive discussions about the "framework", the GOJ agreed that it would be feasible to foster greater USJ R&D cooperation within GOJ's national laws and regulations and without having to obtain Diet approval. Based on this judgment, the GOJ agreed to use the USG draft as the basis for creating a revised, more equal draft agreement and the drafting work on this agreement began during this period. It appears the principal negotiators on the US side (from the Legal Office in State) and from the Treaty Bureau in MOFA had earlier worked together on the Space Station agreement.[76]

Notwithstanding this apparent agreement about the "policy framework", the sparring between the two delegations about the meaning and intent behind the usage of words like comparable, symmetrical, or balanced access, reciprocity in the exchange of personnel, information and access to R&D facilities continued. The GOJ felt particularly stung by the insistent use of Japanese researchers at NIH as an example of the imbalance. They felt very strongly about the unfairness and the USG's utter disregard of the facts of the case. This contributed greatly to the deepening atmosphere of acrimony.

The GOJ agreed to the concept of including IPR stipulations in the new agreement. Although the IPR working group continued its negotiations on the side, the IPR remained a thorny issue especially the idea that all rights would remain with the institution which was receiving a scientist/engineer. To the GOJ this did not seem equitable.

On December 30, 1987, the *Sankei* and *Yomiuri* newspapers of Tokyo carried articles on USJ S&T relations. The former, an economically conservative paper, reported that the GOJ had decided to open a series of national R&D projects to foreigner researchers as part of the GOJ program to respond to criticism that its R&D facilities were closed to non-Japanese. These projects which had hitherto not included foreign researchers, included the fifth generation project being conducted at ICOT, large scale industrial development project, basic technology projects under S&TA. It claimed that many of these projects surpassed the U.S. "both in the industrial field and technology". The GOJ also planned to provide housing and insurance for the anticipated foreign researchers.[77]

The latter article disclosed the GOJ plan to be announced by Prime Minister Takeshita when he visited Washington in January 1988 to give 100 fellowships to foreigner researchers, 50 of which would be from the U.S. The foreign researchers would be assigned to national laboratories and national universities in FY1988 for six months to one year. The article commented that the USG asserted that the imbalance in personnel studying in the U.S. was an indirect cause of the trade imbalance. The USJ negotiations were "having hard sailing" because of the USG demand for participation of foreign researchers in Japanese R&D facilities including those in the private sector. There is no mention of problems with management structure, the dual use technology problem, and national security obligations. The latter points, if known to the Japanese press, would undoubtedly have stirred up a hornet's nest of opposition to the USJ STA. Finally, it commented that because of these GOJ suggestions and actions, the government was aiming to reach a settlement with the USG about the STA.[78]

*ii. **Session 4, January 5-7, 1988.*** Both delegations went home for their respective Holiday Seasons to regroup, reconsider, and to reconvene again in Tokyo on January 5-7, 1988 to make another attempt to surmount the seemingly impossible hurdles to create an agreement for the two Principals to sign in mid-January. The USG estimate was that the "key issues" now boiled down to 1) management structure, 2) restatement of security obligations and 3) reference to the 1956 patent secrecy agreement (subject to other negotiations).[79] Based on available evidence and interviews, the management structure issue could be solved in time for the summit meeting. Since the USG delegation was operating under EPC guidance and because the negotiations were obviously bogged down in acrimony, the executive secretary of EPC, Gene McCallister was consulted. If the involved agencies agree on the agreement, and the GOJ took the various actions they have been requested to take, then there is no need for further EPC clearance. Since the security obligations (article VI) were not based on specific EPC guidance, use of "common sense" would suffice; but in regard to the 1956 patent secrecy issue, McCallister was noted as "a hawk" and said this might need NSC-type (note that it is not EPC) guidance.[80]

Discussions about all issues continued in the 4th session but special note must be taken about the dialogue surrounding the security obligations, dual use technology, and the implementation procedures for the 1956 agreement in the USG proposal. While talks on these issues had begun in the 3rd session, they became much more focused, intense and introduced a serious element of distrust between the two delegations. The GOJ stubbornly maintained that the GOJ could not classify for national security purposes inventions, information, or devices that might, perchance result from joint research projects which had begun as unclassified open research endeavors. The GOJ explained that the GOJ did not have a legal structure (*hō taikei*) which would permit such classification actions. The GOJ could not prohibit the announcement by a Japanese researcher who would assert academic freedom. They maintained that even though the Coldwar was still in full swing and there were the frayed feelings about the Toshiba-Kongsburg Vaapenfabrik incident, there was no legal way in which the GOJ could insist on classifying certain information that emanated from a joint open research project. The Japanese asserted that the two delegations were discussing, to begin with, a rather improbable situation; it was a discussion about a "just-in-case" occurrence which probably would not in reality occur. In order to accomplish USG's request, GOJ laws would have to be revised with Diet approval to allow such action but that would be nearly impossible under existing circumstances since this was an executive agreement. Such information would have to be handled within the confines of existing laws and regulations. Certain members of the USG delegation insisted with equal vigor that the GOJ had such rules, regulations and laws which would permit such classification. The USG bordered on calling the GOJ delegates liars about their assertion that they had no such classification authority. It was apparently a source of great irritation and frustration to certain members of the USG delegation who were probably never convinced of the validity of the GOJ interpretation. Military classification was accomplished through the power of the U.S.-Japan Security Treaty.[81]

The GOJ delegation queried the USG members on every aspect conceivable concerning the meaning, implication, legality, impact on Japanese researchers who might, perhaps, may be involved in a project which had produced information or a device, or design which the USG deemed of military value and should be classified. The GOJ was adamant that any reference to the 1956 Patent Secrecy Agreement in the new STA and holding the STA hostage to the signing of an agreement on implementation procedures were improper and unacceptable. The GOJ was highly suspicious of the motives behind the USG insistence on including security obligations in an S&T agreement which was supposedly for peaceful joint research collaboration. The USG delegation maintained that the GOJ had an unfortunate perception of USG concerns about the STA and security and insisted the GOJ had a genuine misunderstanding of U.S. objectives. The USG felt the MOFA understood this need but that the other ministry representatives, particularly the MOE delegate, put forth unwarranted opposition to the security obligation clause.[82]

According to one U.S. delegate, the USG delegation conducted "a diplomatic sit-in" at the MOFA, refusing to negotiate, in an attempt to force the GOJ to accept the USG's proposed IPR provisions, the security obligations, and to tying the STA to the signing of the implementation agreement. Eventually the GOJ yielded agreeing that the IPR clauses would be legal in Japan but it was not decided that this was acceptable policy. There was one small glimmer of hope: the USG revised the IPR provisions so as to relate rights to the degree of contribution made by the participating groups. This was acceptable to the GOJ and survived into the final agreement. The term "trade secrets" was changed to business confidential information because of GOJ sensitivity to the word "secret" which might be misinterpreted or misconstrued to be related to military secrets. Since the implementation procedures for the 1956 patent secrecy agreement were being negotiated between the DOD and the JDA and not related to the STA, the GOJ refused to have the STA tied as a hostage to the signing of that agreement.

While it had become clear that there was little chance that there would be an agreement for the Principals to sign in a week, the GOJ -- if articles in the *Nihon Keizai Shimbun* are a reflection of their views -- felt it had compromised on major issues and was still working for an agreement by the Summit meeting in mid-January 1988. The GOJ agreed to the inclusion of a clause restricting the outflow of technology which could be diverted for military use to communist nations; this issue was described as the biggest pending issue in the negotiations at that time. This is the first time that "a security related provision has been brought into a science and technology agreement"; this, in turn, raised the anxiety that restrictions similar to those of COCOM might spread to the basic sciences. Acceptance of this restriction on the flow of information stemming from presumably unclassified projects would result in a fundamental change in postwar Japan's S&T principles that no such restrictions would be placed on such projects. It was feared that the political opposition would strongly oppose the acceptance of this restriction, give rise to further intense debate in Japan about the relationship between S&T and security, and would imply that Japan would become still more deeply involved in President Reagan's defense strategy. This was part of a

U.S. strategy at "gathering in Japan's high technology which is starting to pose a threat to the U.S."

The GOJ "accepted" this restriction because, it is reported, the USG threatened to abrogate the 80STA, and the GOJ recognized that there was a need to prevent the recurrence of the Toshiba-Kongsberg Vaapenfabrik incident where exports were made contrary to COCOM regulations and thirdly, because of the GOJ's felt need to re-studying S&T cooperation in relation to SDI R&D cooperation. The GOJ proposed that this restriction on the outflow of technology be carried out within existing national laws in the two countries and to hold consultations and co-ordination separately concerning the details about how specific technologies would be handled. It was felt that since there were major R&D USJ agreements in a number of fields, such as atomic energy, space medicine and the environment each with their own agreement and management structure, acceptance of this restriction would have little impact or effect on the overall situation. The actual wording, it was reported, was being worked out with the USG which, it said, was inclined to accept the GOJ proposal although this seemed highly unlikely based on the mood in the USG delegation. It also claimed that there were additional agreements on R&D cooperation on superconductivity, acceptance of U.S. researchers in GOJ supported R&D projects, the creation of a method to compare and check S&T in both countries, promoting the exchange of STI.[83]

Based on available evidence, a meeting was held by the Bureau Chiefs (Assistant Secretary level) of the concerned Ministries on January 4, 1988 to discuss the GOJ stance on the proposed STA and the upcoming summit. The MOFA proposed a compromise on security obligations concerning the possibility of restricting the flow of information of certain technologies stemming from joint R&D projects in an attempt to reach an agreement with the USG. Because of MITI's deep involvement in the Toshiba-Kongsberg Vaapenfabrik affair, it felt a need to support a compromise. Apparently, the meeting's decision was somewhat vague which must have given the MOFA the opportunity to think that it might try to work out some compromise with the USG. Apparently a tentative compromise was informally worked out by the MOFA in an attempt to reach a workable compromise with the USG. Surprisingly, this move initially went undetected by other Ministries. It was discovered, however, by other GOJ delegates at the next (5th) session (January 21-23) in Washington although GOJ "acceptance" of a security related clause in the STA had been reported in the *Nihon Keizai Shimbun* (1/7/88, evening edition) and the three Musketeers from MITI, MOE and S&TA (minus MOFA) confronted the MOFA and put a brake on any further movement in this direction, thus the issue of restricting the flow of information was brought back to step one.

The word sparring intensified. The USG delegation returned to Washington with no agreement for the upcoming USJ Summit meeting. The USG delegation had not been able to budge the GOJ on tying the new STA to the signing of the implementation procedures for the 1956 patent secrecy agreement, the security obligation and dual technology provision despite MOFA's secretive attempt to reach a compromise agreement. Even the presumed agreements on IPR had yet to be agreed upon by both sides. As described in the media, a major Japanese newspaper had editorially urged the GOJ to adopt an independent S&T policy, probably a euphemism for stand up to the USG in the negotiations, and even allow the 80STA to lapse at the end of January 1988 rather than hastily agree to an agreement that could not be carried out, comparing this possibility to the 1986 semiconductor agreement. A Japanese newspaper reported that Senator Rockefeller and others warned that if Japan did not specifically agree to 1) measures to restrict the outflow of certain technological information presumably stemming from joint research which may be militarily useful, 2) the equal distribution of research results, and 3) the protection of IPR then there is no reason for the STA and suggested that he would lead a movement in the Senate to oppose the revised STA, even though it would not theoretically require Senate approval.[84] These threats of varying degrees had failed. The USG had not been able to wrap-up the negotiations in one session and could not even rush the GOJ to accept any of the one-sided clauses requiring numerous GOJ actions even for a Summit meeting. It must have been exceedingly frustrating for the DOC, DOD and OSTP to be faced with this failure to "persuade" the GOJ to accept its proposal. Thus, the 4th session was described by U.S. delegates as disastrous, bitter and wasteful. This failure did lead to another major confrontation in the USG and deepened several notches the acrimony and animosity among the USG delegates and between the two government delegations.[85]

The next two short sections are a flash back to focus on a letter which provides another example of the attitude toward the GOJ entertained by selected individuals and the other is a short flash forward to the USJ Summit in mid-January 1988.

The OSTP Letter of January 7, 1988. A few days after the second session, a CISET meeting was convened on December 11, 1987 to discuss 1) the first two negotiating sessions with the GOJ and 2) a proposed letter to heads of agencies about S&T cooperation with the GOJ. The operative part of the letter proposed by OSTP was OSTP's request to all agency heads to refrain "from negotiating or implementing new S&T initiatives with Japan until the renewal of the Presidential Agreement has been successfully completed which I [Graham] anticipate will be in the near future". It repeated the basic USG position that the agreement was to be based on the "explicit commitment by both countries to the key principles of balance, reciprocity, and protection of intellectual property rights" and that the agreement should "provide the policy umbrella to facilitate coordination of the nation's overall S&T relationship with Japan", giving the President's Superconductivity Initiative as an example to be carried out under this framework. It asserted that the USG would be able to maintain its "leverage" at the negotiating table and also "speed that process" if "new bilateral S&T initiatives with Japanese should be held in abeyance".[86]

Interagency discussions about this letter took place just as the GOJ delegation would arrive in Washington for the 3rd negotiating session (December 17-23). Based on the CISET meeting of December 11, OSTP completely redrafted the letter and provided an enumeration of the legislative actions, Presidential Initiatives and executive orders which impinged on how the USG should negotiate S&T agreements with foreign countries. OES's response to OSTP was to have each department and agency submit to CISET "for consideration on a case-by-case basis *new* [emphasis added] S&T initiatives and programs with Japan" until the new agreement would be concluded. Regarding bilateral programs which an agency is currently discussing with Japan they "should be reviewed to ensure policy consistency with the principles and standards embodied in the Head of State Agreement".[87] "Such a step would have serious implications for our bilateral relationship. ... A general approach such as you [science adviser Graham] recommend, is not desirable, especially in light of the impending visit of Prime Minister Takeshita. Moreover, it would not offer measurable leverage for our [U.S.] negotiators at this late stage in the process" and withdrawal of this request was requested.[88]

Dr. Graham replied to State/OES A/S John Negroponte (although he had been assigned to NSC by that time) on January 7, 1988 saying that because of State's concerns, OSTP was incorporating the recommended language but he stressed and warned State about the possible need to take some action to "bring this Agreement to a successful conclusion":

> ... we should not lose sight of the considerable leverage we possess, particularly in view of the Government of Japan's disappointing performance to date, their continued rejection of the basic principles of balance and reciprocity as fundamental tenets of our relationship in science and technology, and their apparent intent to use Prime Minister Takeshita's visit to maximum political advantage.
>
> Should the fourth round of negotiations, now underway in Tokyo not produce significance results [and this session was a dismal failure from the USG point of view], we may find it necessary to strengthen further our negotiators' position in order to bring this Agreement to a successful conclusion and, in so doing, achieve the policy goals and objectives the Administration has established for our science and technology relationship with Japan.[89]

OSTP Graham seems to have assumed all the leverage was with the USG all the time. As stated earlier, the GOJ members did feel that the leverage in the negotiations was initially on the USG side. But that leverage was not stable; it shifted back and forth. The style of USG negotiating and the determination of the GOJ not to accept certain provisions taken as given by the USG forced the USG either to try to bluff their way in demanding GOJ acceptance of the USG proposal or continue negotiations. In this case, time would be used to GOJ's advantage. Graham apparently also seemed to assume that the GOJ was to accept "the policy goals and objectives the Administration has established" for governing USJ S&T relations. Indeed the GOJ felt that it was the USG that was immovable, unwilling to compromise and was determined to ram the new agreement down the GOJ's collective throats.

Dr. Graham signed a letter to all USG Departments and agencies on January 7, 1988, the day that the 4th session was completed and described by the U.S. side as disastrous, bitter and wasteful. The original idea of holding all discussions with the GOJ concerning R&D projects in abeyance was toned down to "please review programs your agency is discussing with Japan to ensure policy consistency with the principles and standards" in the USG proposed draft agreement, and inform CISET "of proposed *new* [emphasis added] programs with Japan for review on a case-by-case basis" until the new agreement is signed.[90] This was indeed a far cry from OSTP's original aggressive intent to hold discussions in abeyance. Judging from what had transpired at the four sessions, such a move of holding project discussions "in abeyance" would not have increased USG's leverage but would have exacerbated GOJ stand on numerous issues and a stiffer confrontation would surely have resulted.

The Reagan-Takeshita Summit Meeting. The flash forward is only a week to the Reagan-Takeshita summit on January 13, 1988. Takeshita Noboru was chosen as Prime Minister in November 1987. Whenever a new Prime Minister is chosen in Japan, a pilgrimage summit meeting with the U.S. President is immediately set in motion. As described above, the GOJ and USG each tried to maneuver the other into accepting a version of the STA which the other party would not accept so that a new STA could be signed by the Principals on January 13. This was a failure and the feeling of some in the USG delegation, e.g. DOC, DOD, OSTP, was particularly bitter. The lack of an agreement was elevated as a topic for discussion between the Principals and comment in the joint communique.

The NSC and EPC held a joint meeting on January 12, 1988, the day before the Summit Conference. In the first of nine items and selected other items under the leading first topic, "Japan's Role in Global Strategy", Japan is assessed in the following manner:

> Japan's increasing technological competence, vitality and managerial knowhow have created a giant economic power and awesome competitor in all fields of endeavor. ... It is in our interest to continue to benefit from that energy through access to Japanese technology, knowledge, ... a close and cooperative relationship and solidarity on international issues are essential to [the] welfare of the American people ... Japan invested some $40-60 billion in the U.S. last year [1987] ... helping finance domestic investment as well as the government deficit. ... [Japan provided the] most generous host national support of our allies – over $40,000 per serviceman. ... Japan has no intention to become a major military power, threat to U.S. or neighbors; wishes to work with US ... We expect [a] gradually expanding role, closely tied to us.

The key objectives of a successful meeting were described as "a viable demonstration that the President and the [Japanese] Prime Minister as the leaders of the world's two most dynamic powers can work together to resolve the bilateral and global issues facing our two nations, ..." and specifically 1) the continued expansion in our security cooperation, 2) continued solidarity and support on strategic issues, 3) continued cooperation in key areas (Philippines, Persian Gulf) and 4) continued efforts to resolve economic and trade issues.

While S&T was not specifically mentioned among these objectives, the new STA was the only specific item singled out for attention. It began by saying that "one of the most difficult problems we face vis-a-vis Japan is how to compete yet cooperate for our mutual benefit" without specifically mentioning S&T. It described the principles and framework to govern our S&T relations which the USG was trying to persuade the GOJ to accept. Our goal was "to find ways of cooperating in areas of high national priority in both countries while protecting our national security and commercial interests." About the status of the negotiations, it said that while we had not been able to prepare a new STA for signature, "still, we are encouraged by the progress that has been made and we are continuing our efforts to conclude the new agreement by the end of the month [January 1988] [and] "our S & T relationship is crucial to both countries and we must make every effort to nurture and improve it." Even documents at the highest internal level in the USG did not describe the sorry state of affairs in these negotiations. Understandably we did not air the USG's own serious dissensions, but silence about our unhappiness with Japan's refusal to accept our demands quickly was perhaps surprising. Yet even here if such were described, this process itself would bring to the fore USG's own dissensions and widely differing interpretations of the negotiations.[91] At the end of the Summit meeting, President Reagan, among other issues, commented that

> we affirmed our determination to conclude a new science and technology agreement with equitable and expanding research benefits for scientists of both countries. I also expressed appreciation for Japan's initiative to provide more than four million dollars in science fellowships to American researchers.

Prime Minister Takeshita also stated that "the President and I agreed on the importance of enhancing the cooperation in the field of science and technology." Although this support was somewhat passive these comments were tantamount to more than an exhortation, basically an imperative, to the working level staff to re-double their efforts to reach an agreement.[92]

e. The Final Push

The last three sessions, the 5th, 6th and 7th, were filled with impatience, frustration, threats and counter-threats, interminable discussions about the meaning of words, phrases, the rights of inventors, a final political wrap-up, interspersed with a crisis of confidence among senior members of the USG concerning the extent to accept the promise of the GOJ and whether to continue the S&T negotiations. Yet, throughout this wrangling and bickering, a new agreement bit by bit, word by word gradually began to take shape.

Before the 5th session (January 21-23, 1988) began, two articles on these negotiations appeared in the *Asian Wall Street Journal* and the *Sankei Shimbun*. Perhaps the former could be described as reflecting more or less USG feelings and the latter those of the GOJ.

In the *Asian Wall Street Journal* (reporting from Washington), the emphasis was on the U.S. frustration in being unable to persuade the GOJ to be more generous in sharing the fruits of Japanese research laboratories without specifics and that this "impasse could spark U.S. retaliation". The retaliation would be to invoke a section in the 1986 Technology Transfer Act to prevent Japanese scientists from visiting U.S. national labs. The GOJ was asked to publish more research in English and allow U.S. participation in product development work financed by MITI. By this time, there had actually been a number of Japanese offers to increase U.S. researchers in Japanese government laboratories. The NSF maintains that the GOJ is making a serious effort to open even corporate labs; "our [U.S.] problem is finding American scientists who can read Japanese and handle the difficult cultural experiences in Japan." Obviously the USG was not unanimous in its assessments. The U.S. also wanted a new arrangement about IPR, and national security. It reported that a U.S. negotiator opined that GOJ opposition to U.S. demands was "so stiff it would be a miracle if the accord could be completed" before the January 31 expiration date (*The Asian Wall Street Journal,* 1/21/88).

In contrast, the Japanese newspaper item reported that negotiations would now be hastened because of the Reagan-Takeshita meeting which endorsed the STA but that there are "two very big points" -- who owns intellectual property rights, and the control of information stemming from joint projects. Japan strongly objected to the notion that all IPR should belong to the host country research institution. In regard to the USG's persistent demand that the preservation of secrets, and controlling the out-flow of information be specifically included in the agreement the GOJ response was that this would then require the present STA executive agreement to be submitted to the Diet for approval. This would result in "strong repulsion" by the opposition leftwing parties. Obviously both parties can play the Congressional/Diet card in jousting with the other party. While hoping that a new agreement could be worked out in accordance with the USJ Summit accord, it noted that "the two countries are showing the attitude of not easily changing their respective assertions" and that the outcome of the negotiations was totally unclear and that, according to some hard-liners the lapse of the 80STA may not be avoided. (*Sankei Shimbun* 1/20/88).

Thus the "final push" begins with an ominous atmosphere, bleak prospects, and an abundance of black clouds.

i. Session 5, January 21-23, 1988. The USG report on the fifth session (January 21-23, 1988 in Washington) is sufficiently sanitized to give the impression that some progress was made; yet one U.S. delegate described the same session as a wasteful meeting. Despite the publicly declared intention to work toward a new S&T agreement, there was precious little headway made in the session following the Summit meeting. Let us summarize the session as follows:[93]

The concept of balance was still being "intensively negotiated". But agreement was reached in small steps. The USG proposal had read

> "... to strengthen that relationship on the basis of equity, mutual benefit and overall balance" was changed to read
> "... to strengthen that relationship the two governments will conduct their S&T relationships based on the principles of shared responsibilities, mutual and equitable contributions and benefits, commensurate with their respective scientific and technological strengths and resources."

It would seem that this improvement was not significant but indeed much more detailed about equitable contributions and commensurate benefits.

Management mechanism. No significant progress.

Joint Initiatives. Considerable informal discussion took place and a consensus was reached on the goals of the actions but no agreement on wording.

IPR. This was "exhaustively" discussed. The USG proposal to have the receiving institution receive all the rights to an invention was based on the assumption that the probability might more likely arise with substantial numbers of Japanese scientists and engineers coming to U.S. institutions and few Americans going to Japan. The GOJ questioned the USG delegation "at length" about the basis behind the USG proposal and the possible effects on visiting scientists. The USG proposal reduced the incentive and would hinder the exchange of scientists because the USG "scheme" removes "all rights from the visiting scientists". The USG delegation must have known that NSF was not pleased with the USG proposal in this regard. [Note that this subject will also be raised by NSF and eventually a compromise worked out.]

Security Obligations. "Some progress" -- details not specified -- was reported. A DOD representative described this as one of two sticking points -- the other being the patent secrecy issue. He described the GOJ group as understanding the issues but being unable to internalize them. Since the GOJ Team had studied U.S. laws and regulations about security they asked hard questions and thus "were difficult to deal with". The USG group tried to persuade the GOJ that there was nothing insidious about security, it was important to both the U.S. and Japan. This same person described the talks as cordial and frank but understandably frustrating and acknowledged that the GOJ passed laws including those for handling classified materials in accordance with the U.S.-Japan Security Treaty of 1960. The USG was asking Japan to accept and undertake something quite unprecedented, especially for such Ministries as MITI, S&TA, MOE. The GOJ felt this issue could leave a long shadow, as indeed it did, that sparks could fly, depending on how the USG acted. In their view, "defense secrecy" could be interpreted to include ultimately mathematics, artificial intelligence, computer science; it reflected the Reagan Administration's tendency to attempt to control science. The GOJ preferred "national defense" to national security and objected to such words as "classified" in light of the lack of such regulations on the civilian side.

Defense Patent Secrecy. Although this issue was never actually a genuinely S&T issue and had not been part of the EPC instructions, the three U.S. musketeers from DOD, DOC and OSTP, injected it into the USG proposal and caused considerable confrontation with the GOJ and within the USG. The GOJ consistently and persistently refused to accept the tying of the patent secrecy issue to the new STA. Although this issue was "addressed twice during the plenary session", the GOJ insisted that this issue had been addressed in the patent secrecy discussions and would not budge -- "there was no GOJ movement on this issue". During this negotiating session, the GOJ was "under the strong impression -- quite justifiably as will be shown later -- that the patent secrecy issue had been solved between the JDA Minister (Kawara) and the DOD Secretary (Carlucci). "Each time" this occurred the USG "corrected" the GOJ Ambassador Chairman declaring the patent issue "one of the key issues" and stressed that it very much remained an integral part of our [USG] S&T negotiation package and threatened the specter of "possible Congressional and other consequences if this obligation is not met by them."[94] This adamancy greatly irritated the three musketeers' U.S. agencies and only fostered their suspicions and distrust of the GOJ and even of the U.S. Department of State.[95]

Frustrations among certain elements of the USG delegation and their respective agencies were reaching the boiling point. The seeming lack of acceptable progress in pushing forward with an agreed upon textual agreement -- although there appear to have been some agreements for selected parts -- was, of course, causing the USG to think that perhaps a new agreement might not be possible -- at least by January 31, 1988 was now impossible.

The DOC/USTR/OSTP Letter of January 25, 1988. In stark contrast with the above report by a State Department staffer, Secretary of Commerce Verity, U.S. Trade Representative Yeuter and Science Adviser Graham sent a strong, caustic and threatening letter to SecState Shultz on January 25, 1988, only two days after the 5th session had ended, predicting dire consequences for U.S. national interests. It maintained that 1) the two governments were "still far from agreement on key points mandated by the EPC", 2) the "conclusion appears unavoidable that the Japanese are not prepared to conclude a new agreement on terms which meet the criteria agreed to by the EPC last July [1987]", 3) continuing the current agreement would "perpetuate a science and technology relationship with Japan that is fundamentally unbalanced and ... [would] adversely affect critical national interests in the areas of commerce, trade, and technology, 4) Japan had refused for 32 years to implement the 1956 Patent Secrecy Agreement and Japan's "earlier pledges to honor the 1956 agreement have proved, in fact, to be a string of broken promises". Based on available documents and information, there is no evidence that 1) USJ had earlier agreed on implementation not once but repeatedly and then Japan had reneged on its pledges and 2) the GOJ refused to implement a patent secrecy system while the NATO partners did agree and 3) the GOJ "refused over the past 32 years" to implement the 1956 agreement. It concluded that DOC would concur in extending again the

80STA but said it was "not prepared to conduct any further negotiations with the Japanese unless the Japanese Government immediately implements the 1956 accord" and threatened that if the patent issue was not resolved by the end of January 1988, "this issue will become a matter of high public profile and Congressional scrutiny and cast serious doubts on the reliability and integrity of our relationship with Japan" and called upon the SecState to make the strongest representation to the Government of Japan.[96]

Another Extension of the 80STA, the fourth. Immediately after the fifth negotiating session, a CISET meeting was convened on January 25, 1988 -- the day the blistering letter described above was sent to SecState -- to assess the status of negotiations and to devise some method to pressure the GOJ to sign a new S&T agreement including all the revisions desired by the USG. It recommended that the 80STA be extended for two months (February 1 -- March 31, 1988), established three policy objectives and created a working group to draft "a menu of options" to achieve USG policy goals. The afore-mentioned policy goals were

1. To ensure, during the negotiating process, that U.S. Government departments and agencies maintain policy consistency with the principles and standards set forth by the Economic Policy Council's guidance and embodied in the U.S. draft agreement. This clearly acknowledged what was well known and mentioned previously that the Chairman was unable to control the more vociferous delegates and coordinate a unified USG position in discussions with the GOJ.

2. To increase U.S. negotiating leverage and pressure on the GOJ.

3. To undertake alternative steps and unilateral actions to implement U.S. policies and objectives, if a new agreement was not possible.

It was obvious that frustrations and irritations were mounting. After all, the USG expected to conclude negotiations in one session of three days. We were now faced with the 4th extension of the 80STA. Earlier short two-month extensions had been intended to put pressure on the GOJ to sign the USG proposal. This pressure tactic had not worked its magic; it had, in reverse, probably made the GOJ delegates more determined to pursue every possible implication, meaning of every word and phrase in the USG draft agreement. Again, the Circular 175 authority papers recommended granting a two-month extension and increase U.S. pressure on the GOJ to 1) expedite the negotiating process and 2) underscore the need to reach agreement on outstanding policy issues consistent with the objectives and guidelines set forth by the EPC.

Through an exchange of diplomatic notes between the USG and GOJ, the 80STA was extended two months, February 1 - March 31, 1988.[97]

ii. The Defense Patent Secrecy Crisis.[98] The implementation procedures for the 1956 Agreement for Defense Patent Secrecy were being negotiated between the USG and GOJ by DOD (and State) and JDA. When the DOD representation changed on the interagency group creating USG policy for the renewal of the 80STA, DOD inserted an article into the USG draft proposal which made the proposed S&T agreement hostage to USJ agreement on implementation procedures, assuming that this would pressure the GOJ to sign the implementation agreement. Apparently certain USG delegates vigorously pushed for the inclusion of this article in the new STA and the GOJ with equal vigor rejected its inclusion in the new STA. The defense patent secrecy issue was a totally different set of issues, had no substantive relationship to the unclassified S&T agreement, and was being negotiated by completely different agencies of both governments. Linking the two agreements at the insistence of the DOD, DOC, OSTP and USTR, and pushing this linkage relentlessly led to a major crisis of trust and confidence among senior members of the USG and how each group viewed a commitment by the GOJ as evidenced by the above-mentioned letter of January 25, 1988 to SecState Shultz.

Just before the USG delegation was to leave for Tokyo in mid-February 1988 for the sixth session, the USG was at an impasse with itself.

DOC, OSTP and USTR contended that negotiations for a new STA "should stop until the GOJ implements the implementation procedures of the 1956 agreement". These agencies obviously assumed that stopping negotiations would provide leverage to pressure the GOJ to implement the 1956 agreement, i.e. sign the implementation procedures agreement. They wanted the implementation agreement signed according to their schedule as a matter of principle and, in theory, to protect several dozen defense-related patents which otherwise may lose their patent protection in Japan. Secretary Verity of Commerce, USTR Yeutter, and OSTP Graham sent the afore-mentioned joint letter of January 25, 1988 to the SecState demanding that the negotiations be stopped. It should be pointed out that the DOD Secretary was not a signatory to this letter. Science Adviser Graham had met with Under Secretary of State Derwinski on February 10, 1988 to emphasize the need to stop the negotiations with the GOJ. Trade issues with Japan had now "spilled over into the S&T area with negative effects". Heretofore, science and technology discussions had been kept separate from security and trade issues; each of these areas had lived their own lives. Under the Reagan Administration there was a distinct and determined effort to link all these different areas together as part of an integrated national posture. Obviously, while there may be political agreement on the principle involved here, there was from the beginning a major schism on how to implement this overall policy.[99]

This demand by the three agency heads was at odds with the CISET interagency recommendation to extend the 80STA yet another time in order to continue negotiations with GOJ toward a new STA. The DOC, USTR and OSTP were fully represented on the CISET committee, indeed the OSTP Director was chairman. Furthermore, it would be politically embarrassing to President Reagan and Prime Minister Takeshita for the STA to be allowed to lapse after they had at the January summit meeting expressed a determination to conclude a new STA. The new STA would not produce any results, even if signed, before the Defense Patent Secrecy agreement was expected to be signed in April 1988.

It should be noted here that the DOD and PTO supported by State had completed negotiations on December 16, 1987 "on all substantive issues and reached agreement on the text of the implementing agreement."[100] Concerning the remaining issue of when the implementation agreement would be signed, an OES memo stated that the DOD Secretary had accepted his Japanese counterpart's commitment to an April signing and had, therefore, removed the patent secrecy clause from its MOUs with Japan.[101] DOD Secretary Frank Carlucci had agreed to an arrangement with JDA Minister Kawara to accept a confidential GOJ commitment to an April 1988 implementation of this agreement but the USG delegation had "rejected the GOJ proposal for an April and June 1988 two-stage implementation and demanded a January 1988 signing without conditions."[102] In addition, the USG had not been similarly insistent with the GOJ about the defense patent secrecy agreement in USG's negotiations with the GOJ concerning the Space Station and SDI --"thus, there is no longer a unified USG position on this issue". DOC, USTR and OSTP, however, doggedly pushed the defense patent secrecy issue when DOD itself had already made a separate arrangement with JDA -- an amazing and sad disarray at the highest levels of the USG.[103]

According to an analysis by OES, about 60% of patent applications filed receive patents, and 1-2% of those are commercially viable, hence 0-1 commercially viable patents may be threatened. Assuming this analysis to be correct, the demand to postpone negotiations would appear to be a high price to pay to protect 0-1 commercially viable patents. Between April 12, 1988 when the agreement on implementation procedures was signed by the USG and GOJ, and October 12, 1988, a six month period, 30 requests were made to the U.S. PTO to register their secret patents, presumably in Japan based on this new implementation agreement. This is far fewer than the "dozens" which were reportedly eagerly waiting to register their secret patents in Japan.[104]

The GOJ "cannot understand" why the USG was insistent on immediate implementation after 31 years of inaction by both sides, and especially since, based on the Carlucci-Kawara agreement, the GOJ had already agreed to sign the agreement in April, 1988. The new Prime Minister Noboru Takeshita's major concern in early 1988 was to pass the budget in the spring 1988. The defense patent secrecy agreement would be politically sensitive in Japan. Signing the agreement in January or February 1988 would create adverse reactions domestically, and could threaten the passage of the 1988 budget. A smooth passage of the budget was essential for Takeshita's leadership as the new Prime Minister. From the GOJ's point of view the issue had been satisfactorily arranged with Secretary Carlucci and was a settled issue.

The SecState responded to the joint letter from DOC, USTR and OSTP on stopping the USJ negotiations urging that the USG continue negotiations with the GOJ about a new STA. This was felt by State staff to be inadequate in light of the refusal of the three agency heads to budge, so a telephone call from the SecState to Sec/DOC was proposed; this was converted into a breakfast. In the meantime, SecState had written to GOJ Foreign Minister Uno strongly urging GOJ to take expeditious action on the implementation agreement. As a pro forma statement, the letter reiterated that the USJ S&T relationship was "one of our most important and we wish to see it expanded in an equitable manner." It strongly stressed that delaying the signing of the implementation agreement until April 1988 "may disadvantage some U.S. firms holding potentially valuable patents [which] will expire in April [1988]" and called upon the GOJ to make every effort to find a means of providing patent secrecy protection in Japan for these patents. This message was conveyed with some urgency to the Prime Minister's Office and to the MOFA Minister. Uno's reply affirmed the GOJ intention of signing the agreement in early April 1988. The Japanese Ambassador also confirmed that the agreement would be signed "two or three days" after the 1988 budget was passed by the Diet. In a coordinated effort to impress the GOJ, DOC Secretary called in the Japanese Ambassador, State Under Secretary and the science adviser called in the Japanese charge to demand quick and immediate GOJ action.

At a CISET meeting on February 17, 1988, where agency representation had been upgraded to State's Under Secretary Wallis and Deputy USTR Mike Smith, OSTP's Graham broke ranks with the three agency heads and urged that the USG resume negotiations with the GOJ; in contrast, USTR and DOC reported that they still opposed continuation of the negotiations until the implementation agreement was signed.[105] Notwithstanding these flurries of activities within the USG and across the Pacific Ocean, State commented that the "GOJ has not budged" except to give additional assurances about future implementation.[106]

The outcome of this high level flurry of activity was a face-saving arrangement -- an acceptable bureaucratic compromise -- for the agencies and the egos of the personalities involved. Because of the four agency protest and their refusal to participate in further negotiations with the GOJ, the USG delegation's departure for Japan for the sixth session was delayed; it had apparently been planned for three days starting February 21, 1988.[107]

Based on interviews, the SecState had to put his personal credit and confidence on the line with other agency heads in order to obtain their support for continuation of the USJ negotiations. The negotiations would be resumed based on the following scenario.

1. The GOJ would be informed that the USG's intent is to conclude the S&T negotiations by March 30, 1988 which was believed to be in keeping with the Reagan-Takeshita summit commitment.

2. The USG is prepared to "accept reluctantly" the GOJ assurances about an early April 1988 implementation (i.e. signing) of the patent secrecy agreement a) within 2-3 days following the Diet approval of the GOJ FY1988 budget or within the first ten days of April 1988 (the GOJ budget year begins on April 1) and b) that this will include the whole arrangement, both military and commercial patents and c) that the GOJ will accept the U.S. defense secrecy related patent applications as of the date of signature.

3. The USG will not sign the S&T agreement until the defense patent secrecy agreement is signed. Even if the S&T agreement were negotiated by March

31, 1988, both governments would need to find an opportunity when the two Principals will be meeting together -- hardly another bilateral since they had just met in January 1988, more likely a multilateral summit -- in order to sign this high level agreement for which so much energy and emotion had been expended.

In order to emphasize their distrust, antipathy and reluctance about this process, DOC, USTR and OSTP "made known their intent to pursue filing a self-initiated 301 action" which, "of course, [had] not been vetted through the interagency process." If an S&T agreement was not worked out as planned by March 31, new cabinet or presidential guidance would be sought concerning further extension of the current agreement, and/or continuing the negotiations, and/or considering other options.[108]

With this scenario in place, the stage was set AGAIN for the resumption of negotiations.

iii. Session 6, March 3-8, 1988. At this point there was less than a month to conclude a new agreement before the 80STA again expired on March 31, 1988. The Embassy report to Washington on the specifics of the negotiations and the *Nihon Keizai Shimbun* report on the same session are in stark contrast to each other.

"Much remains to be resolved" said the Embassy cable on this session. The USJ are still debating a fundamental issue on the scope of application of the principles enunciated in article I: do these principles govern all USJ S&T relations or only those agreements under the new USJ agreement? -- AND this was not resolved at the 6th session. The USG was, according to its own notes "reorganizing the document [the STA] to reflect the dual nature of the document, as a framework and as an umbrella agreement.[109] The GOJ wanted every suggested action described bilaterally, references should be only to government or national facilities, did not feel the need for committees to monitor JSTI and access to and flows of researchers, and the deletion of the word "comparable". It did not want any annual reports, nor the creation of task forces to monitor JSTI and access to R&D facilities, nor for committees to have rigid schedules, and it did not wish to develop work plans for the JHLC, merely inform the JHLC of operation of the agreement.

Notwithstanding these differences, the GOJ did agree on the three tiered management structure and terms of reference of the three levels of committees, the joint working level committee, high level committee and high level advisory panel. The security obligations article was again a major stumbling block. While "some progress" was made in defining the issues and defining "classified information" since the GOJ again insistently maintained that it did not have classification laws like the U.S. except through the U.S.-Japan Security Treaty for military classifications and the civil service law which technically is not a classification law. The GOJ also tried to assure the USG that the COCOM regulations would be ever more faithfully enforced following the Toshiba-Kongsburg Vaapenfabrik Incident.

According to a DOD delegate, they spent two hours one Saturday during this session answering questions from several GOJ Ministry representatives about security. Notwithstanding this effort, this issue was accurately described as the one issue "that may be one of the last to be resolved."[110]

The IPR provisions concerning the distribution of rights to the receiving/host institution was unsatisfactory not only to State and NSF but to the Japanese. While it has never been clear why this proposal was made by the USG, one can speculate that those fearing the larger number of Japanese researchers in the U.S. than those in Japan assumed that there may be more patentable creations by the not so creative Japanese visiting U.S. facilities than in reverse. Even before the EPC meeting of July 24, 1987 and even before the bilateral negotiations had begun, State/OES had objected to this concept "as not in the best interest of the U.S."[111] During the latter part of the negotiations, the NSF asserted, however, under these conditions, those inventions created when U.S. researchers visited Japanese facilities would give Japanese institutions and inventors access to the U.S. market, whereas the reverse would allow the U.S. institution access to the Japanese market, much smaller than that of the U.S. "This is especially so when the aggressive practices of the Japanese to patent every minor finding is taken into account". On the assumption that the number of U.S. researchers in Japan would rise, dramatically it was hoped, over the years, that would mean giving up in advance any benefits from joint research. Thus NSF urged that IPR be shared on the basis of contribution to the finding. The GOJ also had reservations about the lack of any rights accruing to its researchers in U.S. facilities. In this way, the more equitable principle of sharing IPR based on contribution replaced the original USG proposal.[112]

The IPR negotiations were described as posturing by both sides, followed by quite efficient and professional negotiations with the Japanese expert making numerous constructive suggestions to the English version. The Japanese were described as feeling they were being singled out and punished with these detailed IPR clauses and highly skeptical of USG intentions until they received an analysis and description of U.S. laws and the mutual need for IPR.

In contrast with the cautious optimism of the USG report, the Japanese press reported, for the first time, political threats from an unnamed U.S. government official and another commentary urging the GOJ to be flexible in its negotiations with the USG. First the threats. This USG official is reported to have warned that

- if the GOJ "does not change its attitude, as is, it will become a political problem" between the two countries on the same level as agricultural products and public works projects.

- the new 1986 Technology Transfer Act could be invoked to exclude Japanese scientists and engineers from U.S. R&D facilities because the GOJ was not allowing U.S. researchers to equivalent Japanese facilities. (This official must have known that was, at that time, erroneous information)

- Senate minority leader Dole who was then campaigning in the presidential election and other congressmen were "showing strong interest in this problem". According to available information, there is no record showing that Dole had expressed an interest in the USJ S&T agreement.

- the 80STA cannot be extended beyond March 31, 1988. It would be difficult to persuade Congress on this point. Extensions had been made by the Executive Branch (specifically, the State Department with the concurrence of other agencies).

- if no agreement is reached by March 31, 1988, then the USG will have no choice but to abrogate the agreement.[113]

This sudden outburst of irritation, impatience, political threats seemed to imply a possible subtle change in the dynamics of the negotiations. It was interpreted by the GOJ delegation as a USG weakness. The USG was looked upon as anxious to push for an agreement. (The Japanese term used was "*aseru*" which means in a nervous haste, but it does have an implication that one side has lost one's cool, and nervously jumpy to reach an agreement at all costs.) As the negotiations began in October 1987, six months previously, the USG and the GOJ both felt the advantage lay with the USG. By the end of the 6th session, this advantage seemed to be subtly shifting toward the GOJ. All the various parts of the agreement seemed to be moving in the direction of a solution, but the security article was running into a Japanese brick wall. While the GOJ would not want negotiations to founder and collapse, especially since it involved the Prime Minister, they were not under such specific policy imperatives (i.e. policy guidelines approved by the EPC). The USG had politically decided to push for a new agreement; in contrast, the GOJ was passive, if we must have an agreement then let us meet and discuss the various possibilities. The USG pushed for all or nothing in preparation for the Summit Meeting in mid-January 1988 and did not succeed. If it failed again to reach an agreement in time for another summit looming in May at the UN General Assembly or at the G-7 Summit starting on June 20, 1988, this would probably result in yet another confrontation within the USG and a new political problem with Japan. The USG had -- with GOJ consent -- been insistent on setting deadlines, 2 or 3 months at a time assuming this would pressure the GOJ into accepting USG demands but this failed, the negotiations dragged on. A failed negotiation would reflect negatively on the USG delegates.

The other Japanese press commentary described some strong feelings in the GOJ to "confront the USG" and for the GOJ to act logically with a resolute attitude. It noted that the security clause was a major stumbling block, that the USG "maintains a stern posture" and harbors a deep-rooted view that outflow of technology to the Communist Bloc must be prevented at all cost. It called upon the GOJ to rethink its posture and consider how Japan could resolve the present controversy with the USG in light of how Japan could contribute to increasing world science and technology in the 21st century. It reminded the government that Japan's postwar recovery owed much to the U.S. There was no need for the GOJ to accept all USG proposals and that even the USG was subject to "strong criticism" for its recent posture to "fence in the results of research." (Referring to the summer 1987 Conference on Superconductivity in Washington during the Reagan Administration which excluded foreigners.) On the other hand, the GOJ Ministries clashed over the inclusion of a security clause in the new STA and its presumed implication that the freedom of science and technology in Japan would now be circumscribed. The MOFA, because of its concern for USJ relations and MITI's concerns and anxieties over the Toshiba-Kongsburg incident, appeared to favor an accommodation with the USG. The MOE, in contrast, carried the torch for continuing the principle of unrestricted S&T which had become a deep seated tradition and belief in Japanese academia in the postwar period -- initially strongly encouraged by the U.S. Occupation as a counter to the governmental guidance and control for nationalistic purposes. The implications of these policies fostered and cultivated over several decades now collided with the perceived political-security demands of the Coldwar as envisioned by the USG. It was felt that such "potential" restrictions on S&T would be inconsistent with the stated peaceful purposes of the STA.[114]

Thus, the GOJ was being buffeted by USG demands and admonished by comments in an influential conservative business newspaper (often compared to the *Wall Street Journal*). But the depth of the chasm between the two governments was well reflected in these reports and the dynamics of the negotiations. Both were now digging in their heels. Perhaps only a "high level political settlement" at the next meeting would break the impasse in these negotiations -- as reported in a small article in the *Nihon Keizai Shimbun* (3/9/88, p. 5). Emotions were running high on both sides as the final session convened.

iv. Session 7, March 21-30, 1988. As the final session began meeting in Washington on March 21, 1988, for what turned into a ten day session, both sides were still talking about the placement of words, e.g. "peaceful purposes", the use of "comparable" to describe the bilateral S&T relationship in the future, the scope of the policy framework (i.e. whether it would apply to all arrangements or only those under this agreement), the definition of the distribution of IPR and security obligations. There were, in other words, many "bracketed items", items which were momentarily set aside for future discussion and resolution; this had been, in the eyes of many delegates, the failure of the negotiating style of the USG and GOJ chairmen. Much was postponed for later resolution, i.e. many topics were thus pushed toward the final deadline for hopeful resolution. Even though under these conditions, a so-called crisis could erupt at any time, one Japanese delegate was optimistic enough to tell the USG that two days could wrap up the discussion of these issues and the remaining time devoted to "making trades".[115] Nevertheless, considering that session 7 was the longest session, there is, except for a few known facts, a mistiness as to what actually was discussed, in what tone, and how the jigsaw pieces suddenly came together as a whole.[116]

There are other developments which turned out to be crucial to these S&T negotiations. The chairman of the Japanese delegation for this session was significantly Nomura Issei, the DCM in the Japanese Embassy in Washington; the fortuitous subsequent arrival in Washington of Ozawa Ichirō, then Deputy Chief Cabinet Secretary, for talks with the USG on access by

U.S. companies to public sector construction in Japan played accidentally an important political role in breaking the impasse. Minister Nomura was more forceful in organizing a Japanese consensus in the first place and more skillful in maneuvering, manipulating, persuading and controlling and even perhaps cajoling -- according to some interviewees -- certain members of the USG delegation in order to drive toward a USJ "consensus" before the expiration of the fourth extension of the 80STA. He strongly believed that at this juncture, it was politically absolutely necessary to work out a compromise with the USG on science and technology, that the 80STA could not be abandoned. That he was able to resolve the many "bracketed items" within a limited amount of negotiating time was, in the eyes of the Japanese delegation, well accepted by both delegations. The Japanese Ambassador, also played a role at the very end of the negotiations. The Japanese delegation felt that the USG chairman, Ambassador de Vos, had played a constructive and important role in the final stages in "pulling things together" (*matome yaku*) in contrast to their earlier assessment of him.[117]

Open Research, Security Obligations and a Side Letter Compromise. Instead of including in article I the principle of maintaining an open basic research environment and the widest dissemination of the results of joint research where other principles of the USJ S&R relationship were enunciated, these were included in article VII on the relationship between joint research and the theoretical possibility that, in an unusual instance, some information or equipment might become, in the USG's opinion, defense related and should be classified. This was debated and argued from every possible angle by both sides but a resolution evaded the delegates. In each of the private papers on the negotiations, the relation of openness of research results and security (and the access to STI and R&D facilities in Japan) were singled out for special emphasis. It was not without reason, therefore, that these topics stayed unresolved until the final negotiating session.

The Japanese had, they thought, informally worked out a compromise wording with a DOD delegate concerning the principle of openness for R&D information at the 6th session but this was later rejected by the USG. This is reminiscent of the MOFA attempt to work out a compromise on security obligations with the USG but this attempted compromise when later discovered by other Japanese delegates was also rejected by the GOJ delegation.

A week prior to when the 7th session was to convene, the USG gave the GOJ another version of the security obligations article which came very close to the finally adopted version in the agreement. The revised version can be summarized as follows:

1. both governments support the widest possible dissemination of the information or equipment created in the course of cooperative activities.

2. in furtherance of the principle to maintain an open basic research environment, no information or equipment classified for reasons of national defense will be exchanged under this agreement. But if classifiable information or equipment does result from this research then it will be protected in accordance with applicable national laws and regulations. According to the USG, the latter sentence means that "for Japanese origin work Japanese laws apply; for U.S. origin work, U.S. laws apply. There is no intent to 'reach out' and impose U.S. legal restrictions on Japanese origin work". But it does leave unclear how such information/devices/techniques classified in the U.S. under a joint project may be used in Japan and vice versa.

3. use of export controllable information or equipment or the prevention of unauthorized transfer of such information and equipment will be handled in accordance with applicable national export control laws and regulations.[118]

The rationale for the revision was, in USG words, to respond to the GOJ desire to cast this provision in a positive manner and to avoid reference to "classified in accordance with applicable national laws and regulations" of either government and by narrowing the definition of classified information to "national defense" the USG legal requirement was met and relieves the GOJ, in effect, from any obligation to accept such information from the USG under this agreement. From the GOJ's point of view this maintained the status quo and would not require the GOJ to revise any existing laws which would require Diet approval.[119]

According to Japanese sources, the USG asked for inclusion of the above principle in Article I. While seemingly reasonable as mentioned above, inclusion of this principle in Article I would clearly highlight the theoretical relationship between open research and national defense, in contrast to inclusion of this concept in Article VII, one might say "buried" toward the end of the agreement. This USG request/demand would re-open at the 11th hour a basic concept which had been laboriously debated from the beginning as part of the security obligation article. This set off alarms bells anew in the GOJ delegation; this would require referring back to Tokyo MOFA for instructions. Fortunately, however, Ozawa was in Washington for talks on access to Japanese public sector construction projects. He had been earlier a parliamentary vice minister for science and technology and thus had a special interest stemming from this experience and also had a sense of felt need to reach a political compromise. He called for a meeting and, apparently to the surprise of the USG, raised the S&T agreement talks with Deputy Secretary Whitehead. This gave the USG another opportunity to press its case at a high political level in the GOJ. While Japanese compromise language on this linkage was unacceptable to the USG, Ambassador Matsunaga commented that it would be inappropriate to include new language and concepts into the STA which are not in the existing U.S.-Japan security treaty and that this would lead to difficult political discussions in Japan. This, according to Japanese sources, persuaded the USG to agree to the wording described above in a side letter signed by the U.S. Secretary of State and the Japanese Minister of Foreign Minister -- apparently proposed by the GOJ -- stipulating that if information or equipment that "may be" classified is unexpectedly created it "may be" protected by applicable laws and that "neither government is obligated to modify existing laws and regulations or to create new ones". Thus the USG demand for a linkage while still existing in this agreement has been, through the use of "may be" and no need to revise or create laws and regulations, diluted and less formidable.[120]

A Second Side Letter. At the suggestion of the USG, another side letter was accepted by Ozawa which stipulated that the GOJ would take steps to identify and increase substantially opportunities for American scientists and engineers to engage in research and study in Japan and that the USG will also encourage and support American researchers to take advantage of such opportunities. The GOJ prevented such unilateral actions being described in the agreement -- an original demand by the USG -- but privately agreed to such in a side letter.

These side letters, according to State's Legal Office, are not, as a legal matter, an integral part of the Agreement; they are documents associated with the Agreement, may be circulated with the Agreement and will be sent to the U.S. Congress with the signed agreement. These letters, however, are not included in State's Legal Office's treaty files on this agreement. They have not been translated into Japanese (while, of course, the agreement and annexes have been translated and co-equal with the English text) and not formally submitted to the Japanese Cabinet for approval.[121] Thus, while they have no legal binding effect they serve as expressions of political intent and interpretation which could later be used with effectiveness in political discussions about the USJ S&T relationship.

With one day to spare, the USG and GOJ announced on March 31, 1988 that they had reached agreement in substance *in the evening of March 30* on a new science and technology agreement designed to enhance USJ S&T cooperation, stressing comparable access for researchers and ushering in a new era in S&T cooperation. The agreement was signed by President Reagan and Prime Minister Takeshita on June 20, 1988, when both were attending the G-7 Summit meeting in Toronto. The agreement was hailed as a major achievement of the Reagan Administration although the psychological aura for a new beginning which would have accompanied a USJ bilateral summit meeting was overshadowed by a multilateral meeting of the leading industrial powers.[122]

DOC letter of March 22, 1988. The day after the seventh and final session of the negotiations for a new STA had begun, the DOC Secretary, William Verrity, created the "Commerce Committee on International Scientific Agreements" and instructed all agencies in DOC that "any pending or proposed agreement with any Japanese government research agency or Japanese government-funded research institution" shall be submitted to this Committee "before negotiations begin with that entity." In issuing these instructions he cited the Dole-Rockefeller Amendment to the Technology Transfer Act [presumably of 1986] that there must be reciprocity of access between U.S. and Japanese institutions, protection of IPR and safeguarding classified information. This Committee shall review any proposed or pending international agreements, shall coordinate policy inputs to develop a unified DOC position, shall be a coordinating mechanism not a policy body for this purpose and it shall consist of representatives from the various Under Secretaries for Economic Affairs, International Trade Administration (ITA), Oceans and Atmosphere, PTO, General Counsel and Assistant Secretary for Communications and Information and chaired by the representative from ITA. This instruction, specifically aimed at Japan, must reflect the frustration of the DOC. It was probably rather unpublicized at the time. While access and IPR and classified information were being negotiated, there was, at the time, no known access, IPR or classified information issues of contention between the U.S. and Japan about public sector R&D over the past 40 years. While the Toshiba-Kongsberg problem was a different and separate issue, its existence and the ill-feeling created on both sides of the Pacific probably was a major factor in issuing these instructions. While, on the surface, it would seem a useful, appropriate and cautious administrative measure by an alert DOC Secretary, it is difficult to image what would be achieved by this committee at this stage in the negotiations when a new agreement of some kind was, by this time, a foregone conclusion, except to emphasize DOC's antipathies toward Japan. It is yet another symbol of the depth of the frustrations and annoyances in some part of the USG about its inability "to persuade" the GOJ to accept the USG draft STA.[123]

Japanese Media Response to the Agreement in Principle. A few days after the "final agreement in principle" on March 30, 1988, three Japanese newspapers, *Mainichi Shimbun* (4/2/1988), *Asahi Shimbun* (4/5/1988), and the *Japan Economic Journal* (4/16/1988 in English) printed commentaries on the negotiations, their backgrounds and the implications of the resultant compromise.

The *Mainichi* reported that Ozawa's plan, which he had been working on since January 1988, was to find a solution which would satisfy both the "format" needed for U.S. acceptance and the "substance" needed from the GOJ's point of view. He accepted the U.S. demand that "security considerations" be included in the agreement but then hedged this inclusion by obtaining the inclusion of the principle of open research and that military R&D would not be conducted under this agreement as a neutralizing counter to the security clause. This clause was described as having a "smoldering burnt smell" in English, where there's smoke there must be a fire or a potential for fire. Notwithstanding the hedging and seemingly neutralized language, there was an underlying fear that the security clause would develop a life of its own, i.e. become uncontrollable and dominate the R&D projects under this agreement. The *Mainichi* commentary reported that it was felt (presumably by government officials) the U.S. demanded the inclusion of the security and the IPR clauses as a strategy to "control" Japanese science and technology.[124]

The *Asahi* commentary was somewhat more extensive and blunter in its characterizations. The first and major point of the article was the description of U.S. motives behind the draft agreement. The USG strategy as reflected in the new STA was to react strongly to Japan's economic challenge to the U.S. The language used says Japan pushed the U.S. into a tight corner economically and, quoting a Japanese government official, that the USG was seeking a platform (*ashiba*) through the draft STA from which "to control" (*kontorōru* in katakana was used) Japanese science and technology. As a reflection of this strategy, the second point was the inclusion of the security clause. Again, Kondō Jirō, President of the Science Council was quoted as saying that "the net of secrecy" (*kimitsu no ami*) could be thrown over even basic mathematics research which could be interpreted as related to computers, related to military technology to classified research. The third point was the IPR provision to use the case-by-case approach instead of the initial USG formula for the receiving institution to receive all benefits (copyrights and patents). The Japanese were probably unaware that the original USG position was being revised due to NSF argumentation inside the USG. Even Uenohara Michiyuki, then VP in NEC and for many years prior to that with Bell Labs, expressed his skepticism about this IPR arrangement until he could see the actual wording in the agreement. Fourth, even though technologies are listed in the STA as the areas where the US and Japan were to carry out joint projects, most experts doubted that much would come of these "competitive cooperative projects" (*kyōgō*). They were just merely listed in the agreement. Lastly the fifth point evaluated the three tiered management structure as the means to strongly pressure the Japanese if the expected results and cooperation in high technologies were not sufficiently forthcoming from the USG's point of view.

The third commentary from *Nikkei's Japan Economic Journal* centered on the reactions of selected senior Japanese to the new STA. Again, the first preoccupation was opposition to the security clause as totally inappropriate for inclusion in an S&T agreement which stresses peaceful basic science research. Uchida Moriya, director of Teijin Ltd, surmised that active private sector participation in governmental projects under the STA was unlikely. Already there were complaints from Japanese high technology industries that the U.S. was "enclosing and absorbing" selected technologies, such as charge coupled devices, image sensors used in VTRs as military technologies but also used by Japanese in civilian products. Shimura Yukio, managing director of Kgōyō Chōsakai Publishing Co., a major Japanese think tank, opined that technology friction will increase in superconductors, battery storage devices and electromagnetic ship propelling systems where the U.S. considers them military technologies. Japan, he said, was in a weak position because of its special status of being under the military umbrella of the U.S. The American Embassy in transmitting this article to Washington, said it typified "opposition which has begun to emerge and [that] it is difficult to estimate the extent of opposition from these examples.[125]

The three examples signify unfortunately a well of deep distrust of U.S. motives in demanding the creation of the three tiered management structure, the inclusion of the security clause and the IPR provision even though the decision would be on a case-by-case basis. The distrust mirrors the obsessive USG distrust of Japan's reliability and strength of commitment to adequately control the flow of high technology information. The USG attempt described above, to try to insert the concept of open research and security in Article I at the 11th hour must have deepened Japanese interpretations about USG motives about the agreement itself. Japanese anxieties were not assuaged by MOFA's assurances that nothing has changed, that Japan does not need to create any new laws about security classification nor change any laws and regulations because of the new STA.

v. What was achieved? Let us now turn to an assessment of what was achieved based on a comparison of the USG proposal and the signed agreement.

A completely new agreement -- not a revision as originally envisioned -- was created based on a much more specific set of principles of cooperation and policy framework which it was hoped would govern all USJ future S&T relations.

The "umbrella-ness" of the agreement. This was not popular with USG and GOJ technical agencies. The USG did make a major attempt to centralize the control of bilateral USJ S&T policy and project management relations but it did not succeed. Those projects which had their own management structure would be expected to include the principles enunciated in the 88STA. That was a far looser arrangement than originally planned and would reinforce turf issues among competing technical agencies. The coordinating agency OES at State and the MOFA Science Affairs Bureau do not have the manpower to exert their authority to micro-manage the wide ranging USJ S&T relations.

The Presidential nature of the STA. It was reinforced by the USG approach to the negotiations with the GOJ and thus has now highly politicized future bilateral S&T relations in contrast to the 40+ years since 1945 when these relations were the province in the main of technical experts, scientists and engineers.

The USG proposed three tiered management structure. This structure was reluctantly accepted by the GOJ. It is cumbersome and time consuming to administer. This structure gave the USG what it had sought for some time, a platform where it could raise policy issues about GOJ science policy, expose problems in the Japanese structure which were not to the liking of the USG, and it was also, in a more positive tone, a platform to have a genuine dialogue with the Japanese about the coordination of S&T policy. This structure gave the USG three opportunities -- at the Joint High Level Committee (Ministerial level), the Joint Working Level Committee at the Assistant Secretary/Bureau Chief level and the High Level Advisory Panel of academic and private sector representatives -- to raise and pursue S&T issues with the GOJ; the GOJ also had the same opportunity but generally did not use it. The USG has been singularly unsuccessful through these kinds of negotiations to persuade the GOJ to commit large financial resources into USG S&T projects without itself investing in large Japanese S&T projects.[126] It is particularly significant that only the USJ STA contains this three-tiered management structure, a fact which would tend to confirm the above interpretation as to the purpose of this bureaucratically cumbersome structure.[127]

USG Demands shifted to JWLC. In the opening sections, Annex I and II of the USG proposal, there were many items where the language clearly was intended or actually said the GOJ would take numerous specific actions. For example, there was a need [for Japan] to create more opportunities for interaction, where "at times one country [Japan?] may need to make greater commitments in particular components of the R&D process", GOJ "should increase its commitments to our shared responsibilities" in selected areas, "the two nations will discuss their respective S&T policies, activities, plans, the balance of reciprocity ... identify [Japanese] impediments to balanced opportunities, access and benefits" to overcome them, the GOJ will facilitate U.S. participation in GOJ government supported or sponsored projects. In Annex II on Mechanisms, there was a list of what GOJ will establish, will actively support, will provide annually. Many of these ideas were folded into the responsibilities of the JWLC but stated as neutral bilateral responsibilities and many of the imperative "will" were changed to the more permissive "may". In the Annex of the 88STA, many of the functions of the JWLC were to make recommendations “as necessary”, thereby providing considerable leeway and discretion.

Comparable Access. Article I.1.D and in Article VII stated that there would be free exchange of STI, comparable access to GOJ and USG R&D facilities with equitable contributions and shared costs. The concept of “access” used in the STA is reminiscent of trade discussions and agreements. This may be described as symbolic of the changed nature, role and character of the USJ S&T relationship. Annex III.2.H and I stipulated that a task force would be established, “as necessary” on access to R&D facilities, but a task force on STI would be created. The “as necessary” condition for the former was totally meaningless since these two task forces were promptly created and were still meeting in 1996-97. Annex II, paragraphs B, C, D, E and F enumerate how both governments must improve language training, provide government opportunities and fellowships for foreign nationals with adequate allowances for accommodations and other needs, disseminate information about these opportunities, and urge their S&Es to take advantage of these possibilities. All these are now stated as bilateral responsibilities but the USG draft described them as GOJ's responsibilities. Because USG's major emphasis during the negotiations was not satisfied with the emphasis in Annex II, it proposed a side letter which was accepted by the GOJ re-emphasizing the GOJ's commitments “to take steps to identify and increase substantially opportunities for American scientists and engineers to engage in research and study in Japan.” Through the agreement text and side letters many of the USG demands were incorporated into the 88STA.

IPR. Detailed intellectual property rights that might stem from joint R&D projects under this STA were included. But an important revision took place in the negotiations stemming from the logic of the Japanese position and from the NSF inside the USG. While the USG had attempted to establish an inflexible principle that all inventions made by a visiting S&E in the course of joint activities would accrue to the hosting/receiving facility, this was modified importantly by stipulating that where the hosting facility was “expected to make a major and substantial contribution to the programs of the cooperative activity” then the host has the right “to obtain all rights and interests in the invention in all countries”. This detailed formula concerning rights to inventions will need to be adjusted to and coordinated with the principle enunciated in Article I.1.c that there will be “adequate and effective protection and equitable distribution" of IPR in the entire process of collaboration.[128] In actual practice, therefore, these provisions will be applied on a "case-by-case basis" as advocated by the GOJ.

Security & Export Control. The USG successfully persuaded -- or rather strong-armed -- the GOJ to accept inclusion of an article on security and export control of technology but this issue remained a source of endless argument to the bitter end and was resolved only at the last moment at a high political level. The relevant section in the USG proposal called "Annex V Security Obligations" read like a clause in a military security treaty on the assumption that classified information and equipment would be created, not an agreement on civilian R&D. At the GOJ's request article VII began with the assertion that "both parties support the widest dissemination of the information or equipment in the course" of joint projects (a principle already enumerated in Article I.1.D) and "in furtherance of the principle of maintaining an open basic research environment, both parties confirm that no information or equipment classified for reasons of national defense will be utilized" in joint activities. Since the GOJ held out (*nebaru ni nebatta*) for "a politically acceptable solution" and the USG wanted a resolution of this impasse, the momentum swung to the GOJ's advantage when it suggested that a side letter be signed by the SecState and MOFA Minister specifying that if information or equipment with national security implications were "unexpectedly created" it "may" -- note not shall -- "insofar as permitted by applicable laws and regulations be protected from unauthorized disclosure" and that "neither government is obligated to modify existing laws or regulations or to create new laws and regulations." While the spirit of the USG proposal may have survived, its substance was compromised. It would now seem difficult for the USG to press the GOJ to create new regulations, let alone laws, concerning classified information under this side letter arrangement.

Similarly, the transfer of export-controlled information and equipment stemming from joint programs would be done "in accordance with applicable national laws and regulations."

Potential technical areas of cooperation. Again, this was a USG initiative and included a list of seven technical areas of potential cooperation which by definition implies that the scientific and technical levels of both countries in these areas were more or less comparable.

9. The Media

Unlike some other USJ issues such as trade, the media both in Japan and in the United States played a rather low key, almost non-existent role during the STA negotiations; the Japanese media was perhaps more a mirror of GOJ attitudes with some embellishments. To the best recollection of this writer, there were no TV news reports in the U.S. about the USJ STA but a few articles in two U.S. magazines. In contrast, there were a substantial number of reports in the Japanese press considering that there were only seven negotiating sessions and these articles appeared as editorials and regular articles in all the major dailies in Tokyo, thus a much broader spectrum of the media reported on this issue in Japan, often in prominent locations in the newspapers.

a. The U.S. Media

The articles appearing in the U.S. media were in specialized journals such as *C&EN* (published by the American Chemical Society) and *Science* (published by the American Association for the Advancement of Science); they were more or less straight news stories with seemingly no "message" for the GOJ. Also the *Asian Wall St Journal* reported at least a couple of times from Washington, but never from Tokyo. There was one substantive article on USJ S&T relations published in the *C&EN* on April 11, 1988, soon after the final agreement in principle on the 88STA had been reached. As the Japanese press generally commented with pessimism about the new agreement and how it was concluded, this article begins with an assessment that

> Despite feverish economic strife and vast cultural gaps Japan and the U.S. are finding themselves bound on a joint course into a technological future. Conflicts may be hiding the commonalities, but most Americans who have visited Japan and worked within the Japanese technical establishment come away with feeling that the two countries now share a common path. ... The ties with Japan can be used to invigorate and complement our vast resources. One can liken the situation to a muscle cramp. The worst antidote is to tighten up" (p. 13 and 21).

No major U.S. daily newspaper analytically commented on the in principle agreement except again for an article in the *C&EN* of 7/25/1988 and *Science* of 7/1/1988. The two articles had a general tone of optimism about them. *Science* quoted the Science Adviser Graham as proclaiming the 88STA as "trail blazing" and that "we did bridge a gap between our structural differences." While similarly somewhat upbeat, *C&EN* pointed out that the IPR and national security clauses "could prove troublesome if pressed by federal agencies hostile to Japan's technological might."

b. The Japanese Media

In contrast to this more cheerful note in the U.S. media, the articles in the Japanese media were, in most cases, obviously based on interviews with MOFA or other officials, were perhaps intended as "messages" to the USG negotiators. They were not happy optimistic reports about a constructive relationship but reports that reflected a frustration and displeasure with the negotiations and a dark and skeptical interpretation of USG intentions to "attempt to control Japanese R&D" through the S&T mechanism. They were apparently read carefully by the Tokyo American Embassy and interpreted periodically as conveying a particular Japanese policy response, outlook or assessment by the GOJ toward earlier USG proposals and discussions at negotiating sessions. The Embassy felt these articles could be of use to the U.S. Team in preparing their next round of negotiations with the GOJ.

There was one case, and only one, where the chief Washington correspondent of a major Japanese newspaper, the *Asahi Shimbun*, interviewed the A/S OES, Ambassador Negroponte, about the outstanding S&T issues between the two countries. This gave him an unusual platform to convey some USG requests, hopes, dissatisfactions and objectives to the Japanese public. But it is doubtful that this one time attempt had much impact on the Japanese public or the government.

All the major Tokyo daily newspapers, *Asahi, Mainichi, Nikkei, Sankei, Tokyo, Yomiuri,* carried a total of at least 22 news articles and ten editorials. The newspaper with the largest number of articles was the *Nikkei* with 12 including two editorials. *Asahi and Mainichi* were next with 6 news items each with 4 and 3 editorials each, respectively. This media interest is in sharp contrast to the U.S. media where there were no editorial comments. It seems that as a pattern there was a news item before and after almost all the seven negotiating sessions in one of the Japanese major daily newspapers. Although some of the reported specifics may not be completely accurate -- after all these talks were supposed to be "quiet" talks between the two bureaucracies and also confidential according to the Japanese prime minister's explanation to the Diet -- the general trend was correct. The way the negotiations were reported, the tone of the language used and assessments reported probably draw a fairly reasonable picture at least as viewed from the GOJ's side, but, based on the available record, little attention, if any, was given to these Embassy reports by the USG delegates.

Before the USG had given the GOJ a draft STA in late August 1987 and before the negotiations began in October 1987, the *Tokyo Shimbun* and the *Nihon Keizai Shimbun* reported that the aim of the U.S. through it's draft agreement was to achieve superiority over Japan in high technology and that the U.S. was calling for a completely open exchange of all S&T information and acceptance of U.S. researchers into Japanese research institutions (by implication both governmental and private industry). The USG was said to be demanding "symmetrical access", an expression which, it seems, was derived from a report prepared by the National Academy of Sciences and given to the Japanese in early Fall 1987. The Japanese feared that if these ideas were completely accepted, Japanese S&T would come under the control of the U.S. Additionally, the head line of the *Nikkei* was "the chasm deepens between U.S. and Japan over technology friction". This was an ominous beginning to the bilateral negotiations.

Although the GOJ reportedly accepted the U.S. demand to include a security clause in the new agreement, as described in an earlier chapter, it was not possible to arrive at a compromise in time for the Reagan-Takeshita summit meeting in mid-January 1988. A *Nihon Keizai* article (1/7/88) described this new demand as potential "waves of control" flowing over Japan's basic postwar research principle and policy of open research, and expressed the fear that the U.S. would now attempt to expand its control over other technical fields. On the eve of the summit meeting, the *Nihon Keizai* editorialized on January 9, 1988 that 1988 was a year of S&T diplomacy, that the U.S. and Japan for their mutual interest should cooperate in S&T, that a new set of IPR understandings should be worked out carefully in order to avoid future problems. It cautioned Prime Minister Takeshita not to buckle under to U.S. pressure to an agreement which cannot be carried out, which, in turn, would lead to demands for U.S. retaliation against Japan just as was occurring under the 1986 semiconductor agreement. It opined that the U.S. objective was "to bind [or tie up] Japanese S&T" through this agreement in some clear form. If a hasty agreement is reached it will leave "roots of calamity" unresolved into the 21st century. Therefore, even if the 80STA were to be allowed to expire at the end of January 1988, specific problems should not arise since specific and special cooperation could be arranged.[129] The implied threat of allowing the STA to expire or to threaten an abrogation (as apparently the U.S. side also tried) did not result in a lapse of the STA since the USG ultimately agreed to another two-month extension.

Soon after this extension, the *Asahi Shimbun* and *Mainichi Shimbun* both editorialized about these troublesome negotiations on February 8 and February 1, 1988, respectively, and published commentaries on the negotiations. They each urged the government to listen carefully to the USG positions on exchange of personnel, IPR, the security clause and greater access to Japanese R&D facilities. They thought the USG position had some validity about the exchange of personnel and access facilities, but did admonish U.S. researchers to study Japanese as Japanese had studied English in order to read the technical literature. They did urge the government to stand fast on its principles about IPR and the need to resist any attempts to restrict the flow of scientific information stemming from joint projects. The *Mainichi Shimbun* did, however, raise the U.S.'s apparently strong stand and demands to Japan, and again raised the issue of the U.S. intent to try to control, restrict or tie-up (*shibaru*) Japanese R&D implicitly because the U.S. greatly feared that Japanese applied research results would leak to Communist countries.

The *Nihon Keizai Shimbun* (3/9/1988) reported (after the 6th session) on the "threats" made by the USG to the Japanese at the 6th session in early March to "politicize the S&T negotiations" -- as if that had not already been done -- like the discussions over agricultural products and public works construction and to deny access by Japanese scientists and engineers to U.S. national laboratories as a new way of putting pressure on foreign governments under the 1986 Technology Transfer Act -- if the GOJ did not agree to a revised S&T agreement.

After the two governments announced on March 30, 1988 that they had reached in principle agreement on the outstanding issue on the new STA, most of Tokyo's major dailies printed editorials on the new Agreement over the next six months: the *Mainichi Shimbun* (4/1/1988 and 8/16/1988), *Asahi Shimbun* (4/8/1988 and 9/15/1988), *Sankei Shimbun* (6/22/1988), *Nihon Keizai Shimbun* (6/22/1988). Two just after the March announcement but before the detailed agreement had been made public, two immediately after the signing and two sometime later.

The *Mainichi* editorial first pointed out that some parts of the U.S. positions were not reasonable (*datō de nai*). The earlier STA was a symbol of friendship but the U.S. now demands to make the "simple" (*awai*) relationship into a thick and complicated (*koii*) relationship. The demand for the security clause was the most puzzling to the Japanese because any modern high technology can be used for military purposes. It called upon the GOJ to make very sure that the security clause was not left ambiguous so that it could develop a life of its own. It noted that the Human Frontier Science Program proposed by Japan, the joint fusion project being carried out by the U.S., Japan, Europe and the Soviet Union are all based on open research without the shackles of classified information and called upon the U.S. to keep this approach in finalizing the agreement with Japan.

The *Asahi* editorial, in the beginning, quoted Dr. Tanaka Shōji, the Japanese scientist of worldwide fame in superconductivity. He lamented that the U.S. fears the Soviet Union and now contemplates completely (*kanzen ni*) controlling Japanese S&T and is trying to create a frame (*waku*) for this purpose. It was his judgement that we have now entered a fearful era (*erai jidai*). Because of the insistent U.S. position, Japan felt obliged to accept most of its demands. As a result, there are two important questions that have to be raised.

1. hiding behind the new agreement, the net of U.S. security classification system may be thrown over Japanese technology since any advanced technology can have a dual purpose. There has been too much military emphasis in U.S. R&D in contrast to the Japanese stress on civilian uses. The U.S. should seriously consider moving in the direction that Japan is following.

2. the great possibility that politics will creep into S&T issues especially in the implementation phase.

Both editorials were polite but left no doubt about what they thought were U.S.'s motives and objectives in creating the three tiered management structure and demanding the inclusion of a security clause.

The *Sankei* editorial of a politically rather conservative newspaper acknowledged that there was somewhat of an imbalance in the researchers going to Japan in comparison to those Japanese researchers going to the U.S. and called upon Japan to prepare for receiving U.S. researchers. The issue of the security clause was given a passing reference and IPR was not even mentioned. It was a weak, ineffectual editorial of little substance.

The *Nikkei* editorial, in contrast to commentaries cited in this chapter, struck a positive note of hopefulness by mentioning six areas where cooperation can be emphasized: information will be open except where it contravenes the domestic laws of each country, specifies technical areas of cooperation, that each country will facilitate the exchange of personnel, the creation of a three tiered management structure, IPR will be decided on a case-by-case basis, that research involving military secrets would not be conducted. It concluded that the new agreement does not include any factors which would obstruct future U.S.-Japan S&T cooperation and calls for ever closer bilateral relationships in the future.

The occasion for the next two editorials by the *Mainichi* (8/16/88) and *Asahi* (9/15/88) was the convening of the first Joint High Level Committee created by the 88STA. The *Mainichi* editorial was bland and innocuous and hoped for the best. While it cited the MOFA and S&TA as being satisfied that the basic position of the GOJ had been sustained, it cautioned that it is still possible that Japanese science and technology might be "taken for a ride" because the agreement espouses a system of open R&D, the opportunity to participate in government sponsored/assisted R&D, especially by the private sector. Unless both the U.S. and Japan genuinely share the results of their basic research in an open and free manner, the Joint High Level Committee could easily become the battleground for the confrontation or S&T war between the two countries. Already there was substantial interchange in S&T among the universities, the private sectors and national laboratories of both countries. It concluded by calling upon the parties to pour good *sake* into the "new leather bottles," the 88STA.

The *Asahi* editorial, in good *Asahi* tradition, was more stringent yet concluded with an appeal for Japan to push forward with its idealistic conception and policy of R&D for peaceful objectives minimizing the military/-security needs to the greatest extent possible. It pointed out the dilemmas facing scientists and engineers in weighing their own independence and the need of national security in light of the increasing dual use nature of science and technology. Despite the seeming assurances in the STA about security needs being executed within the confines of each country's domestic laws, Japanese scientists and engineers cannot rest easily and lightly dismiss the security clause. It should be kept in mind that the objective, intent (*nerai*) of the USG demanding this clause was "to control" (*kontorōru* in *katakana*) Japanese S&T from the [U.S.] national security point of view -- repeating and thereby re-emphasizing a phrase used in a commentary on April 5, 1988 cited earlier. Depending on how the STA was implemented there was a fear that the open environment in Japanese S&T might be lost. It is conceivable that Japanese scientists might invoke "self control" because of the possible consequences of prosecution when visiting the U.S. if there is a difference between the U.S. and Japan in handling certain S&T information and data. It called upon the GOJ to reconfirm the openness of science and technology and put a brake on (*hadome o kakete*) any possibility that the U.S. might attempt to emphasize the security aspect of S&T information at the upcoming Joint Working Level Committee and the Joint High Level Committee. "Technology for Peace" (*heiwa gijutsu*) it asserted seemed to be assuming a greater role over military technology in recent years and probably into the future. It said Japan could take pride in being in the forefront of this change.

A *Yoimuri* article (4/2/1988) wrote about a U.S. net under the guise of national security being thrown over Japanese R&D, thereby losing its freedom, that the USG was, through the 88STA, attempting to create a foothold to control Japanese R&D, and that the bilateral process left smoldering embers (*hidane*) behind, implying they could ignite at any time. All the newspapers, however, emphasized that the future relationship between the two countries depended on how the new agreement would be implemented. The *Nikkei* (6/22/1988), *Tokyo Shimbun* (6/24/1988) and *Sankei Shimbun* (6/22/1988), all conservative newspapers, were far more generous and sympathetic to the USG in interpreting the future possibilities of the 88STA.

These editorials and articles expressed Japanese hopes and anxieties for a closer USJ S&T relationship especially after the rather acrimonious negotiations which, in turn, were a reflection of the extremely unpleasant confrontations within the USG over future S&T relationships with Japan. Nevertheless, there was close to the surface a strong undercurrent of suspicions about the motives and intent of the USG in pushing for and demanding the inclusion of the security clause and detailed IPR stipulations in the new agreement. In summation, the editorials sketched the possibilities of a new but cautious era in USJ S&T relationships.

10. Post-Mortem Exercises

After rancorous negotiations, a new science and technology <u>cooperation</u> agreement was signed on June 20, 1988 as a sideshow at the G-7 Summit in Toronto, Canada. It should be kept in mind that while there were only 7 "negotiating" sessions from October 1987 to March 1988, actually the process of formulating an S&T policy towards Japan, new or otherwise, by the USG began about summer 1983 in the State Department, four years of internal discussion, negotiations and bickering. One might fairly describe this entire process as the "science and technology wrangle" between the U.S. and Japan.[130]

Each side returned to home base, Tokyo and Washington, and at appropriate times and places proclaimed their respective "victories" at the negotiating table. In a way, they are both correct if one chooses to ignore certain demands made by the USG and certain hopes by the GOJ to avoid acceding to certain demands by the USG. One side held a public Congressional hearing and flamboyantly and aggressively proclaimed victory and determination to pursue the Japanese "to the last yen"; the other held a low-key round table discussion (a *zadankai*) by the four musketeers as they were reassigned to new positions.

The USG preceded its negotiations with the GOJ with a Senate subcommittee hearing; the GOJ held no such hearing or meeting. There was a short interpellation in the Diet about the Defense Patent Secrecy issue and this discussion tangentially touched upon the 88STA negotiations. The U.S. side, in a symmetrical fashion, held a House subcommittee hearing in July 1988, less than a month after the Agreement was signed to apprize Committee members of the newly signed agreement. There was no such Japanese hearing; none would be considered necessary since it was an executive agreement not requiring Diet approval. An English-language Japanese publication organized a short round table discussion about this agreement for worldwide distribution. These two "post mortem exercises" are a contrasting reminder of how the two governments operate.

a. U.S. Congressional Hearings-II, July 1988

On July 13, 1988, a joint subcommittee hearing in the House of Representatives was co-chaired by Congressmen Doug Walgren and Ralph M. Hall to hear from the DOC, State, NSF and OSTP about the 88STA.

The USG witnesses formed a panel of representatives from the most mutually antagonistic agencies, yet naturally they were a model of comportment and cooperation at the hearings. They naturally strove to prove how effective they had been in negotiating a new type of STA, a model agreement, which included a policy framework for future S&T relations not only with Japan but also for use with other countries, a detailed set of model IPR arrangements, R&D and national security precautions, technical areas of potential collaboration, and a three tiered management structure. We do get a glimpse of how the USG believes the IPR distributive arrangements give an advantage to the U.S. They assert naturally that the USG secured from the GOJ what it wanted in accordance with the EPC guidelines -- except that assertion would be subject to further scrutiny in the next chapter. There is a fragmentary reference at both the Congressional hearing and the afore-mentioned Japanese round table discussion to the protracted negotiations and strained atmosphere but no one on the subcommittee pursued this issue in any manner. Since this was an executive agreement not requiring Senate approval it was a pleasant and friendly informational hearing with no anticipated action by the committee. It is nevertheless part of the legitimizing process in the USG.

The subcommittee hoped to find out 1) how the comparable access issue would be implemented, 2) how research which may impinge upon national security would be handled, 3) how IPR would be apportioned, 4) about the nature, coverage of the 88STA as an umbrella agreement and 5) how the Japanese will be financing more basic research, and increasing publication of the results of their research.[131] The witnesses were Frederick W. Bernthal A/S OES, Peter de Vos, chairman of the U.S. delegation, Maureen Smith of DOC, Deborah Wince-Smith of OSTP, Charles T. Owens of NSF. DOD and USTR representatives were not present. The two most vocal and vigorous U.S. delegates during the negotiations were from DOC and OSTP. Much of what was said at these hearings has been stated in a variety of forms in this study. Nevertheless, it is important how certain statements, how certain parts of the agreement would be pursued, and the kind of English language used to describe how these objectives would be pursued.

A/S Bernthal of State naturally gave a most positive description of the agreement describing it as "a model international" umbrella agreement where the world's two "leaders in science and technology" have joined "in a new partnership based on equality, reciprocity and comparability" and established a framework of principles for our overall Government to Government S&T relationship. It is a model of an S&T agreement between "two highly advanced and capable countries". The new agreement will produce "significant gains" for both countries based on shared cost of collaboration taking into account respective risks, benefits and management accounting. It is also described as a "hallmark" as it "sets out specific IPR provisions for cooperation."[132] He stressed that the 88STA applied to government-funded research whether at a national, university or industry laboratories, and industry funded R&D whether applied or basic was not subject to this agreement. Bernthal suggested that automation and information technology would be two areas where the USJ would have "a very active and fruitful exchange". He pointed out, however, that such organizations as the dual use technology consortia, SEMATECH, which is DOD funded would exclude Japan from cooperative R&D in this project. While the USG can easily classify dual use technology under the national security rubric without challenge from any country, and thus deny foreign participation, the GOJ in forming similar consortia cannot make such a similar claim in the name of national security since it does not have appropriate classification laws for this purpose. In this sense, therefore, the GOJ is at a distinct disadvantage when it comes to sharing in this kind of domestic U.S. cooperative research activities.

Under the principle of distributive IPR when junior researchers participate in joint projects all IPR will go to the sponsoring organization, not the junior researcher. The USG felt that since "far more researchers from Japan visit the U.S., than vice versa, ... the U.S., as a nation, gains from the distribution of invention rights contained in the agreement". From the sheer numbers point of view, statistical probability, the assumption that Japanese researchers are as effective and creative as their U.S. associates and that Japanese researchers in the U.S. will continue to substantially outnumber U.S. researchers in Japan, then this assessment may be correct. But we should remind ourselves that we have been preaching that Japanese scientists/engineers are not as creative as our own. This deficiency is made up with much larger numbers of visiting researchers, or we have conveniently decided that Japanese are, after all, probably as creative as our personnel -- especially as we are officially saying that the Japanese and the U.S. are the two scientific and technology leaders in the world.

Despite the size of the Space Station project -- and probably the SSC if it had materialized -- its management structure and control was placed outside the "umbrella-ness" of the 88STA. But it was stressed that the principles outlined in article I would apply to all S&T agreements between the U.S. and Japan whether specifically under the 88STA or not.

Since Japanese government-funded research is mostly carried out in industrial laboratories, in contrast to the U.S. where much of it is done in universities, the issue of comparable access to this research by foreigners, especially U.S. researchers, was identified by Maureen Smith of DOC as one of the key aspects of the agreement. "We had", she said, "not several, but numerous and very protracted negotiations or discussions" and she asserted "it is certainly understood that it is our objective to obtain access to that portion of Japanese Government-funded research which is conducted in a private or proprietary setting" and "we define government-funded in the Japanese context as a single yen". Because of the Japanese governmental system, and the role that the GOJ plays in coordinating, guiding, sponsoring R&D projects even though the government itself provides only a small part of the funding, the issue naturally arose whether "government-supported and government sponsored" projects come under the rubric of this agreement. Bernthal maintained that even projects that did not have government funds in it but were supported or sponsored by the government would be subject to the "comparable access" principle. This interpretation would seem to give the USG the rationale to push with vigor, determination and relentlessly for opening all kinds of Japanese laboratories. This is an enormously wide ranging assertion that would be unprecedented if, in reality, the GOJ would agree to such an expanded interpretation of "comparable access".

This broad interpretation was later considerably circumscribed -- one might say repudiated or almost completely reversed -- in a written response (included in the hearing as part of Appendix I) to specific questions from the Committee chairman: the 88STA does not "require access to these programs and facilities, but rather makes them eligible for access. ... The agreement does not compel each side to provide access to such facilities, but it does provide for a forum to discuss cases in which access has been denied". Each side may still deny access to a specific program and/or facility, but "Japan must increase tremendously the overall level of access it provides to its government-sponsored or government-supported facilities if it wishes to continue to enjoy its current degree of open access to such U.S. facilities". This is an implied threat to the GOJ although no instance of denial was mentioned in this hearing or other documents.

Congressman Walgren asked for a description of the "sticking points" in the negotiations. Bernthal prefaced the answer to this issue by asserting that there was "quite broad agreement on the general principles". Chief negotiator de Vos said the "broad principles" were agreed to early in the negotiations, "clarifying the specific points that took the time"; we needed to explain in the IPR section what "each of the little paragraphs" meant. What was the point of greatest satisfaction?": "a new recognition of the significance and important of protection of IPR." Ambassador de Vos described his most satisfying aspect to be spelling out what access means and being able "to bridge the gap in the structure between our two countries" and maintained that the USG "got everything what we asked for ... we're happy with it".

When asked for her assessment, Deborah Wince-Smith of OSTP observed:

> that our science and technology cooperation is really not a zero sum game and that in order for it to continue as an integral, healthy part of our relationship, both sides have to come to the table as equal players and they have to put something on the table of equal value. Clearly, the perception in the past that our system was open and our system was there to be exploited is one

> that the Japanese recognize they could no longer go forward with without having some repercussion from that. ... in a parallel negotiation in the OECD which ... in May [1988] established a new framework for principles for international collaboration and that accomplishment also set forth in the international arena the idea that all advanced countries in science and technology have to come to the table as equal players and the concepts of equity and fairness and at the end of the day, there's some overall balance in contributions.

In contrast to the protracted negotiations and the rather strident proclamation of success by the government witnesses, Chairman Hall concluded this hearing, rather ironically, by recognizing the Japanese Embassy's science counselor, Ikeda Kaname, describing him as a friend of the House Science, Space and Technology Committee and praised Japan as "good people for us to cooperate with in the space program and certainly on the Supercollider" and that they have "more genuine interest" than the Europeans in putting up their share of the costs in developing science and technology.

In a more private setting of an interview, I was told by Dr. Graham, the President's science adviser that 1) this was a <u>first</u> major agreement in science and technology by the Reagan Administration, 2) that it set an important precedent, i.e. a model agreement, for future STAs, 3) we need to emphasize the principles enunciated in the new STA, reciprocity, IPR and dual use technology. A senior member of USTR characterized the 88STA as one of the most significant achievements, a landmark agreement, of the Reagan Administration, that the STA has leveled the playing field, has integrated the U.S. approach to international S&T, that it was commensurate with a Presidential level agreement, that both, the U.S. and Japan, negotiated as equals. He could not avoid a critical comment on the OES: that it was obstinate and backward. An OES staffer commented that the STA had given "given purpose and direction never before provided". An NSF senior official described the process as a "pulling and tugging that resulted in a symmetrical agreement". An OMB observer commented that Japan got 70% of what it wanted and a legitimation of free exchange while the U.S. got 30% of its objectives.

b. "Four Musketeers" Discussion Group, July 1988

The Japanese post-mortem was a short get together by "the four musketeers" sponsored by *Look Japan* to discuss the just signed 88STA. When the round table discussion (*zadankai*) was held the foursome had become former musketeers, former Division Directors. Hinata Seigi of MOFA, for example, was sent to Harvard University as a Fellow at the Center for International Affairs for a year's sabbatical; in June 1989 he prepared a report on the institutional features of the U.S. Negotiating System.[133] In contrast to the formal Committee hearing where the USG witnesses were trying to convince the subcommittee members that it had achieved its stated objectives, this round table discussion was short, a mere two page report in the magazine.[134] Only a brief summary will be given here of this discussion.

- The same type of executive agreement was signed which is subject to the domestic laws of each country, not a treaty which supersedes national laws and would require the approval of each country's national legislature.

- Clear guidance for cooperation was given, describing the scope and format of joint activities and even suggesting fields of joint activity. That the agreement made explicit the cooperative nature of all future joint ventures was described as one of its most far-reaching implications.

- Technological relationship between the U.S. and Japan was described as the most important part of the 88STA. Future projects would be shared responsibilities with mutual and equitable contributions. It was asserted that Japan was "only too willing to comply" with the USG demand that Japan should make a commitment to a more balanced contribution provided there was comparable access.

- Both governments must exert themselves to provide a systematic organizational framework to accommodate foreign and Japanese researchers.

For public consumption purposes, especially in an English language magazine published in Japan but distributed worldwide, the tone was upbeat in a limited way, optimistic and forward looking. Except for one former Musketeer who clearly stated that the negotiations had been "long and arduous", that the greatest difficulty had been a communications gap between the two teams, one would get the impression that the negotiations had been smooth. This member went on to point out that the GOJ team felt that the issue of cooperation had previously been agreed upon, whereas the USG demanded an overhaul of the "entire framework" of USJ cooperation in S&T. It had taken months to overcome this gap in expectations and communications. This discussion was a pale reflection of their feelings, impressions and images left behind by the negotiations.

Kondō Jirō, President of the Science Council, immediately after the agreement had been agreed to in substance, i.e. end of March 1988, issued a statement opposing the new STA, insisting that it would now become more difficult to publish the achievements of basic science, e.g. in mathematics and physics, because of the security clause and the maintenance of military secrecy.

These Japanese opinions in semi-official dialogue, and non-governmental organizations are a far cry from the seemingly exuberant and aggressive USG posture about its negotiating achievements and future plans.

11. Act IV, A Legacy of Smoldering Embers?

The USG demanded and obtained GOJ acceptance of almost all its major points. The USG had attempted "to rush" the GOJ into signing the 80STA in a short time, and in one session of three days for the 88STA. These tactics clearly reflect not a sign of respect of an equal but a disrespect, even contempt, for the GOJ; the USG took the GOJ for granted while at the same time flattering and extolling the Japanese as technological equals in words to try to persuade it to accept its demands. But in the process of persuasion the USG was put through the wringer and the hoops by the GOJ to its annoyance, frustration and even anger. The GOJ, insofar as it accepted the USG proposal as the basis for the negotiations, necessarily played a rear-guard action by forcing the USG Team to explain and re-explain -- probing and pushing as Michael Blaker described this process -- the philosophical framework and intent of the USG proposal, the meaning of words, phrases, paragraphs, whole sections, yet acceding to the USG in the final text with some diluting modifications.[135]

The GOJ counter proposal was never a serious alternative and was a mere flicker on the negotiating horizon; it was based apparently on slight revisions of the 80STA which the USG adamantly rejected (although it was basically the author of that agreement). It seemed to be put forward as a proforma exercise in negotiating and an empty gesture of self-respect in this process.[136]

The GOJ obviously aimed in the agreement-writing process to make the new STA "a neutral balanced document", a bilateral document from a unilateral document with no blatant unilateral burdens and responsibilities on the USG and GOJ. The issue of "security obligations" was pushed to the negotiating deadline and was solved through a political compromise (*seiji--teki ketsudan*) between the Deputy Secretary of State John Whitehead and Ozawa Ichirō, Deputy Chief Cabinet Secretary. While accepting the basic details of the framework proposed by the USG, the GOJ was successful in diluting USG demands, whittling down U.S. demands to a manageable level, for example, in the security area.

While the USG "succeeded" in securing a new STA incorporating most of its demands and accepting some GOJ dilutions, a high price was paid in lingering feelings in both bureaucracies about USJ bilateral negotiations. One U.S. delegate described it as his most bitter, emotional and unpleasant experience. A Japanese delegate said there normally is an exchange of greeting and sense of something achieved and well-done and mutual congratulations. But there was no clinking of wine glasses -- no *kampai* -- to mark the occasion.

When I completed my interviews and document research up to the signing of the 88STA, I too felt sadly that the bitterness and poisoned atmosphere created within the USG and between the two governments in the negotiating process had indeed left smoldering embers which were not dying but still quite hot and ready to burst into flame but simultaneously felt that the 88STA although a major departure from the past did open up new possibilities, and a new relationship which I hope could be fostered and strengthened and could become a positive watershed between the two countries for the sake of world science.

Notes

1. The data for this chapter came from interviews of more than 40 people and materials provided through the FOI process. Generally speaking, I found the State Department interviewees helpful in their responses. A similar comment could be made about the cooperation of the staffs of other agencies and Departments, e.g. Treasury, Energy, DOC's NOAA and Patent and Trademark Office, DOD, and OMB. OSTP was reluctant and provided minimum support. The U.S. Department of Commerce was the most belligerent and downright antagonistic. They were insistent on not commenting on or discussing the U.S. negotiating strategies for fear of revealing such to the GOJ – as if the GOJ did not know.
2. For additional details see sections IV.6.c.ii and IX.4.b and IX.8.e.ii on the Defense Patent Secrecy Agreement. In light of the postwar tradition in Japan of openness of all research and development outside industrial R&D, the restrictive stipulations of Annexes IV and V would be major points of contention with the GOJ.

3. This policy of exclusion was, it was reported, at the direction of the Science Adviser, William Graham, whose office was spearheading the push against the GOJ. Graham himself led a USG delegation to Japan in February 1987 to push the GOJ for openness of R&D facilities, the opposite of the White House conference. (*Far Eastern Economic Review*, 3/31/88:1958) The above was based on several press releases: The President's Superconductivity Initiative, 7/28/1987; text of remarks by the President to the Federal Conference on Commercial Applications of Superconductivity, 7/28/1987; press briefing by Science Adviser to the President Dr. William Graham, 7/28/1987; and three articles in the *Washington Post*, Superconductivity Conference to Exclude Foreign Officials: Scientists Warn U.S. decision could backfire on Nation (7/25/1987), Commercializing Superconductors (7/29/1987), The Superconductor Follies: Chauvinism, Defense Contracts and Secrecy won't beat Japan by Robert L. Park, director, Washington Office of the American Physical Society (8/2/1987). Earlier *The Washington Post* had printed a series of articles, 5/17-20/1987, Super-Challenge: The Race to Exploit Superconductors. These articles were symbolic of the great anxiety as to whether Japan would reach the marketplace first.
4. There were other cases where techno-nationalism was invoked: As mentioned earlier, the invitation to the Japanese (Fujitsu) in 1986 to consider taking over the ailing Fairchild semiconductor manufacturer was withdrawn by Fujitsu because of pressure by the U.S. for so-called national security concerns, although Fairchild itself when it made this invitation was owned by a French company. The USG had not objected to French ownership but did to Japanese. When the Japanese government Research & Development Corporation offered in August 1987 to support some basic research at Rockefeller University in New York City it was rejected. When MIT chose to buy a Japanese supercomputer in October 1987, it was forced to withdraw its offer because a U.S. funded R&D project should not use a Japanese supercomputer. Yet the USG pressured the GOJ to purchase many Cray supercomputers for its agencies and Ministries on Japanese government projects. These cases were mentioned in the *Far Eastern Economic Review*, 3/31/1988:58.
5. Over the following decade, technoglobalism and strategic alliances appear to have become more the norm than techno-nationalism except in a few technical areas such as supercomputers. In 1997 an informal memo was prepared in the Office of Technology (DOC) on U.S.-Japanese high-tech strategic alliances based on four articles on this subject: Elizabeth Garnsey and Malcolm Wilkinson, Global Alliance in High Technology: a trap for the unwary, *Long Range Planning* 27:6(1994):137-146; Tōru Sasaki, What the Japanese have learned from Strategic Alliances, *Long Range Planning* 26:6(1993):41-53; Peng S. Chan and Dorothy Heide, Strategic Alliances in Technology: Key Competitive Weapons, *Sam Advanced Management Journal* Autumn 1993:9-17; Terutomo Ozawa, Technical alliances of Japanese firms: an "industrial restructuring" account of the latest phase of capitalist development, *New Technology Policy and Social Innovations in the Firm* (chapter 7) London. Pinter, 1994. Also, a *JEI report*, Japanese Research and Development: an era of "technoglobalism", *JEI Report* 7/24/1992:28A, 20 p. provides yet another viewpoint on the this spreading phenomenon.
6. Two OES memos (both 9/22/1987) on the A/S OES meeting with Senator Rockefeller on September 22 [sic, should be September 23], and Background paper and talking points for meeting with the Senator on September 23. The briefing paper was undoubtedly given to the Senator although that is not specifically stated.

7. U.S. Congress. U.S. Senate Hearing before the Subcommittee on Science, Technology and Space of the Committee on Commerce, Science and Transportation, October 15, 1987, 38 p. S Hrg, 110-359. The three government witnesses were: Ambassador John D. Negroponte, A/S OES, Hon. Bruce Smart, Acting Secretary of Commerce and Under Secretary of Commerce for International Trade, Department of Commerce, Dr. William R. Graham, Science Adviser to the U. S. President. The other two witnesses who were generally regarded as highly critical of Japan were William C. Norris, chairman emeritus of Control Data Corp., and Clyde Prestowitz, formerly Counselor in DOC during the first Reagan Administration.
8. *Washington Post*, 10/28/1994.
9. OES memo (7/10/1987), John D. Negroponte to Science Adviser Graham, *Renewal of U.S.-Japan 'Non-Energy Agreement': State Comments on Draft CISET Paper for the EPC.*
10. *Japanese Scientific and Technical Information in the United States*. Report of a workshop at MIT January 1983. Report prepared by National Technical Information Service, U.S. Department of Commerce. March 1983. p. iv.
11. The following comments are derived from Tokyo 20495 (10/2/1985)
12. *The Survey of Doctoral Recipients* has been conducted biannually since 1973. The Survey was initially sponsored by NSF but by 1989 it was supported by the National Endowment for the Humanities, NIH, USDA, DOE. I was given data and reports on the 1987 and 1989 surveys. The released data was organized by surveys and each was organized by many factors, gender, race, with and without foreign research experience and by fields but not by country in a meaningful manner. The data on Japan about hindrances was specially prepared for me by the Fogarty Center. These documents are: *Doctoral Scientists and Engineers* (Draft) [Survey of 1989]. Fogarty International Center, NIH, 1/3/1992; *Advisory Board: Summary Minutes of 11th meeting*, 1/24/1989. Fogarty International Center, NIH; and letter (6/1/1989) to me from Dr. Coralie Farlee, Assistant Director for Planning and Evaluation, Fogarty International Center, NIH.
13. Memo, 3/19/1987, from NSF International Study Group Director to Deborah Wince, Assistant Director, OSTP on "Revised Background Paper on Foreign Participation in the United States and Japan."
14. This information was given to me during an interview with William H. Tallent, Assistant Administrator, Cooperative Interactions, Agricultural Research Service, USDA. One of the so-called anecdotal stories was that a Japanese S&E sent to Peoria as a post-doc returned to Japan with data from a federal laboratory and developed a new fermented soya product.
15. John P. Stern, Executive Director of U.S. Electronics Industry Japan Office to Joseph P. Allen, Acting Director, Federal Technology Management Division, Office of the Under Secretary for Economic Affairs, DOC. 5/25/1988. In an interview (11/30/1990) with Dr. Joseph Clark, Deputy Director, National Technical Information Service, DOC, he said that as a polymer chemists for 25 years he never had any problems of access to Japanese research laboratories.
16. Tokyo 17218 (9/8/1986).
17. The following paragraphs are a distillation of these six cables: Tokyo 03669 (3/3/1987), Tokyo 03863 (3/9/1987), Tokyo 03855 (3/5/1987), Tokyo 03933 (3/6/1987), Tokyo 04159 (3/10/1987), Tokyo 04994 (3/23/1987). Only those on reciprocity and access to R&D facilities will be discussed here.

18. NSF Tokyo Office Memorandum no. 92 (1/31/1986), *Directory of Japanese Company Laboratories Willing to Receive American Researchers.* Summarized in Tokyo 04994 (3/23/1987) This was later updated and distributed as NSF Tokyo Office Memorandum no. 124, *Survey of Japanese Corporations Hosting American and other Foreign Researchers.* Summarized in Tokyo 10341 (6/12/1987).
19. Access to Japanese Laboratories. Memo (3/5/191987) prepared by NSF International Division, Untitled memo (3/6/1987) with recommendations for the draft USG S&T agreement.
20. The above information was summarized from Tokyo 05008 (3/23/1987), Tokyo 06076 (4/7/1987). U.S. scientists headed the labs for bioelectronic materials and non-linear optics and advanced materials and the Frenchman, the molecular regulations of aging. Dr. Kevin M. Ulmer was then Director of the Center for Advanced Research in Biotechnology, a joint project of the National Bureau of Standards and the University of Maryland; Dr. Anthony Garito is professor of Physics at the University of Pennsylvania. The French scientist, Dr. Gabriel Gachelin, is an expert in the pathology and biochemistry of diseases of the aged such as lupus and senile dementia (Alzheimer). According to Tokyo 14391 (8/13/1987) Riken planned to provide housing on RIKEN grounds, travel expenses, stipends for 100 foreign scientists. The Embassy also reported that MITI had been conducting for three years (JFY1985-87) a feasibility study on superconducting power generation systems and MITI plans and hopes for international cooperation on superconductivity power generation and electronics (Tokyo 12227 (7/13/1987).

    Dr. Frank Press, President of the National Academy of Sciences, stated that the NAS had agreed with "Japanese intellectuals to promote a special project for expanded exchange ... [to begin] on a full scale in one or two years." (Tokyo 17465 (9/30/1987).
21. Tokyo 04957 (3/20/1987).
22. The formal name of the 1956 agreement is Agreement between the Government of the United States and the Government of Japan to facilitate Interchange of Patent Rights and Technical Information for purposes of Defense, signed March 22, 1956. This section is based, in the main, on an interview with the DOD official responsible for the early discussions with the GOJ about implementation procedures, and a binder of materials on this subject and other materials gathered by the writer.
23. The countries are Australia, Belgium, Canada, Denmark, France, the Federal Republic of Germany, Luxemburg, Greece, Italy, Japan, The Netherlands, Norway, Portugal, Sweden, Turkey and the United Kingdom. Implementation procedures to carry out these agreements had been worked out with each of these countries, except Japan.
24. These weapon systems are howitzers (BMY Co), F15J Interceptor Fighter (McDonnell Douglas), P-3C patrol aircraft (Lockheed), and the Hawk, Sparrow, Sidewinder and Patriot missiles (Raytheon).
25. At an interpellation session of the House of Councilor's Foreign Affairs Committee on April 19, 1983, then Prime Minister Nakasone Yasuhiro expressed unequivocally that the government did not plan to propose an anti-espionage law and that it would continue to use the National Civil Service Law and the Criminal Code to combat spying activities in Japan (*Sangi-in Gaimu Iinkai Kaigiroku Dai-7-gō/4/19/1983:16-19)*. Another attempt was made in mid-1985 when an anti-espionage was submitted to the Diet but was abandoned (*haian*) in December 1985, and not enacted.

26. The GOJ representative also pointed out at these hearings that another good point in the 1956 DPSA was that technical information owned by either Governments (USG or GOJ) will be available without charge to the GOJ, i.e. technical information on advanced weapon systems owned by the USG will be made available to the GOJ without charge. This was stressed as one of the important benefits of the DPSA. "This will make possible for the first time the obtaining of new and advanced knowledge, instead of the stale and outdated things up till now", said the GOJ representative at these hearings.
27. According to an interview with Stephen Piper (then with DOD), it had been a pleasure working as part of this small interagency group on the implementation agreement, in contrast to the reported bitterness and distrust which permeated the other USG working groups on the USJ S&T agreement.
28. The number of pending applications was based on materials given me for this section and do not completely coincide with those provided by the U.S. PTO as mentioned earlier. That there was an issue of some indeterminate value is undoubtedly correct. The reason for it becoming an emotional and intractable issue for a while in the USG is not clear. A GOJ delegate aware of these developments, speculated that perhaps some one in the USG (probably DOD or U.S. PTO) had made a personal commitment to one or more U.S. companies to press for early GOJ agreement on the implementation procedures agreement.
29. This comment is based on discussions with officials on the Japan Desk, Department of State and the U.S. Patent Office which is part of DOC; I was assured that a written note by the Secretary of State did exist but could not be shown or a copy given to me.
30. A substantial part of the implementation procedures are classified; only the unclassified portion was released to this writer by the DOD.
31. Because "the NIH Issue" pervaded so many of my interviews, I decided to meet with as many NIH (National Institutes of Health) officials as possible in 1988 who could shed light on this issue. They are listed in appendix 7 under NIH.
32. These descriptions are based on NIH pamphlets on these various programs.
33. Data on foreign scientists is derived from the official NIH report, *National Institutes of Health Annual Report of International Activities*, FY 1986 and 1987, and various NIH memos: FIC Director Craig K. Wallace, to Director on Opportunities for U.S. Biomedical Researchers to work in Japan, 3/16/1987, on NIH Cooperation with Japan, 3 p. 10/28/1987, Remarks by NIH Director, James B. Wyngaarden to NIH Alumni Association in Japan Symposium, Tokyo, 6/30/1987, letter from Alexandra Stepanian, FIC to Cecil H. Uyehara on 8/31/1988 and 10/11/1988. It is basically impossible to make all the statistics completely jive with each other. The overall percentages, however, are all in the same range.
34. Undated memo (soon after June 1988 when the new USJ STA was signed by FIC Director, Craig K. Wallace, The National Institutes of Health's Role and Interest on Biotechnology in the Pacific Basin, 1 p.
35. The NIH Director congratulated the Japanese on the formation of the NIH Alumni Association in Japan on June 30, 1987 in conjunction with the 100th anniversary of NIH. The Japanese Alumni Association, he noted, was the first such organization outside the U.S. But he did not avail himself of this opportunity to press for greater Japanese support of U.S. scientists in Japan. Obviously, it was not a high priority concern in NIH.

36. According to an email report to me on 3/12/1998, ironically Japanese researchers at NIH had increased in actual numbers above those of the 1980s but had decreased in percentage of the total: FY95 (10/1/1994-9/30/1995) of 2782 visiting scientists, 417 (15%) were Japanese, FY96: 2834:454 (16%), FY97:2615:479 (19%).
37. USG cables as released to me report on who said what, when in a somewhat dry manner, they do not provide how the GOJ institutionally and its staffers individually felt about the atmosphere surrounding these negotiations. The GOJ does not have an FOI process through which one could obtain the release of MOFA documents; the GOJ releases documents on foreign relations decades after the events take place. As is often the Japanese practice, senior members of a GOJ delegation prepare their own notes and interpretations of events to which this writer had access. These private notes are sensitively held since they were shared with me in confidence. These two sets of notes together with interviews provide the basis of understanding and appreciation of the GOJ position. These two documents and interviews will be referred to in this chapter without specific attribution. I was given access to these two documents with the understanding that the writers' names would not be divulged.
38. This non-paper, entitled *Initial Comments on the U.S. Proposed Text*, was handed to the acting science counselor at the American Embassy and given to the State Department by the Japanese science counselor at the Japanese Embassy according to Tokyo 17402 (9/29/191987). It appears these comments were prepared in English by MOFA. In addition, the positions of the GOJ officials as reported in several cables reporting on the "briefing session" and meetings with the GOJ leading to the early October meeting were deleted when released under the FOI process. Related cables on this subject are: Tokyo 15915 (9/8/1987), Tokyo 16459 (9/17/1987), Tokyo 16883 (9/22/1987), Tokyo 17198 (9/25/1987). Tokyo 17914 (10/7/1987). This MOFA document was found among a collection of State documents assembled for the "briefing" session and subsequent negotiations without attribution and date but based on the contents and English style it can be safely assumed to be a MOFA document.
39. Tokyo 17402 (9/29/1987).
40. These interpretations were obtained from interviews and personal notes. For example, the first three items were emphatically volunteered by a Japanese delegate at the beginning of the interview "as important points" to understand as background to the USJ negotiations. He had served in the Japanese Embassy in the U.S. in the 1980s, was an active member of the Japanese delegation.
41. According to a *C&EN* article (7/25/1988:20), Japanese delegates regarded the negotiations as "simply another example of America's popular sport of 'Japan bashing.' Some sources familiar with U.S.-Japan relations say the Japanese still feel that political, national security, and patent issues should have played no role in scientific cooperation between the two countries."
42. Tokyo 14510 (8/17/1987). There was also another article in the *Sankei Shimbun* of 8/13/1987 on this subject. According to the former cable, the Embassy believes these articles stemmed from interviews given by Hinata Seigi, Director, Scientific Affairs Division, MOFA, after his visit to Washington.
43. Tokyo 16127 (9/10/87).
44. The following discussion is based on the proceedings for these two subjects for these two days as reported in *Dai Hyaku Jūni-kai Kokkai, Shūgiin Yosan Iinkai Gijiroku Dai 11-gō, Dai 1-rui, Dai 15-*gō, p. 30, 30-34, same source but from *Dai 16-gō, p. 49, 10-12; Sangiin, Yosan Iinkai Kaigiroku, Dai 8-gō,* 3/16/1988, p. 14-22.

45. *Defense News* (1/18/1988) by Daniel Sneider.
46. Circular 175 Authority to renegotiate the 80STA with Japan for five years was granted on 9/10/1987. The request by Ambassador de Vos stated that the draft text would be given to the GOJ after the authority was granted; in reality the text had been given to the GOJ on 8/28/1987. An additional authority was granted on 10/22/1987 to extend the 80STA from November 1987 for three months to 1/31/1988.
47. State-250739 (8/13/1987).
48. Tokyo 15915 (9/8/1987) from meeting with Seiji Hinata of MOFA.
49. Diplomatic Note and Annexes and commentaries, about September 1987, p. 11.
50. ibid. p. 37.
51. ibid. p. 12.
52. ibid. p. 17.
53. In an OES memo (10/13/1987) to A/S Negroponte about a meeting on October 14, 1987 with Dr. John Moore, NSF is described as "the most successful at promoting collaboration between scientists of the [U.S. and Japan. Dr. Bloch [NSF Director] is especially "concerned about the need to maintain the separation between public and private sector activities". He does not believe that private sector activities should be affected by the S&T agreement.
54. ibid. p. 7.
55. Tokyo 15915 (9/8/1987).
56. Talking Points, 9/10/1987; "We expect negotiations to begin in Washington later in October and have indicated [to the GOJ] that we would like to conclude them in a single session". One session was defined as three days to be set aside for these negotiations during the week of October 12,[1987]. State 287801 (9/16/1987).
57. Memo, OES/SCT to OES/Negroponte, 9/22/1987, "Your meeting with Dr. Graham September 22, 10 a.m." OSTP wanted to know how, with de Vos as chairman, the U.S. delegation would respond to each and every GOJ concern, who was going to be involved in and how it was going to be decided that these negotiations are "unproductive." The memo seemed to exude with distrust between State and OSTP on the eve of the USG briefing and bilateral negotiations.
58. Tokyo 17193 (9/25/1987).
59. For example, Deborah Wince (later Wince-Smith) represented OSTP on the delegation. In the Bush Administration (1989-1993), she became assistant secretary for science policy in DOC. Prior to joining the OSTP she was a program officer for Eastern Europe at NSF. In commenting on a consortium proposed by Japan in 1989 to establish international manufacturing standards, she is quoted as saying that Japan will use this program to raid U.S. technology. She also posed the question: "I frequently ask people to give me an example of a significant technology transfer from Japan to the U.S., where they gave us a strategic technology advantage in a world-class product or service ... no one ever has", she says. (*Insight* 6/4/1990:36-7). According to a *C&EN* article (7/25/1988:19), the "U.S. Team was sharply divided" on linking trade, security and basic research, but the "trade/security faction won out". It was this philosophy that underlay the approach of OSTP and probably DOC and others in the negotiations with Japan about future S&T relationships.
60. According to a *C&EN* article (7/25/1988:21), "Indeed, one American close to the negotiations complained of 'racist' attitudes among some of the more 'hawkish' negotiators. 'That was very, very sad to see.'"
61. *Newsweek,* 2/20/1989:30.

62. This description of the role of the Embassy and the U.S. Teams was based on an interview with the U.S. minister-counselor in Tokyo 1986-1990. It underscores the accounts that appeared in the *C&EN* articles quoted earlier.
63. Tokyo 255558 (12/17/1984).
64. In a study of MOFA in the 1970s, Haruhiro Fukui had already described the rise of a restless, younger group of officials who were "impatient, defiant [with a] cocky posture of self assertion, with heavy nationalistic overtones" (p. 35). The manner in which these musketeers dominated the negotiations, well underscored Fukui's findings that in MOFA (and, I would add, other ministries too) *kachō* are given "considerable discretion in decision making", thus "A powerful man" whose true test of effectiveness was "his ability to work with and not against his bureau director" (p. 13). Policy-Making in the Japanese Foreign Ministry, *The Foreign Policy of Modern Japan,* Robert A. Scalapino, ed. University of California Press. 1977.
65. Glen Fukushima, formerly with USTR, The Nature of United States-Japan Government Negotiations, in *United States Law and Trade*, ed by Valerie Kusuda-Smick, Transnational Juris Publications, Inc. These are from a paper given at a conference in Tokyo in 1988. These comments, therefore, probably applied to the USJ S&T negotiations which were proceeding in the first half of 1988.
66. Tokyo 18692 (10/19/1987). The *Nikkei* article also reported that the NAS had given the GOJ a draft plan to improve the balance of exchange, i.e. symmetrical access by 1) creating an organization to make public all research findings and share them equitably, 2) re-assessing systems for copyright, patents, IPR which are obstacles, and 3) creating a data network for access immediately on demand. I tried to verify the existence of this "plan" with the NAS but could not do so with certainty partly in light of NAS's rule that project related materials cannot be released for 25 years. So much for openness.
67. U.S. - Japanese S&T Negotiations, 11/6/1987. This was a short note on the status of negotiations for a senior State official's visit to Korea and Japan.
68. This point is mentioned in the private papers and was also mentioned to me in interviews on a number of occasions. While the fairness or accuracy of this observation cannot be ascertained, it was obviously a quite strongly felt perception of the situation which acted as a hindrance in negotiations.
69. Tokyo 20938 (11/25/1987).
70. Tokyo 21015 (11/27/1987) provided a translation of this article.
71. For example, a series of nine memos and papers were prepared as follows: the impact of the expiration of the S&T agreement, an overview of the government-to-government S&T relation, a brief description of USG supported R&D in the private sector, a description of GOJ supported R&D in the private sector, a rationale of the USG choice of priority areas, a list of the recent GOJ steps in opening their R&D to foreigners, U.S. programs to teach Japanese to scientists and engineers, what are the relative benefits of each side paying the salaries of American researchers at Japanese labs, impact of the U.S.-Japan S&T agreement proposals on the private sector, active projects under the US-Japan S&T Agreement (11/1/1987).
72. Tokyo 20762 (11/20/1987).
73. "Opening Statement" (12/2/1987) p. 2 and 5.
74. Tokyo 21596 (12/7/1987), unnumbered cable Tokyo to Washington, 12/8/1987 on Japanese morning press highlights, December 8.
75. Quotes are from Tokyo 21703 (12/8/1987).

76. The US treaty lawyers are given substantial credit by the GOJ for playing a constructive role in creating the new agreement which was ultimately signed in June 1988. The USG and GOJ staffs worked quietly together to craft this new agreement. The USG staffer felt more nervous in negotiating with emotional fellow U.S. delegates. While the GOJ questioning deepened the suspicions of the USG delegation, the legal representatives pointed out that the GOJ needed to be given consistent interpretations, not incoherent, inconsistent and emotional responses.
77. Tokyo 22988 (12/30/1987).
78. Tokyo 23033 (12/30/1987).
79. Memo, U.S.-Japan Science and Technology Agreement, 12/29/1987, one page.
80. Memo, OES/SCT (S. Libicki) to OES (Ambassador de Vos), Telecon with Gene McCallister of the EPC regarding the U.S.-Japan S&T Agreement, 12/31/1987. one page.
81. This writer was informed that the USG had made a study of GOJ laws and regulations concerning classification, but no one volunteered to give me a copy (even by those who accused the GOJ of bad faith), nor did a copy of this presumed study ever appear through the Freedom of Information process.

    As mentioned earlier, Japan does not have an anti-espionage law. Article 100 of the National Civil Service law requires those in and out of the Japanese government who possess secret information to maintain the secrecy of that information and not divulge it without the permission of the government. Various examples of the kinds of information that would need to be kept confidential and secret are information concerning weapon systems capabilities, secret codes, certain kinds of financial information related to national defense, diplomacy, international economics, tax, medical, and financial data on individuals, data on government investigations, the courts, contracts to be let, activities that must be kept secret for a specific length of time, etc. No examples are given for the need to keep information secret stemming from R&D projects within Japan or joint projects with other countries. If such projects were funded by the self-defense forces, then the results would be governed by procedures emanating from the USJ Security Treaty. But to classify information or equipment stemming from a civilian funded project may theoretically be classified but the process to succeed in classifying such information or equipment would undoubtedly become a highly sensitized political issue. Based on the above, Japan does have a law, the national civil service law which covers "secret or confidential information", but it is not intended to cover the results of joint R&D projects; thus, Japanese can legitimately assert that Japan does not have a legal system (*hōtaikei*) to handle such cases. This interpretation is based on a reading of Himitsu o mamoru gimu, *Hōritsugaku Jiten (3rd edition)*. Tokyo. Yūhikaku, p.1200, Himitsu o mamoru gimu, *Chikujō Kokka Kōmu-in Hō*. Gakuyō Shobō.1988. p. 830-841.
82. The Japanese chairman insisted, contrary to the USG description, that the MOFA did not agree to the inclusion of this clause in the agreement.
83. The above was based on AmbEmb translation of two *Nihon Keizai Shimbun* articles in the evening edition of 1/7/1988 (p. 1 and 3) in Tokyo 00389 (1/11/1988) and 1/9/1988 in Tokyo 00490 (1/12/1988). In the eyes of the GOJ, the USG, in order to obtain an agreement for the USJ summit only a week away, took the high ground, refused to compromise, and adopted an all or nothing stance in demanding the GOJ accept the USG proposal.

84. If such a threat was made it would not be an idle threat. The FSX agreement negotiated by the Bush Administration with Japan in 1989 was revised at the insistence of the U.S. Congress. *Nihon Keizai Shimbun*, 1/8/88 Eve ed. p. 3.
85. A rather inexplicable newspaper account on the alleged withdrawal of the security clause by the USG appeared in the *Tokyo Shimbun*, February 14, 1988, a week before the next session starting on February 21 but which was postponed to early March. This article, based on an interview with a GOJ official, reported that the USG had withdrawn its request for the inclusion of a security clause at the 4th session in early January 1988. This was described as "a big step forward in the direction of a settlement" suggesting that an agreement in general outline could be reached at the next round of negotiations to be held in Tokyo. This occurred as a result of GOJ explanations of its positions and because the USG showed "understanding" of Japan's position concerning the lack of a law on classification. There was, however, no confirmation of this withdrawal on the USG side and indeed this issue remained contentious to the very end. This is cited here to show the apparent misunderstanding that had occurred during these negotiations. The article was translated by the Embassy and sent to State without comment by Tokyo 02781 (2/18/1988).
86. Memo (12/8/1987) OSTP (Wince) to ten USG agencies on Japan negotiations.
87. Letter (12/18/1987) OES (de Vos) to Wm Graham, OSTP. This sentiment was also supported by the technical agencies of the USG.
88. Letter (12/28/1987), John C. Whitehead, Acting Secretary of State to Science Adviser Graham.
89. Letter (1/7/1988) OSTP Graham to John [Negroponte].
90. Letter (1/7/1988) from Dr. Graham to Heads of Departments and Agencies of the USG. The replies to OSTP were largely a formality assuring OSTP that they would abide by their wishes. DOC's National Bureau of Standards ended their letter with "NBS has always found Japanese labs open to our staff." USDA's reply raised a question of interpretation of the USG draft in connection with comparable emphasis in cooperative projects in one section (1.6.2) and its possible inconsistency with the preamble where it stated that one country (meaning Japan) may need to make greater commitments in particular components to the R&D process, etc. This could imply that the strength of one country in a particular technology could not be balanced off with another technology in the other country. It gave the example of engineering scale-up of bio-processing in Japan with new laboratory-scale biotechnology procedures in the U.S. It also raised some question about selected parts of the copyright clauses in the proposed agreement. NSF informed OSTP that it was proceeding with a Circular 175 authority to allow 30 U.S. researchers to be assigned to AIST labs in Japan. The Secretary of Housing and Urban Development took this opportunity to warn that "During the next decade, the U.S. home building industry will have to respond to Japanese manufactured housing techniques as these approaches inevitably begin to show up in the market."
91. The quotations in the preceding two paragraphs were from a 5-page document, *Talking Points for January 12 [1988] NSC/EPC Meeting*. Under normal circumstances the President would preside at this meeting. The list of attendees was not provided as distinct from denied under the FOI process. Another shorter 3-page version (presumably earlier version) included fact sheets on 30 topics on our relations with Japan, including, of course, the STA, but these sheets were not available. Often such appendices were not available, had become detached from the main document.

State's memo (1/6/1988) to NSC's Colin L. Powell on *Talking Points for the President's use in his meeting with Prime Minister Takeshita on January 13, 1988,* described the negotiations as having made progress although difficult issues remain and that a new STA would be "an important achievement for both nations."

92. When he visited Washington, Takeshita also announced a plan to create "Japan Science Fellowships" for U.S. researchers, to mollify the U.S. and to respond to U.S. demands for greater access to Japanese R&D facilities.
93. The following is based on a cable drafted by S. Libicki, OES AAAS Science Fellow. It was probably sent to the Tokyo Embassy but the numbered copy was not provided to me.
94. State 38907 (2/9/1988).
95. The *Nihon Keizai Shimbun* (2/9/1988) reported in a fairly extensive article that there were two sets of parallel USJ negotiations being conducted on revising the 80STA and implementation procedures of the patent secrecy agreement of 1956 with apparently no knowledge that the USG was trying to make the STA a hostage to a GOJ agreement to implementation procedures for this agreement. The article did not point out that the USG was very eager to prevent any leakage of technology to the Communist countries and for this purpose was doing its best "to throw controls" over Japanese R&D results and their dissemination.
96. Letter (1/25/1988) sent from SecCommerce, USTR, Science Adviser to SecState Shultz. As mentioned, a letter or letters were sent to the Japanese Foreign Minister and he replied but these letters were not released through the FOI process.
97. The above short account is based on the OES memo (January 25, and revised January 29, 1988 requesting Circular 175 authority to extend the 80STA an additional two months. The working group's menu of options was not made available through the FOI process.
98. The following is based on a cable drafted by S. Libicki, OES AAAS Science Fellow. It was probably sent to the Tokyo Embassy but the numbered copy was not provided to me.
99. 1) Memo (2/12/1988), OES to Whitehead, Deputy Secretary of State, Calling Secretary Verity regarding U.S.-Japan S&T cooperation. This memo includes the above analysis of viable patents and explains the positions of DOC, USTR and OSTP and the GOJ position on the implementation procedures agreement. The tabs to this memo included copies of the joint letter to the SecState but these letters were blotted out so completely in the State FOI release process that not even the date, the senders' names, nor that of the recipient were allowed to stand. A few years later, the DOC released the three agency letter of 1/25/1988 to SecState Shultz in entirety.
    2) Memo (2/2?/1988), OES/EAP/EUR to Derwinski, Your meeting with Dr. William Graham at 2:30 on February 10 [1988] to discuss the U.S.-Japan S&T Agreement and the U.S.-Soviet Basic Sciences Agreement. This set of documents like those in 1) above was extremely heavily deleted even though the rationale for these viewpoints had been fully released.
100. Tokyo 22225 (12/16/1987).
101. OES/EAP memo(1/29/88) to the Secretary, Reply to Joint Letter from Varity, Yeutter and Graham on S&T and Patent Secrecy Agreements with Japan. The purpose of the memo was to seek the Secretary's signature on a response to DOD/OSTP/USTR's joint letter of January 25, 1988.
102. ibid.

103. ibid.
104. The number of patents was provided by the office in the U.S. PTO which handles defense patent secrecy orders.
105. OES-Whitehead (Deputy Secretary) memo (2/22/1988), Update for your breakfast with Secretary of Commerce Verity Regarding Japan S&T Agreement.
106. OES-Whitehead (Deputy SecState) memo of 2/12/1988 on calling the Secretary of Commerce about the USJ S&T negotiations.
107. This data was obtained from an article in *Tokyo Shimbun* February 14, 1988 on these negotiations.
108. The scenario description was based on State 060048 (2/26/1988), U.S.-Japan S and T negotiations. That the USG delegation's departure was delayed derives also from an OES weekly highlights report to the SecState of February 11, 1988 which states that "despite your [the SecState's] response to Verity, Yeutter and Graham that, because of the recent Reagan-Takeshita commitment to a new S&T agreement, you believed we should continue the negotiations and asked for their support, they have refused". OES memo (2/11/1988) OES Highlights, February 8-12 [1988]. The other comment about the SecState was based on an interview with a State staffer who also showed me a SecState letter to this effect but would not give me a copy.
109. Based on notes prepared for a USG Delegation Meeting on the USJ S&T Agreement on February 25, 1988. This was an interesting memo only part of which was released to me; the other parts were not denied, they were just in that folder, so to speak. The released part of 3.5 pages of bullet notes concerned a USG interpretation of the GOJ position and how it differed from the USG.
110. Tokyo 04212 (3/9/1988).
111. Memo (7/15/1987), OES/SCT Reifsnyder to OES/OCT Prochnik, Draft IPR Annex for Japan S&T Agreement.
112. The NSF analysis was based on January 29, 1988 and February 26, 1988 NSF memos to the USG delegates on the legal logic of this issue. A draft CISET report of June 1987, listed the U.S.-PRC STA of 1979 as a model to be considered in formulating the USG approach to Japan. The IPR annex to the U.S.-PRC agreement was signed in May 1991, seven years after the expiration of the this agreement; in comparison to three years in Japan's case. The IPR Annex stipulated that disagreements would be settled by binding arbitration. The stipulations were similar but less detailed; like the USJ agreement was also basically on a case by case basis. All rights and interests stemming from joint research would remain in the researchers home country and those in third countries would be determined in the implementing arrangements.
113. Tokyo 04458(?)(3/11/1988), Tokyo 04818 (3/17/1988) both transmitted same translation of an article in the *Nihon Keizai Shimbun*, 3/9/1988, evening edition. In the latter cable the Embassy said it was unable to identify the U.S. source who had made these comments.
114. This interpretation was derived from the *Nihon Keizai Shimbun* article (3/15/1988) on the Japanese inter-ministry conflicts.
115. Tokyo 04787 (3/17/1988).
116. There is a dearth of reasonable materials for this period. State has not released any documents for this last session which describe what transpired. There are, of course, press releases about the USJ agreeing in substance on a new agreement. To a large extent, this section has relied on interviews and the private notes of two officials involved in these negotiations, and newspaper articles.

117. According to one of the MOFA delegates, the text of the new STA was the result of quiet backstage discussions and negotiations among mid-level staff from MOFA, State's OES and Legal Office while the senior delegates thrashed out the "big and thorny" issues. A few working level staffers on both sides worked together to transform the draft USG agreement into its finally agreed upon format; this resulted "in working around" certain U.S. delegates.
118. State 081506 (3/16/1988).
119. According to the Japanese chairman, Ambassador Endō, there was, nevertheless, considerable anxiety about the implications of the possibility of the U.S. side classifying certain information or equipment stemming from joint R&D work. This classification would not be valid in Japan. The Japanese engineer/scientist involved could find himself/herself subject to prosecution if he/she visited the U.S. To prevent this from happening a "Record of Discussion" was prepared jointly on March 30, 1988 specifying that such Japanese researchers could return to Japan, publish this information but would not be subject to U.S. prosecution if this researcher visited the U.S. in the future. This is based on a conversation in January 1993 with Ambassador Endō. This "record of conversation" could not be found in the State Department's treaty files on this agreement. The GOJ also has not produced this record of conversation. Although it has not become available it is a logically necessary document in light of how the final agreement came to be drafted.
120. According to one Japanese interpretation, the USG was unable to counter this usage of the U.S.-Japan security in the STA negotiations which led them to accept the side letter idea. According to a State document dated "2/3/88" which could mean February 3 or March 2, the idea of a side letter and a proposed text had been discussed earlier but the wording was not agreed upon and thus delayed. The content was changed from not providing classified information or equipment for projects under this agreement to the idea that any such information developed from cooperative activities would be protected according to each nation's laws and regulations.
121. State Legal Office to OES, 7/18/1988 on the recently signed agreement. This was State's legal response to OSTP's assertion that the two side letters "constitute an integral part" of the 88STA. (Memo OSTP's Wince to State/OES 6/29/1988 concerning the public release of the 88STA).
122. The 80STA was extended for the fifth time for three months from April 1 to June 30, 1988 to allow time for legal editing of the agreement and to find an opportunity for the Principals to sign the agreement.
123. This short section is based on an internal DOC memo from the Secretary to Heads of all [DOC] Agencies on the "Agreement between the United States and Japan on Cooperation in Research and Development in Science and Technology", 3/22/1988.
124. The Japanese word used was *kensei*, which is translated in the *Kenkyusha J/E dictionary* as check, restrain, curb, constraint, contain, rein in.
125. Tokyo 06679 (4/14[sic]/1988).
126. To the best of my knowledge the GOJ has not asked for USG investment in their projects, large or small.
127. This observation was confirmed by the foreign service officer covering STAs in the OES and a personal perusal of many STAs between the US and other countries.
128. In an interview with a patent attorney, the IPR provisions were described as unduly detailed for a President-level international agreement. It read like a business licencing agreement. A Chemical & Engineering News article (7/25/1988, p. 21) quotes sources

in the State Department and NSF did not think that the national security and particularly the IPR clauses were necessary. The rhetorical question was raised: How many patents have ever come out of any these agreements? Negotiating the intellectual property clauses in the agreement was a waste of time.

129. *The Sankei Shimbun* (1/12/1988) reported that a Japanese government source also felt that it would not be "a big problem" if the 80STA was allowed to lapse, just a "few obstacles" to overcome in the exchange of researchers and cited the USG demand for a security clause in the STA as the principal cause for this attitude.
130. This description was adapted from the title of another book called *The Textile Wrangle* by I.M. Destler, et al. Cornell University Press. Ithaca, NY. 1979.
131. The two subcommittees are 1) Science, Research and Technology, and 2) International Scientific Cooperation of the House Committee on Space and Technology. The chairman of the latter subcommittee was Ralph M. Hall. This subsection is based on the *Hearings, U.S./Japan Science and Technology Agreement*, No. 148. This writer attended this hearing in person to savor the give and take and the atmosphere of the hearing.
132. Bernthal in providing responses to additional Congressional questions stated that the IPR language in the 88STA was considered "to be a good example" of how IPR could be handled in STAs. (P. 39 of these hearings).
133. Hinata, Seigi. *Institutional Features of the U.S. Negotiating System*. Center for International Affairs, Harvard University. June 1989. 64 p. The table of contents reflects the kinds of issues with which Hinata was confronted in negotiating the 88STA: U.S. Presidential authority to negotiate treaties, international agreement other than treaties, the discretion to decide whether or not an international agreement constitutes a treaty, locus of negotiating authority within the Executive Branch, and Presidential capacity to preserve the negotiated agreement which had been agreed upon. Subsequently, Hinata became DCM at the Japanese Embassy in Mexico.
134. *Look Japan*. October 1988. p. 42-43.
135. Michael Blaker, Probe, Push and Panic: The Japanese Tactical Style in International Negotiations (p. 55-101), *The Foreign Policy of Modern Japan*, University of California Press. 1977.
136. The GOJ proposal was never released or specifically denied to me in the FOI process. It appears as "a phantom document" which was commented on in USG documents but "the body" (the proposal itself) was never in evidence. If it had appeared, probably the State Department, as a courtesy to the GOJ, would have denied release of the document.

X. Act V, Implementation and Renewals

Before the ink had hardly dried on the new STA -- i.e. the day after the Principals had signed the agreement -- the U.S. science adviser, William Graham, called for meetings of the three-tiered management structure, the joint working level committee, the joint high level committee (the so-called Ministerial level meeting), and the joint high level advisory panel, to move forward in the materialization of the objectives of the 88STA. He took a strong and threatening stance toward Japan about the implementation of the agreement. He asserted that it is necessary to devote more energies to the materialization of cooperation particularly in the selected fields itemized in the 88STA and promptly. He warned Japan that if it did not cooperate, move to correct the imbalances that this problem might resurge in the U.S. Congress -- a standard ploy to pressure Japan and other countries into submission -- and that, although "there is no enforcing or compulsory mechanism per se in the STA", there are several levels of committees which would be watching Japan's conduct and pressure her to comply. In addition, he called upon the GOJ to 1) work with the USG to carry out cooperative R&D in the various technical fields specified in the 88STA, such as new materials and the life sciences, 2) take the lead (*shudōken*) in international R&D cooperation and reach a final decision on planning the GOJ proposed Human Frontier R&D program based on the economic declaration at the summit meeting and 3) reach an early decision to support the U.S. big science project, the Superconducting Supercollider (SSC) with substantial funding assistance.[1] In light of the decidedly contentious and bitter negotiations in which Graham's office, the OSTP, played a large role, this was an ominous beginning to the implementation period. A more benign and philosophical approach with a biblical twist was given by Ambassador Endō, the GOJ delegation chairman: the new STA is "merely a new leather bag. This issue is what kind of wine will be put into the bag."[2]

1. Implementation

The 88STA created a complicated series of interrelated committees, task forces and liaison groups to implement the S&T agreement. The Joint Working Level and the Joint High Level Committees met for the first time on September 14-15, 1988 and October 11-12, 1988, respectively. They endorsed the creation of the Task Forces on Science and Technology Information, and Access to major government-sponsored and government-supported R&D and six liaison groups on advanced materials, manufacturing and automated process control, life sciences, joint data bases, information science and technology, and geoscience. The STA actually suggested a slightly different array of seven technical areas in Annex I which could be modified by mutual agreement. The newly appointed Joint High Level Advisory Panel met on January 17-18, 1989 in Tokyo. Let us recap the principal functions of the three-tiered management structure, as described in Article IV.

The Joint High Level Committee (JHLC). It will serve as the annual forum to review and discuss, under the policy framework of the STA, S&T policy issues between the U.S. and Japan and cooperative activities under the Agreement, prepare an annual report to both governments, and set forth steps and new initiatives for the next year.

Joint Working Level Committee (JWLC). It will provide the staff support for the JHLC, prepare recommendations for the JHLC, prepare an annual report to the JHLC, report on projects outside this agreement and appoint task forces on access to STI and to R&D facilities.

The Joint High Level Advisory Panel (JHLAP). It will advise the JHLC on the overall science and technology relationship between the U.S. and Japan, make recommendations concerning major advances in R&D, the mechanisms for S&T cooperation and their effectiveness, and on how to access research and training opportunities, facilities, expertise, data and results.

According to the Agreement, each of the three tiered committees was to meet annually starting from the Joint Working Level Committee and Joint High Advisory Panel to the Joint High Level Committee. While this arrangement basically held for three rounds (1988-91), the cycle from the JWLC to the other two began to stretch out from one year apart to two to three years apart as shown in the table (in appendix 4) on the cycle of committees meetings. The concept of the annual meeting was central to maintaining the tempo and pressure of the S&T dialogue -- especially on the Japanese -- now cut both ways. Senior officials of both governments would have to find the time and sense of priority to set aside and prepare for and attend these meetings. Obviously, the lack of time, priority and the felt need for such an intense dialogue became quickly evident. Pursuit of such intensity soon became a farce and unsustainable from an administrative point of view and probably quite unnecessary from a substantive point of view also.

a. JWLC/JHLC Activities

Before the STA was renewed in June 1993, the JWLC had met 5 times, the JHLC four times. Based on a reading of all the reports of these two committees, it was clear that the real work was conducted by the JWLC and its actions and recommendations ratified by the JHLC. The following commentary was derived from the reports of these two committees.

Based on the sanitized "annual reports" of these committees, the atmosphere was cordial not confrontational, in a way self-congratulatory, and in the JHLC it was an elaborate political "show'n tell" operation. It seems a substantive part of the JHLC was spent in explaining each other's S&T policies, programs, projects and various S&T activities, and ratifying the recommendations made by the JWLC. The initial meetings were a frenzy of activities being done and set to be accomplished in the next year -- with most of the suggested activities coming from the U.S. side. For example, some of the listed activities were how to facilitate access to gray literature in Japan and the U.S.[3], ways to expand U.S. access to JSTI in the long term [but the reverse was not mentioned], ways to increase U.S. utilization of the existing Japan Fellowship program [again, the obverse was not mentioned]; evaluation and assessment of various categories of cooperation under the STA (government-to-government, agency-to-agency MOUs, activities not covered by written agreements, multilateral agreements, non-governmental cooperation). Gradually this pace slowed as the JHLC met over the years; the reports were filled with such phrases as: presentations were given ..., so-and-so stressed the importance of ..., we discussed the issues of ..., it welcomed the proposal to do ..., heard a report on ..., appreciated the significance and candidness of ..., i.e. one could not find evidence that the JHLC had initiated or even approved an original major policy departure or accomplishment, or a Presidential/Prime Ministerial program of significant national import to both countries. The work of the JWLC did not slow down; it remained fairly relentless as it obviously tried to meet as close to an annual basis as possible.

Generally speaking the Japanese side was passive in its presentation; it did not touch on trade issues and S&T and their political interconnectedness, IPR, access to R&D and STI. It was in general the U.S. side that expressed its concerns about these kinds of issues. As early as the second meeting in 1990, the U.S. science adviser, Dr. Alan Bromley, commented that the two countries had the greatest opportunity and challenge in coordinating and managing big science. The USG stressed that the research from experience gained from such projects as the SSC, the Space Station, Human Gnome Project, ITER, Global Change would add to the knowledge base of all countries, that both sides may wish to consider a format for approaching large scale science initiatives, that those interested in participating in these projects should be involved from "the very early stages so as to eliminate potential problems" (The U.S. had not involved Japan or the Europeans in the early stages of SSC planning, but later pressed the GOJ for funding support).[4] The USG called for a discussion of this vexing issue at the 1991 meeting. At the 3rd JHLC meeting in 1991, the GOJ raised the unreliability of the U.S. in supporting large scale international S&T projects.[5] Ambassador Ōta Hiroshi, the Japanese chairman, announced at the 4th JHLC (5/93) that it would be difficult for Japan to support the construction costs of the SSC because Japan at the urging of foreign countries meaning the U.S., had decided to improve the infrastructure of basic science in Japanese universities. He expressed GOJ's expectation that the USG would continue to do its best to secure stable funding for the success of the ITER international project.[6] At the 6th JHLC (1994) meeting there appeared to be gentle sparring by the GOJ and USG as seen through the sanitized reports: the GOJ pressed for early participation from the beginning especially on budget issues and the importance of the OECD's Megascience Forum where Japan could more easily work with the Europeans vis-a-vis the USG, in diplomatic language the GOJ warned the U.S. that depending on how Russia joins the space station program, it could have "negative effects on Japan" and asked the USG to "make maximum efforts for steady progress of this program [meaning I believe adequate consultations and stable provision of the U.S. financial contribution] lest the confidence of the partners is damaged". The USG defensively expressed its appreciation for Japanese patience and assured that Russian participation would "give technical and political benefit to all parties." The big science issue continued to loom large in these high level meetings.[7]

The JWLC/JHLC committee meetings obviously served as useful fora to discuss S&T policy issues, coordinate actions, listen to brief reports from the various committees, task forces and liaison groups, ratify the recommendations of the JWLC, but the time required to manage the JWLC, preparations for the JHLC meetings and the time necessarily consumed by these meetings seriously impacted on the schedules of the senior officials of both governments. They were indeed a far cry from the July 1988 House hearings where it was determinedly, vigorously, aggressively and pointedly stated that the USG would pursue the GOJ "to the last yen" spent in its national R&D programs. The various meetings appear to have been in the beginning a vortex of frenzied activities at the strong urgings of the USG, this seems to have gradually changed and receded so that the USG was finding itself more on the defensive especially in the arena of the management and funding of big science projects. It was a polite discussion, especially at the JHLC meetings where it was recorded that both sides congratulated each other on the accomplishments achieved in the one-day sessions.

b. The Joint High Level Advisory Panel

The concept of obtaining "private sector" constructive and useful input into the USJ S&T relationship was an important expectation of the USG in insisting on creating this Panel. The Panel was to consist of senior corporate officials and academicians from the U.S. and Japan. In 1996, the U.S. panel consisted of six from industry and two academics; the Japanese panel had three from industry (but two were formerly with the government), two from academia, two from research institutes (each were formerly with the government and academia). The makeup of the two Panels was thus distinctively different. The U.S. panel appears to be more genuinely "private sector" represented; the Japanese Panel had a clear and distinct connection with the GOJ, so it was far less "private sector-ish" in appearance and mode of dialogue than the U.S. Panel. It should be remembered that the GOJ felt that since this STA was a government-to-government agreement, private sector involvement was not necessary. In any case, the GOJ and the private sector are known to cooperate and coordinate their activities to a greater extent than in the U.S. With that philosophy in mind, the composition of the Japanese Panel is understandable. The Panel was supposed to meet annually; it met fairly regularly, four times before the STA was renewed in 1993 and twice since (end of 1996) in various cities in the U.S. and Japan. It always met in between the JWLC and before the JHLC with the hopes that its ideas would be reflected in the deliberations of the JHLC.

The Panel is "supposed to" (in the eyes of the State Department) 1) identify important issues in the USJ S&T relationship, 2) review major advances in R&D from which to suggest new areas for collaboration, 3) review the USJ S&T mechanism for effectiveness and strengthening, 4) identify how comparable access to STI and R&D facilities might be enhanced and make recommendations in these areas. The first meeting in 1989 recommended programs in global environment, advanced materials in high temperature superconductivity, advanced ceramics and polymer-based materials, safety in aircraft, railroads, and nuclear plants and a joint study in science and engineering education in Japan and the U.S.[8] One of the outstanding recommendations of the Panel was the creation of the "Summer Institutes" which have been most successful and well received. These recommendations were "acted upon to some extent". Five years later, the MOT and DOT signed an agreement on investigating safety and six superconductivity projects were listed among the 150 projects under the 88STA in 1995. It is difficult to prove any connection between the recommendation and the specific projects.

At the second through fourth meetings (1990-1993), the Panel members generally briefed each other on their respective S&T policies and discussed access to STI and researchers in Japan and megascience issues. As was the case in the JHLC, the Japanese "complained about the U.S. being an unreliable partner in large scale projects" at the 4th meeting in 4/1993; this feeling was re-emphasized by the US cancellation of the SSC in Fall 1993. Based on a reading of the JHLAP reports it was clearly a pleasant almost social gathering where no major issues were discussed with energy nor any major recommendations of great import were made -- the above mentioned paper noted the difficulty to keep track of the execution of the Panel's suggestions. With the renewal of the 88STA in 1993, the USG attempted to revitalize the Panel by giving it an "unofficial role" to advise the USG on suggestions for projects which would be of benefit to the U.S. private sector, its access to Japanese technologies, specific comments on particular technology or even institutions. This new role appears to stem from U.S. Panelists' feelings that the dialogue was not a private sector-to-private sector dialogue but one tinged with a GOJ voice. The USG was now going to transfer the substantive dialogue to the JWLC and JHLC, introducing the USG voice to the so-called "private sector dialogue". This small evolution implied that the JHLC would be a more effective indirect venue to involve the U.S. private sector in discussions with the GOJ; the implication was that the JHLAP had not proved a route for effective dialogue with the Japanese.

The Panel members obliged by providing a four-page nine point memo to the science adviser on December 22, 1994. Its first and foremost request was that the USG be "sensitive" to ... competitive concerns in the design and implementation of all cooperative S&T programs with Japan."[9] The first and foremost concern here was that if the Japanese proposed a joint project for the National Information Initiative (NII) that the U.S. representative to the JHLC clearly explain to the Japanese that U.S. companies have access to the Japanese telecommunication market on par with Japanese companies. If that does not happen and discrimination occurred -- as judged by the U.S. -- then that "would likely adversely affect cooperation in this field and could also lead to renewed trade tensions". The panel also called for the JHLC to seek assurance that "Americans are invited to participate in all Japanese NII-related standards and protocol settings to insure inter-operability, compatibility and the ability of U.S. firms to compete for supply contracts." Based on available documents, this is the first time, that such a close and tight commercial connection was drawn for U.S. participation in a potential joint USJ R&D project. This same concern about IPR stemming from joint R&D projects with Japan is stressed again in two other items in the report, under cooperation with Japan on "global climate change, the environment, etc ... The Japanese do understand this dimension and will act accordingly." It sounds almost that U.S. companies are not sensitive enough to this issue and urgently need the backing and support of the USG to protect them from the Japanese. In connection with IPR it also, under another item stated that "no opportunity be missed to emphasize [to the GOJ] the importance"of the IPR issue "in the JHLC context" and commends the USG to improve IPR rules when U.S. companies are involved in joint projects, such as the Civil Industrial Technologies project stemming from the Common Agenda of the USJ Summit meeting.

While the Panel supported the increased Japanese effort in basic research, one panel member in the biotech industry saw "a barrier in the Japanese university support system which makes it impossible for a U.S. company to provide research support for a Japanese university laboratory in return for intellectual property rights to any resulting inventions."

The Panel "got the impression that there is a great unsatisfied need and demand in the private sector" for JSTI but candidly admitted that its own attempts to verify this did "not lead to the same conclusion". It pointed out that much savings could be achieved if it is true that the current collective effort is "as vast as it appears and the impact as small as our simple survey suggests." (see the subsection on access to STI).[10]

In regard to access to R&D facilities in Japan, the Panel found that such programs as the Manufacturing Technology Fellowships administered by the DOC and the NSF post-doctoral fellowships funded by Japan "are greatly under-subscribed by Americans" and commented that large companies were not interested and that the benefits for smaller companies were not great enough to justify the costs. It will be remembered that access was one of major points of USG dissatisfaction with Japan, yet six years after the 88STA was signed the U.S. could not find enough researchers willing to stay in Japan for a long period. On the other hand, shorter-term sojourns, like the Summer Institutes, were apparently popular. In such short periods of time, one can only get a flavor of the country (Japan), the cursory concept of the status of research but probably a good appreciation of Japanese S&T. Probably substantive research in three months is rather difficult. The Panel urged the USG to make a greater effort to publicize these programs, assess the impact of the Japan-sojourn on the researchers.

Concerning megascience projects, it unequivocally recommended that Japan be "approached as early as possible and treated throughout as a fully equal partner in the program" and that "past history of dealing with Japan in this regard has not always been exemplary." Judging from habits and traits of handling Japan in the past, it is not unreasonable to say that this would be quite difficult for the U.S. to carry out.

The number of joint projects had increased from about 50 to over 150 in 1995. The Panel felt that in light of budget pressures there should be a "broad assessment ... with Japan" about their results, mechanisms of cooperation, management and efficiency of these programs. When that idea of a cooperative joint assessment was proposed in the mid-1980s, it should be remembered, it was pushed aside and forgotten and a confrontational approach about future S&T relations with Japan was adopted by the USG.

In this connection and in conclusion, the Panel made an important observation about the "political" aspect of USJ S&T relations. It said that "many of the present joint programs with Japan grew out of political initiatives to strengthen the 'U.S.-Japanese relationship.'" (emphasis made in the report) While I have discussed this characterization in a later subsection, this depiction is not mentioned nor discussed in almost all accounts. The Panel called for considering how much weight should be given "to this political dimension in what will be increasingly harsh competition for research monies." It significantly opined that "the broad array of commercial alliances, joint ventures, and cooperative industry programs ... could already well exceed the numbers involved in cooperative government programs." The implication here is clear, let us -- USG and GOJ -- put into clear and appropriate perspective the governmental S&T relationship in the totality of the bilateral S&T relationship. This is an important observation that should be taken seriously and the intense governmental S&T relationship reassessed.

This short report reiterated the U.S.'s continued extraordinary anxiety about the combined effect of joint R&D projects and trade, with IPR and budget pressures and in the context of the total S&T relationship with Japan; and through this report the character and mission of the JHLAP appears to have shifted to include the role of "political" adviser on USJ S&T relationships and the originally intended channel of communication with their Japanese counterparts and, in my opinion, became another channel, another instrument, to exert pressure on Japan to achieve USG's political, trade and financial goals, rather than on genuine joint R&D cooperation. This shift was described in the talking paper for the September 1994 HLAP meeting as follows: the meetings of the JHLAP with the Japanese will not be the most important, rather the most important will be those informal recommendations the U.S. Panel members make privately to the USG. Notwithstanding this apparent shift, the characterization of the past role of the JHLAP by a recent chairman was that those meetings were a series of pleasant events of polite dialogue with little effect and impact. Perhaps that will change in the future.

c. Access to STI

The creation of the Task Force on [Scientific and Technology] Information Access was mandated by the 88STA and it held meetings annually from 1989 through 1992 but like the others began to slip into a 15-18 month cycle from 1994 onwards. Apparently, the atmosphere of the meetings was cordial, constructive and forward looking. The first meeting began with a mutual education program of each other's STI program, their dissemination, and utilization, access to gray literature in Japan which was later (at the 2nd meeting) dropped. The task force pointed out that the two governments were "exchanging a fairly large amount" of STI at this time -- just after the STA was signed and that Japan is "more enthusiastic in collecting information from the U.S. than the U.S. is in collecting information from Japan." The task force's goals were established as follows:

- improve awareness and understanding of organizations and systems established to improve the use of scientific and technical information (Mutual Education)

- increase the quantity and quality of STI

- reduce impediments, if any, to the flow of STI (Japan Reprographic Rights Center, similar to the U.S. Copyright Clearance Center)

- increase the translation of scientific and technical information[11]

The "bottom-line" of this Task Force's mission was to persuade U.S. researchers and companies to study and use JSTI in a much more continuous, aggressive and constructive manner. After many years of steady cooperative work, a Reprographic Rights Center was created in 1991, an agreement was reached in 1996 between the GOJ and DOC for the latter to use in the Machine Translation (MT) Center in the DOC a Japanese/English machine translation system developed with Japanese software, and annual conferences have been held to explain and persuade U.S. researchers and companies to use STI to their advantage. These annual conferences were held in various cities in the U.S. and jointly sponsored by the National Technical Information Service (NTIS) and the Japan Information Center for Science and Technology (JICST).[12] Separately, at the suggestion of the Library of Congress, the GOJ, through the Center for Global Partnership under the Japan Foundation (funded by GOJ), assisted in creating and maintaining the Japan Documentation Center in the Library to collect Japanese documents to assist, in reality and in the main, the Congressional Reference Service in the Library in its policy studies vis-a-vis Japan for the U.S. Congress.[13] The periodic Task Force reports are replete with what the GOJ has done and will do to facilitate the distribution of JSTI and an expression of USG appreciation for the increased budget allocation of the equivalent of $1 million for this purpose by JICST. Nineteen projects were listed in the 1996 report all conducted by Japanese organizations, including JICST, NDL, Agency for Cultural Affairs, Science and Technology Agency, National Center for Science Information Systems, the Agency for Industrial Science and Technology under MITI, the Communications Research Laboratory (MPT) the Geographical Survey Institute and the Public Works Research Institute (both MOC), the Institutes of Physical and Chemical Research (RIKEN). Many of these organizations have been providing documents to U.S. organizations for many years. It is indeed noteworthy that there were no projects listed for which U.S. organizations were responsible. Furthermore, this list of activities to provide more JSTI to U.S. organizations was in line with what was called for in the draft given to the GOJ in August 1987. Japan was providing more funding for this endeavor and was satisfying the USG demand for "better access to JSTI" without it being included in the 88STA as a one-sided demand by the USG.

It is extremely difficult to ascertain whether, with all these substantial efforts by the GOJ to provide more JSTI through all these means and the annual conferences in the U.S. to persuade U.S. researchers in academia and companies to utilize/exploit the data in JSTI thus provided, there had been a quantum jump among U.S. researchers to use this material. Based on random interviews, the type of people attending those aforementioned annual conferences (mainly providers of JSTI not the users of STI), there had not been a great surge by U.S. researchers and companies in the known usage of JSTI notwithstanding all the efforts by the GOJ.

d. Access to R&D Facilities

This was the second Task Force appointed by the JWLC to improve and increase the access of U.S. and Japanese researchers to each other's R&D facilities, but in reality this meant Japanese R&D facilities for all intents and purposes. This was the area where, on the U.S. side, there was the greatest perception of imbalance between the U.S. and Japan in numbers of researchers in each other's country. This was the area where both Japanese and U.S. delegates agreed that there was an imbalance but where there was a wide chasm between them as to why there was this imbalance. This is the area where a DOC witness at the July 1988 hearings stated with gusto and a glint in her eyes that the Japanese would be pursued "to the last yen" to assure that the U.S. is given access to R&D facilities where R&D projects are government supported or sponsored. The USG demanded that the Japanese open more research facilities to U.S. researchers -- implying without using specific words that the Japanese had not opened their facilities, particularly industrial research laboratories, were not willing and perhaps there was a conspiracy not to open these facilities. This was the area where the U.S. Embassy in Tokyo through a series of cables attempted to persuade Washington that contrary to the perception of some, indeed there were many opportunities to work at Japanese R&D facilities.[14]

It would be useful to provide a brief statement on NSF plans to increase opportunities for U.S. researchers in Japan. In the late 1980s, NSF obtained $1 million in new appropriations for such purposes with a plan to gradually increase this investment to five million in 5 years. These funds were intended for small scale R&D study missions to Japan, short and long-term language training, and development of technical Japanese curriculum for implementation about 1988. But as part of the 1988 Summit goodwill, Japan offered almost $5 million to be administered by the USG (NSF was given this responsibility) to send researchers (not undergraduate or graduate students) to Japanese national laboratories, universities, and industrial laboratories with no institutional restrictions. This Japanese grant of funds to the USG caused the NSF appropriations for this purpose "to disappear." This relatively straight forward and simple program was replaced by a broad array of programs funded by the GOJ and the USG as described below.

The Task Force on Access began meeting in this rather charged atmosphere inherited from the negotiations. It first met in 1989 and thereafter annually alternately in Washington and Tokyo until the renewal of the STA in 1993 when its meeting dates began to slip into a more or less 18 month cycle like the other groups under the 88STA. At the first meeting, it was agreed that both sides would conduct surveys of the foreign researchers in their midst for over two months (indeed a very short time for research work in a foreign country) at "government-sponsored or government-supported programs" and facilities, exchange the results of these surveys for future reference and remedial action, and handle complaints by researchers concerning their research tours would be handled expeditiously. In order to facilitate a greater number of U.S. S&Es to become interested in an R&D sojourn in Japan, NSF/Tokyo updated its Report Memorandum #92 (1/86). It found that 154 laboratories said they were willing to receive U.S. S&Es at their laboratories. This provided another source of opportunities in Japan for foreign researchers.[15] In addition, the U.S. side proposed and the GOJ agreed -- to creating a "summer program" to assist U.S. students in Japanese language training, familiarization with Japanese R&D facilities for future exchanges.[16]

Some rather important findings emanated from these initial endeavors:

1. Both U.S. and Japanese researchers identified language as a major difficulty more for daily life in the community not so much as a problem in the workplace. This, of course, was not surprising. This was followed by hous-

ing, employment of spouses and schooling. "...the consensus was that one could still have a good research experience using only English".[17]

2. Neither side noted any major access complaints. This is important in that there was an implication in the USG presentation that there was an access problem. The very raison d'etre for this Task Force was to overcome presumed access problems.

3. According to a separate survey of U.S. and European ST&A (Science and Technology Agency, GOJ) Fellows, Europeans tended to evaluate their experiences in Japan, whether of housing or research facilities, etc, as similar to or better than their own, while U.S. researchers tended to note more of the differences.

4. The first Summer Institute Program was begun in 1990. Announcements were sent to 324 major U.S. research universities, 89 applications received, 25 selected. While this was regarded with "great satisfaction" as the first trial phase, it cannot be said that there was an overwhelming number of acceptable applicants chafing at the bit not to do research but to go to Japan for language training and help them establish a basis for future exchanges. It was regarded by the Task Force as "fewer than expected."

5. No "specific complaints" (presumably other than access issues) had been made to the complaint mechanism.

6. Concerning the controversial issue of Japanese students in the U.S., in 1988/89 24,000 Japanese students in all fields; this compares to 68,724 ethnic Chinese (from China, Taiwan, Hongkong). This is 6.6% of total foreign student population. 11.9% (1987-88) of the Japanese students were in science and engineering with science having 3% and engineering 4.2%. The U.S. granted 20,257 PhDs in science and engineering in 1988, foreigners receiving 5,933, and Japanese receiving 100 (less than 2% of total number of PhDs received by foreigners and less than 0.5% of all PhDs) and most of them were more inclined to stay in the U.S. instead of returning (presumably a plus for the U.S.) This would not seem to constitute a Japanese threat to the U.S. The U.S. sent 3,633 students to Japan.

7. What about the U.S. respondents evaluations of Japanese host institutions and their research: Of the 55 U.S. respondents 5 felt research in their field

was better than that in the U.S., 21 that it was comparable to that done in the 5-10 best U.S. institutions, 24 that it is comparable to that in standard U.S. institutions, 3 felt the Japanese research was less advanced, and two no answers. Concerning those who would benefit from research in a Japanese institution, the post-doctoral and junior researchers, and professor and senior researchers believed their groups would most benefit.[18]

Since the 88STA was renewed in 1993, it would be meaningful and useful to report on the September 1992 Task Force Report on Access.

The GOJ had created five programs to facilitate the research of U.S. scientists and engineering in Japan and were funding the costs of the Summer Institutes researchers in Japan.[19] Through these programs, there had been a quantum jump in the number of U.S. researchers in Japan, from, for example, about ten long term visits through NSF to almost 300. One of the "most significant" accomplishments - and one of its most popular -- of this Task Force was the gradual increase in the number of students under the Summer Institute Program (25 in the first year, 49, and 58 in the next years), breadth of coverage to include national university and corporate labs, and also those in the Tokyo areas from the original Tsukuba Science city orientation.[20]

The above recitation shows clearly that however neutrally and even handedly the report text may be presented, the direction of the Task Force's efforts had been on what Japan had accomplished with very little on what the U.S. side had accomplished. This was an unfortunate symbol of how R&D access was approached and reported on. However, a 1994 NSF report on American researchers in Japan provided a more complete and thorough presentation of the complicated and varied series of programs for American S&Es in Japan. It reported, for example, that 809 S&Es studied in Japan for one week to a year or more in FY1994 (10/93-9/94). Nearly half were for three months or more, one fourth for a year or more. About 300 (mostly bachelors and masters degree students) were sent by the U.S.-Japan Industry and Technology Management Training Centers supported the U.S. Air Force Office of Scientific Research. About half were, in essence, "get acquainted short-term visits for a hopeful future." Almost 600 others were supported by ten programs; half of these programs have long histories, the other half established between 1988 and 1992, after the 88STA came into effect.[21]

In relation to barriers which had been presumed to have existed because of the large imbalance in the number of students in each country the 1992 report to the JWLC

> "... acknowledged that no hidden barriers have been found to inhibit the exchange of researchers to government-sponsored and/or government-supported research facilities. ... that the complaint mechanism which has been built into the TFA structure since its inception, has never been used.[22]

In addition, the NSF itself conducted an analysis of the first ten years of "US-Japan programs for postdoctoral fellows and graduate students." It noted that 27,000 S&Es, mostly senior-level researchers from the U.S. and Japan who had participated in the oldest bilateral Cooperative Science Programs for both countries since 1961, had participated in jointly funded research projects and joint seminars. Concerning the fellowships and summer programs initiated since 1988, it had the following comments to make about the trends:

- the level of interest among U.S. S&Es in long term fellowships (post doctoral fellows and visiting researchers) has been declining since 1991, averaging 52.3 per year for 1989-92, 1993-1996 dropped to 45.8, in 1997 the number slipped to 29, the lowest since 1988.

- in contrast, Japan has had no difficulty filling these positions with S&Es from Asian and West European countries, the "decline in interest is peculiar to the U.S."

- the attractiveness of short-term fellowships remains relatively "relatively strong" among U.S. S&Es.

- the summer institute programs, popular at one time among U.S. S&Es, dropped to 69 in 1998, well below the available slots of 110.[23]

This conclusion of the 1992 Task Force report and the 1998 ten-year survey by NSF put to rest the need to pursue the GOJ "to the last yen" as the DOC witness vowed to do at the 1988 Congressional Hearing on the 88STA. This perception was completely demolished. The Task Force whose creation was based on the assumption that the Japanese supported hidden barriers and there were many complaints to pursue with the GOJ, fortunately conducted its proceedings in an atmosphere of cordiality, cooperation and understanding, with obviously much more modest goals.[24] Another major revelation -- but to many just a confirmation of what was expected and, therefore, un-surprising -- was that in many cases the assigned quotas for U.S. S&Es

for specific R&D opportunities in Japan went unfulfilled, i.e. there was no long line for S&Es to study in Japan.

e. Technology Liaison Groups (TLG)

The TLGs were created to manage and maintain liaison among the scientists and engineers concerned with the various projects which were carried out under Article II of the 88STA and Annex I. Seven technical areas are listed in Annex I where cooperative activities could be conducted; of course, this grouping could be revised by mutual agreement.[25] There were liaison groups for life sciences, information science and technology, manufacturing technology and automation and process control, global science and environment, advanced materials and superconductors; joint database development was handled by the Task Force on STI. The JWLC and JHLC both focus, in the main, on high level S&T policy matters, the JHLAP provided a variety of recommendations on bilateral S&T issues to the JHLC, the two Task Forces focused on ways of increasing the flow of STI in general. The Liaison Groups come nearest to doing "S&T work" but even here it is a liaison and management function, one-step removed from R&D benchwork.

One of the major objectives of the 88STA -- as with its predecessor -- was to intensify, broaden, and deepen the USJ S&T relationship. One approach has always been to encourage the S&Es of both sides to propose R&D projects. When the 88STA was signed there were about 50 projects under the old 80STA; there were still only 50 in September 1991. By October 1995, however, there were 187 projects, a complete explosion of proposals. Practically every GOJ Ministry and USG Department and their national laboratories and institutes and the principal universities were involved in these projects. By far the largest groups of projects were in global geoscience and environment (73 projects), biotechnology (71 projects) and advanced materials and superconductors (23 projects), almost two-thirds in only three technical areas. This shows obviously where the principal interest, direction, and emphasis lay in the minds of the S&Es of both sides.[26] All of these projects were worthwhile since they had been accepted by both sides, but none of them, except perhaps one, can be regarded as a major big science project, i.e. Presidential level projects. The one exception is the Human Gnome Project which was not bilateral but multilateral.

This explosion created a management and coordination problem for the liaison groups and the governing policy committees. The USG had, for some time, been urging the Japanese to take a more active role in international affairs, including S&T. For the first time, Japan proposed a series of projects in 1978-79 which resulted in the U.S. agreement on energy research. The Japanese took the U.S. at it's word: a great part of the above explosion came from proposed Japanese projects; the exact proportion is difficult to ascertain. According to a USG analysis, "the majority of U.S. proposals for cooperative projects come from the bottom up, from individual researchers at the technical agencies ... as a consequence of this process of bottoms up decision making, the U.S. ends up making far fewer proposals ... than Japan." Based on the conventional wisdom on the Japanese system, they also use the bottoms up system, apparently more effectively. The "language barrier" -- a barrier which the U.S. side has permitted to become a barrier, not because of the Japanese -- was singled out as the major culprit. But it did admit that "Japanese researchers seem to be more aware of what is going on in their fields in the U.S. than U.S. researchers are of the work being done in their specialties in Japan", thus more projects from Japan. "The bottoms up process may not be serving us [the U.S.] well". We appear to have a sense of being overwhelmed by the torrent of Japanese R&D proposals. The JHLAP was appealed to for assistance on how to stimulate, create, etc cooperative projects from the U.S. side.[27] The JHLAP suggested that priorities could be assigned to these projects, but because funding control in both countries lies in the agencies involved not a central authority, that may be difficult. Perhaps, in light of budget limitations in both countries, there should be more concentration on careful project management and more direct communications between researchers themselves.[28]

If numbers have an implication, then the Liaison Groups and the number of projects and depth, width, character, and direction of USJ R&D cooperation would appear to be highly successful. But the sheer numbers of projects presented a management problem not previously envisioned, and definitely not within the personnel limitations of State/OES, OSTP, to manage as a whole. Yet an alternative system had not been devised by the end of 1996. The operations of these several liaison groups, the number of agreed upon projects -- though perhaps unwieldy for management purposes -- was a manifestation of the possibilities and potentialities of USJ S&T cooperation. Apparently from the Japanese point of view, these projects and exchange of ideas which take place through these many projects are important: " ... to let loose misgivings over individual issues and spoil the tightly intertwined grass-roots cooperation at the researcher level [as exemplified by these projects] would be gross folly."[29]

f. The Techno-Growth House (TGH)

The draft agreement given to the GOJ in August 1987 specified that the GOJ would provide improved facilities for foreign researchers (meaning basically those from the U.S.); while one-sided stipulations were negotiated out of the agreement, the GOJ, in a side letter to the 88STA, confirmed that it would "take steps to identify and increase substantially opportunities for American scientists and engineers to engage in research and study in Japan." To this end, the GOJ created the "Techno-Growth House" (TGH) in Tsukuba science city in 1995. While this was not directly related to "access to R&D facilities", it was an important response by the GOJ to facilitate the sojourn in Japan of foreign researchers; it is a facility which indirectly makes the researchers stay in Japan that much more appealing.[30]

The GOJ had planned to create some kind of facility to facilitate the stay in Japan of foreign researchers. Under USG prodding to create such a facility it was put on the Common Agenda, a summit document itemizing activities, projects, programs to be accomplished, thereby putting the GOJ under pressure to accelerate its construction. It was built for two billion yen and opened in spring 1995. It provided 40 rooms at very modest cost. TGH has a wide range of facilities, a conference room for 20 people, library, eight office spaces with electronic facilities. It was an extension of the existing adjacent dormitory (the Educational and Training Center) for foreign researchers.

The USG was not pleased with the TGH being a wing of the existing MITI Guest House, not an independent facility. It, therefore, inserted itself in the decision-making process in the planning for the TGH by persuading MITI/AIST to accept five U.S. representatives from the government and U.S. businesses in Japan on the committee overseeing the creation of the TGH.[31] The USG was concerned that there would be insufficient rooms for U.S. visitors, that the operating budget would be insufficient, that the facilities would be inadequate and not up-to-date, that MITI would not provide support for obtaining visas for foreign visitors, that a mentor program be created requiring Japanese researchers to assist foreign visitors with introductions to appropriate contacts. The USG felt that the dialogue at the management board with U.S. representatives present would "help ensure that the facility is first rate and will appeal to [foreign] researchers"; the implication here being that without U.S. participation, prodding, demanding, and pressure the Japanese would not be capable of creating a "first rate facility." It is difficult to imagine that a similar management oversight arrangement would be even seriously entertained let alone approved if such a facility were being constructed in the U.S. (No such facility, to the best of my knowledge exists in the U.S.)

While the TGH was financed, constructed and managed by the GOJ, the decision concerning the acceptability of applicants resides, extraordinarily, NOT with an agency of the GOJ, the TGH but, in the first instance, with a foreign government. This was done at the request of the USG. U.S. applicants must submit applications initially to the U.S. Department of Commerce. Obviously, it wanted to control who, among U.S. researchers, had access to these facilities, not let the GOJ have this prerogative as a sovereign nation. The appropriate applicants were recommended by the DOC (or any other government) or nominated to the GOJ for acceptance. Except in the most unusual circumstances TGH, acceptance is a mere formality. This is another example where the GOJ made the actual STA neutrally balanced, but accommodated the USG separately under a side letter, separate funding, and unusual management structure. To date by far the largest user of the TGH have been U.S. researchers but the utilization rate of TGH has been low.[32]

2. Special Issues

a. Presidential/Prime Ministerial Level Projects

The lack of Presidential level projects under the 80STA was a major point of USG dissatisfaction with that agreement. Presumably such prestigious projects would be bilateral since they would need to be included under the purview of the 88STA to be counted. The big science projects that have been mentioned or discussed at JHLC meetings have been multilateral. The USG pressed the GOJ repeatedly for support of the SSC project with billions of dollars of contribution. The Intelligent Manufacturing System originally suggested by the GOJ but looked upon with great distrust by the Office of Technology, DOC during the Bush Administration as a Japanese ploy to secure U.S. technology in a back-handed fashion, became, after much travail, a multilateral project. ITER, the Human Gnome Project, and the others are multilateral big science projects. There was a fair amount of discussion at the JHLC meetings on the future handling of megascience projects but with no obvious conclusions. In any case, probably JHLC meetings were not the principal forum for such discussions. One might say that the USG's long time demand, expectation, and hope for prestigious bilateral projects was overtaken by the increasing internationalization of science and the absolute need for international cooperation if these large scale projects were to be brought to fruition. To the best of my knowledge, there are no bilateral projects under the 88STA of such stature to be characterized as Presidential/-Prime Ministerial; rather, in a more constructive interpretation, there are projects, e.g. Disaster Prevention and Global Change, which are of distinct mutual benefit though not Presidential in size and stature.

b. Umbrella Nature of the 1988 STA

To make the 88STA "an umbrella agreement" for USJ S&T relations was a major objective. But this concept was not frontally challenged by the technical agencies of either governments which did not want this concept in its pristine form adopted in any manner; USG agencies went along with the OSTP demand and the inclination of the State Department. Nevertheless, the original concept was always compromised and became something less than all-conclusive in the final agreement. Those projects/programs which had their own management structure, meaning the large projects, e.g. Space Station, ITER, Human Gnome project, were exempted from being under the 88STA; but, all bilateral S&T agreements and MOUs, were "expected" to reference the basic S&T cooperation principles and the IPR provisions of the 88STA. In this way, the umbrella aspect was weakly upheld in implementation by reference.

c. Security Issues

Because of the deep-seated distrust of Japan and its presumed inability to prevent the leakage of critical civilian technologies developed by Japan to the then Soviet Union notwithstanding its tightened regulations concerning such possible exports, the USG insisted to the bitter end, and the GOJ dragged out the discussions on this point to the bitter end and, was, in the end, finessed by a revision of the wording and the side letter signed by the Secretary of State and the Minister of Foreign Affairs. During these negotiations, it was clear that the Japanese scientific groups, specifically the Japan Science Council, and some elements of the GOJ did not trust the USG either. It believed that the USG would, under cover of this clause, attempt to classify unjustifiably open basic research resulting from joint R&D projects and attempt to control Japanese S&T and R&D in this manner. This clause did not, as the Japanese feared, "develop a life of its own" (*hitori aruki*) to their disadvantage. Since

1988 the Japanese have not been accused of leaking critical technical information to the Soviets (later the Russians) nor did the U.S. appear to be attempting to control the Japanese R&D agenda. Their mutual suspicions were thus unwarranted and inaccurate; it was a non- issue in the implementation phase. While it would appear to be a non-issue, the clause remained dormant and ready to be invoked at the slightest cause by either side; in light of the past history concerning this clause more probably by the USG rather than by the GOJ.

Unlike the STA itself and the IPR provisions, the security clause was not, to the best of my knowledge described as a model to be used in other STAs. A similar clause was nevertheless, used in a few STAs such as those with Venezuela, Korea, Italy, but none, it seems, with any major European country.

d. "Political" Science and Technology

In recent years, several aspects of USJ science and technology relations have been handled on three planes: implementation of the 88STA, the Common Agenda created during the Clinton Administration, and the "technology basket" in the perpetually vociferous trade dialogue. The 88STA has been implemented under Presidents Bush and Clinton and each has included their own special "bilateral programs" with an S&T component in each case for a communique on Summit accomplishments. The general understanding of the USJ bilateral relationship involved the military and the economic and trade pillars; the S&T relationship was broad and deep both on a government-to-government basis and between the two private sectors but traditionally has not been the source of discussion for most summit meetings. This changed with the 88STA under recent Presidents. Under President Bush there was the Global Partnership Plan of Action and under President Clinton the Common Agenda. Each has a specific S&T section although some of the items in other sections, e.g. environment, could be included under S&T. It is politically significant that neither bilateral plans even mention the 88STA as an overall agreement for S&T relations; in contrast, the trivial matter of future meeting dates of selected working groups were mentioned. One can only get the impression that the S&T part was included to make a rounded and fulsome bilateral plan emanating from the summit meeting with multiple political, financial and research purposes in the background.

The S&T section in the Global Partnership Plan of 1992 focused on Global change research, major international projects (SSC, ITER, IMS,

Space Station, Human Frontier Science Program, Optoelectronic/OE technologies). These are all multilateral projects except the last one on OE, which depends on close international cooperation and the provision of stable annual funding from the participating countries. Only the SSC mentioned a specific method: establish a joint working group to consider how "this project can be formulated as an international project to enable Japan to participate in the project, i.e. how to persuade the GOJ to commit billions of dollars (it naturally did not say that the USG project was in danger of cancellation which actually happened in 1993). As would be expected in a summit level plan, the language used was general with no hard and fast commitments of funds except to cooperate, reaffirm, ensure success, and support various activities. Most of the listed projects were already taking place and there could be no reason to object to their inclusion. The 1993 Common agenda for Cooperation in Global Perspective included reference to USJ cooperation in transportation, telecommunications, civil industry, road technologies and prevention of disaster analysis. Most of these projects were short to mid-term projects. There were no financial commitments by either side, direct or implied except, of course, those funds to carry out their already agreed upon commitments.[33]

The January-June 1995 and June 1996 Common Agenda joint reports for the USJ Summit contained a section on "Bilateral Cooperation on Advanced Technologies"; there are other technology projects under forests, conservation, oceans, earth observation, environment, road technology, energy efficient technologies, and national information infrastructure.[34] The Advanced Technologies section dwelt on minutia about working groups finalizing plans for cooperation in the above mentioned technologies, the opening of Techno Growth House in Tsukuba science city to house foreign researchers. It must be remembered that these projects are just a tip of the numerous R&D projects already being carried out under many agency-to-agency MOUs and other arrangements. This overlap "initially led to some confusion as to their roles [the Framework and Common Agenda and the USJ 88STA], but these have been clarified over time." [35] There was an inherent management problem in placing projects under the Common Agenda: because the Common Agenda does not have an established management structure to monitor long term development of projects, and because "only the UJSTA [88STA] has an established set of guidelines for dealing with IPR" broader longer term projects would need to be transferred to 88STA management with an appropriate Implementing Arrangement.[36] It appears there have been no project transfers to the 88STA except for Road Technology. There is another aspect to including selected projects in such documents as the Common Agenda. The inclusion of specific projects tends to make it easier for GOJ Ministries to secure necessary funding for these projects from the Ministry of Finance. On the other hand, NASA under President Clinton was perhaps able to prevent substantial budgetary cuts of the Global Observation Information Network by a GOP majority in the U.S. Congress (1994-96) which regarded this program as a private sector kind of effort.[37] To list just these specialized projects in either the Global Partnership or the Common Agenda gives the wrong impression that these are the major items in U.S. S&T relations when, in reality, they are not. This kind of impression is difficult to avoid since they emanate from a U.S.-Japan Summit Meeting. The USG would request/demand that selected items be included in the Common Agenda in order to get the attention and action of the GOJ, i.e. like the TGH. The 88STA and Common Agenda would, in the eyes of some, provide areas of synergy and potential synergy between the two groups. The Common Agenda was a short term statement, while the STA would provide the long term perspective. Listing these specific projects compromises the perceived and presumed political value of the 88STA, a Presidential agreement about which the USG was, at one time, under a different Administration, most exercised about and insistent upon. The inclusion of an S&T section in the Common Agenda and the Global Partnership was clearly "political" science and technology and rounds out a bilateral Summit coverage of "all or most important bases" for political and public relations purposes.[38]

e. 1988 STA as a Model Agreement

In July 1987, the Economic Policy Council under the Reagan Administration, provided guidance and instructions to the Executive Agencies involved in USJ S&T relations to negotiate an S&T agreement which could serve as a model for future use. In principle this would appear to be an excellent idea, especially since S&T issues were coming more and more to the fore in relations between the U.S. and some foreign countries. It should be kept in mind that these were the instructions of one Administration; when the Reagan Administration departed Washington, the EPC files were promptly retired, ignored, forgotten by the next Administration, though a Republican Administration, and ne'er heard from again and became difficult to locate.[39]

Based on a search in the State Department of STAs signed after June 1988, there were no STAs which contained the three tiered management structure used in the 88STA.[40] Indeed, a similar overall kind of STA with any of the European countries except Italy could not be found but there were

STAs with PRC, Venezuela, New Zealand, Mexico, Korea and Poland. By 1992, it had become clear to the USG that this complicated management structure, while seemingly quite rational and systematic was, in reality, cumbersome, time consuming and a burden on the limited number of government officials and private sector advisers in the JHLAPs who would be required to invest a substantial amount of their time to only one country. In any case apparently the priorities of the latter years of the Bush Administration and the Clinton Administration did not call for pursuing such a cumbersome S&T arrangement with foreign countries. No standard STA was prepared by the Bush or Clinton Administrations.[41]

Japan was thus again treated differently, a special case. Furthermore, and most interestingly, none of these S&T agreements were signed by the U.S. President or by his foreign counterpart but by the Secretary of State or Foreign Minister or some other officials. Inevitably the question arises as to the political need and motivation for such a special treatment of Japan. No documents are known to exist which specifically address this issue since such agreements are negotiated quite often by a different set of people each time. Was this approach a cynical ploy to persuade the GOJ that the 1987-88 negotiations were to create a new bilateral STA but also to create a model STA for future negotiations with other countries. After all, the earlier STA and arrangements had been publicly described by the USG as a model arrangement to be emulated with other countries. Perhaps this concept of a model was a sincere aim of the 1987-88 USG negotiations; only that it became unwieldy in implementation and other countries would not agree to such a management structure.[42] One can address this issue of differentiated treatment only by drawing conclusions or deducing interpretations from available data and these interpretations have been presented elsewhere in this study.

f. 1988 STA IPR Section as a Model

In addition to the total STA as a model to be used, the IPR clauses in the 88STA were also presented as a model for other agreements. Under the Bush Administration, using the IPR in the 88STA as the basis, a more flexible and less rigid IPR standard annex was prepared for use when agencies of the USG negotiated an STA with a foreign government. The new model IPR annex was to be included in any Circular 175 request for interagency review. It was felt that one side of a personnel exchange under the 88STA was too much in favor of the R&D facility receiving a researcher, i.e. too one-sided, and that the 88STA IPR clauses tried to describe too rigidly various types of cooperation and method of allocation. The standard IPR annex approach was used in several STAs between the U.S. and the PRC, Venezuela, New Zealand, Mexico, Korea and Poland. A new "model" IPR was developed in early 1992.[43]

3. Implementation Wrap-up and Renewal

a. Implementation Wrap-up

Overall. Despite a seemingly ominous beginning in implementation, the 88STA created a forum to discuss (*tōron/giron-suru ba*) S&T policy issues though tepidly but continuously, tamely and quietly, contributed importantly to a smooth and cooperative S&T relationship, and engendered a better atmosphere for bilateral and multilateral S&T cooperation. But, on the other hand, it seems that implementation of the 88STA became absorbed in process -- a bureaucratic process -- of convening committees, task forces and liaison groups, all emanating from the mandates of the 88STA. One might say that it had little time for major controversy and contention, something that both governments wanted to avoid in light of the particularly contentious and unpleasant bilateral negotiations for the STA. There were, however, lingering doubts, if not an underlying *suspicion* in the USG about the implications of results of joint R&D projects on trade issues and on the GOJ side a politely articulated *lack of confidence* that the USG would be a reliable partner in big sciences especially in relation to a reliable commitment to provide adequate funds over a number of years. This feeling does not stem from doubts about the serious personal commitments and intentions of government officials but from the systemic problem facing the U.S. in providing the necessary funds for an international big science project.[44]

Exchange of Information and Personnel. Based on available reports, the USJ S&T dialogue in the various fora under the 88STA has probably resulted in 1) a genuinely better understanding of each other's S&T culture, R&D systems, requirements, etc; 2) persuading the Japanese to create a number of R&D study programs in Japan for foreigners including U.S. researchers. Surveys have shown that language is not as much a barrier as envisioned, that the complaint mechanism had not been used, that PhDs granted to Japanese were a mere 0.5% of all U.S. PhDs, hardly a threat to the U.S., that a majority of U.S. researchers found Japanese R&D facilities comparable to standard U.S. institutions or even the 5-10 best U.S. institutions, that while new programs for foreigners to study and do research in Japan were created after 1988, half of all programs had long histories of bilateral exchange, that short term programs like the Summer Institutes programs were most popular. Notwithstanding almost all the projects listed in the Task Force reports are markedly unbalanced: practically all for Japan to carry out.

The imbalance in the number of S&Es from Japan in the U.S. and U.S. S&Es in Japan still persisted. But based on surveys and studies the myths and perceptions about access to R&D facilities were found to be incorrect. Through an MOU the USG had a virtual monopoly on the use of the Techno Growth House, yet the utilization rate was low. The USG has had difficulty in filling almost all its assigned slots for researchers in Japan in the various programs. In the JSTI area a similar gap persists; again, it seems the problem lies more in the lack of interest on the part of U.S. S&Es in JSTI than in the unavailability of information from Japan, even when it is in English.

The management structure. Even IF the motive originally had been to influence, persuade and even attempt to control Japanese R&D, the management structure did establish a forum to press the Japanese for internal changes which the USG perceived to be to the U.S.'s advantage, has been found to be too cumbersome and has persuaded the USG to relax on the tight management structure without actually revising the STA itself. The formal meetings have been made "more informal" to accommodate the stress on high officials schedules, budget squeeze on travel and personnel funds and telescoped into basically a two-day session for the JWLC, JHLC and JHLAP once a year. It does not seem feasible that a meaningful discussion of high policy in S&T for both countries can adequately be discussed in one day without the meetings deteriorating into perfunctory and non-substantive fora. While it may be a small point, originally the JHLC prepared "reports to the Governments of the U.S. and Japan"; these devolved into mere press releases. The Task Force on Access produced reports through 1994, none thereafter. Whatever the rationale for these modifications, this is a downgrading of the felt need to have such a complicated, tiered and intensive high level and time consuming interaction and relaxation of the apparent need to pursue the Japanese as originally envisioned. This could clearly imply that the original underlying assumptions and anxieties about Japan S&T relations were excessive, misplaced and unwarranted.

Desire for GOJ financial contributions. In 1979-80, the USG pressured the GOJ for funding of USG R&D projects but did not succeed. Again, the USG pressured the GOJ for funding contributions to the SSC; again, it was not successful and would not heed the signals that funding was not forthcoming.

Joint R&D projects. The substantive R&D work of the STA lies in the numerous joint R&D projects -- 187 as of October 1995 -- carried out in a systematic yet necessarily plodding manner away from the limelight, slowly, cumulatively without major scientific technological breakthroughs in a cooperative not contentious manner as scientific and technological equals, each having a sense of contribution and receiving benefits. This achievement is a continuation of a trend and tradition started decades ago but which, it seems, has been deepened and broadened based not on political, economic, trade or military imperatives or needs, but based on S&T-R&D needs, requirements and curiosity of both U.S. and Japanese researchers.

Umbrella-ness of the STA. The perennial attempt to make the USJ STA a truly umbrella agreement was not completely achieved. It was agreed that each joint R&D arrangement "should" -- not "will" -- include by reference the IPR article; in reality, however, this will probably result in including it in all such arrangements. The GOJ insisted on making such arrangements appear to be permissive on a case-by-case basis but, in reality, agreed to the inclusion of IPR by reference.

The security clause. It was not invoked nor discussed but remains a sleeping clause which could theoretically spring to life at any moment.

b. Renewal-I

The renewal process on the U.S. side began when the USG informed the GOJ at the JWLC in July 1991 that it was beginning the USG internal review process concerning the renewal of the STA. After considerable movement back and forth among the interested agencies, the USG -- 14 months later -- gave the GOJ "a renewal strategy paper" on September 1, 1992.[45] An internal USG paper justified the decision to renew the STA by stating that "opening the agreement up for any changes in the IPR annex would threaten to derail an agreement that serves to benefit the USG more than the GOJ. There are a variety of cooperative activities associated with this umbrella agreement (including IMS, MTI, RWC) that could be jeopardized with any renegotiation, which would last for months, if not years." It should be in mind that as mentioned previously, a model IPR had been prepared in early 1992. Attempting to apply this new model IPR would indeed have opened up past history and probably "derail the agreement". But this model did surface in the second renewal. It is significant that the USG did not feel that the 88STA was a benefit to both countries. One might have thought that in this case even in an internal document the USG could have felt comfortable saying the 88STA was of benefit to both countries. The benefits were not listed nor even hinted at. It is unfortunate that the USG could not feel sufficiently confident enough to justify the renewal to itself and possibly to the Congress as beneficial to both countries. This stated lack of mutual benefit is significant as an indication of how the USG feels compelled to view this S&T relationship, especially in light of its glowing public stance, while, at the same time, it pursues the GOJ to rectify this and that imbalance and lack of availability of this and that opportunity.

The USG paper stated that it has "determined that a renegotiation of the U.S.-Japan Science and Technology Agreement is unnecessary at this time", but that its implementation should be improved and streamlined.

1. The 88STA should become "a true umbrella agreement" by having the JHLC conduct comprehensive reviews of the entire USJ S&T relationship, that each government agency briefly review it's bilateral S&T relations, and that all bilateral S&T arrangements include references to article I of the 88STA. Interestingly and ironically this interpretation comes very close to the original State Department idea put forward in the mid-1980s. The technical agencies in both governments were not in favor of this "centralization" for fear that their specially developed S&T relationships would be subject to State and MOFA jurisdiction and implicit review. The "umbrella" aspect of the 88STA was thus always left ambiguous and partial.

2. Since the three tiered management structure was "costly in terms of government resources", it suggested that there be one meeting, the JHLC, where all agencies, instead of each of the three tiers meeting separately each year. The "smaller JWLC" would continue to be chaired by the same arrangement with the same mandate but it would meet more informally and act "more like a secretariat." This was an attempt to stretch and re-interpret the Agree-

ment's rather clear mandate for annual meetings of the three tiers.

3. Concerning the task forces, they have "in fact been very useful" but they should continue to meet to identify and accomplish "further objectives", and like the three tiered committees they also should meet on a more informal basis with a smaller numbers of officials participating. Since the access to R&D facilities had been a source of considerable friction, it is important to note that the paper reports (based on an aforementioned study conducted by the Task Force on access) that

> "we [presumably the USG] now understand better that our mutual access situation is not mainly a matter of specific barriers, but rather one which reflects the asymmetries between our respective R&D systems."

4. The liaison structure. The liaisons are "points of contact" who should represent more than the agencies they come from; they are to be a dedicated group who have "the time to devote to this responsibility and the willingness to coordinate activities". Not all cooperative projects must be initiated or reviewed by the liaison structure but nevertheless, they are to remain "an important contact and review arena for keeping abreast of developments and possibilities for expanded cooperation". The crux of this ambiguous seeming cutback stems from the "given that the USG has no earmarked funds for expanded cooperation under this agreement" and "would like the joint [liaison group] meetings to be a forum for discussing trends and developments in their areas, not solely a meeting where proposals are discussed for expanded cooperation, a forum for "dialogue between technical and policy experts from both sides and [to serve] as an informal mechanism".

5. The JHLAP. The JHLAP should report directly to the JHLC which presumably it had been doing from the beginning and its members should have frequent and "often informal" contact with their Japanese counterparts and exchange papers and correspondence. In light of the emphasis placed on the U.S. conceived need for a specific role for the private sector in the USJ bilateral S&T relationship this was a rather tepid and nondescript encouragement.[46]

The clear and unequivocal message from the USG paper was to down-size the original frenzied activities into a more controllable management structure, stressing "informality", due to lack of USG travel funds and the drain on officials' time and resources. This was an important re-interpretation of the 88STA and a major departure from the rigid management structure clearly stipulated in the 88STA. I do not believe this was meant to down-grade the importance of the USJ S&T relationship but rather to attempt to manage it with less resources, financial and human, especially since the relationship was not fraught with all manner of barriers and impediments on the Japanese side.

Let us now turn to the GOJ response to the above USG proposals.[47]

1. The GOJ "still strongly wants the S&T agreement renewed without change and for five years" and feels that it has been "operating smoothly". Presumably this reflects the GOJ assessment that the STA had been beneficial to Japan and *perhaps*, in their eyes, the U.S. also.

2. The GOJ believes the bilateral S&T relationship could be strengthened by inviting all government agencies on both sides who are doing "non-military research" not only to report on their activities, but also report on R&D areas where "there is not yet cooperation" implying that through this approach both sides could find additional areas where new technical cooperation would be possible.

3. There were two cautionary notes sounded by the GOJ: 1) while the JHLC was a useful forum to discuss such issues as big science projects and multilateral projects, "we should be careful not to control S&T-related activities outside the framework by taking them as activities under the policy framework" of the STA and 2) the felt need by the USG to make the 88STA an umbrella agreement by reference was not "necessarily useful" for every agreement and should be decided on a case-by-case basis. This is probably a typical Japanese preference to have specific arrangements for specific topics, not automatically apply universal rules, regulations and principles. 3) This particularistic approach is applied to the JWLC by suggesting that only those thought to be appropriate should attend the JWLC which is the approach already used by the Japanese side.

4. The two task forces were regarded as having made good contributions and called for the completion of separate five year reports by the task forces. The GOJ raised the administrative issue of the rank and status of the chairmen of these task forces.

5. The liaison structure was "playing a useful role in promoting S&T cooperation in the seven major areas" listed in the 88STA and the creation of a liaison office to manage those projects outside the seven areas.

6. The JHLAP. Both this Panel and JHLC should inform each other of the kinds of policy issues which are of concern to the bilateral relationship, i.e. make the relationship and the JHLAP more meaningful.

The GOJ was reported as strongly desiring the renewal of the STA, was suggesting ways to increase S&T cooperation but was cautious about having the JHLC involved in big science and multilateral science projects and attempting to bring these projects under the 88STA purview. While the USG wanted to make the management structure more informal, the GOJ appeared to want to maintain the present structure in light of the diversity of the US bilateral S&T relationship. Both sides agreed that the JHLC would be the mechanism to discuss S&T policy issues of interest to both countries. It is noteworthy that there was no sign of any rancor, hostility, suspicions on either side, only slight disagreements on relatively minor points and that both sides were interested in extending the STA for another five years.[48]

The 88STA was renewed for another five years through an exchange of letters on June 16, 1993; a record of discussion was signed on June 10, 1993 on steps on implementation. These steps were a compromise with a tilt in the USG direction for more informal meetings, but bilateral agreements "should" -- rather than "will" which the USG wanted -- reference the IPR as applying to all bilateral S&T activities (including task force and liaison activities). There had been no problems over the controversial security obligations article, nor had there been any IPR problems. In light of the negotiations leading to the 1988 STA, the subsequent implementation experience, and political requirements of the bilateral relationship and the apparent emphasis placed on S&T in the regular USJ Summit meetings, this renewal was inevitable, understandable, uncontroversial, and a renewal without any modifications of the text of the STA.

c. Renewal-II, Preparatory Studies

Several years after the 88STA was signed, and when the anxieties about the Japanese Challenge were still rife and rampant, the Defense Authorization Act passed in 1991 for FY1992-FY1993 called for the Secretary of Defense:

> ... to commission an independent study ... by the National Academy of Sciences ... [to] analyze the strengths and weaknesses in Japanese science and technology and present a framework for pursuing U.S. interests through scientific and technological relations with Japan in the future.

In light of the prevailing moods in and out of the government at that time about this Challenge – while the first evidence of the possibility of an incipient bursting of the bubble in Japan could barely be detected – the Challenge continued to loom large on everyone's radar screen; thus, this request for a study and action plan was a natural and prudent action; nevertheless yet another study to fathom Japanese S&T. While the studies requested were not, on the surface, related to the renewal of the 88STA, one would expect them to become important background papers for the 1998 renewal process, the tenth anniversary. Two task force reports were prepared, for Defense and Competitiveness (the latter included the combined conclusions of both reports) supplemented by two additional working papers. These publications began appearing in 1994 and each year through 1997.[49] It is not insignificant that these studies were funded by an array of USG Departments, DOD, State, DOE, DOC, NSF and the Rockefeller Brothers Fund. Undoubtedly these reports will reflect many if not all of the issues and approaches the USG would consider in its interagency discussions about the renewal of the 88STA. These studies were reminiscent of the procedures and timing followed when the USG began considering the renewal of the 80STA. These were begun in 1983 with a contractor study and followed a rocky path in the USG and difficult bilateral negotiations until the 88STA was finally signed in June 1988. These studies for the June 1998 renewal were much more elaborate, systematic, continuous, voluminous and substantive in comparison to the initial efforts in the mid-1980s.

The Office of Japan Affairs in the NRC was given the task of conducting these "overview studies" under the rubric of "maximizing U.S. interests in science and technological relations with Japan ... to resolving contentious issues emerging as a result of deeper U.S.-Japan competition and collaboration in science and technology." Based on the analysis under implementation in this Chapter these on-going "contentious issues" under the 88STA were basically two, access to JSTI and Japanese R&D facilities. I am convinced that the GOJ did not feel that "contentious [S&T] issues" stemming from the deeper USJ competition and collaboration existed in the mid-1990s between the U.S. and Japan, unless, of course, one insisted on putting SSC funding issue in this category. The presumed authority of these reports was closely tied to the position of the NAS/NRC in the U.S. S&T structure. While the NRC under the National Academy of Sciences is, in theory, a non-official institution, it conducts numerous studies at the behest of and funding by the USG and other institutions, its findings and reports are generally highly respected. Under these circumstances, it would be useful to comprehend the analysis and suggestions of the two working papers because they are more specific and tend to indicate more clearly the kind of thinking and mind set that underlay the generalized 1997 Report.

The 1994 Working Paper. It was concerned with JSTI. It should be kept in perspective that when the 88STA was being negotiated in 1987-88 – several years before these studies were mandated – the "hot item" was definitely the pursuit of increasing U.S. researchers studying in Japan and also – almost in the same breath but not quite – the effective and broader utilization of JSTI by U.S. researchers. For these purposes two Task Forces under the JHLC, one for access to R&D facilities, and another for JSTI were created. The imbalance in U.S. researchers in Japan and Japanese in the U.S. could not be corrected, of course, by definition because for one reason the U.S. did not have enough suitable candidates to fill its allotment of slots for researchers, the incentives for young U.S. researchers to study in Japan were and still are not a plus for their careers. Based on a survey, the joint USJ Task Force on Access to R&D facilities admitted the anticipated barriers were not found; thus, the cause for pursuing the Japanese in this regard had basically evaporated. Similarly, the Joint Task Force on STI – meaning of course JSTI – worked assiduously on improving access to STI with the various measures to be taken, again all on the Japanese side. Annual meetings were held in the U.S. to acquaint potential U.S. users about JSTI, yet the utilization of JSTI was not as high as it was hoped it would be, "not optimal" as the 1994 working paper reported on p. 8. Although the availability of JSTI has been focused on since the first Congressional Hearings on JSTI in 1984, this report incredulously declared that:

> Over the long term, U.S. government could work with U.S. industry, particularly small companies, to better define and specify their needs as input to bilateral discussions [i.e. a euphemism for the GOJ to take further actions] ... A growing number of U.S. high technology companies, both large and small, appear [in 1994] to be growing aware more of the need to monitor developments in Japan. ... It is an appropriate [time] to reexamine JSTI access and utilization issues. ... it still appears that access [in

Japan] and utilization [in the U.S.] are not optimal (p. 4-8).

Even after a decade plus of intensive bilateral efforts through the Task Force on STI and all the efforts previously conducted in this regard the U.S. is still groping for this definition. Half a dozen measures were suggested as possible new approaches on the U.S. side beginning with "an inventory of government [USG] JSTI programs, expanded role of consortia of university-industry-nonprofit organizations to articulate and define needs, more effective use of MT."[50] While recognizing that "Japan has taken steps" – of course, never enough to satisfy the USG or industry – "barriers to timely access to a range of important and potentially useful information remain. Some barriers reflect inherent characteristics of Japan's 'information culture' that Americans certainly need to understand but are likely to change only in the long term" – in our direction.[51] It then suggested half a dozen measures which could be taken up "through the mechanism established under" the 88STA, i.e. up through the JHLAP, JWLC, JHLC and if this was not sufficiently persuasive at the bilateral summit. The GOJ could make its information available on-line, could expand access to existing information services, the USG could pronounce yet another evaluation of the performance of the Japan Reprographic Rights Center, the USG/GOJ and private sectors could work to develop common standards for indexing and cataloging information, the USG and U.S. industry could articulate information needs, especially those of small companies, and the USG/GOJ could collaborate on MT with JSTI as a targeted application area.[52]

The 1996 Workshop report on strategies to achieve U.S. objectives. It listed several key findings in USJ S&T relations:

1. [Understandably] "U.S.-Japan cooperation in science and technology is not an end in itself. ... [and for this purpose the U.S.] will need to continue efforts to access Japanese resources, scientific and technological capabilities, markets and skilled human resources." One of the other objectives mentioned is moving from a "defense assistance paradigm" with Japan to "a new partnership in which sharing of technologies, engineering capabilities and risks with Japan will allow the United States to leverage constrained resources [small budgets] for defense R&D and procurement." Another is "to advance U.S. commercial and economic interests' through S&T cooperation with Japan. Still another is to cooperate in basic science to gain "expanded insight ... due to exposure to Japanese research."

2. There are "a number of long standing barriers to achieving U.S. objectives: accessing the Japanese market, low Japanese funding for basic research in open non-proprietary settings, legal or attitudinal factors related to cooperation in specific fields." This section states that the U.S. "Has not fully risen to the challenge of Japan opportunities" due to budget constraints and insufficient incentives for individuals and organizations to obtain the necessary expertise and pursue opportunities." This was an important recognition by the workshop. It clearly stated that "future U.S. access to a variety of Japanese assets are required in order for the United States to derive significant benefits from the relationship", such as financial assets. There was a recognition here that "effective planning, coordination and continuity of commitments on the U.S. side will be needed to ensure maximum leverage for U.S. efforts and resources." Greater access is needed to information in regulatory and legal affairs, trade data, patents and results of government sponsored research. Under "structural and ideological barriers" Japanese interpretation and implementation of "Japan's arms export principles have served to constrain cooperation in defense technology and equipment." IPR and underlying concerns about academic freedom were mentioned "as barriers to cooperation between U.S. companies and Japanese universities." Access to and for people "appear to have been resolved."

3. Other findings were: Streamlining the 88STA management structure, leveraging S&T relations to achieve U.S. objectives, centralized critical review of projects and programs has been difficult to achieve, a new element has been introduced with the recognition that R&D and innovation in Asia is growing rapidly requiring the need for increasing U.S. attention to this phenomenon while continuing to focus on Japan.

Possible Priorities and Strategic Options in four areas

Streamlining. Because of tight budgets and squeeze on staffing, it proposed "to eliminate or deactivate joint task forces or committees working in areas where major issues appear to have been resolved, such as the Task Force on Access." In light of the stress on and inadequacies of access described above this is truly surprising. It should be remembered that it was the USG which had insisted on this management structure which it was now proposing to slim down. But how then does one slim down and pursue the Japanese to do this and that for the U.S. based on this report to maximize U.S. interests. Surprisingly, the Japanese are now described as "the main barrier to such streamlining." Concerning the implications of re-opening the bilateral discussions about streamlining the management structure, the report stated that there was "some reluctance to risk another protracted renegotiation process" like that which happened in 1987-1988. On the other hand, it was pointed out that

> it was the possibility that the agreement [80STA] would not be renewed that brought the Japanese government to the negotiating table in the mid-1980s, and that it is sometimes necessary to take such extreme positions in negotiations with the Japanese government to achieve any worthwhile results.

This was an ominous implied threat and unfortunately an incorrect interpretation of what is known of the historical facts: the GOJ reluctantly agreed to negotiating a new STA, was extremely upset about the manner in which the negotiations were conducted by the USG and both sides used implied threats that they would be willing to let the 80STA lapse unless this or that happened. Refer also to Ambassador Mansfield's cable to the SecState mentioned in Chapter XI. Once begun, both sides did not want the negotiations to fail. The belligerent attitudes underlying the thinking in the 1996 report was a danger to the renewal process and, if adopted, could easily unduly complicate the renewal both within the USG and with the GOJ.

Linkage and Leverage. The report asserted that "an informal linkage already exists on a specific as well as on a more general level" between S&T and other aspects of the bilateral relationship and that "this linkage appears to have increased in recent years." Based on a reading of the available documents this "informal linkage" was barely discussed and was not a major issue. The U.S. delegation of the JHLAP did comment in a report to the Science Adviser and Chairman of the U.S. delegation, that the U.S. should raise the issue of cooperative research in telecommunications and concomitant access to the Japanese telecommunications market. But it did not appear as a major and contentious issue. Even while the two countries argued over the sale of supercomputers in Japan and the U.S., there were about half a dozen joint USJ R&D projects concerning supercomputers. It opined that "linking improved access to specific Japanese markets to government supported U.S.-Japan cooperation in important areas of research" and that the U.S. could "decline to renew the U.S.-Japan S&T Agreement unless specific objectives were met" should be considered as alternatives. It also, at the same time, raised a number of questions which it admitted could not be answered in a one-day session: would S&T sanctions really have an impact on U.S.-Japan trade relations particularly if more direct trade remedies were available? Can S&T be used effectively to achieve more open access to the Japanese market? Should the "existing information linkage" between S&T and trade be formalized? Does this informal linkage inhibit the freer flow of S&T? While it is perfectly legitimate to consider such alternatives, approaches and options as a matter of considered strategy, it is nevertheless symbolic of the mind set on the U.S. side and a clear and distinct reminder of the approach adopted in the 1987-1988 bilateral negotiations.

Improving U.S. Coordination and Implementation. Three points were outlined: 1) That USG representatives "speak with one voice". While this seems to be stating the obvious, this was a problem for the USG Team during the 1987-1988 negotiations with the GOJ so it needs to be stated. 2) Oversight of government-to-government cooperative R&D programs is a difficult issue for both USG and GOJ. The technical agencies enjoy their relative freedom to conduct cooperative programs but there is a constant tension between the felt need by some in State, OSTP to evaluate and review the bilateral programs for their worthiness, needs and reciprocity in a centralized manner. So long as there is inadequate staff and financing of OES and OSTP *for this purpose,* this wish will remain largely a mirage. 3) Effective communication between the government and the private sector. The USG has pushed for this linkage within the U.S. context and with the Japanese. Based on the known record it's effectiveness has been spotty to say the least. When, as the report admitted, most bilateral R&D between universities and industry occur "outside this official structure" it is difficult to mobilize the private sector, except for specific and specialized purposes.

Rising Asian Capabilities and Impact on the USJ Relationship. The inclusion of this issue is more than the bilateral S&T relationship. It is in this section that *for the first time* that "mutually beneficial" was used: "The major opportunities and challenges in structuring beneficial U.S.-Japan cooperation in addressing global problems that lie ahead." It pointed out with some concern that Japanese investments and education of Asian students might contribute to building an Asian infrastructure and innovation that is tightly linked with the Japanese system and less permeable than the American system. In this connection it urged that the U.S. pay more attention to Asia BUT not at the expense of paying attention to Japan.

This report also interestingly included in Appendix B a list of ten goals for USJ S&T cooperation which had been gleaned from a variety of sources. It is the most comprehensive and useful list that I have seen since it goes beyond the 88STA and brings together in one place a list of the hopes, ideals and potentialities of this bilateral S&T relationship – if one just believed that only a part of the listing were sincerely felt by both sides or could be implemented, it would be impressive.[53] One more could be added from Article VII of the 88STA calling for "widest possible dissemination of the information or equipment created in the course of cooperative activities. Since the focus, drive and objective or the overview exercise was to "maximize U.S. interests" in this relationship there was no discussion nor recognition of these goals toward which the bilateral S&T relationship should aspire; they were, of course, sadly ignored. While maximizing U.S. interests is perfectly legitimate, ignoring these goals which the study took the trouble to assemble must be interpreted that the USG goals in this exercise – the early steps in the USG consideration of the future of the 88STA – have little or no relationship to pursuing the more lofty goals of the much proclaimed USJ S&T relationship. At one time, much was made of the need to create bilateral R&D projects of Presidential/Prime Ministerial stature which was emphasized so strongly by the USG in the 1988 negotiations. These objectives, these hopes, these potentialities were never discussed nor even alluded to in the report. It is a sad commentary and distinctly disappointing that such a distinguished, senior and high level study group would not, or could not transcend the narrow issues of maximizing interests and discuss and propose a challenging, aspiring joint program for the greater good of the world – because after all, we have said over and over that the two technological superpowers could together make such a contribution. Perhaps this is the sad reality of the bilateral S&T relationship, that neither side has sufficient drive and will power to invest the necessary manpower, resources and time for these grander purposes and it may also and probably does reflect the underlying distrust – as exemplified in the inordinate stress on maximizing U.S. interest to the exclusion of other more positive possibilities – which inevitably negates any drive toward more ambitious bilateral R&D projects. On the other hand, surely a perusal of the goals enumerated in the footnote would give many opportunities and mandate to both the USG and GOJ to propose substantial major cooperative projects for the betterment of mankind; yet this was not done. It was the USG that was critical of the 80STA for its lack of Presidential/Prime Ministerial level prestigious bilateral R&D projects, and so insistent that such projects should result from the 88STA, yet none have been proposed so far by either country.

The 1997 Report on Maximizing U.S. Interests in Science and Technology Relations with Japan. In many ways this is a unusual report: it attempts to look at USJ S&T relations in a more comprehensive manner and in a broader context than those in the past; its provides what is calls "A Framework for Maximizing U.S. Interests..." The word "framework" was a term used with great controversy during the 1987-1988 bilateral negotiations; because of the changed circumstances between its inception and its final report which the report fortunately recognizes, the report also includes for the first time a discussion of the "Key challenges" emerging in Asia despite it's recent and continuing financial travails. The Report analyses S&T innovation before and after WWII in Japan and the United States, a statistical analysis of this relationship, the USJ competitiveness trends in automobiles, advanced materials, biotechnology and health care, semiconductor manufacturing equipment, information industries, key lessons for the U.S. In this sense this is an important document in further understanding U.S. attitudes and recommended policies through the lenses of this semi-official NAS/NRC Task Force. It is a useful reference document. Let us select some observations from this report which would symbolize their interpretations of the context:

> ... the U.S. economy and U.S.-based companies appear to be very strong today, while Japanese advantages have diminished and in some cases dissipated. ... The emergence of new high-technology competitors based in Asia is affecting U.S. and Japanese strategies, prompting a shift in focus from bilateral to multilateral relationships. ... [But many USJ] traditional asymmetries in exchange and cooperation, have not fundamentally changed. ... Many U.S.-based companies have improved their response to challenges from Japan-based competitors. Through investment in U.S. research and development and manufacturing, Japan-based companies have begun to contribute to overall U.S. capability to innovate. ... Japan is increasing government R&D investment aimed at strengthening its own fundamental research base and will move ahead of the United States in absolute non-defense government R&D spending early in the next decade (2000) if current trends continue. U.S. entities and the U.S. innovation infrastructure may derive increasing benefits from monitoring, tapping, and leveraging Japan's expanding capability; being second is not a familiar position for the United States. ... it is very possible that potentially harmful concentrations of high-

> technology production and capability [presumably to the U.S.] will emerge in Japan or in other countries in the future. ... Japan will be the most important country in this process [of accessing and utilizing a global science and technology base] for some time to come ... Although there is currently a growing segment of opinion that would write Japan off as a technological and industrial power, the task force reports document Japan's resiliency and continuing strengths in technology and innovation. ... The U.S.-Japan [S&T] relations both past and present can be considered a leading indicator of issues and questions that may emerge in relations with other countries, particularly the emerging techno-industrial powers of Asia (p. 2-8).

In this context, the 1997 Maximizing Report made 11 recommendations which go far beyond the narrower interpretation of just USJ S&T relations calling for the creation of an integrated policy for the USG relating to USJ S&T relations, patents and copyrights, worldwide trade policy, increased cooperation among US industry, universities, and government briefly described below.

1. Removing Barriers and Protecting U.S. IPR. The USTR should identify patents pending in Japan covering major scientific and technological advances, identify barriers to participation in the Japanese markets, investments and competition policy.

2. Keeping Up with Japanese S&T developments. Here the USG is urged to provide funds to train "Japan-capable" S&Es, maintain support for efforts to obtain, translate and disseminate Japanese S&T, business and policy information, press the GOJ to make available electronically regulations, laws, administrative guidance and other policy-relevant documents that impinge on the market and technology, and for companies and industry associations to expand access to Japanese markets and technology and a greater role in Japanese and U.S. policy debates. These are not particularly new ideas; they require determination, funds and staff and a sense of need and urgency. It cannot be said that there was an abundance of this drive during the past 15 years in this direction.

3. Use the 88STA to encourage effective program management, develop "a simple set of metrics" in areas such as personnel exchanges in publicly funded programs where lack of willing U.S. S&Es has been the problem.

4. USG, university, and industry should increase investments in S&T and new collaborative mechanisms, particularly focused on important commercial technologies linked with agency or broader national needs, such as SEMATECH. This is clearly a controversial industrial policy type recommendation.

5. Use of lessons learned from the USJ experience in S&T relations with the emerging nations, especially in Asia, where S&T is beginning to play to a major role.

It is significant that many of the more specific suggestions mentioned in earlier reports were subsumed into more generalized recommendations. While this report's recommendations have been sanitized to one higher level, I feel that the greater specificity in the other reports more clearly reflect the real intentions of the USG. In light of past experience, the lack of funds, staff and senior official time in the USG to invest in adhering to even the management structure of the 88STA would imply that the probability of the USG pursuing the above ideas with vigor and determination seems somewhat improbable.[54]

1998 Senators Roth/Bingaman Senate Resolution. On July 30, 1998, Senators William Roth (Republican, Delaware) and Jeff Bingaman (Democrat, New Mexico) submitted resolution S262 on USJ S&T relations as the bilateral negotiations to renew the 88STA were beginning. The purpose of the resolution as stated in the title was that "the Government of the United States should place a Priority on Formulating a Comprehensive and Strategic Policy with Japan in Advancing Science". As the negotiations for the 88STA began in Fall 1987, the Senate, under the guidance of Senator Rockefeller, held hearings about the USG position in these negotiations, to provide "an informal backing" to the USG Delegation and by implication a message to the GOJ. In light of this precedence, it is hardly coincidental that this time just as another set of negotiations are beginning a Senate Resolution is proposed – instead of a hearing.

Among the eleven "whereas" phrases, we might single out several of special interest: the GOJ's 1996 Basic Plan for S&T made S&T a higher priority area for investment for the GOJ, the GOJ FY1998 Supplemental Budget will increase Japan's S&T budget by 21%, the flow of science and technology from the U.S. to Japan is still larger than the reverse "due partly to barriers Japan has erected to the outward flow of STI as well as barriers to the inward flow of foreign investment and foreign participants in industrial organizations such as consortia and associations". This is quite amazing, self-deceptive and disingenuous since during the past ten years, Japan has been declared not guilty of these practices by two joint USJ task forces. Furthermore, it should be kept in mind that major U.S. consortia such as MCC and SEMATECH, when created steadfastly excluded Japan in particular and all other foreigners from this USG supported cooperative R&D effort. The resolution proposed three points as the "sense of the Senate":

1. the USG should place priority on formulating a comprehensive and strategic policy of engaging and cooperating with Japan in advancing S&T for the benefit of both nations as well as the rest of the world,

2. among other goals, that policy should aim to promote strategic cooperation in areas that further U.S. policy interests in S&T; more balanced flows of STI and personnel between the U.S. and Japan; more rigorous application of scientific methods in the development of standards and regulations to promote efficient technological progress and mitigate trade problems; and more equitable IPR,

3. the USG should integrate this strategic policy into current and future STAs with the GOJ.

This is probably about one of the few times that the USG is urged to advance S&T with Japan "for the benefit of both nations as well as the rest of the world". To rake up yet again the STI and Access issues is indeed unhelpful, if not a deviation from reality. That both Senators called for a "more equitable" IPR where questions "have existed far too long and should be rectified" is extraordinary since the detailed clauses in the existing STA were basically derived from USG proposals and there have been no IPR issues or problems under the 88STA or its predecessors over the past decades. In an accompanying statement to the proposed resolution, Senator Roth called for Japan to "move toward greater emphasis on [S&T] cooperation" as though this had not been the basis for the past decades since 1945. Senator Bingaman provided a more balanced assessment: he listed eight major projects where Japan and "nearly every department and agency" of the USG participated and all have "prospered as a result of the USJ STA" and will grow even more with the renewal of this Agreement. He did urge that ways should be found "to leverage increasingly scarce funds" to "solidify the cooperative linkages that exist between two countries. In the past that has been the code word for Japanese monies, an approach that has been notably unsuccessful. He did declare that we should take pride in this development ["interaction between our two countries exists on a scale far beyond what many once considered possible"] and that the current STA "to be an interactive arrangement of the highest importance." Senator Bingaman enumerated three criteria which the renewed STA should satisfy:

1. it must recognize that serious structural and procedural asymmetries still exist between the two countries and that they must be resolved,

2. it must provide freedom for S&Es to interact and complete their research as free as possible from government interference [to the best of my knowledge there has not been any problems on this issue under the many bilateral USJ projects over more than 40 years,]

3. it must recognize that the results that derive from USJ S&T cooperation has the potential to alleviate many of the problems we face in the world today and, as such, should be easily diffused into the international community.

The resolution was referred to the Senate Foreign Relations Committee but was not acted upon by the 105th Congress. Nevertheless, it was another factor that was taken into consideration to an unknown extent by the U.S. executive branch and the GOJ in negotiating another renewal.

The GOJ Stance. In contrast, the GOJ was satisfied and had become well accustomed to the 88STA arrangement as it had been with the 80STA. Consequently, from their standpoint, did not feel the need to conduct elaborate studies and surveys to assess and maximize GOJ interests in the USJ R&D relationship or even to propose ambitious big science projects. The GOJ philosophy about and approach to the 88STA was to rely on the initiative of the working S&Es to propose and discuss with counterparts in the U.S. or in any other country R&D projects of special interest to them – unless, of course, there were special projects in which the government needed to become involved, e.g. the IMS, the Genome Project, ITER, etc. Basically they were passive and waited for the USG to propose revisions – and the USG always seems to oblige. It repeatedly commented on the unreliability of the USG for providing and living up to its international R&D commitments; it felt that an international multilateral project would probably fare better in persuading the USG to make a commitment which it would probably carry out. The GOJ's approach is understandable, especially in light of how some of it's international R&D proposals have been viewed, in the first instance justifiably or not, with suspicion and distrust by the USG. Nevertheless, it was highly regrettable that in light of its insistence and advocacy of S&T and R&D for peaceful purposes and concurrence with the USG that combined efforts, resources, and determination the two technological powers could and should make contributions to the world, the GOJ appears reluctant to play a more pro-active role in pushing S&T for these purposes.

4. Act V, An Anti-climactic Renewal, 1999

As in the past, the GOJ requested that the 88STA be extended another five years without any changes as was done in 1993. The USG response was to request a re-negotiation of selected parts of the 88STA but also, as in the past, the proposed changes were given to the GOJ about two months before the expiration date of June 20, 1988. This naturally and automatically necessitated an extension in order for the GOJ to consider these proposals; this was reminiscent of the earlier approach used by the USG. Instead of the niggardly two months extensions, as in the past, the 88STA was extended initially for nine months to March 31, 1999. Because of gentlemanly but cautious disagreements the STA was extended a couple of more times and was then renewed on July 20, 1999 for another five years through a protocol amending and extending the 88STA.[55]

Based on available information, there was apparently, unlike the mid-1980s negotiations, a sense of trust among and between the two delegations and oral agreement that the discussions would focus on STA issues not other topics. This understanding was an obvious attempt to avoid the debacle of the 1988 negotiations where the USG injected issues not directly related to the problem at hand thereby greatly complicating the negotiations. The USG's proposed changes centered around administrative relaxation and IPR/copyrights. Since there had not been any IPR problems under the 88STA, there was understandably a deep seated skepticism by the GOJ why it was so necessary to change the existing already rather complicated IPR clauses – keeping, in mind, of course, that it was the USG which had sold itself and the GOJ that the IPR in the 88STA was to serve as a model for future STAs between the US and other countries. This model was, however, as mentioned, modified somewhat after 1988 and another "model" created in

1992. It was used as a partial basis for revising the IPR provisions of the 88STA. The GOJ cogitated over these proposed IPR revisions for months. Judging from these revisions, the elaborate studies described earlier appear to have had little impact on the reality of revising the 88STA. The revisions proposed by the USG were given to the GOJ in April 1998 through the AmEmb in Tokyo; subsequently three "negotiating sessions" were held alternately in Washington and Tokyo.[56] The Embassy was used more than during the 1988 negotiations; this was possible because of the type of revisions being discussed and because this required less travel costs and time of senior personnel. While these detailed revisions were seemingly quite innocuous were they really of such importance that they had to be included in such a high level agreement which resulted in a delayed signing – of 13 months – for a mere five year extension. As in the past, a USJ Bilateral Summit was used as a lever to persuade the GOJ to accept USG proposals. A "substantive agreement" was announced at the Obuchi-Clinton Summit meeting in early May 1999.[57]

They -- referred to in the summit communique as "The Leaders"-- also called for the preparation of a joint report by representatives from industry, academia, non-governmental and government sectors "toward the Spring of the new millennium" (March 2000?) on how S&T "can most effectively contribute to our societies and the global community and to identify areas in which enhanced cooperation would be desirable." Even though it is a much belated decision, it is a heartening step in the right direction for both countries hopefully to focus on how to accomplish the many objectives enumerated in fn 53 in this Chapter. Actually, the 88STA itself in Annex I already includes a list of seven areas of S&T cooperation. It is, to the best of my knowledge, the first time that such a determination was made to make this kind of joint analysis.[58]

The agreed upon revisions consisted of four parts: 1) the preamble, 2) reports, meetings and future plans, and 3) IPR and 4) copyrights.[59]

The Preamble. The existing 88STA began with nine paragraphs, recalling, confirming, stressing, desiring numerous understandings and expectations but did not include an overt statement linking S&T and industrial and trade issues. This linkage has been an objective and desire of the USG since the mid-1980s. A short statement has now been added "Recognizing that science and technology play more important roles in social and economic issues, including industrial and commercial issues, in this era of rapid innovation", making a total of ten "Whereas" paragraphs. The original USG proposal had included "trade" but this was too sensitive for the GOJ and was dropped. This addition stemmed from the recommendations of the Joint High Level Advisory Panel and perhaps was also a nod in the direction of the aforementioned un-passed draft Senate Resolutions 262. Since these preamble paragraphs have not become controversial in the past, neither side has invoked them for specific action, hopefully these paragraphs will continue to create an atmosphere without controversy.

Reports, Meetings, etc. The strict specifications for annual reports and annual meetings in the 88STA which had become a major administrative burden particularly for the USG were relaxed so that annual reports became "a report", the JHLC annual forum requirement became merely "a regular forum", the JWLC which was to meet at least annually became "in principle annually" and future plans for next year became "for next or the following year." This was merely a recognition of reality which had already begun in practice some years previously as described earlier in this chapter.

IPR. The third part on IPR gives the distinct impression that some USG lawyers were trying to cover some additional bases and contingencies which could or may arise in the future. The proposed IPR revisions were the most difficult to negotiate between the U.S. and Japan. It should be noted that the USG proposed that "IPR allocation will be determined by the U.S. and Japan on a case-by-case basis"; GOJ (and U.S. NSF) had advocated in 1988 that such arrangements should be on a case-by-case basis.

If an invention is made by an S&E visiting an institution in Japan or the U.S. the rights to that invention would be received by the inventor with an important proviso. The 88STA, in principle, gave those rights to the institution where that S&E was visiting. This seeming change is immediately hemmed with a fine print proviso that the inventor would receive these rights, benefits, bonuses and royalties "in accordance with terms and conditions of a standard arrangement between the Receiving Party and the Inventor". By convention and in reality these rights are almost always retained by the Receiving Party.

Under the 88STA, rights from an invention resulting from joint research would be based on a "mutually agreed upon disposition, on an equitable basis of rights" but the revised version states that the benefits shall be derived from an agreement made prior to beginning the joint research and shall take into account "the relative contributions of the parties to the creation of the Invention, the benefits of licensing by territory or for fields of use, requirements imposed by the parties' domestic laws and other factors deemed appropriate." If this agreement cannot be worked out then the joint research "will not be initiated". This provided more detailed specificity to achieve the similar goals as stated in the 88STA.

Copyright. The fourth part on the revised copyright section, like the IPR, contained details about assuring that the parties concerned "shall be given the opportunity to review the translation [of a copyrighted work] prior to its public distribution", publicly distributed copies of copyrighted work prepared under joint research would indicate the name of the author, and each party to cooperative activities would do their best to obtain a non-exclusive, irrevocable, royalty-free license in all countries. Is it truly Presidential to burden this agreement with such minutiae of etiquette. This issue of "translation etiquette" was not derived from the aforementioned model IPR.

Other Forms of Intellectual Property. The objective of this revision was to cover presently unknown kinds of IPR and not protected by the laws of one Party's country. If this occurs then the cooperative activity from which this potential IPR stems "will be suspended" while consultations take place; if no agreement is reached then the joint activity may be terminated.

Lastly, it must be remembered that a great-to-do was made by the USG that a Presidential agreement must be revised in an appropriate manner for appropriately high level R&D cooperation, yet the second five year extension – unfortunately not even ten years – was signed without fanfare on July 20, 1999 by a mere acting secretary of state for OES (for the U.S. Secretary of State) and the Japanese Ambassador to the U.S. with additional details on various contingencies that MAY arise in IPR and Copyrighted works, quite an anti-climactic conclusion to a major Presidential/Prime Ministerial S&T agreement between two major technological superpowers.

Notes

1. Interview of William Graham in *Nihon Keizai Shimbun* 6/23/1988. Though a small item on p. 7, it was indicative of the approach and intent of those in the USG who conceived and negotiated the agreement.
2. The principal sources of information for this chapter are the reports (and some tabs attached to the reports) of the several committees and task forces operating under this agreement. The chapter will be divided into 1) the implementation of the new STA as can be gleaned from the various reports from 1988 to the renewal of the STA and 2) special issues (such as political "science and technology") under Presidents Bush and

Clinton and the 88STA, the 88STA as a model S&T agreement, the IPR clauses in the 88STA as a model for negotiating an STA with other countries and 3) the renewal of the 88STA in 1993 and 1999.

3. Based on my experience obtaining gray literature in the U.S. on U.S.-Japan S&T issues was as much a problem as it was in obtaining similar literature in Japan. One simply had to know what literature existed, whom to ask for a copy and when to ask for a copy.
4. The interest of the USG in trying to persuade the GOJ to be financially more forthcoming, especially in support of the SSC, is exemplified in the preparation of a talking points paper of nine points on big science initiatives for discussion at the 1990 JHLC. Some of the other points were: the USG is exploring new tools for dealing with international cooperation in big science to overcome national rivalries to share costs and benefits equitably, such as to average the share over a number of different megaprojects, the need to develop more stable and credible financial arrangements (aimed at itself because of foreign criticism of the USG for its unstable budgetary support of such projects).
5. The "reliability issue" was not merely raised by the GOJ, it was a problem recognized in several recent USG and other studies, such as *International Partnerships in Large Science Projects*, U.S. Congress, Office of Technology Assessment, 1995. 131 p., *Megaprojects in the Sciences*, President's Council of Advisors on Science and Technology, 12/1992. 28 p., *The United States as a Partner in Scientific and Technological Cooperation: Some Perspective from Across the Atlantic* by Alexander Keynan, 6/91, 88 p, *Megascience Policy Issues*, OECD, 1995,199 p. specifically a chapter by Albert H. Teich and *Future Potential Science and Technology Cooperation between Japanese and the United Sates* by Shigeo Yoshikawa and Cecil H. Uyehara, 3/96, 117 p. Teich of the AAAS wrote pessimistically about the U.S.'s inability to make long-term financial commitments to international projects. The OTA study while admitting that the cause of this sense of unreliability among U.S.'s partners stems from a few big projects, it explained that this occurred because of changes in Administrations and policies, budget pressures, and have tended to be exceptions to the U.S. record in international collaboration (p. 18).
6. In January 1992, the Bush-Miyazawa summit meeting established "a joint working group to examine technical and other essential aspects of the SSC project and to consider how this project can be formulated as an international project to enable Japan to participate in the project." The working group met twice in 1992 and the technical subgroup met eight times in 1992; the GOJ suspended this process in November 1992 after the U.S. Presidential election. In early 1993, the GOJ linked its participation in the SSC to a satisfactory conclusion of the redesign of the Space Station Freedom, a firm commitment from both the Administration and the Congress that the project would be completed but also advised that the request for $1 billion plus contribution was unrealistic and that the project needs to be appropriately internationalized. Some more meetings were held in 1993 but ultimately the GOJ respectfully declined to support the SSC project.

   Ambassador Ōta published a book on science in the U.S., *Kuzureyuku Gijutsu Taikoku* (The Waning of Technological Leadership: How the U.S. sees itself [Ōta's English title]. The Simul Press. 1992. 242 p. Ōta later became Japanese Ambassador to Saudi Arabia. This book is about the anxieties of the U.S. and its position in the S&T world in the 1980s and concludes with a scenario for the revival of the U.S. as a technological leader.

7. The depth of GOJ concern about U.S. reliability is well-expressed in an article in *Look Japan* (9/1993, p. 22-23) by Yamada Yoichirō, Assistant Director of Scientific Affairs, MOFA: "... the withdrawal of support by one partner can drive an entire project to a halt." When the Clinton Administration in its first months in office in spring 1993 announced its intention to redesign the Space Station Freedom, "shivers went down the spines of international partners, Canada, the European Space Agency and Japan all of whom had already invested large amounts of money in the project." It "was a relief" when President Clinton announced on June 17, 1993 after consultations with its international partners that the project would continue "based on a scaled-down version of the original design." But when the House of Representatives endorsed the manned space station on a 216-215 vote, again the international partners "worried" about how reliable is the U.S. The MOFA writer expressed the hope that the U.S. would "introduce a system that allows for long-term financial commitments to major international cooperative projects."

   The January 1994 Discussion Paper on Formulating U.S. Research Policies within an International Context, by the Government-University-Industry Research Roundtable endorsed "Greater international cost-sharing for research as enhancing the breadth of U.S. research and trade policies" (Proposition 4), p. 10.
8. The Panel was obviously unaware that the topic of the fourth U.S.-Japan Science Policy Seminar in October 1986 -- only 2.5 years earlier -- was a study on engineering education in Japan and the U.S. The quotations cited are from an OES paper, *Presentation to the Joint High Level Advisory Panel of the US-Japan S&T Agreement* [5th meeting 9/1994].
9. Comments cited are from Report of the U.S. Members of JHLAP to Dr. John Gibbons, U.S. Chairman of the JHLC, Washington, D.C., December 22, 1994. This report was for U.S. members only. A separate joint report with the Japanese Panel was being prepared.
10. This suggestion has been mentioned a number of times in various reports but little of substance was accomplished. Part of the problem lies in the copyright issue which would prevent the USG from distributing translated copyright materials from the Japanese.
11. According to a senior U.S. member of the STI Task Force, the S&TA urged the USG to exert outside pressure (*gaiatsu*) on the Japanese to stimulate the creation of the Reprographic Rights Center, the four objectives were quickly established in one afternoon without contention, implicitly the issue revolved around the flow of STI from Japan to the U.S., the reverse was never raised. This same person also commented that U.S. individuals' use of domestic U.S. STI data was "not up to snuff" although it was improving, and that their knowledge of JSTI was indeed meager.
12. A similar set of JSTI Conferences have been held in Europe starting in UK (1989), Germany (Berlin 1991), France (Nancy 1993). The fourth conference was scheduled for Sweden in 1995 but not held. The next Conference was held in July 1997 in the U.S. -- perhaps this will be the last such conference. I attended two in 1991 and 1993; they were indeed enjoyable but I was highly skeptical about the continuing value of these meetings since there was a tendency for approximately the same group of people to attend these delightful get togethers of the faithful.

   In a way, the MT Center in the DOC is the culmination of an effort to obtain an MT capability in the U.S. by Congressman Doug Walgren and myself in 1988 -- but the objective then had been to develop a genuine U.S. capability not a borrowed system. The DOC sponsored a national conference on Japanese/English computer--aided MT at the National Academy of Sciences in December 1989. A National Science Foundation funded study was also conducted in late 1990 to study and assess the status of MT capabilities in Japan. The DOC-JICST Agreement was signed May 1, 1996 with an accompanying contract whose objective was "to evaluate the performance and the applicability of the machine translation system developed by JICST for use by researchers, engineers and scientists in the U.S."

13. The Japan Documentation Center (JDC) was created in 1994 in the Library. Its mission was to collect difficult to obtain Japanese documents relating to policy issues and to conduct research on selected topics requiring the use of these documents for both governmental and non-governmental organizations. So far, JDC has received about 900 inquiries per year. The JDC was abolished in March 2000.
14. It should be pointed out that the eleventh report of the Science and Technology Council, chaired by the Prime Minister and which sets the basic R&D policies for Japan, stressed the role of basic research, creativity, access to new knowledge from abroad, Japan's obligation to the rest of the world as a contributor to new knowledge, the strong encouragement of the internationalization of Japan's R&D enterprise, laying the policy foundation for and leading hopefully, thereby to an increased number of foreign researchers in Japan. The question as always was, would Japan be sufficiently attractive for European and particularly U.S. S&Es to invest their time and effort in a long term R&D sojourn in Japan.
15. *Directory of Japanese Company Laboratories willing to Receive American Researchers*, 3/1/1991, 157 p. It surveyed 553 Japanese company laboratories with more than 30 researchers; 284 responded, and 154 said they were willing to receive U.S. researchers. This Directory of R&D surveyed opportunities in all major Japanese companies and in 25 industrial fields: transportation (automobile, shipbuilding, etc), ceramics, cement and glass, chemical (fat, oil, detergent cosmetics), chemical (inorganic/organic), chemical (paints/surface coating/ink), chemical (plastics), chemical (others), construction, electrical/electronics, food beverage, information service, iron & steel works, machine manufacturing, machine products, manufacturing (other), non-ferrous metal (mining and smeltering), petroleum/coal products, precision machinery manufacturing, precision machinery products, research, rubber, textiles, utilities.
16. In March 1992, the GOJ (MITI/AIST) issued a report, *Interim Report on International Exchange of Researchers* [translated title]. The report was originally published in Japanese and later translated into English by FBIS as JPRS-JST-93-012 (3/23/1993), 125 p. The report was prepared by the "Society for Studies Toward New Development in Researcher Exchange". This Society consisted of five senior Japanese company officials at the managing director level (e.g. Hitachi, Mitsubishi Petrochemical, etc), two professors (Hiroshima Institute of Technology, Tokyo University), and the chairman, formerly with MITI/AIST and significantly the science counselors from the British and U.S. embassies, the economic counselor from the French Embassy. This report either in the Japanese or English versions was never referenced in various committee/Task Force reports, etc. Some 1500 survey questionnaires were sent to private companies, national laboratories, and universities.
17. This new finding only confirmed what had been known for a quarter century. The USJ co-chairmen's statement commemorating the 25th anniversary since the initiation of the 1961 USJ Science Cooperation Program, noted that "While differences in language and separation by [the Pacific Ocean] pose severe difficulties, it is surprising how fully

such difficulties can be and have been overcome by scientists willing to make a special effort to do so. ... English has become sufficiently common among Japanese scientists that there are now a great many academic departments and laboratories in Japan where American scientists with only a minimal command of [Japanese] can work effectively with their Japanese colleagues." (From a special commemorative report published in 1986).

18. The figures and quotations in these six items are from the 1990 Report of the Task Force on Access and appendices. These numbers may differ from other cited statistics because of the different periods covered and compiling institutions. The same kind of results were obtained in subsequent surveys in 1991, 1992 and reported in the annual Task Force reports.
19. These programs are 1) the Japan-U.S. Fellowship Fund from the GOJ to USG for supporting long term research stays in Japan, 2) Post-Doctoral Fellowships of the S&TA, 3) the Post-Doctoral Fellowships of the Japan Society for the Promotion of Science, 4) the Post-Doctoral Fellowships of AIST/MITI and agreement between AIST and NSF for up to 30 American researchers per year in AIST laboratories, and 5) Japan Foundation's Center for Global Partnership to encourage cooperation between the U.S. and Japan across a broad spectrum. *(Task Force on Access Summary Report,* 9/1992, p. 2).
20. The Summer Institute program did not consist of full blown research projects for individuals; rather, it was intended "to introduce U.S. graduate students to Japanese science and engineering research laboratories and to initiate personal relationships which will better enable students to collaborate with Japanese counterparts in the future." Indeed a far cry from the type of Presidential project which was supposed to be created under the new STA. This program notwithstanding its modest scale was an example of close and cooperative efforts of the agencies of both the USG and GOJ. (A similar program was also sponsored by the MOE in conjunction with NSF and the Deutcher Akademischer Austrauschdienst of Germany starting in 1993).
21. NSF Report Memorandum #94-7 (11/15/1994). 64 p. This report divided the varied and extensive series of programs for U.S. S&Es to study in Japan into four groups. This report described only the breadth and depth of training projects which naturally excluded all the other R&D projects between the two governments.

    I: Those involving the NSF: U.S.-Japan Cooperative Science Program (started 1961, number of projects in 1994, 63), Japan Society for the Promotion of Science Post-doctoral Fellowship Program (1988, number of researchers in Japan in 1994, 56), Science and Technology Agency Post-doctoral Fellowship Program (1988, 40), NSF-Center for Global Partnership Fellowship Program (1992, 27), Summer Institute Programs (1990, 60), MOE Research Experiences Fellowship for Young Researchers (1993, 3), NSF Dissertation Enhancement Program (1993, 3),

    II: Programs Administered by Japanese Organizations: JSPS Fellowships for Foreign Scientists (1959, 75), Japan Foundation Dissertation and Research Fellowships, (1972, 51), ERATO (1981, 4), Japan Trust Project of the Japan Key Technology Center (1985, 0, so far 5), RIKEN Frontier Research Program (1986, 9), Agency of Industrial Science and Technology (1988, 1), International Superconductivity Technology Center Fellowship and ISEC Scholarship (1990, 1),

    III: U.S. Air Force Office of Scientific Research. The U.S.-Japan Industry and Technology Management Training Program was begun in 1991 with funds from the National Defense Authorization Act. Such programs were created at the Universities of

Michigan (1991, 37), Vanderbilt (1991, 2), Pittsburgh and Carnegie Mellon (1992, 59), New Mexico and Texas (1992, 54), California (Berkeley (since 1988, but USAF funding since 1992, 5), Washington (1991, 4), Stanford (NA), Weber State College/Brigham Young, Wisconsin-Madison (1990, 61), University City Science Center (1993, 18) and MIT (1981, 60),

IV: Other programs: Stanford University Center for Technology and Innovation (1988, 35), U.S.-Japan Education Commission (1952, 52), and Manufacturing Technology Fellowship (Through 1996, 70, not as many as originally hoped for). Fifty Japanese companies were participating.

22. *Task Force on Access Summary Report,* September 1992, p. 3. This issue of "barriers" was also focused on in a memo, *Notes on the Extended-term International Mobility Issue* (2/12/1987), by the CISET Working Group on International Infrastructure, Education and Facilities. It said the nature of this problem was more complicated than it first appears. It pointed out that even when residency abroad was permitted with some programs (e.g. NSF's plant science, engineering and mathematics programs) for many years virtually all recipients for those fellowships choose to spend their tenure in the UK or Canada, suggesting that language was a significant barrier to long term mobility. The low stipend of $1600 per month, not competitive with U.S. salaries, in the highly competitive U.S. job market, concern at the post-doctoral level that chances of gaining tenure depend on day-to-day contact with senior faculty, concern about the loss of laboratory space, and the perception that the best research is going on in the U.S. in any case, perceived difficulties in acclimating to a foreign culture both in and out of the laboratory, the prevalence of two-career families, and the barriers to the foreign employment of spouses were other inhibiting factors. It also pointed out that a survey by an NIH contractor indicated that these concerns were frequently reinforced by senior medical school faculty and deans. The attractiveness of studying abroad even for short periods would require a change in both management philosophy, practices and culture of the sponsoring organization. This study did not single out any country for criticism, or that any country (including Japan) was deliberately keeping out U.S. scientists explicitly or implicitly. This is an important finding in light of how the issue of mobility and access to Japanese R&D facilities were handled in the bilateral negotiations.
23. NSF Report Memorandum 98-08 (4/16/1998), *US-Japan Exchange Programs for Postdoctoral Fellows and Graduate Students: Accomplishments and Trends*, 4 p.
24. It must be emphasized that this Task Force focused naturally only on S&T opportunities in various forms in Japan. A broader study was presented in *American Students in Japan: Educational Imbalances and Opportunities* (JEI report 3/3/1995, 17 p.) where it describes "options around every corner" to study in Japan in all fields of intellectual endeavor. Another report on this subject is *Foreign Students in Japan: An update* (NSF Report Memorandum #96-8, 4 p.). Between 1994 and 1996, this Task Force held one formal meeting but did not produce a report. Letter, Larry Weber, NSF to Cecil H. Uyehara, 3/18/1997.
25. This list of technical areas was suggested by the USG during the negotiations. In a [1996] GOJ document outlining the structure and activities under the 88STA, the GOJ lists five main points. The fifth point is "the list of the seven technical areas which are of high interest to the U.S. side". This description is interesting as an example of how some parts of the 88STA are perceived by the GOJ even eight years after the agreement was signed; it is referred to not a joint list, but a list of special interest to the

USG. This choice of technologies would imply that the USG felt that there as something to be gained by the US from Japan in these areas.

26. These data derive from 50 Agreed Upon Projects, 9/1991, 2 p. and Summary List of Agreed Projects prepared by the GOJ probably STA, bilingual, as of 10/12/1995.
27. Presentation to the Joint High Level Advisory Panel of the US-Japan S&T Agreement, undated but prepared for that Panel's 5th meeting in 9/1994.
28. *Joint Conclusions and Recommendations to the JHLC of the Joint High Level Advisory Panel on US-Japan S&T Relations*, Seattle, 10/19-20/1995.
29. Science Protocol, Yamada Yoichirō, Assistant Director, Scientific Affairs Division, MOFA. *Look Japan*, 9/1993, p. 23. On the U.S. side, a more systematic approach was attempted in measuring benefits in *International Cooperation in Research and Development: An Inventory of U.S. Government Spending and a Framework for Measuring Benefits* (Critical Technologies Institute, Rand Corp. MR-900.0-OSTP by Caroline S. Wagner). Interestingly this report chose earthquakes and seismology as the case study to test the framework for measuring benefits. This choice included one of the oldest and long standing cooperative joint USJ R&D projects. It identified three measures which could be reliably used to measure benefits: bibliometrics, survey of participants, and expert judgment on standards development. The number of papers increased through joint research and when written by two authors, Japanese researchers were the most likely collaborators with a U.S. author. Almost one half (47%) of the projects were supported in greater amounts than the U.S.; 35% were equal in contributions. The highest leverage of funds with research projects was with Japan. These joint projects helped U.S. researchers to stay at the state-of-the-art (p. 33). At the very least, this tentative attempt to measure the benefits of international R&D cooperation was positive.
30. Information on the TGH is derived from "*Techno-Growth House*", managed by the Japan Industrial Technology Association under commission by the AIST/MITI, and NSF Report Memorandum #96-7 3/7/1996) on the TGH and several briefing papers, *Techno-Growth House*, prepared for JWLC and JHLCs.
31. The expression used in USG documents on this subject is "We [the USG] ... are pleased that MITI has allowed U.S. representatives to be on its management board [of the TGH].
32. While the DOC/USG is no longer involved in the *direct* management of the TGH by being on a board of management, it signed an MOU on June 5, 1995 with AIST/MITI on "a cooperation plan" on the TGH. Keep in mind that the TGH was wholly funded and is managed by the GOJ, yet this MOU specifies that eight office facilities, accommodation rooms, a lounge, a seminar room and a library are available for foreign users of TGH; basically this means that the USG has a first call on these facilities. If no foreign user with his government's support applies for the use of an accommodation room, it can be released for no more than two weeks. Except for the accommodation rooms the number of which are not specified, the enumerated facilities constitute the entire TGH. This kind of foreign control would be unimaginable in the U.S. for a USG funded and managed facility. According to the TGH, French researchers have used it for 12 days, Italians for 2, and the U.S. for 126 days out of 142 days from the opening to January 17, 1997. I am informed that the Guest House and TGH are fully used almost all the time by other foreign S&E s working with Japanese S&Es.

33. After the first Clinton-Miyazawa summit meeting in July 1993, a Joint Statement on the U.S.-Japan Framework for a new Economic Partnership was issued. It contained the "Common Agenda" plan. In 1994 and 1995, however, no joint statements or communiques were issued; a statement was issued in April 1996. In the meantime, joint reports were issued for two meetings in 1995, and again in June 1996.
34. It should be remembered that no joint statements were issued after the 1994 and 1995 USJ Summit meetings, presumably reflecting difficulties in preparing such agreed-upon joint statements. The several joint reports, therefore, are concerned with progress and developments measured against earlier joint statements. They could be interpreted as an expression of non-confidence in the discharge of the Task Force's responsibilities. It was another venue used by the USG to pursue the GOJ on "access" to technology which conceptually accessed both to STI and R&D facilities and showed how the GOJ was "challenged" on S&T issues from different direction from differs parts of the USG.
35. An undated paper, *U.S.-Japan Science and Technology Relationship*", prepared for the 5th JHLAP in 1994, p. 2.
36. ibid. p. 2.
37. This comment is from an interview with a science counselor at the Embassy of Japan in Washington, October 1996.
38. Based on available documents, the "technology basket" under trade discussions appears to have started vigorously in 1993 and continued to 1995 but appears to have petered out like a comet in the sky. It apparently stemmed from the State's Economic Affairs Bureau which felt constrained that its trade discussions should include a technology aspect. It was apparently not meant to displace the S&T issues in the Common Agenda or the Task Force for STI under the 88STA or even be an expression of non-confidence in the discharge of the Task Force's responsibilities. It was just another venue used by the USG to pursue the GOJ on "access to technology" which conceptually includes access both to STI and facilities. It is included here even though it faded away, because it is yet another example how the GOJ was "challenged" on S&T issues from different directions from different parts of the USG. Based on a draft cable (12/21/93),the main focus of access to technology in this instance was on the involvement of the private sector. "The numbers suggest that technology is less accessible in Japan"; thus the emphasis on how this situation "can be improved within the context of existing R&D Budgets and other conditions" but the USG chair observed that neither governments would change their R&D expenditures sufficiently to effect the ratios cited – then why raise a useless issue? The GOJ delegation included some MOFA officials who were quite willing to take on the USG officials and challenged their assertions, their statistics, their assumptions, and called upon the U.S. research to do more for itself. The US chairman said, however, that "achieving a balance would require more of an effort from the U.S. side. The USG also proposed the creation of technoparks in Japan to help foreign (i.e. U.S. small and medium) enterprises (SME) to set up offices in Japan at subsidized rates. Does the USG do this for other foreign SMEs? The GOJ asked how the USG was implementing the metric system, etc.
39. I understand they were retired to the Reagan Library. Just before this happened I met with a sympathetic staff person with the EPC who was also en route to being "retired"; he provided me with a detailed oral description of the important July 1987 EPC meeting the minutes of which were then classified. Subsequently, I was given a copy of this report through the FOI process.
40. This was confirmed in an interview with an OES staff officer in Spring 1996.

41. Based on a conversation with Jeff Schweitzer, Senior Policy Analyst, OSTP in January 1992 and with Larry Huffman, State/OES in 1996.
42. This was also confirmed by OES staff in interview in 1996. The survey of State files on STAs was conducted in 1994.
43. A model IPR Annex was given to me by a senior policy analyst in the OSTP in January 1992. The standard annex consisted of 3 pages with seven pages of guidelines for the use of this new standard annex. According to this analyst, the felt need to prepare a model IPR was based on experience gained from IPR negotiations and implementation particularly regarding the following two problems: 1) in personnel exchanges the receiving side or party would receive all rights and benefits, a too one sided arrangement and 2) the description of types of S&T cooperation and definition of allocation of rights for each type of cooperation were too rigid. The 1992 model IPR was intended to revise this approach and make the conditions somewhat more flexible. Although this model IPR was available the USG did not raise the issue of revising the IPR section in the USJ 88STA during the 1993 renewal negotiations; undoubtedly, such a move would not have been welcomed by the GOJ in light of the difficult negotiations only five years earlier. But the model IPR's applicability was raised by the USG during the second renewal process in 1998-1999. See section 4 An Anti-climactic Renewal, 1999 toward the end of this chapter.
44. In 1996, the American Chamber of Commerce of Japan made a study and evaluation of trade agreements between the U.S. and Japan from 1980 to 1996 (*Making Trade Talks Work: Lessons from Recent History*, Tokyo. 1996. 188 p. One of the "trade agreements" was the 88STA. The results brought about by the agreement "are broadly positive. ... A team of several individuals rated the agreement a seven on a scale of one to ten. Within the team there is some diversity of opinion in a few areas, but the consensus is favorable." (p. 56) There was no indication or elucidation who constituted this small group of evaluators and what were their qualifications to make a judgment.
45. *U.S.-Japan S&T Agreement (STA) -- History on Renewal Discussions,* lists 20 dates concerning various exchanges of memos among USG agencies about the renewal. In earlier internal discussions OSTP seemed to procrastinate and not attend inter-agency meetings. This time it was the USTR. This absence was particularly noted under "suggested talking points" of the same paper.
46. The above is based on State 284123 (9/1/1992) and the identical document, *U.S. Paper: Renewal of the U.S.-Japan S&T Agreement*, September 2, 1992, 4 p.
47. The original GOJ response in Japanese was not available. The description of the GOJ response was based on the American Embassy translation as reported in Tokyo 15541 (9/18/1992).
48. *Look Japan*, 9/1993 (p. 22-23) published an article on the renewal of the STA by Yamada Yoichirō, Assistant Director of the Scientific Affairs Division of the MOFA.
49. 1994. National Research Council (NRC). Committee on Japan (COJ). *Japanese Scientific and Technical Information: U.S. Industry Needs, Access and Policy Issues*, 9/26/1994. 30 p. A working paper of the Competitiveness Task Force.
    1995. NRC, COJ. *Strategies for Achieving U.S. Objectives in Science and Technology Relations with Japan.* 19 p. Report of a Workshop.
    1996. NRC. COJ. *Maximizing U.S. Interests in Science and Technology Relations with Japan.* 126 p. Report of the Defense Task Force.
    1997. NRC, COJ. *Maximizing U.S. Interests in Science and Technology Relations with Japan.* 140 p. Committee on Japan Framework Statement and Report of the Competitive-

ness Task Force.

50. Inventories of USG JSTI activities have been made several times in the past with no known effect, yet this report made a recommendation for yet another report. It said 26 USG agencies were found to be collecting and disseminating JSTI. Wide dissemination of these materials would clearly involve copyright issues with Japan which was recognized in the 1994 working paper.
51. P. 14. The assumption underlying this assessment was that the U.S. treats information "as a public good", in contrast the Japanese regard information "as a proprietary source" (p.10). This is a highly debatable assumption and judgment upon which to base this comment.
52. P. 15-16. Concerning MT, the DOC and MITI agreed in 1996 that GOJ would and later did provide an MT system in the DOC MT Center to expedite availability of JSTI.
53. They were selected from various parts of the 88STA, the USJ Framework for a New Economic Partnership (including the Common Agenda for Cooperation in Global Perspective) and a number of agency-to-agency agreements.

    1. USJ sharing responsibility in contributing to the world's future prosperity and well-being by strengthening respective national research and development policies.

    2. Sustaining long term investments in basic research and creating dynamic R&D environments with a view to generating fundamental new knowledge.

    3. Ensuring the protection of intellectual property rights so as to preserve the value of innovations derived from joint collaboration.

    4. Providing for the smooth application of new technologies.

    5. Nurturing and expanding the next generation's human resources in S&T.

    6. Ensuring long-term mutually beneficial international S&T collaboration through long-term partnerships between scientists of different nationalities, performance of joint research and development at each other's facilities, education and training of each other's promising students, and publication of joint research and development results in international journals.

    7. Affirming commitment to equitable contributions and comparable access to each nation's R&D systems.

    8. Sharing responsibilities and mutual and equitable contributions and benefits commensurate with the two nation's respective S&T strengths and resources.

    9. Promoting international competitiveness.

    10. Promoting S&T cooperation in areas of global concern.
54. In addition to these studies, there is also as mentioned earlier, the Report of the Defense Task Force, *Maximizing U.S. Interests in Science and Technology Relations with Japan*. This study was intended to provide one of the inputs into the final report covering the multifaceted USJ S&T relations. The recommendations of the Defense Task Force report were as follows:

    1.The DOD should pursue technology reciprocity in the defense relationship with Japan as a major goal.

    2. The USG should seek to reduce or eliminate barriers to technology flow that result from Japanese policies. The U.S. should seek from the Japanese government a) clarification of the arms export principles and a public statement that export of items embodying substantially commercial technology that undergo minor modifications for defense applications are not restricted and b) change to the 1983 exchange of notes stating that Japanese military technologies transferred to the U.S. are exempt from retransfer restrictions, with changes addressing legitimate Japanese concerns and including

provisions for the payment of royalties.

3. DOD should develop a new mechanism for facilitating technological collaboration between U.S. and Japanese companies to address common defense needs. One promising approach would be a program to fund U.S.-Japan industry R&D on specific enabling technologies – including the adaptation of commercial technologies – targeted at applications in future weapons systems.

4. The U.S. and Japan should institutionalize an enhanced comprehensive security dialogue featuring an integrated discussion of the political-military, economic, technological and other aspects of the relationship.

5. DOD should ensure a coordinated approach in future collaborative defense programs with Japan. One approach that might be adopted as a minimum is designating a single authority with the responsibility for coordinating strategies toward major systems in which collaboration with Japan is under discussion.

6. DOD, in cooperation with the DOC and other appropriate agencies, should continue to build capabilities to monitor and manage dependence on foreign sources of critical technologies, with the goal of ensuring U.S. access.

55. The STA was extended on 6/16/1998 for nine months to 3/31/1999, on 3/19/1999 to 5/31/1999 and on 5/19/1999 to 7/31/1999.

56. The dates of these meetings were 12/18-19/1998, 2/17 & 19/1999 and 3/9-10/1999.

57. Using Summit meetings as a Lever. In an unclassified "sensitive cable" from Acting A/S OES, Melinda Kimball, to the [U.S.] Ambassador [in Tokyo], the latter was urged to mention in discussions with the GOJ to persuade it to accept our revised proposal that "While we take the point raised by your side [GOJ] that this delicate negotiation should not be driven by high-level meetings, the fact is that the President and the Prime Minister will meet early next month and we [USG and GOJ officials] have no choice but to inform our leaders as to the progress of our negotiation." (State 071525 4/17/1999)

U.S.-ROK S&T Relations. A revised U.S.-Republic of Korea (ROK) STA was signed on July 2, 1999. A cursory comparison of this STA and the USJ STA will reveal major divergences in the two arrangements. While the USJ STA has ten "whereas paragraphs" to underpin the relationship, the US-ROK STA has none. The original US-ROK STA of 1992 created a "Joint Committee" for "coordinating and facilitating cooperative activities". This was only a few years after the USJ 88STA was signed when USG said to itself and the GOJ that this STA was to be model. This Joint Committee was similar to the Joint Committee in the USJ 80STA which had been criticized as disadvantageous to the U.S., vilified, denounced as inadequate by the USG. Instead the USG demanded the three tiered management structure to govern USJ S&T relations. No such tight oversight was imposed on the South Koreans. The US/ROK Joint Committee would be "composed of representatives designated by the Parties." and shall conduct "A joint review of activities under this Agreement every two years". There is no mention of "translation etiquette" to be followed in the US-ROK agreement. Perhaps the Koreans are more polite and considerate. In revising the 1992 STA, IPR provisions were also a stumbling block for the Koreans. The "allocation of rights" is similar in detail to the 1999 revised USJ STA. The USG insisted that as part of its planned model STA structure, that a "policy framework for the overall science and technology relationship [between USJ]" must be created, that "principles and provisions" must be established, enumeration of steps to be taken to "strengthen the overall [USJ S&T] relationship" must be clearly stated in the STA; under each of these items numerous steps are listed for actions to be taken presumably by both Japan and the United States. None of these kinds of detailed binding and obligatory actions are included in the US-ROK STA. The latter agreement does include an Annex on "security obligations" similar to the USJ STA but there are apparently no side letters as there were in the USJ STA. Overall, the US-ROK STA is far looser than the USJ STA; Japan originally requested a similar kind of agreement in 1988 with the U.S. but the USG demanded serious consideration of only its draft agreement and the GOJ bowed to this demand.

Record of Discussion. A "Record of Discussion" of July 16, 1999 consisting of seven points, was created and became a part of the USJ STA agreement's record. My copy was obtained from the U.S. Department of State's Legal Office; interestingly the Japanese Embassy said it could not release this "Record", symbolic of how differently even simple unclassified documents are handled by the two Governments. It had not been translated into Japanese. The intent of first item on the important role of S&T in social and economic issues was to note that this "simply describes the fact" of this linkage and that only and was "not meant to prejudge the positions of either side in any future discussions". I would maintain that the inclusion of this explanation was to avoid or prevent the USG from attempting to raise this linkage for purposes other than those directly related to the 88STA. Item two stated that the JHLC could take up S&T issues "other than cooperative arrangements under the Agreement and including general issues that may be the subject of other arrangements"; this sounds like a carte blanche for either side to raise almost any issue. Item 3 was rather obvious stating that the seven areas of possible technical cooperation were not intended to be exclusive. Item 4 stated that the representatives of the USG and GOJ would be the equivalent rank of the Science Adviser to the U.S. President or the Japanese Minister of Foreign Affairs; this would avoid GOJ appointing a designee of less rank than the Science Adviser. Item 5 is a back door arrangement to assure that when the JHLC meets it would "facilitate the contributions of the JHLAP to the overall S&T relationship". The JHLAP which supposedly was to be a non-governmental group, was envisioned by the USG as the channel through which the ideas/demands/assertions of the U.S. private sector could be injected into S&T policy discussions between the two countries; this was indeed a long standing objective of the USG. Item 6 concerned the need to notify their respective authority of the date when a request was made for consultation under a new subparagraph 6.B under "other forms of intellectual property" of Annex IV on IPR. Since neither side could envision what potential IPR might be involved, this requirement was merely a "just-in-case" precautionary measure. Item 7 confirmed that the standard IPR arrangement was not limited to a formal agreement but may include a written policy which may require an assignment to be made to the Receiving Party of any Invention made by the Inventor of the Assigning Party. The latter items are involved with esoteric points in IPR. Most of these points reflect not a relationship of complete trust but of lingering concerns.

U.S.-German S&T relations in cursory comparison. In contrast to Japan, Germany, another major R&D powerhouse, does not have an overall S&T agreement with the USG; there is, therefore, no three tiered management structure to that relationship. The German federal constitution restricts the federal government's role in cultural and scientific affairs. Because of this restriction, the German Federal Government has not entered into an overall S&T agreement with the USG. Instead, the German Federal Government does have about 25 separate agreements about the space station, with NASA, on health research with NIH, etc. The duration of these agreements range from 5 to 25 years. There are no known overall studies by the German Federal Government about the US-German S&T relationship and no studies similar to the aforementioned four USG studies during

1994-1997 on how "to maximize U.S. interests in its S&T relationship with Germany." or vice versa.; even such academic studies, I am informed, do not exist. Although I was informed by OES that the U.S. S&T agreements with the German Federal Government would come under the general purview of the recently signed EU/USG STA, I was informed by the German Embassy that is definitely not the case. This is just one example how the USG does not attempt to use or impose it's model IPR and STAs on all countries, only selected ones.

58. Appendix 8 (1 and 2), Japan and U.S. to explore role of Science and Technology in Society into the new millennium, and the Japan-U.S. Science and Technology Agreement, 5/3/1999. The seven areas of cooperation mentioned in Annex I of the 88STA are: life sciences, including biotechnology, information science and technology, manufacturing technology, automation and process control, global geoscience and environment, joint database development and advanced materials, including superconductors.

    The joint analysis task was given to the U.S. JHLAP committee and a special Japanese group chaired by Dr. Imura Hiroo, former President of Kyoto University. Each group had eight members. The two groups held video conferences and were expected to prepare a report by mid-April 2000; this report would then be put on the agenda of the JHLC meeting in May 2000. Under these circumstances, this report might survive administration changes in the U.S. in 2001 and continue as a viable project into the future.

59. The discussion of the revisions was based on following cables: State 084587 (5/11/1998), State 071525 (4/17/1999) and the protocol revision the 88STA.

XI. An Afterword

After "living with" the 88STA for more than a decade, from the internal USG preparations in the mid-1980s to the signing of the agreement and its implementation, and its renewal and then another renewal for five years, 13 months after the tenth anniversary, it is somewhat anti-climactic to talk about significant conclusions about USJ political S&T relations. Rather, it is perhaps more useful to provide some overall observations and note a few hopes for the future. As it was before the 80STA so it is as the 88STA moves forward, that the benchwork scientific and technical work between the two groups of S&Es is constructive and useful but with no spectacular breakthroughs, and moves forward steadily and unabated within an atmosphere and ambiance of equality, respect and mutuality. IPR problems, security clauses and lack of Presidential/Prime Ministerial level R&D projects are of little apparent concern. It appears that project level research has fortunately not been affected or contaminated by the mischief indulged in at the political policy level. The mutual and genuine cooperation that has evolved since 1945 in S&T/R&D among the practicing S&Es has been an outstanding victory for scientific and technological cooperation between Japan and the U.S., but one cannot, based on the known record, truthfully assert that the political S&T record of treatment of a major ally, Japan, has been exemplary. On the political level, there is a continual long standing implicit lack of confidence in the USJ S&T relationship (which was highly resistant to change), in particular by the U.S. -- thus creating an apparently genuine disconnect between these two levels of the S/T-R/D relationship.

The USG was determined to persuade, cajole, force – whatever verb one wishes to use – the GOJ to accept a specific framework, a three tiered management structure, a new set of IPR, security obligations and much more. The USG was, thus, eminently successful in achieving these goals.

The USG had created a structure that gave it a multilayered platform to raise with the GOJ ANY issues concerning science and technology in the broadest sense about which, in the eyes of the USG, the GOJ "performance" was not acceptable and to pursue and demand the GOJ to take remedial actions. The Joint Committee under the 80STA had not been used by the USG (or the

GOJ); based on this past precedent perhaps that would happen again under the new three tiered management structure.

Let us inject, at this point, the comments of the U.S. Ambassador to Japan, Mike Mansfield, about the Washington-inspired brand of negotiations as sketched above and described in some detail in earlier chapters. While he was highly respected in Japan as a senior U.S. political leader, his standing in Washington after many years in Japan was somewhat down-graded as a strong aggressive advocate of U.S. interests. Immediately after the USG had given its draft to the GOJ, he sent a cable DIRECTLY to the Secretary of State expressing his strong concern for Washington's approach to the upcoming negotiations for a new STA.

> I am disturbed by what appears to be a rather heavy-handed approach in Washington towards renewal of our science and technology cooperation agreement with Japan. ... [it] provides us the opportunity to work with the Japanese to gain more access to the exciting and leading-edge R&D work going on here. ... Some Washington agencies appear to insist that the Japanese should be expected to agree to our proposed text after only one negotiating session. This approach ignores the fact that the many delays in beginning negotiation of the agreement have all been on the U.S. side. ... due to [the USG's] inability to fashion an interagency agreement on a negotiating text to propose to Japan. ... Given the one year it took us to come up with our own proposals, common courtesy and diplomatic practice would dictate that we provide the GOJ sufficient time for its own interagency deliberations and for full negotiations with us. I urge that the De Vos Delegation come to Japan as scheduled to explain our text to the GOJ. I also urge that we give due consideration to an interim extension of the agreement, as will likely prove necessary. It is in the best interests of the United States to have a full and balanced S and T relationship with Japan. All indications are that the GOJ is prepared to work with us to facilitate more balanced access to R&D work going on here. A heavy-handed approach on our part is counter-productive, and threatens to undermine this willingness to cooperate.[1]

The admonition about a heavy-handed approach was completely ignored by the U.S. delegation. The earlier cables from the Embassy about R&D opportunities in Japan were also ignored completely. The Ambassador was correct in anticipating the need for an extension of the 80STA and the need to give the GOJ more time to conduct their own interagency deliberations and the Japanese interest in cooperating with the U.S. in S&T. Each time, however, some Washington agencies only grudgingly agreed to the extensions and to giving the GOJ a respectable amount of time. The role of the American Embassy in Tokyo was minimal based on the records, and even the role of the U.S. Ambassador in a direct cable to Secretary ultimately seems **to have had little real impact** on the conduct of the negotiations.

One of deepest and abiding impressions I have from this research was the depth of DISTRUST of some agency officials of another Department of the USG, and the apparent willingness to tolerate, or the lack of political will to prevent, a small group to poison the negotiations both within the USG in its negotiations with Japan with a deliberate distortion of the facts in one aspect at the expense of all others. Differences of opinions, proposed policy approaches, strategies, which will naturally occur, are understandable, and to be expected; these disagreements would stem from individual philosophical and departmental turf issues and jealousies, but to slip over into distrust, suspicions, implicitly questioning the motives of one's fellow compatriots in whom we have invested a trust is a treacherous psychological minefield in which to maneuver. Normally an STA need not be brought to a domestic economic policy council; normally it would be more a matter of NSC concern. But the determination to force the STA issue in mid-1987 to the EPC, chaired in theory (at the time) by the President, stemmed undoubtedly from complete distrust of the State Department and its approach to Japan. Because one Department was so distrusted, the attempt here was to lessen the negotiating leeway of the U.S. delegation to the greatest extent possible. It should be remembered that originally the USG had entertained the idea and conveyed many times to the GOJ the concept of having a mutual, friendly and constructive joint review of the entire USJ S&T relationship. To listen in the interviews to the words of antipathy and to the tonal expressions of distrust used was, in the beginning, shocking and as the implications of these words gradually sank in it became saddening in the extreme and indeed tragic. This is in the starkest contrast possible to the publicly avowed policy of the USG in the 1980s that the USJ relationship was the most important, bar none. I wonder if such feelings of distrust in the USG existed in the 1980s and were so frontally stated about the Soviet Union; the distance between the public record and some of the internal governmental discussions concerning Japan were truly difficult to comprehend, rationalize and appreciate.

Another target of distrust was, of course, clearly Japan and this was intertwined with the mutual distrust in the USG. In each case where this was manifested the doubting group in the USG was proven incorrect. The USG had entertained the idea of a joint mutually friendly and constructive review of their S&T relationship but this was later transformed into a rather one-sided draft agreement which was basically "thrust" (*tsuki-tsukeru*) upon the GOJ which was expected to sign the draft in one long three-day session -- a rather contemptuous attitude reminiscent of mid-1945. It is understandable that the GOJ would privately question USG motives when the GOJ had already agreed to the bilateral review. Another example was the blatant attempt to try to hold the STA hostage to force Japan to sign the defense patent secrecy implementation agreement. The Secretary of State was pressured into signing a letter assuring the doubters that Japan would sign the agreement as they promised by a certain date even though there was one part of the USG which had already accepted Japan's word, the Department of Defense. Yet another example symbolic of the distrust of Japan was the three-tiered management system creating a forum where the USG could at least try to keep track of Japanese R&D, try to influence its direction and if necessary pressure Japan to do this or that in S&T. In reality, this did not occur and the USG called for a relaxation of the management of the system to make it "more informal" rather than being so intense and formal as originally conceived by the USG.

Even behind the "sanitized" committee reports, one can detect, as mentioned before, that there is an underlying sense of distrust, not strongly stated but implicit in the demands and issues the USG tends to raise with the GOJ with varying degrees of enthusiasm and determination. The USG briefing books are full of ideas for actions the GOJ could take. In contrast there are hardly any for the USG; since these briefing books contain the issues to be taken up at any one meeting, this tendency would seem to imply that the GOJ fails to list actions which the USG should take. This is yet another example of the general passivity of the GOJ. On the other hand, there appears to be a generalized lack of confidence in the GOJ that the USG, while asking its international partners to make long term commitments in human and financial terms for megascience projects, cannot itself make such commitments or even devise a system within the U.S. governmental structure to provide satisfactory and convincing assurances.

Japan may have been regarded as a technological challenger, a threat, even a technological superpower with the U.S. -- as the latter's rhetoric has constantly indicated -- but a technological superpower amenable to USG wishes, and the means towards achieving this objective would probably be through the three tiered management structure which was used only in USJ S&T relations (to the best of my knowledge). We told ourselves at the Economic Policy Council that the STA (including this management system) was going to be used "as a model" for STAs with other countries but that was never done.

The USG used quite an aggressive stance in the bilateral negotiations occasionally using threats -- invoking the name and threat of Congress. Whether these threats were effective is difficult to gauge. My intuition -- and a judgment in this regard can only be intuitive -- is that such threats did not carry much weight in the bilateral negotiations. Although the historical circumstances were quite different, these threats were reminiscent of Admiral Perry's threats to use military action in the mid-19th century in opening Japan to the outside world. In the 1980s instance of the STA, one of the main USG objectives was to force open the presumably less open Japanese R&D facilities, make JSTI more available presuming it was not sufficiently available, and increase foreign participation in the Japanese R&D process under threat and presumed fear of the "awesome" U.S. Congress.

When GOJ support was not deemed sufficient, strong enough, or wanting in some way, the issue at hand had a way of arriving on the Summit meeting agenda -- in recent years called The Common Agenda. There have not been enough U.S. applicants to fill the slots available to the U.S. and no complaints were lodged against the Japanese thus disposing of a presumed barrier in this area. Despite annual meetings sponsored by the GOJ and USG, there is a presumed persistent gap in the utilization of JSTI by U.S. researchers. The GOJ surprisingly agreed to set aside almost a whole facility in Tsukuba Science City for the support of foreign researchers meaning basically U.S. researchers through an MOU with the DOC. I am tempted to believe that the Japanese sign such agreements half believing that the U.S. researchers will not use their allotted slots and not use the JSTI adequately and efficiently, so why not accommodate the USG; that way it will be less hassle. After expressing great satisfaction with Japanese performance -- and the USG is always in a posture of judgment and evaluation -- there seems to be an unrelenting drive/pressure/demand for the Japanese to do something more; and the USG urges industry "to keep up the pressure on Japan."

Access to R&D facilities and JSTI were not even on the horizon for the 80STA, and again not in the initial purview of the STA review exercise in the mid-1980s. The access issue became a USG obsession during the negotiations. In the implementation period, it was investigated by two task forces with basically no case found against Japan. After a decade of working on the issue of access to JSTI, a 1994 working paper still most surprisingly intoned that "A growing number of U.S. high technology companies, both large and small, appear to be growing more aware of the need to monitor developments in Japan, are keeping closer tabs on Japanese capabilities and are perhaps less likely to be technologically 'blinded-sided' by Japanese competitors. ...[Nevertheless, it] is an appropriate time [1994] to reassess [after how many similar efforts?] to reexamine JSTI access and utilization issues ... [and how] to improve U.S. industry utilization of JSTI and enhance U.S. competitiveness".[2] There is nothing, it seems, the Japanese can do in any one area that will genuinely satisfy the U.S. side with no more requests for yet another action.

For many years, trying to loosen Japanese purse strings to support U.S. R&D projects was a major U.S. objective and each time the GOJ negotiated their way out of financial commitment to the great chagrin of the USG.

The USG steadfastly refused to take seriously GOJ intimations that the new STA could not be negotiated in one session, could not bring itself to believe that the GOJ would, in this case, drag these negotiations out for six months and, in the end, basically accept the USG plan but with many modifications. This tends to encourage such U.S. tactics which then become the source of additional unpleasantness.

IPR became a major issue between the two governments whereas it was not a concern for the 80STA. While no major leaks from federal laboratories were discovered, there was a persistent fear that the Japanese "would steal" the crown jewels from U.S. taxpayer funded labs. While negotiating the IPR was difficult, they were not bitter. It was a source of internal USG friction, and the IPR philosophy in the draft agreement given to the GOJ was based on a condescending attitude toward the Japanese. The premise upon which the initial IPR was based was erroneous. Probably because of NSF criticism, eventually the IPR were revised to a more equitable case-by-case basis (basically the GOJ position), thus an unnecessary self-imposed retreat by the USG.

In light of the success, cordiality, trust and a sense of equality in R&D cooperation during the previous decades, with no IPR conflicts, no security problems for example, GOJ was surprised and frustrated, dismayed and angry -- and probably felt quite betrayed -- that the USG would insist on such a detailed Presidential/Prime Ministerial STA.[3] The USG had already given the GOJ many examples of using various platforms to pressure, pursue through relentless dialogue to change GOJ policies and approaches, e.g. to loosen GOJ purse strings to support and invest in U.S. research projects. Inevitably, the USG gave the distinct impression – whether true or not, whether fair or not – that it was trying to find a way to influence, if not control the Japanese R&D enterprise through IPR and national defense classification. These GOJ suspicions of U.S. motives were never so baldly stated – to the best of my knowledge – in bilateral negotiations but did appear in Japanese newspaper commentaries and editorials. It appears that the GOJ was probably resigned to accepting the USG draft agreement in some form; the question was how to de-tooth, defang it, lessen the impact of imperative clauses in the eyes of the GOJ bureaucracy, not the people of Japan because it was too specialized for their attention and was regarded as the responsibility of the presumably responsible and effective and dedicated civil service.[4]

They conducted a rearguard action by persistently questioning the USG delegation about every word, phrase and paragraph in the draft agreement, the need for IPR and particularly security obligations in a non-military S&T agreement, to the great annoyance and frustration of and ultimately anger of the USG delegates. In the negotiations, the GOJ weakly put forward it's alternative agreement which was summarily dismissed since it was merely a tinkering of the 80STA and not responsive to what the USG demanded in its draft. The GOJ's weak effort was replaced by an effort to neutralize the USG draft in form and strive to prevent the inclusion of too many universalistic principles and endeavor to obtain agreement on particularistic stress on the case-by-case decisionmaking process.

By making the free flow of STI the major emphasis and thrust of the STA, the GOJ in a way, finessed, neutralized and subordinated the USG demands to this overarching principle underlying the 88STA. Through this emphasis on the free flow of STI, it could be said the GOJ succeeded in preserving its option to maintain its long standing goals and policies of indigenizing, diffusing, and nurturing the development of S&T in Japan almost without the encumbrances of national security considerations and restrictions.[5]

Eventually, the agreement was written as a neutral document with most of the "wills" changed to "may" and in other ways finessed by the GOJ. However, in side letters the GOJ agreed to carrying out a fair number of the items the USG wanted in the first place. The 88STA, it must also be admitted, has not produced any radical changes comparable to the great expectations of some in the USG. The USG was the one to call for streamlining the implementation structure which it had imposed on the GOJ; the GOJ will accept this streamlining, although it has a tendency to get used to a procedure which it is then somewhat reluctant to revise or give up. Remember also that it was the USG that demanded and obtained the creation of a Joint Committee under the 80STA and then did not ask for nor push for its effective utilization.

When the 88STA was considered for renewal, the USG said to itself that the STA had been more advantageous to the USG than to the GOJ. It is surprising that in light of its own rhetoric that both are technological superpowers capable of cooperating in R&D as equals and making major contributions to the world in S&T that the USG could not have had greater self-confidence and magnanimity and suggest that the 88STA has been to the mutual benefit of both countries. The gnawing question remains: why does the USG feel so pressed, so determined, so obsessive about Japan and Japan only in such a detailed S&T relationship? Is it inertia of the past when most initiatives concerning R&D came from the U.S. side, a genuine desire to cooperate with Japan, yet intermingled with fear of Japanese S&T accomplishments and their potential? This element of fear was decidedly not the case from 1945 to about 1980; perhaps since about that time the U.S. became less sure of its own position in S&T. Perhaps U.S. consciousness that its dominance in leading edge technologies was now, for the first time, seemingly threatened not by Europe but by an erstwhile Asian enemy who was not supposed to have such qualities and capabilities. That would make the sense of the challenge and threat that much more acute.

That the NAS/NRC at the behest of the DOD conducted a series of studies over several years, 1993-97, to ascertain how "to maximize U.S. interests in science technology relations with Japan" is an example of the persistent underlying concern with Japan's rise as a technological superpower. While it is legitimate as statecraft and the governmental function to conduct such studies, it is somewhat unusual to be so blatant in the use of language and, to the best of my knowledge, Japan is the only country in the 1990s about which we make such studies. These studies would seem to strongly underscore the opinion of GOJ officials during the negotiations of the 88STA that USG's underlying purposes – notwithstanding assertions even actions to the contrary – were to try to manage and control the Japan S&T enterprise as much as possible. This should have been an obvious mirage to the USG for the past decade but a series of actions appear to send the message that this is still perhaps the USG's hidden agenda. The USG made numerous assessments of USJ S&T relations in the past especially during the 1980s. Yet even in the late 1990s was the U.S. still so unsure of its bilateral S&T relations it had to conduct these studies to tell itself, yet again, how it could maximize it's interests in USJ S&T relations between two technological superpowers and discover weaknesses in Japanese S&T.

Just as the renewal negotiations began in early summer 1998, Senators Roth and Bingaman introduced a draft Senate Resolution about the USJ S&T relationship and expectations and criteria for judging success. While this was a merely un-passed Senate resolution, it was reminiscent of the October 1987 Senate hearings just as the negotiations began on the 88STA. The draft provided moral support to the Executive and an implied message to the GOJ but, in all probability, with no impact.

The 88STA was ultimately renewed for five years from July 20, 1999 with a relaxation of requirements on reports and meeting frequency and a few more detailed stipulations about IPR and copyrights in order to cover additional legal possibilities that might occur. These revisions took a leisurely 13 months to conclude, probably because of Japanese skepticism on why they were needed in the first place and to try to fathom what were the USG motives for calling for these specific revisions.

The GOJ is heavily responsible for the USG approach of constant demands on the GOJ; it's style of response encourages, facilitates and stimulates the USG to make yet another request of the GOJ since the latter will seemingly debate the merits of an issue and then in the final phase (*saigo no dotan ba de*) capitulate to U.S. demands. One might venture to say that in the ultimate sense the GOJ appears to be somewhat obsequious (*koshi ga hikui*). The GOJ and Japan need a stiffer backbone. A harsher judgment would be that an air of cooperation will persist so long as Japan does most of U.S.'s bidding with yet another agreement which will keep the Americans happy and off their backs for a while. There is a slowly growing tendency for some postwar generation GOJ officials to be more assertive and raise issues that the USG is forced to answer in the bilateral dialogue.[6]

There is a distinct need for the USG *to come to terms* with the rise of R&D Japan as a technological superpower, transcending its long standing narrow suspicions and accept and treat Japan with genuine respect, not condescension, and equality befitting its position as a technological superpower. Two actions would enormously enhance this relationship: for the USG to invite Japan to conduct a joint review as suggested earlier, and offer

Presidential/Prime Ministerial projects which would require investments by each country in the other country.

And in the same breath, it behooves Japan *to come to terms* with its responsibilities and obligations as a technological superpower, to conduct itself with more assurance and self-confidence, take a more positive, active and constructive forward looking role in its worldwide S&T relations, particularly those with the USG, and recognize that its behavior has fostered, encouraged and stimulated the USG to take a more high-handed and condescending attitude toward the GOJ.

Notes

1. Tokyo 17308 (9/28/1987). It was marked "Ex Dis" signifying a most limited distribution cable and for the Secretary's personal attention.
2. *Japanese Scientific and Technical Information: U.S. Industry needs, access and policy issues*, A working paper of the Competitiveness Task Force, Committee on Japan, National Research Council. 9/26/1994. p. 7.
3. This was one of the strongest, longest lasting but surprising impressions I received from my interviews with Japanese officials. It was a near unanimous expression on their part; while details were no longer clearly remembered they felt there had been a deliberate and conscious distortion of data by selected members of the U.S. delegations. It must be kept in mind that such issues as the Toshiba-Kongsburg issue and FSX were technically, legally, administratively separate and distinct from the S&T cooperation of the past decades but these very issues created the critical confrontation, an anxiety loaded atmosphere and political overtones spilled over into and importantly colored these negotiations.
4. The sense of betrayal might be interpreted as deriving from the Japanese notion of *amae* in its relations with the U.S., but I do not feel this way. I regard the Japanese reaction as quite a human reaction, not uniquely Japanese. Nevertheless, I would suggest Peter Berton's essay, Understanding Japanese Negotiating Behavior, *Intercom* (18:2/11/1995:1-8), International Studies and Overseas Programs, University of California, Los Angeles.
5. These goals were cited by Jonathan Lewis is his paper, Undermining Techno-nationalism: Japanese Policymaking for the International Space Station" at the 1998 Association for Asian Studies meeting.
6. A senior DOC official told me that probably in the near future the USG will conduct continuous and probing discussions and negotiations with the PRC. That may be so, but I doubt that the response will be as convenient as it has been with the GOJ; the Chinese will be far less accommodating.

   Justin Bloom made a most insightful observation about USJ S&T relations: ... the U.S. is more comfortable with a relationship in which the U.S. dominates the scene and offers largess freely and in great amounts. When another country [e.g. Japan] approaches the competence of the U.S., fears of competition may arise to force a drawing away from what is perceived to be an unwelcome threat. Bilateral Cooperative Programs: A case study – the United States and Japan, in *Journal of the Washington Academy of Sciences*, 77:3(9/1987):96.

Appendix 1

1980 U.S.-Japan Science and Technology Agreement

An Agreement between Government of the United States of America and the Government of Japan on Cooperation in Research and Development in Science & Technology

The Government of the United States of America and the Government of Japan;

Desiring to create a truly productive partnership and further strengthen the bonds of friendship existing between the two countries;

Believing that the future welfare and prosperity of mankind requires untiring efforts in the fields of scientific and technological research and development and that such efforts are not only beneficial to the country involved but also to the entire world;

Realizing the great benefits derived by the two countries from their long and highly successful scientific and technological relationship;

Recalling that the Agreement between the Government of the United States of America and the Government of Japan on Cooperation in Research and Development in Energy and Related Fields (hereinafter referred to as "the Energy Research and Development Agreement") was signed at Washington on May 2, 1979; and

Desiring to promote cooperative activities in the scientific and technological areas not covered by the Energy Research and Development Agreement and to establish the foundation for a new era of mutually beneficial cooperation between the two countries characterized by a recognition of equality and by a common perception of research and developments needs;

have agreed as follows:

Article I

The two Governments shall develop cooperative activities in scientific and technological research and development in such fields as may be mutually agreed for peaceful purposes on the basis of equality and mutual benefit. Such fields shall exclude those agreed upon under the Energy Research and Development Agreement.

Article II

Forms of the cooperative activities in research and development under this Agreement may include:
a. Conduct of joint projects and programs, and other cooperative projects and programs;
b. Meetings of various forms, such as those of experts, to discuss and exchange information on scientific and technological aspects of general or specific subjects and to identify research and development projects and programs which may be usefully undertaken on a cooperative basis;
c. Exchange of information on activities, policies, practices, and legislation and regulations concerning research and development;
d. Visits and exchanges of scientific, technicians or other experts on general or specific subjects; and
e. Other forms of cooperative activities as may be mutually agreed.

Article III

Implementing arrangements setting forth the details and procedures of the specific cooperative activities under this Agreement may be made between the two Governments or their agencies, whichever is appropriate.

Article IV

1. The two Governments shall establish a Joint Committee, the functions of which shall be:
a. Exchange of information and views on the science and technology policies of the two Governments and other issues relating to the implementation of this Agreement;
b. Review of the cooperative activities and accomplishments under this Agreement;
c. Provision of advice to the two Governments with regard to the implementation of this Agreement.

2. The Joint Committee shall meet alternately in the United States of America and Japan at mutually agreed times.

Article V

1. Scientific and technological information of a non-proprietary nature from the cooperative activities under this Agreement may be made available to the public by either Government through customary channels and in accordance with the normal procedures of the participating agencies.

2. The two Governments shall give due consideration to the equitable distribution of

industrial property resulting from the cooperative activities under this Agreement and of licences thereof and to the licensing of other related industrial property necessary or the utilization of the results of such cooperative activities; and shall consult each other for this purposes as necessary.

Article VI

Nothing in this Agreement shall be construed to affect the arrangements for cooperation between the two Governments or the official frameworks for cooperation between the two countries which exist at the date of signature of this Agreement. However, such arrangements and frameworks may be incorporated into the framework of this Agreement as may be mutually agreed.

Article VII

1. Activities under this Agreement shall be subject to budgetary appropriations and to the applicable laws and regulations in each country.

2. Costs for the cooperative activities under this Agreement shall be borne as may be mutually agreed.

Article VIII

The termination of this Agreement shall not affect the carrying out of any project or program undertaken under this Agreement and not fully executed at the time of the termination of this Agreement.

Article IX

1. This Agreement shall enter into force upon signature and remain in force for five years. However, either Government may any time give written notice to the other Government of its intention to terminate this Agreement, in which case this Agreement shall terminate six months after such notice has been given.

2. This Agreement may be extended by mutual agreement of the two Governments.

Done at Washington on May 1, 1980, in duplicate in the English and Japanese languages, both being equally authentic.

FOR THE GOVERNMENT OF THE UNITED STATES OF AMERICA	FOR THE GOVERNMENT JAPAN
JIMMY CARTER [U.S. President]	MASAYOSHI OHIRA [Prime Minister of Japan]

Appendix 2

Comparative Analysis of Three Versions of the 1988 U.S.-Japan Science and Technology Cooperation Agreement

Draft March 25, 1987	Proposal to Japan, August, 1987	1988 Agreement
The Government of the United States of America has the honor to refer to the Agreement between the Government of the United States of America and the Government of Japan on Cooperation in Research and Development in Science and Technology, signed at Washington on May 1, 1980 and extended for a period of two years by an exchange of notes on May 1, 1985 (hereinafter referred to as the Agreement) and proposes a further extension of the Agreement, pursuant to Article IX, for a period of five years from May 1, 1987. In implementation of Article I of the Agreement, the Government of the United States of America proposes that the two sides agree upon the following principles and	The Embassy of the United States of America presents its compliments to the Ministry of Foreign Affairs of Japan and has the honor to refer to the Agreement between the Government of the United States of America and the Government of Japan on Cooperation in Research and Development in Science and Technology, signed at Washington on May 1, 1980, and extended for a period of two years by an exchange of notes on May 1, 1985, and further extended for a period of six months by an exchange of notes on April 30, 1987 (hereinafter referred to as the Agreement) and proposes a further extension of the Agreement, pursuant to Article IX, for a period of five years from November 1, 1987. In	AGREEMENT BETWEEN THE GOVERNMENT OF THE UNITED STATES OF AMERICA AND THE GOVERNMENT OF JAPAN ON COOPERATION IN RESEARCH AND DEVELOPMENT IN SCIENCE AND TECHNOLOGY The Government of the United States of America and the Government of Japan (hereinafter referred to as "the Parties"); Recalling the purposes of the Agreement on Cooperation in Research and Development in Science and Technology, which was signed by the President of the United States of America and the Prime Minister of Japan and entered into force on May

objectives:

The Importance of Science and Technology

The United States and Japan have long recognized that the future prosperity and global security of Mankind are driven by the world's ability to generate new scientific knowledge and translate new discoveries into operational technologies and commercial applications. The U.S. and Japan have shared responsibilities in this effort.

Scientific progress and technological innovation underpin our nations' economic growth, high standards of living, and security.

Today, more than ever before, the

implementation of Article I of the Agreement, the Government of the United States of America proposes that the two governments agree upon the following principles and objectives:

The Importance of Science and Technology

The United States and Japan have long recognized that the future prosperity and global security of mankind are driven by the world's ability to generate new scientific knowledge and translate new discoveries into operation technologies and commercial applications. The U.S. and Japan have shared responsibilities in this effort. Scientific progress and technological innovation underpin our nations' economic growth, high standards of living, and security.

Today, more than ever before, the rapid advancement of scientific and technological know-how, coupled

1, 1980 (hereinafter referred to as "the previous Agreement");

Recognizing that the two countries derive great benefits from their long and highly successful scientific and technological relationship;

Believing that the future prosperity and well-being of mankind depend upon the world's ability to generate new scientific knowledge and translate new discoveries into operational and applied technologies;

Affirming that the United States of America and Japan, sharing responsibilities in contributing to the world's future prosperity and well-being, should make further efforts to strengthen their respective national research and development policies;

Stressing the importance of sustaining long-term investments in basic research and creating dynamic research and development environments with a view to generating fundamental new knowl-

rapid pace of scientific advance and technological know-how, coupled with the internationalization of the R&D enterprise, provide unique challenges and responsibilities for countries such as the United States and Japan, who are working at the frontiers of research and who possess the national assets and determination to profit from the unparalleled opportunities of the 21st century.

To maintain leadership in science and technology and contribute to the world's well-being and future security, advanced countries such as the U.S. and Japan should establish national R&D policies that sustain long-term investments in basic research and create dynamic research environments that advance the following broad goals:

(1) Generate fundamental new knowledge to expand the world's pool of science and technology understanding;

(2) Swiftly transfer new technologies

with the internationalization of the R&D enterprise, provide unique challenges and responsibilities for countries such as the United States and Japan, which are working at the frontiers of research and development and which possess the national assets and determination to profit from the unparalleled opportunities of the 21st century.

To maintain leadership in science and technology and contribute to the world's well-being and future security, advanced countries such as the U.S. and Japan should establish national R&D policies that sustain long-term investments in basic research and create dynamic R&D environments that advance the following broad goals:

(1) Generate fundamental new knowledge to expand the world's understanding of science and technology;

(2) Protect intellectual property rights so as to preserve the value of

edge, ensuring the protection of intellectual property rights so as to preserve the value of innovations derived from joint collaboration, providing for the smooth application of new technologies, and nurturing and expanding the next generation's human resources in science and technology;

Convinced that long-term mutually beneficial international science and technology collaboration is built upon long-lasting partnerships between scientists of different nationalities, performance of joint research and development at each other's facilities, education and training of each other's promising students, and publication of joint research and development results in international journals;

Affirming their commitment to equitable contributions and to comparable access to each nation's research and development systems;

Determined to strengthen the overall science and technology relationship and applications to the marketplace; and

(3) Nurture and expand the next generations' talent base.

Cooperation in International Science and Technology

Acknowledging that in many recent cases the world's most important discoveries and technological applications are the direct result of successful partnerships between colleagues of different nationalities; and recognizing that long-term mutually beneficial international S&T collaboration is built upon long lasting partnerships between scientists of different nationalities, performance of joint research at each other's facilities, education and training of each other's promising students and publication of joint research and development results in international

innovations derived from joint collaboration;

(3) Provide for the swift transfer of new technologies and applications to the marketplace; and

(4) Nurture and expand the next generations's talent base.

Cooperation in International Science and Technology

Acknowledging that in many recent cases the world's most important discoveries and technological applications are the direct result of successful partnerships between colleagues of different nationalities;

Recognizing that long-term mutually beneficial international S&T collaboration is built upon long-lasting partnerships between scientists of different nationalities, performance of joint research and development at each other's facilities, education and training of each other's promising students, and publication of joint

based on the principles of shared responsibilities and mutual and equitable contributions and benefits, commensurate with the two nations' respective scientific and technological strengths and resources;

Affirming their commitment to further enhance cooperation in science and technology; and

Desiring to set forth the policy framework for the conduct of the overall science and technology relationship between the Parties and to strengthen that relationship for peaceful purposes;

Have agreed as follows:

journals;

The Governments of the United States and Japan affirm their commitment to enhance cooperation in science and technology through acknowledging the need to create more equitable opportunities for interaction and to work for a balance of contributions and benefits.

In order to achieve balanced collaboration in science and technology in which both countries derive equitable benefits from the relationship, the Governments of the United States and Japan acknowledge that at times one country may need to make greater commitments in particular components of the R&D process, such as basic research and infrastructure, such as graduate and postgraduate education and training at universities and national research institutes, research facilities and S&T information dissemination, commensurate with its scientific and technological strengths.

R&D results in international journals;

The Governments of the United States and Japan affirm their commitment to enhance cooperation in science and technology through acknowledging the need to create more equitable opportunities for interaction and to work for a balance of contributions and benefits.

In order to achieve balanced collaboration in science and technology in which both countries derive equitable benefits from the relationship, the Governments of the United States and Japan acknowledge that at times one country may need to make greater commitments in particular components of the R& D process and infrastructure commensurate with its scientific and technological strengths. In this regard, the Government of Japan should increase its commitments to our shared responsibilities in such areas as open, basic academic research support, graduate and post-graduate education, training at universities and national research

	institutes, research facilities and the published dissemination of information.
<u>Revitalization of the U.S.-Japan Agreement on Research and Development in Science and Technology</u>	<u>Revitalization of the U.S.-Japan Agreement on Research and Development in Science and Technology</u>
The Governments of the United States and Japan recall that the Agreement was signed by the President of the United States and the Prime Minister of Japan and concur that the scientific and technical content of the agreement should match the commitments of its Head of Government signatories.	The Governments of the United States and Japan recall that the Agreement was signed by the President of the United States and the Prime Minister of Japan and concur that the scientific and technical content of the Agreement should match the commitments of its head of Government signatories.
The Governments of the United States and Japan concur that this agreement sets forth the overall policy framework and national goals in the science and technology relationship between the two countries.	The Governments of the United States and Japan concur that this agreement sets forth the overall policy framework and national goals in the science and technology relationship between the two countries.
Accordingly, the two countries intend to carry out under the agreement	Accordingly, the two countries intend to carry out under the Agreement cooperative science and technology

cooperative science and technology projects and programs of the highest national priority. The cooperative activities undertaken are expected to provide new knowledge and technology of importance to each country into the twenty-first century.

Under the policy framework of this agreement the two governments will discuss the two nations science and technology policies, priorities, activities and plans. They will also examine the balance and reciprocity of the overall cooperative relationship between the United States and Japan and identify impediments to balanced opportunities, access and benefits, with a view to overcoming those impediments.

The Government of the United States of America further proposes that the following annexes which are an integral part of this note, shall operate as implementing arrangements under article III.

1. Major initiatives and priority areas

projects and programs of the highest national priority. The cooperative activities undertaken are expected to provide new knowledge and technology of importance to each country into the 21st century.

Under the policy framework of this Agreement the two governments will discuss the two nations' science and technology policies, priorities, activities and plans. They also will examine the balance and reciprocity of the overall cooperative S&T relationship between the United States and Japan and identify impediments to balanced opportunities, access and benefits, with a view to overcoming those impediments.

The Government of the United States of America further proposes that this note and the annexes appended thereto shall become an integral part of the Agreement. The annexes are:

1. Major Initiatives and Priority Areas

2. Mechanisms, initiatives and areas for joint review
3. Management of science and technology cooperation
4. Intellectual Property Rights

In this exchange of notes the Parties acknowledge that there is nothing herein, nor in the attached annexes, that is inconsistent with the Agreement signed in Washington on May 1, 1980.

If the foregoing is acceptable to the Government of Japan, the Government of the United States of America has the honor to propose that this note together with the reply of the Government of Japan to that effect shall constitute an agreement between our Governments to renew the Agreement for a period of five years from May 1, 1987, adopt the principles and objectives included in this note, and approve the four annexes appended hereto.

Please accept the renewed assurances of the highest consideration of the

2. Mechanisms, Initiatives and Areas for Joint Review
3. Management of Science and Technology Cooperation
4. Intellectual Property Rights
5. Security Obligations

In this Exchange of Notes the Parties acknowledge that there is nothing herein, nor in the attached annexes, that is inconsistent with the Agreement.

If the foregoing is acceptable to the Government of Japan, the Government of the United States of America has the honor to propose that this note and the annexes appended hereto, together with the reply of the Government of Japan to that effect, shall constitute an agreement between our Governments to renew the Agreement for a period of five years from November 1, 1987.

Please accept the renewed assurances of the highest consideration of the Government of the United States of America.

Government of the United States of America.

Annex I - Major Initiatives and Priority Areas

Article I: Major International Initiatives and National R&D Programs

(1) When the United States and Japan agree to collaborate in major international R&D initiatives and large-scale national projects, such cooperation will be pursued under the policy framework of the Heads of Government Agreement in keeping with the functions of the Joint Committee detailed in Article II of Annex III. If such projects have separate management mechanisms they would not fall under the technical and management review of the Joint Interagency Executive Committee established in Annex III, Article III.

(2) Consistent with Article III of the Agreement, interagency memorandums of understanding and

Annex I - Major Initiatives and Priority Areas

I. Major Bilateral Initiatives and National R&D Programs

(1) This Agreement provides a basis for the United States and Japan to collaborate in major international R&D initiatives and large-scale national projects, and to pursue such cooperation under the policy framework of this Heads of Government Agreement and in keeping with the functions of the Joint Committee detailed in Article II of Annex III. Projects that have separate management mechanisms shall not fall under the technical and management review of the Joint Interagency Executive Committee established in Annex III, Article III.

(2) Consistent with Article III of this Agreement, interagency memoranda of understanding and other inter-

Article I

1. This Agreement establishes the policy framework for the overall science and technology relationship between the Parties, including collaboration in large-scale projects and major research and development initiatives. To strengthen that relationship, the Parties will conduct their science and technology relationship based on the principles of:

A. Shared responsibilities and mutual and equitable contributions and benefits, commensurate with the two nations' respective scientific and technological strengths and resources;

B. Comparable access to major government-sponsored or government-supported programs and facilities for visiting researchers, and comparable access to and exchange of other intergovernmental instruments may be negotiated by appropriate lead agencies in both countries to determine the specific terms and provisions of collaboration.

(3) When the U.S. and Japan agree to collaborate in major international R&D initiatives and large-scale basic research projects that require significant, new investments in infrastructure and state-of-the-art facilities, the two countries will share the costs equitably.

(4) The U.S. has proposed major scientific and technology initiatives such as the Manned Space Station in which participation of Japan would be of great mutual benefit. The recently announced Super-conducting Super Collider will also fall into this category when it has been formally proposed. Many other large scale U.S. or international R&D programs (including such potential programs as Mapping of the Human Genome and global environmental initiatives) could have a US-Japan cooperative governmental instruments will be negotiated by appropriate lead agencies in both countries to determine the specific terms and provisions of collaboration.

(3) When the U.S. and Japan agree to collaborate in major R&D initiatives and large-scale research projects that require significant, new investments in infrastructure and state-of-the-art facilities, the two countries will share the costs in proportion to their respective risk, benefit and management shares, and considering the balance and reciprocity of their overall S&T relationship.

(4) Collaboration will be undertaken subject to the appropriate laws and regulations of each country and to the availability of funds and personnel resources.

information, in the field of scientific and technological research and development;

C. Adequate and effective protection and equitable distribution of intellectual property rights created in the course of collaboration and adequate and effective protection of intellectual property rights introduced in the course of collaboration;

D. Widest possible dissemination of information consistent with applicable national laws and regulations, including those related to security; and

E. Shared costs of collaboration taking into account their respective risks, benefits and management shares.

2. Under this policy framework, the Parties will discuss matters of importance in the field of science and technology relationship between the two countries.

component.		
(5) Japanese international initiatives such as the Human Frontiers Science Program, the International Frontier Research System at RIKEN and the Japan Trust of the Key Technology Center, as well as national programs such as ICOT (the Institute for New Generation Computer Technology) ERATO (Exploratory Research for Advanced Technology) and institutes such as those of AIST also have great potential for cooperative activities between the United States and Japan.		
Article II: Priority Areas for Science and Technology Cooperation between the United States and Japan	II: Priority Areas for Science and Technology Cooperation between the United States and Japan	Article II
(1) For an activity to be included under this Agreement, the activity should meet the criteria described below. Those activities under the current agreement that do not meet these standards of priority will be placed under other existing bilateral agreements or will be dropped.	(1) For an activity to be included under this Agreement, the activity should meet the criteria described below. Those activities under the current Agreement that do not meet these standards of priority will be placed under other existing bilateral agreements or will be considered for	1. This Agreement also sets forth the principles and provisions for cooperative activities under this Agreement. Thereunder, the Parties will undertake cooperative activities for peaceful purposes in such areas of science and technology of national importance as may be mutually agreed.

	termination.	2. The main areas and the forms of the cooperative activities under this Agreement are provided in Annex I, which is an integral part of this Agreement.
		3. Implementing arrangements for the cooperative activities under this Agreement are provided in Annex I, which is an integral part of this Agreement.
		4. A cooperative activity under this Agreement will be initiated by mutual agreement and should meet the following criteria:
(2) The criteria are: -- Each partner should possess strong, complementary or counterbalancing research capabilities, adequate resource bases, and appropriate centers of excellence to engage in joint ventures.	(2) The criteria are: -- Each partner should possess strong, complementary or counterbalancing R&D capabilities, adequate resource bases, and appropriate centers of excellence to engage in joint ventures.	A. Each party to that cooperative activity should possess strong complementary or counterbalancing research and development capabilities, adequate resource bases, and appropriate centers of excellence to engage in that cooperative activity;
-- The subject areas should reflect the national R&D priorities of both countries and contribute to an equitable distribution of investment and pay-off to each partner's national	-- The subject areas should reflect the national R&D priorities of both countries and contribute to an equitable distribution of investments and pay-off to each partner's national	B. The subject area of that cooperative activity should reflect an area of importance to both countries;

needs.

-- Bilateral cooperation in the areas chosen should have the potential to accelerate the rate of scientific progress compared to what is achievable now and offer tangible contributions to the world's knowledge base.

(3) The following general scientific and engineering fields are priority areas for cooperation under the Agreement in implementation of Article 1 thereof. Identification of specific areas for actual implementation of cooperation and definition of the precise terms and mechanisms by which this cooperation is to be undertaken will be decided by the Joint Interagency Executive Committee which will establish working groups under the guidance of lead agencies in individual scientific areas to implement the agreed cooperation. This list of fields may be modified by mutual agreement of the two sides.

needs.

-- Bilateral cooperation in the areas chosen should have the potential to accelerate the rate of scientific progress compared to what is achievable now and offer tangible contributions to the world's knowledge and technology base.

(3) The following general scientific and engineering fields are priority areas for cooperation under the Agreement in implementation of Article I thereof. Identification of specific areas for actual implementation of cooperation and definition of the precise terms and mechanisms by which this cooperation is to be undertaken will be decided by the Joint Interagency Executive Committee which will establish working groups under the guidance of lead agencies in individual scientific areas to implement the agreed cooperation. This list of fields may be modified by mutual agreement of the two sides.

C. The results of that cooperative activity should be expected to contribute to an equitable distribution of benefits to each Party; and

D. That cooperative activity should have the potential to accelerate the rate of scientific and technological progress and to offer tangible contributions to the world's knowledge and technology base.

5. With regard to the cooperative activities under this Agreement, the Parties or their agencies, as appropriate, may allow the participation of researchers and organizations from all sectors of the research establishment, including universities, national laboratories, and the private sector.

6. The Parties or their agencies may include their respective major government-sponsored or government-supported research programs in the basic and applied research areas listed in Annex I as part of the cooperative activities under this Agreement when these programs and cooperative ac-

		tivities meet the criteria set forth in paragraph 4 of Article II. 7.This Article will be implemented subject to the applicable laws and regulations of each country. Annex I Main Areas and Forms of the Cooperative Activities
(4) The general fields agreed upon as priority areas for cooperation are: a. Materials science and engineering (particularly electronic materials) b. Life sciences (particularly bio-technology) c. Information science and technology d. Automation and process control e. Global geosciences and environment f. Joint database development g. Joint standardization and nomen-clature development	(4) The general fields agreed upon as priority areas for cooperation are: (to be identified during the negotiations)	1. The following may be included as main areas for the cooperative ac-tivities under this Agreement: A. Life sciences, including biotech-nology; B. Information science and technol-ogy; C. Manufacturing technology; D. Automation and process control; E. Global geoscience and environment; F. Joint database development; and G. Advanced materials, including superconductors. This list may be modified by mutual agreement.
(5) Under this agreement, the in-	(5) Under this Agreement, the in-	2. Forms of the cooperative activities

stitutional performers of the collaborative research developed in these priority area may include all sectors of the R&D enterprise in both countries, i.e., universities, national laboratories, and private sector research centers.

(6) Each government agrees that under this Agreement, research programs eligible for bilateral cooperation include those private sector programs, exclusive of national security projects, that meet the criteria described in Article II (2) of this Annex and that receive government support or government subsidies.

stitutional performers of the collaborative R&D activities in these priority areas may include all sectors of the R&D enterprise in both countries, i.e., universities, national laboratories, and private sector research centers.

(6) Given the structural differences between the R&D systems in the United States and Japan, the Japanese Government will facilitate U.S. participation in those national research programs in basic and applied sciences and engineering that meet the criteria described in this article and that are under government sponsorship and receive government funding. Such national R&D programs include, but are not limited to, the Japan Key Technology Center, ICOT (the Institute for New Generation Computer Technology, ERATO (Exploratory Research for Advanced Technology) the Next Generation Research Program as well as national R&D programs conducted by MITI's Agency for Industrial Science and Technology. Such

under this Agreement may include:

A. Conduct of joint projects and programs and other cooperative projects and programs;

B. Meetings of various forms, such as those of experts, to discuss and exchange information on scientific and technological aspects of general or specific subjects, and to identify research and development projects and programs which may be usefully undertaken on a cooperative basis;

C. Exchange of information on activities, policies, practices, laws and regulations concerning research and development;

D. Visits and exchanges of scientists, engineers or other experts on general or specific subjects; and

E. Other forms of cooperative activities as may be mutually agreed.

1. This Agreement supersedes the previous Agreement. Implementing

	national R&D programs in Japan, supported by MITI, as the Science and Technology Agency and Monbusho, are comparable to ongoing U.S.-government supported activities conducted at leading U.S. universities such as MIT and Caltech and U.S. national laboratories such as NIH and Oak Ridge in which there is significant Japanese participation.	arrangements and cooperative activities undertaken under the previous Agreement are hereby incorporated under this Agreement, except that the Parties may agree that those arrangements and activities under the previous Agreement that do not meet the criteria set forth in paragraph 4 of Article II will be placed under other bilateral agreements.
		2. This Agreement does not otherwise legally modify existing bilateral science and technology arrangements between the Parties or their agencies. However, the Parties or their appropriate agencies may amend such arrangements, as may be agreed, in accordance with their relevant amendment procedures, to make them consistent with the policy framework of this Agreement.
Annex II - Mechanisms, Initiatives and Measures of Progress	Annex II - Mechanisms, Initiatives and Areas for Joint Review	Article IV
I. Mechanisms and initiatives to Enhance U.S.-Japan Interactions in Science and Technology.	I. Mechanisms and Initiatives to Enhance U.S.-Japan Interaction in Science and Technology.	With a view to strengthening the overall science and technology relationship on the basis of the principles set forth in paragraph 1 of

(1) The United States and Japan have shared responsibilities to achieve balanced and equitable collaboration in science and technology as agreed in the exchange of notes renewing this Agreement. Each country undertakes to implement appropriate steps in order to establish a fair and equitable relationship.

(2) The U.S. Government will continue to work through appropriate U.S. agencies and organizations, such as the Departments of State and Commerce, the National Science Foundation, the National Academies of Science and Engineering, and others, to encourage and support language training programs for technical personnel at U.S. universities and other institutions to facilitate the participation of American scientists and engineers in Japanese R&D activities.

(3) In parallel, the Government of Japan will establish at Tsukuba University and other key institutions in the public and private sectors, inten-

(1) The United States and Japan have shared responsibilities to achieve balanced and equitable collaboration in science and technology as agreed in the exchange of notes renewing this Agreement. Each country undertakes to implement appropriate steps in order to establish a fair and equitable relationship.

(2) The U.S. Government will continue to work through appropriate U.S. agencies and organizations, such as the Departments of State and Commerce, the National Science Foundation, the National Academies of Science and Engineering, and others, to encourage and support language training programs for technical personnel at U.S. universities and other institutions to facilitate the participation of U.S. scientists and engineers in Japanese R&D activities.

(3) In parallel, the Government of Japan will establish at Tsukuba University and other key institutions in the public and private sectors, inten-

Article I, the Parties will take those steps listed in Annex II, which is an integral part of this Agreement, and such other steps as may be mutually agreed.

Annex II Steps to strengthen the overall science and technology relationship.

In accordance with Article IV, the Parties will, subject to the applicable laws and regulations of each country, take the following steps to strengthen the overall science and technology relationship:

A. Continue their commitment to open research and development systems and international cooperation;

B. Continue to improve foreign language training programs for scientists and engineers to facilitate their communication and participation in research and development and daily life activities;

sive language programs for visiting American scientists and engineers to facilitate communication in R&D environments as well as in daily life activities in Japan.

(4) The Ministry of Education, MITI, STA, and other ministries, will actively encourage the vigorous recruitment and acceptance of more American scientists and engineers, including junior researchers, at R&D facilities under their respective institutions or supported with their resources. In this regard, they will give priority attention to increasing the number of Americans participating in national R&D programs and collaborating the priority areas list in Annex I.

(5) Appropriate U.S. Government agencies and S&T organizations will work with universities and other R&D performers to encourage American researchers to take advantage of existing opportunities to work in Japan. To assist U.S. efforts, the Japanese Government will sive language and cultural programs for visiting U.S. scientists and engineers to facilitate communication in R&D environments as well as in daily life activities in Japan. The GOJ will maintain and accommodate visiting U.S. scientists and engineers consistent with standard practices in other U.S. S&T agreements.

(4) The Ministry of Education, MITI, STA, and other ministries, will actively encourage vigorous recruitment and acceptance of more U.S. scientists and engineers, including junior researchers, at R&D facilities under their respective institutions or supported with their resources. In this regard, they will give priority attention to increasing the number of U.S. researchers participating in national R&D programs and collaborating in the designated priority areas list in Annex I.

(5) Appropriate U.S. Government agencies and S&T organizations will work with universities and other R&D performers to encourage U.S.

C. Provide comparable opportunities for scientists and engineers from the other country to engage in research and study in their respective facilities and major government-sponsored or government-supported research programs in basic and applied research areas;

D. Provide substantial numbers of competitive government fellowships in science and engineering for foreign nationals at their respective centers of excellence, with adequate allowances to cover accommodations and other needs;

E. Promote dissemination of information on the government fellowships and the opportunities for research and study referred to in subparagraphs C and D above;

F. Exert comparable efforts to encourage scientists and engineers to take advantage of the government fellowships and the opportunities for research and study referred to in subparagraphs C and D above;

provide annually to the U.S. Government a comprehensive list, with all necessary particulars, of current opportunities for American researchers to participate and be employed in the Japanese R&D system.

(6) To increase the number of U.S. researchers in Japan and promote the long-term objective of balanced access and development of training opportunities in science and technology, the Government of Japan will establish and widely advertise a significant number of substantial and prestigious "Japan Fellowships" in science and engineering for American undergraduates, graduates and postdoctorals at Japanese centers of excellence. The fellowship subjects should promote interaction in the priority areas listed in Annex I and provide opportunities for the fellowship recipients to participate in the national R&D programs also listed in Annex I.

(7) Under the Joint Interagency Executive Committee established in researchers to take advantage of existing opportunities to work in Japan. To assist U.S. efforts, the Japanese Government will provide annually to the U.S. Government a comprehensive list, with all necessary particulars, of current opportunities for U.S. researchers to participate and be employed in the Japanese R&D system.

(6) To increase the number of U.S. researchers in Japan and promote the long-term objective of balanced access and development of training opportunities in science and technology, the Government of Japan will establish and widely advertise a significant number of substantial and prestigious "Japan Fellowships" in science and engineering for U.S. undergraduates, graduates and postdoctorals at Japanese centers of excellence. The fellowship subjects should promote interaction in the priority areas listed in Annex I and provide opportunities for the fellowship recipients to participate in the national R&D programs also listed in

G. Ensure that scientific and technical report produced by government agencies or through major government-sponsored or government-supported research programs that are not published in readily available professional literature will be made available to researchers of the other country, through central sources such as National Technical Information Service and Japanese equivalents, as well as through expansion of the National Technical Information Service - Japan Information Center of Science and Technology program.

[SIDE LETTER]

June 20, 1988

Dear Mr. Secretary:

Pursuant to the Agreement between the Government of Japan and the Government of the United States of America on Cooperation in Research and Development in Science and Technology and the principles of equitable contributions and

Annex III, the U.S. and Japan will establish a working level task force of appropriate representatives from government, universities, and the private sector, to develop a system to identify, recruit, and monitor American scientists and engineers' access to and participation in Japanese R&D programs. This task force will examine barriers and other structural problems on both sides that impede or inhibit increasing the numbers of U.S. researchers in Japan. The task force will set up a system to obtain accurate, yearly statistical data on Japanese S&T researchers participating in the U.S. R&D and American S&T researchers participating in the Japanese R&D system.

Annex I.

(7) Under the Joint Interagency Executive Committee established in Annex III, the U.S. and Japan will establish a working level task force of appropriate representatives from government universities and the private sector, to develop a system to identify, recruit, and monitor U.S. scientists and engineers' access to and participation in Japanese R&D programs. This task force will examine barriers and other structural problems on both sides that impede or inhibit increasing the numbers of U.S. researchers in Japan. The task force will develop recommendations and actions for the consideration of their Governments. In addition, the task force will set up system to obtain accurate, yearly statistical data on Japanese S&T researchers participating in the U.S. R&D and U.S. S&T researchers participating in the Japanese R&D system.

(8) Recognizing that U.S. scientists and engineers face unique problems comparable access to each Government's research and development system stated therein, I hereby confirm that the Government of Japan will take steps to identify and increase substantially opportunities for American scientists and engineers to engage in research and study in Japan, including in the areas of cooperation listed in Annex I of that Agreement.

I anticipate that the United States Government will also encourage and support American scientists and engineers to take advantage of such increased opportunities.

Sincerely,

Sousuke Uno
Minister for Foreign Affairs of Japan

The Honorable
George P. Schultz
Secretary of State
of the United States of America

[SIDE LETTER]

in gaining access to Japanese scientific and technical information (STI) not encountered in other major STI source countries, an STI committee will be established to examine and develop recommendations on improving STI access. This committee will serve as a forum where STI organizations may raise and resolve concerns. The committee will be comprised of the directors of Japanese and U.S. Government agencies and other organizations which are directly concerned with the availability of STI. In addition, to deal with the specific problem of unpublished literature which is produced as a result of Japanese Government-funded research, the Japanese Government will take steps to ensure that scientific and technical reports produced by GOJ agencies and their contractors which are not published in "open" journal literature will be made available to U.S. researchers through a central source comparable to the NTIS and through expansion of the NTIS-JICST program.

June 20, 1988

Dear Mr. Minister:

I welcome and appreciate the Government of Japan's commitment that, pursuant to the Agreement between the Government of the United States of America and the Government of Japan on Cooperation in Research and Development in Science and Technology and the principles of equitable contributions and comparable access to each Government's research and development systems stated therein, the Government of Japan will take steps to identify and increase substantially opportunities for American scientists and engineers to engage in research and study in Japan, including in the areas of cooperation listed in Annex I of that Agreement.

I hereby confirm that the United States Government will strongly encourage and support American scientists and engineers to take advantage of such increased opportunities.

Annex III - Management of Science and Technology Cooperation

I. The Joint Committee established under Article IV of the Agreement shall be co-chaired by the Science Advisor to the President of the United States and the Minister of Science and Technology of Japan.

II. (1) In carrying out the functions pursuant to Article IV of the agreement the parties agree that the Joint Committee will serve as the forum for discussion of policy level issues related to activities under this agreement. Technical and management issues, including the review of cooperative activities and accomplishments under the agreement, will be dealt with by the Joint Interagency Executive

Annex III - Management of Science and Technology Cooperation

I. The Joint Committee established under Article IV of the Agreement shall be co-chaired by the Science Advisor to the President of the United States and the ------------------- (Japanese equivalent).

II. (1) In carrying out the functions pursuant to Article IV of the Agreement the Parties agree that the Joint Committee will serve as the forum for discussion of policy level issues related to activities under this Agreement. Technical and management issues, including the routine review of cooperative activities and accomplishments under the Agreement, will be dealt with by the Joint Interagency Executive

Sincerely yours,

George P. Schultz

His Excellency, Sousuke Uno,
Minister for Foreign Affairs, Tokyo

Article V

1. The Parties will establish a Joint High Level Committee. The Joint Committee will be co-chaired by the appropriate high-level representatives of both Parties. The U.S. chair will be the Science Advisor to the President. The Japanese chair will be the Minister for Foreign Affairs or his designee.

2. Meeting alternately in the United States of America and Japan, the Joint High Level Committee will serve as the annual forum for the Parties to review and discuss, under the policy framework of this Agreement, matters of importance in the field of science and technology and policy issues related to the overall science and technology relationship

Committee, the functions of which are further detailed in paragraph III below, and	Committee, the functions of which are further detailed in paragraph III below, and	between the two countries, and the cooperative activities under this Agreement.
(2) The Joint Committee will also serve as a forum for regular high level exchange of views between the two governments on matters of importance in the fields of science, technology and engineering. In this capacity it will discuss and negotiate on priority areas for science and technology cooperation between the United States and Japan and related issues of importance to both sides.	(2) The Joint Committee will serve as a forum for annual high level exchange of views between the two Governments on matters of importance in the fields of science, technology and engineering. In this capacity it will discuss and negotiate priority areas for science and technology cooperation between the United States and Japan and related issues of importance to both sides. (3) The Joint Committee also will approve and adopt the annual action plan covering the next review period of the Agreement to implement the provisions of Annex II, Articles I and II.	3. In this context, the Joint High Level Committee will submit an annual report to the Parties. The report will review the operation of this Agreement, including an assessment of major developments with respect to the factors listed in Annex III, which is an integral part of this Agreement. The report will also set forth steps and new initiatives for the Parties to adopt for the next year, and, as necessary, policy recommendations on matters of importance to the overall science and technology relationship between the two countries.
Annex II	[Annex II]	[Annex III]
II. Areas for Joint Review	II. Areas for Joint Review	1. Pursuant to paragraph 1 of Article I and Article V, the Joint High Level Committee will assess major developments with respect to the following
In its examination of the operation of the Agreement, the Joint Committee	In its annual examination of the operation of the agreement, the Joint	

will review the following indicators, which are central to achieving a balanced and cooperative science and technology relationship between the United States and Japan:

a) Developments and trends in each government's policies for the support of science and technology, with particular reference to basic research activities performed at universities, supported research institutions, national research institutes, and the private sector .

b) Efforts by each country to stimulate and achieve a balanced flow of scientific and technical information between the United States and Japan consistent with the provisions of Annex IV, and the concrete actions of each country to ensure that new scientific and technical information generated by their respective S&T establishments comes to the notice of the world scientific community through publication in international refereed journals and other customary practices.

Committee will review the following indicators, which are central to achieving a balanced and cooperative S&T relationship between the United States and Japan:

A) Developments and trends in each government's policies for the support of science and technology, with particular reference to R&D activities performed at universities, national research institutes, and other government supported research institutions;

B) Efforts by each country to stimulate and achieve a balanced flow of scientific and technical information between the United States and Japan consistent with the provisions of Article I-8 of this annex and Annex IV, and the concrete actions of each country to ensure that new scientific and technical information generated by their respective S&T establishments comes to notice of the world scientific community through publication in refereed journals and other customary practices.

factors:

A. Developments and trends in each Party's policies for the promotion and support of science and technology, with particular reference to research and development activities performed at universities and national research institutions, and in major government-sponsored or government-supported research and development programs;

B. Efforts by each Party to stimulate an equitable flow of scientific and technical information between the two countries, and its concrete actions to enable new scientific and technical information generated by their respective science and technology establishments to come to the notice of the world scientific community through publication in open, readily available professional literature;

C. Efforts in each country to promote advanced educational and training opportunities in science and engineering at universities and national re-

c) Efforts in each country to maintain and enhance state-of-the-art educational and training facilities at universities and national research institutes to conduct science and technology and to promote advanced training opportunities for the next generation of scientists and engineers to meet the challenges of the 21st century;

d) Government investments in the establishment and enhancement of world-class R&D facilities in each country, at universities and national laboratories to generate fundamental new knowledge and generic technologies;

e) The levels of flows of scientists and researchers between the United States and Japan to premier educational and research facilities;

f) Japan's initiatives to open research programs and S&T institutions to foreign participation specifically, to encourage American scientists to work in research and development in-

C) Efforts in each country to maintain and enhance state-of-the-art educational and training facilities at universities and national research institutes to conduct science and technology and to promote advanced training opportunities for the next generation of scientists and engineers to meet the challenges of the 21st century;

D) Government investments in the establishment and enhancement of world-class R&D facilities in each country, at universities and national laboratories to genere fundamental new knowledge and generic technologies;

E) The levels of flows of scientists and engineers between the United States and Japan to premier educational and research facilities;

F) Japan's initiatives to open national R&D programs and S&T institutions to foreign participation specifically, to encourage U.S. scientists to work in

search institutions;

D. Each Party's efforts to establish and enhance world-class research and development facilities at universities and national research institutions in its country to generate new knowledge and generic technologies;

E. Flows of scientists and engineers between the two countries to educational and research facilities, and efforts in each country to stimulate and encourage equitable flows;

F. Efforts by each Party to provide comparable access to universities and national research institutions and to major government-sponsored or government-supported research and development programs; and

G. Support in each country for visits and exchanges of and joint projects for, scientists and engineers involving all types of facilities and all levels of seniority and financial assistance.

stitutions in Japan and to participate in research programs in Japan;

g) Actions by each country to encourage increased participation by American researchers in Japanese science and technology programs by providing logistical support, especially language training and cultural familiarization;

h) Support in each country for expanded visits, exchanges and joint projects for scientific and technical personnel in all levels of seniority, types of facilities, and levels of financial assistance;

i) Major science and technology initiatives in each country to generate both new knowledge and generic and applied technologies, and the levels of participation in those initiatives by scientists, researchers and engineers from the United States and Japan.

research and development institutions in Japan and to participate in research programs in Japan;

G) Actions by each country to encourage increased participation by U.S. researchers in Japanese science and technology programs by providing logistical support, especially language training and cultural familiarization;

H) Support in each country for visits, exchanges and joint projects for scientific and technical personnel involving all levels of seniority, types of facilities, and levels of financial assistance.

The Joint Committee shall provide an annual report of its review of the above indicators to its respective Heads of Government.

The Joint Committee also will provide an annual action plan setting for the concrete steps and new initiatives both countries will peruse during the next review period to imple-

	ment the provisions of this annex.	
III. The parties further agree to establish a Joint Interagency Executive Committee to support the work of the Joint Committee at the technical management level. The functions of the Joint Interagency Executive Committee shall be:	III. The Parties further agree to establish a Joint Interagency Executive Committee to support the work of the Joint Committee at the technical management level. The functions of the Joint Interagency Executive Committee shall be:	[Art. V]
(1) to review the progress of cooperative science and technology activities under this agreement;	(1) To monitor, manage, and review the progress of cooperative science and technology activities under this Agreement.	4.The Parties will establish a Joint Working Level Committee at the technical management level to support the work of the Joint High Level Committee. For this purpose, specific functions of the Joint Working Level Committee are provided in Annex III and include a review of the overall science and technology relationship between the two countries under the policy framework of the Agreement and of the cooperative activities under this Agreement, and the preparation of the annual report for the consideration of the Joint High Level Committee.
(2) to resolve routine technical and management issues which may arise;	(2) To ensure that the lead agencies of both Parties keep the Joint Committee informed of the status of the other activities related to the U.S.-Japan S&T relationship, including those large-scale projects endorsed at the head of State level.	
(3) to formulate recommendations on matters related to activities under this agreement which require policy level attention for consideration by the Joint Committee.	(3) To resolve routine technical and management issues which may arise;	5. The Joint Working Level Committee will be chaired by the U.S. Department of State and the Ministry of Foreign Affairs of Japan. Each Party will determine its own representatives, including at least one technical management official from
	(4) To formulate recommendations on matters related to activities under this agreement which require policy	

	level attention for consideration by the Joint Committee. (5) To develop an annual action plan for consideration of the Joint Committee, recommending specific steps both countries will pursue to implement the provisions of Annex II, Articles I and II. (6) To assess periodically and report to the Joint Committee whether other agency-to-agency implementing arrangements are fully consistent with the policy framework of this Agreement and its annexes.	each agency with lead responsibility for a cooperative activity under this Agreement. 6. The Joint Working Level Committee will meet at least annually, alternately in the United States of America and Japan. Meetings of the Joint Working Level Committee will be scheduled with particular attention to its role in supporting the Joint High Level Committee.
IV. The Joint Interagency Executive Committee will be co-chaired by the Deputy Assistant Secretary of State for Science and Technology of the United States and the ---. The membership of the Joint Interagency Executive Committee will be determined by each side but will consist of at least one technical management official from each agency with lead responsibility for a project under the	IV. The Joint Interagency Executive Committee will be chaired by the Department of State and the (Japanese equivalent).The membership of the Joint Interagency Executive Committee will be determined by each side but will consist of at least one technical management official from each agency with lead responsibility for a project under the Agreement.	[Annex III] 2. The functions of the Joint Working Level Committee will include: A. Reviewing and discussing the overall science and technology relationship between the two countries under the policy framework of the Agreement, including an assessment of the factors listed in paragraph 1 [of Annex III.1] above.

agreement.		B. Reviewing and discussing the overall progress of the cooperative activities under this Agreement and preparing recommendations, as necessary, for the consideration of the Joint High Level Committee to strengthen those activities;
V. The Joint Interagency Executive Committee will meet at least annually, alternately in the United States and Japan. Meetings of the Committee will be scheduled with particular attention to its role in supporting and developing policy recommendations for consideration by the Joint Committee.	V. The Joint Interagency Executive Committee will meet at least annually, alternately in the United States and Japan. Meetings of the Committee will be scheduled with particular attention to its role in supporting and developing policy recommendations for consideration by the Joint Committee.	C. Preparing, as necessary, policy recommendations on matters of importance to the overall science and technology relationship between the two countries for submission to the Joint High Level Committee for its consideration.
		D. Preparing and submitting to the Joint High Level Committee for its consideration steps and new initiatives for the next year, including areas of cooperation under paragraph 1 of Annex I.
		E. Preparing an annual report for the consideration of the Joint High Level Committee incorporating the results of the functions enumerated above;
		F. Informing the Joint High Level

Committee of the status of other bilateral science and technology arrangements in relation to the policy framework of this Agreement;

G. Dealing with technical and management issues related to the cooperative activities under this Agreement, except that large-scale projects and major research and development initiatives with separate management mechanisms will not fall under the technical and management review of the Joint Working Level Committee;

H. Establishing, as necessary, a task force to identify and monitor scientists' and engineers' access to and participation in major government-sponsored or government-supported research and development programs and to obtain annual statistical data on Japanese researchers' participation in the U.S. research and development system and U.S. researchers' participation in the Japanese research and development system; and

[Annex III]

VI. The two Parties also agree to establish a Joint High Level Advisory Panel composed of eminent leaders from the scientific and engineering disciplines, representing industry, academia, and government, to advise the two Governments on priority issues in the Science and Technology relationship. The Panel shall conduct annually a joint review of the entire range of U.S.-Japan S&T cooperation. These reviews shall take place alternately in the United States and Japan.

[Annex III]

VI. The two Parties also agree to establish a Joint High Level Advisory Panel composed of eminent leaders from the scientific, and engineering disciplines, representing industry, academia, and government with experience in the proposed area of research, to advise the two Governments on priority issues in the science and technology relationship. The Panel shall conduct annually a joint review of the entire range of U.S.-Japan S&T cooperation. These reviews shall take place alternately in the United States and Japan. A meeting of the Panel will be convened at the request of either side.

I. Establishing a task force to develop recommendations on improving access to scientific and technical information and to serve as a forum where scientific and technical information organizations may raise and resolve issues relating to open access to the results of scientific and technological research.

[Article V]

7. The Parties will establish a Joint High Level Advisory Panel to conduct a joint review of the overall science and technology relationship between the two countries and to advise the Joint High Level Committee on issues concerning that relationship. Specific functions of the Joint High Level Advisory Panel are provided in Annex III.

8. The Parties will each designate members of the Joint High Level Advisory Panel, which will comprise eminent leaders from the science and technology communities of both countries representing academia,

		industry, and other areas.
VII. The Joint High Level Advisory Panel shall be composed of senior scientific leaders representing a cross-section of the science and technology communities of the two countries. Members will be appointed by the President's Science Advisor on the U.S. side and the Science and Technology Minister on the Japanese side, respectively. On each side, a majority of the panelists will be selected from non-governmental institutions, including industry and academia.	VII. The Joint High Level Advisory Panel shall be composed of senior scientific, engineering and S&T application leaders representing a cross-section of the science and technology communities of the two countries. Members will be appointed by the President's Science Advisor on the U.S. side and (Japanese equivalent) respectively. On each side, a majority of the Panelists will be selected from non-governmental institutions, including industry and academia.	9. The Joint High Level Advisory Panel will meet on an annual basis and may also be convened at its own initiative or at the request of either chair of the Joint High Level Committee in consultation with the other chair.
VIII. In appointing members to the Joint High Level Advisory Panel, each side may draw on existing bodies whose mandates encompass the concerns which would be addressed by the Panel in the specific U.S.-Japan context. On the U.S. side such a body is the White House Science Council. On the Japanese side, such a body is the Prime Minister's Council for Science and Technology.	VIII. In appointing members to the Joint High Level Advisory Panel, each side may draw on existing bodies whose mandates encompass the concerns which would be addressed by the Panel in the specific U.S.-Japan context. On the U.S. side such a body is the White House Science Council. On the Japanese side, such a body is the Prime Minister's Council for Science and Technology.	

IX. The functions of the Joint High Level Advisory Panel shall be:	IX. The functions of the Joint High Level Advisory Panel would include:	[Annex III] 3. The functions of the Joint High Level Advisory Panel will include:
(2) to identify issues that are of importance for the S&T relationship between the two countries;	(2) Identifying issues that are of importance for the S&T relationship between the two countries;	A. Identifying issues of importance to the overall science and technology relationship between the two countries and making appropriate recommendations to the Joint High Level Committee;
(1) to review overall advances in S&T in the two countries and recommend to the Joint Committee priority areas for bilateral collaboration, under either private or governmental auspices;	(1) Reviewing overall advances in R&D in the two countries and recommending to the Joint Committee priority areas for bilateral collaboration, under either private or governmental auspices;	B. Reviewing major advances in research and development in the two countries and recommending to the Joint High Level Committee priority areas for bilateral collaboration, under either private or governmental auspices;
(3) to assess ongoing mechanisms of science and technology cooperation for their effectiveness in achieving equitable scientific benefit for both sides, and recommend new mechanisms, as appropriate, to further these objectives;	(3) Assessing ongoing mechanisms of S&T cooperation for their effectiveness in achieving equitable scientific and economic benefit for both sides, and recommending new mechanisms, as appropriate, to further these objectives;	C. Reviewing mechanisms of science and technology cooperation for their effectiveness in strengthening the overall science and technology relationship, and making appropriate recommendations to the Joint High Level Committee; and
(4) to identify and recommend approaches for dealing with structural/-institutional differences that may impede balanced access to research and training opportunities, facilities,	(4) Identifying and recommending approaches for dealing with structural/institutional or policy differences that may impede balanced access to research and training oppor-	D. Identifying and recommending

expertise, data, and results; and	tunities, facilities, expertise, data, and results;	approaches to enhance comparable access to research and training opportunities, facilities, expertise, data, and results, taking into consideration each nation's research and development system, institutions, and policies.
(5) to identify areas and mechanisms for strengthening scientific and technological cooperation between the U.S. and Japan outside the governmental framework.	(5) Identifying areas and mechanisms for achieving symmetry in S&T cooperation between the U.S. and Japan outside the governmental framework;	
	(6) Addressing, at the request of either Party, all other issues, problems or questions pertaining to full implementation of this agreement and making appropriate recommendations to resolve such matters.	
Annex IV - Intellectual Property Rights	Annex IV - Intellectual Property Rights	Article VI Intellectual Property Rights
[None]	[None]	1. Scientific and technological information of a non-proprietary nature arising from the cooperative activities under this Agreement may be made available to the public by either Party through customary channels and in accordance with the normal procedures of the participating agencies.
For purposes of implementation of	For purposes of implementation of	2. The Parties will ensure:

Article V of the Agreement, the Parties agree on the following principles.	Article V of the Agreement, and to provide appropriate stewardship for and equitable distribution of intellectual properties created or introduced under the Agreement, the Parties and Participants agree to the following terms:	A. the adequate and effective protection and equitable distribution of intellectual property rights and other rights of a proprietary nature as provided in Annex IV created in the course of the cooperative activities under this Agreement; and
I. Protection of Intellectual Property	I. Protection of Intellectual Property Rights	B. the adequate and effective protection of intellectual property rights and other rights of a proprietary nature provided in Annex IV introduced in the course of the cooperative activities under this Agreement, in accordance with the laws and regulations of the respective countries and with international agreements to which the United States of American and Japan are or will be parties. The Parties will consult for this purpose as necessary.
The Parties will ensure adequate and effective protection for intellectual property created or introduced under the Agreement and the relevant implementing arrangements, in conformity with their respective laws and regulations and international agreements to which Japan and the United States are or will be Parties.	The Parties will ensure adequate and effective protection for intellectual property created or introduced under the Agreement and the relevant implementing arrangements, in conformity with their respective laws and regulations and with international agreements to which Japan and the United States are or will be parties.	
X. Applicability	K. Applicability	[Article VI]
This Annex is applicable to the Agreement and any implementing arrangements or cooperative activities entered into or performed thereunder, except as otherwise specifically	This Annex is applicable to the Agreement and any implementing arrangements or cooperative activities entered into or performed thereunder, except as otherwise specifically	3. Details and procedures for the protection and distribution of intellectual property rights and other rights of a proprietary nature as referred to in paragraph 2 above are

provided for in individual implementing arrangements. Agency implementing arrangements may elaborate the provisions of this Annex and should explicitly refer to it. This Annex is not applicable to any cooperative activity not conducted under this agreement.	provided for in individual implementing arrangements. Agency implementing arrangements may elaborate the provisions of this annex and should explicitly refer to it.	set forth in Annex IV, which is an integral part of this Agreement. Annex IV is applicable to any cooperative activities under this Agreement, except as otherwise specifically agreed by the parties to the cooperative activities concerned, in individual implementing arrangements or otherwise. Implementing arrangements may also elaborate the provisions of Annex IV.
VIII. Miscellaneous	VIII. Miscellaneous	[Article VI]
Other questions or issues regarding the treatment of information, inventions, discoveries, writings, etc., not covered by this Annex, or any disagreements between the Parties respecting this Annex, will be settled through consultations between the Parties or their competent Government agencies.	Other questions or issues regarding the treatment of information, inventions, discoveries, writings, etc., or any disagreements between the Parties respecting this annex, will be referred to the Joint Committee.	4. Issues that arise between the parties to a cooperative activity regarding the treatment of information, inventions, discoveries, writings, etc., under this Article or Annex IV will be settled, in principle, between those parties. Any such issues which cannot be resolve by those parties may be referred to the Joint Working Level Committee.
II. Protection of Dual Use Technology	II. Protection of Dual Use Technology	[none]
Recognizing that technology having both peaceful and defense uses may	Recognizing that dual use technology may be developed under the coopera-	

be developed or introduced under the cooperation, the Parties agree to accept for filing in their respective patent offices applications classified or otherwise held in secrecy for national security purposes, in accordance with the Agreement between the Government of the United States of America and the Government of Japan to Facilitate Interchange of Patent Rights and Technical Information for Purposes of Defense dated 22 March 1956 (hereinafter "1956 Agreement"). Information classified or otherwise held in secrecy shall not be exchanged under this Agreement before the 1956 Agreement is fully implemented.	tion, the Parties agree to accept for filing patent applications classified or otherwise held in secrecy for national security purposes, in accordance with the Agreement between the Government of the United States of America and the Government of Japan to Facilitate Interchange of Patent Rights and Technical Information for Purposes of Defense, dated 22 March 1956 (hereinafter "1956 Agreement). Such information classified or otherwise held in secrecy shall not be exchanged under this Agreement and this Agreement shall not otherwise be implemented until the procedures under the 1956 Agreement permitting the filing of patent applications classified or otherwise held in secrecy are fully implemented.	
III. Confidential Information	III. Confidential Information	Annex IV Protection and Distribution of Intellectual Property Rights and other Rights of a Proprietary Nature
(2) Information to be protected means information of a confidential nature which is appropriately	[2] Information to be protected means trade secrets, commercial or financial information, or technical	1. Business-Confidential Information A. For the purpose of this Annex,

identified and which meets all of the following conditions:

(a) it is of a type customarily held in confidence by governmental or commercial sources;

(b) it is not generally known or publicly available from other sources;

(c) it has not been previously made available by the owner to others without an obligation concerning its confidentiality; and

(d) it is not already in the possession of the recipient Party or Participant without an obligation concerning its confidentiality.

data if such secrets, information or data are privileged or if their disclosure could reasonably be expected to:

[A] Impair a Government's ability to obtain such information in the future;

[B] Cause substantial harm to the competitive position of the person from whom the information was obtained, either directly or indirectly; or

[C] Harm an identifiable Government interest.

Such protectable information meets the following conditions:

[A] It is of a type customarily held in confidence by government or commercial sources;

[B] It is not generally known or publicly available from other sources;

[C] It has not been previously made available by the owner to others without an obligation concerning its

"business-confidential information" means any know-how, technical data, or technical, commercial, or financial information that meets all of the following conditions:

(i) It is of a type customarily held in confidence for commercial reasons;

(ii) It is not generally known or publicly available from other sources;

(iii) It has not been previously made available by the owner to others without an obligation concerning its confidentiality; and

(iv) It is not already in the possession of the recipient without an obligation concerning its confidentiality.

(1) Any information of a confidential nature, as described below, furnished under the Agreement or its implementing arrangements, shall be protected. Such information shall be introduced and furnished only by mutual written agreement of the Participants and after review by the competent Government agency, or as otherwise agreed in writing by the Parties. Each party, agency and participant shall give full protection to such information in accordance with its laws, regulations, and administrative practices.

(3) Any information to be protected shall be appropriately marked before it is introduced under the cooperation, and responsibility for marking such information is on the Participant who introduces it or asserts that it is to be

confidentiality; and

[D] It is not already in the possession of the recipient Party or Participant without an obligation concerning its confidentiality.

(1) Any information of an unclassified proprietary or business-confidential nature, as described below, furnished under the Agreement or its its implementing arrangements, shall be protected. Such information shall be introduced and furnished only by mutual written agreement of the participants and after review by the competent Government agency, and shall be protected as agreed in writing by the Parties.

(3) Except as otherwise required by law, any information to be protected shall be appropriately marked before it is introduced under the cooperation or, unless otherwise provided in the implementing arrangements, immediately upon being generated, and responsibility for marking such information is on the Participant who

B. Any business-confidential information will be furnished or, when created in the course of the cooperative activities under this Agreement, transferred only by mutual written agreement of the parties to the cooperative activity concerned and will be given full protection in accordance with the laws and regulations of their respective countries.

C. Any business-confidential information will be appropriately identified before it is furnished in the course of the cooperative activities under this Agreement or, unless otherwise provided in the implementing arrangements, immediately upon being created. Responsibility for identifying such information will fall on the party

protected. Unmarked information shall be assumed not to be information to be protected except as required by the laws of the Parties. Implementing arrangements may address in greater detail the provisions for marking, acceptance or refusal of confidential information, and procedures to resolve disagreements as to whether information is to be protected under this Article.	introduces it or asserts that it is to be protected. Unmarked information shall be assumed not to be information to be protected except as required by the laws, regulations and administrative practices of the Parties. Implementing arrangements will address in greater detail the provisions for marking, acceptance or refusal of confidential information, and procedures to resolve disagreements as to whether information is to be protected under this Article.	which furnishes it or asserts that it is to be protected. Unidentified information will be assumed not to be information to be protected, except that a party to the cooperative activity may notify the other party in writing, within a reasonable period of time after furnishing or transferring such information, that such information is business-confidential information under the laws and regulations of its country. Such information will thereafter be protected in accordance with subparagraph B above.
[IV Inventions]	[IV Inventions. Paragraph 2]	2. Ownership of Intellectual Property Rights
Between a Party and its nationals, the ownership of rights will be determined in accordance with the Party's national laws, regulations and practices.	Between a Party and its nations, the ownership of rights will be determined in accordance with the Party's national laws and practices.	Between each Party and nationals of its country, the ownership of intellectual property rights will be determined in accordance with its national laws, regulations and practices.
IV Inventions	IV Inventions	3. Inventions
For purposes of this Article, "invention" includes any invention made or	For purposes of this Article "invention" means any invention made or	A. For the purpose of this Annex, the "Invention" means any invention

conceived in the course of or under this Agreement or its implementing arrangements and which is or may be patentable or otherwise protectable under the laws of the United States, Japan, or any third country.	conceived under this Agreement or its implementing arrangements and which is or may be patentable or otherwise protectable under the laws of the United States, Japan or any third country. "Made" means conceived or first actually reduced to practice.	made in the course of the cooperative activities under this Agreement which is or may be patentable or otherwise protectable under the laws of the United States or America, Japan or any third country.
As to inventions made or conceived under the Agreement or its implementing arrangements, the Parties, their competent Government agencies, and Participants will take appropriate steps to secure rights to implement the following:	As to inventions made under the Agreement or its implementing arrangements, the Parties, their competent Government agencies, and Participants will take appropriate steps to secure rights to implement the following:	B. As to an Invention, the parties to the cooperative activity concerned will take appropriate steps, in accordance with the national laws and regulations of the respective countries, with a view to realizing the following:
(1) If the invention is made or conceived as a result of the exchange of information between the Parties, such as by joint meetings, seminars, or the exchange of technical reports or papers:	[1] If the invention is made as a result of cooperation that involves only the transfer or exchange of information between the Parties, such as by joint meetings, seminars, or the exchange of technical reports or papers, unless provided otherwise in an applicable implementing arrangement:	(i) If an Invention is made as a result of a cooperative activity under this Agreement that involves only the transfer or exchange of information between the parties such as by joint meetings, seminars, or the exchange of technical reports or papers, unless otherwise provided in an applicable implementing arrangement:
(A) the Party whose personnel make the invention (the Inventing Party) has the right to obtain all rights and interests in the invention in all countries;	[A] The Party whose personnel make the invention [the Inventing Party]	(a) The party whose personnel make the Invention (hereinafter referred to as "the Inventing Party") or the personnel who make the Invention
(B) in any country where the Invent-		

ing Party decides not to obtain such rights and interests, the other Party has the right to do so; and

(C) in any country where one Party obtains all rights and interests in an invention, the other party has the right to a nonexclusive, irrevocable, royalty-free license to the invention for the Party, with the right to grant sub-licenses;

(2) If the invention is made or conceived by personnel of one Party (the Assigning Party) while assigned to the other Party (the Receiving Party) during an exchange of scientific and technical personnel:

(A) the Receiving Party has the right to obtain all rights and interests in the invention in its country and in third countries, and the Assigning Party has the right to obtain all rights and interests in its country.

(B) in any country where either Party decides not to obtain such rights and

has the right to obtain all rights and interests in the invention in all countries, and

[B] In any country where the Inventing Party decides not to obtain such rights and interests, the other Party has the right to do so.

[2] If the invention is made by personnel of a Party [the Assigning Party] while assigned to another Party [the Receiving Party] in the course of cooperation that involves only the visit or exchange of scientific and technical personnel:

[A] The Receiving Party has the right to obtain all rights and interests in the invention in all countries, and

[B] In any country where the Receiving Party decides not to obtain such and interests, the Assigning Party has the right to do so.

(hereinafter referred to as "the Inventor") have the right to obtain all rights and interests in the Invention in all countries and

(b) in any country where the Inventing Party or the Inventor decides not to obtain such rights and interests, the other Party has the right to do so.

(ii) If the Invention is made by an Inventor of a party ("the Assigning Party") while assigned to another party ("the Receiving Party") in the course of programs of a cooperative activity that involve only the visit or exchange of scientists and engineers, and:

(a) in the case where the Receiving Party is expected to make a major and substantial contribution to the programs of the cooperative activity:

i. the Receiving Party has the right to obtain all rights and interests in the Invention in all countries, and

interests, the other Party has the right to do so; and (C) in any country where one Party obtains rights and interests in an invention, the other Party has the right to a nonexclusive, irrevocable, royalty-free license for the Party and its nationals.		ii. in any country where the Receiving Party decides not to obtain such rights and interests, the Assigning Party or the Inventor has the right to do so; (b) in the case where the provision in subparagraph (a) above is not satisfied: i. the Receiving Party has the right to obtain all rights and interests in the Invention in its own country and in third countries, ii. the Assigning Party or the Inventor has the right to obtain all rights and interests in the Invention in its own country, and
(3) If the invention is made or conceived as a result of other specific forms of cooperation, such as special joint research projects, the Parties, their competent Government agencies, and Participants shall provide for the appropriate distribution of the rights thereto. In general, each Party should normally determine	[3] Specific agreements involving other forms of cooperation, such as special joint research projects, shall provide for the mutually agreed upon disposition of rights to an invention made as a result of such a special project in accordance with the policies of the participants and their respective Governments.	iii. in any country where the Receiving Party decides not to obtain such rights and interests, the Assigning Party or the Inventor has the right to do so. (iii) Specific arrangements involving other forms of the cooperative activities, such as joint research projects

the rights to such inventions in its own country, and third country rights should be agreed upon by the Parties on an equitable basis.		with an agreed research work scope, will provide for the mutually agreed upon disposition, on an equitable basis, of rights to the Invention made as a result of such activities.
(4) The Party whose personnel made an invention shall communicate promptly to the other Party information disclosing the invention and any patent or other protection it elects to obtain and will furnish the documentation necessary for the establishment of the other Party's rights in the invention. The Communicating Party may ask the other Party in writing to delay publication or public disclosure of such information. Unless otherwise agreed in writing, such restriction shall not exceed a period of six months from the date of the communication of such information. Communication shall be through the competent Government agencies or as designated in the implementing arrangements.	[4] The Participant whose personnel make an invention must disclose promptly the invention to the other Participant and furnish any documentation and information necessary to enable the other participant to establish rights to which they are entitled. The Inventing Party may ask the other Participant in writing to delay publication or public disclosure of such information. Unless otherwise agreed in writing, such restriction shall not exceed a period of six months from the date of communication of such information. Communication shall be through the competent Government agencies or as designated in the implementing arrangements.	(iv) The Inventing Party will disclose promptly the Invention to the other party and furnish any documentation or information necessary to enable the other party to establish rights to which it may be entitled. The Inventing Party may ask the other party in writing to delay publication or public disclosure of such documentation or information for the purpose of protecting its rights or the rights of the Inventor related to the Invention. Unless otherwise agreed in writing, such restrictions will not exceed a period of six months from the date of communication of such documentation or information.

V. Copyrights	V. Copyrights	4. Copyrights
Participants will take appropriate steps to secure copyrights to works created under this Agreement in accordance with their respective national laws, except as specifically provided otherwise in an implementing arrangement. Rights to such works shall be determined in the relevant implementing arrangements, which may also provide for a non-exclusive, irrevocable, royalty-free, license under the copyright to translate, reproduce, publish and distribute such works, for the Parties and, where appropriate, for their nationals. In determining the rights of the Parties, the principles of Article IV may serve as guidance.	Rights to works created under this Agreement shall be determined in the relevant implementing arrangements. Participants must take appropriate steps to secure copyrights to works created under this Agreement in accordance with their respective national laws, except as provided otherwise in an implementing arrangement, rights will be determined in accordance with the policies of the Participants and their respective Governments.	Disposition of rights to copyright-protected works created in the course of the cooperative activities under this Agreement will be determined in the relevant implementing arrangements. The parties to the cooperative activities concerned will take appropriate steps to secure copyright to works created in the course of the cooperative activities under this Agreement in accordance with the national laws and regulations of the respective countries.
		5. Rights to Semiconductor Chip Layout Designs
	None	Disposition of rights to semiconductor chip layout designs created in the course of the cooperative activities under this Agreement will be determined in the relevant implementing arrangements. The parties to the cooperative activities concerned will take appropriate steps to secure rights to semiconductor chips layout designs

		created in the course of the cooperative activities under this Agreement in accordance with the national laws and regulations of the respective countries.
VI. Other Forms of Intellectual Property	VI. Other forms of Intellectual Property	6. Other Forms of Intellectual Property
Rights to other forms of intellectual property, such as mask work registrations, shall be determined on an equitable basis, as set forth in the implementing arrangements.	Rights to other forms of intellectual property, such as mask works, shall be determined on an equitable basis, as set forth in implementing arrangements.	For those other forms of intellectual property created in the course of the cooperative activities under this Agreement which are protected under the laws of either country, disposition of rights will be determined on an equitable basis, in accordance with the laws and regulations of the respective countries.
VII. Cooperation	VII. Cooperation	7. Cooperation
Each Party, its competent Government agency, and Participant will take all necessary and appropriate steps to provide for the cooperation of its authors and inventors which is required to carry out the provisions of this Annex. Each Party and Participant assumes	Each party, its competent Government agency, and participant will take all necessary and appropriate steps to provide for the cooperation of its authors and inventors which are required to carry out the provisions of this annex. Each Party and Participant assumes	Each party to the cooperative activity concerned will take all necessary and appropriate steps, in accordance with the laws and regulations of its country, to provide for the cooperation of its authors and inventors which are required to carry out the provsions of this Annex. Each party to the cooper-

the sole responsibility for any award or compensation that may be due to its personnel in accordance with its laws and regulations, provided, however, that this Article creates no entitlement to any such award or compensation.	the sole responsibility for any award or compensation that may be due its personnel in accordance with its laws and regulations, provided, however, that this article creates no entitlement to any such award or compensation.	ative activity concerned assumes the sole responsibility for any award or compensation that may be due its personnel in accordance with the laws and regulations of its country, provided, however, that this Annex creates no entitlement to any such award or compensation.
IX. Relation to Domestic Law	IX. Relation to Domestic Law	See Article II.7
No provision of the Agreement or this Annex requires either Party to modify its domestic law as to matters covered by the Agreement or Annex.	Except as provided in Article II, no provision of this Annex requires either Party to modify its domestic law as to matters covered by the Annex.	
	Annex V - Security Obligations	Article VII
	I. Protection of Information	1. Both parties support the widest possible dissemination of the information or equipment created in the course of the cooperative activities under this Agreement, unless otherwise stipulated in this Article, Article VI, or Annex IV. In furtherance of the principle of maintaining an open basic research environment, both Parties confirm that no information or equipment
	Both Governments agree that no information or equipment requiring protection in the interests of national defense or foreign relations of either Government and classified in accordance with the applicable national laws and regulations shall be provided under this Agreement. Information or equipment, which is	

known or suspected to require such protection, identified in the course of projects undertaken pursuant to this Agreement, immediately shall be brought to the attention of the appropriate Government officials.

II. Technology Transfer

The transfer of unclassified information and equipment between the parties under this Agreement shall be subject to national export control laws and regulations of each Government. The Governments will take all necessary and appropriate measures, in accordance with the international obligations, national laws and regulations of each Government, to prevent the unauthorized transfer or retransfer of unclassified, export-controlled information and equipment provided or produced under this Agreement. Detailed provisions for the prevention of unauthorized transfer or retransfer of such information or equipment shall be incorporated into all contracts and other arrangements implementing this Agreement.

classified for reasons of national defense will be utilized in the cooperative activities under this Agreement.

2. The transfer of export-controlled information or equipment between the countries in the course of the cooperative activities under this Agreement will be in accordance with the applicable national export control laws and regulations of each country. Each party will take all necessary and appropriate measures, in accordance with applicable national laws and regulations, to prevent the diversion to unauthorized destinations of export-controlled information and equipment provided or produced in the course of the cooperative activities under this Agreement.

[SIDE LETTERS]

The Secretary of State
Washington. June 20, 1988

Dear Mr. Minister:

I wish to confirm the following understanding shared by our two Governments regarding paragraph 1 of Article VII of the Agreement between the Government of the United States of America and the Government of Japan on Cooperation in Research and Development in Science and Technology:

1. In the event that information or equipment may be classified for reasons of national defense is unexpectedly created in the course of cooperative activities under the aforementioned Agreement, it may, insofar as permitted by applicable laws and regulations, be protected from unauthorized disclosure.

2. Neither Government is obligated to modify existing laws or regulations or to create new laws and regulations.

Sincerely Yours,

George P. Schultz

His Excellency, Sousuke Uno, Minister for Foreign Affairs, Tokyo

----

Ministry of Foreign Affairs, Tokyo, Japan. June 20, 1988

Dear Mr. Secretary:

I wish to confirm the following understanding shared by our two Governments regarding paragraph 1 of Article VII of the Agreement between the Government of Japan and the Government of the United States of America on Cooperation in Research and Development in Science and Technology:

	1. In the event that information or equipment that may be classified for reasons of national defense is unexpectedly created in the course of cooperative activities under the aforementioned Agreement, it may, insofar as permitted by applicable laws and regulations, be protected from unauthorized disclosure.
See Paragraph 1, Diplomatic Note	2. Neither Government is obligated

See Paragraph 1, Diplomatic Note

to modify existing laws and regulations or to create new laws and regulations.

Sincerely,

Sousuke Uno, Minister for Foreign Affairs of Japan

The Honorable George P. Schultz, Secretary of State of the United States of America

Article VIII

1. Implementation of this Agreement will be subject to the availability of appropriated funds and to the applicable laws and regulations of each country.

2. Costs of the cooperative activities under this Agreement will be borne by the Parties as mutually agreed, taking into account their respective risks, benefits and management shares.

Article IX

1. This Agreement will enter into force upon signature and remain in force for five years. Either Party may at any time given written notice to the other Party of its intention to terminate this Agreement, in which case this Agreement will terminate six months after such notice has been given.

2. This Agreement may be extended or amended by mutual agreement of the Parties.

Article X

The expiration of this Agreement will not affect the carrying out ot any project or program undertaken under this Agreement and not fully executed at the time of the expiration of this Agreement.

DONE at Toronto, this twentieth day of June, 1988, in duplicate in the English and Japanese languages, both texts being equally authentic.

Ronald Reagan
For the Government of the United States of America

Noboru Takeshita
For the Government of Japan

Appendix 3

Joint Task Force Report

Plan of Action for Proposed Renewal of the U.S. Agreement for Cooperation in Research and Development in Science and Technology

Submitted to the Executive Committee of the FCCSET Committee on International Science, Engineering and Technology by the Joint U.S.-Japan Bilateral Task Force[1]

November 26, 1986

Overview

At its July 11, 1986, meeting, the Executive Committee of the FCCSET Committee on International Science, Engineering and Technology (CISET) agreed that CISET, through its relevant Working Groups, should initiate a review of the United States' scientific and technical relationships with Japan. The Executive Committee noted that the proposed renewal of the U.S.-Japan Agreement for Cooperation in Research and Development in Science and Technology [hereafter the non-energy agreement]. due to expire in 1987, provides an opportune occasion to initiate a dialogue with Japan aimed at a long-term strategy to achieve mutually beneficial goals: first, because the agreement, negotiated and implemented at the Presidential-Prime Ministerial level, has significant political implications for both countries; second, because it is intended to cover all fields of science and technology.

In view of the limited time available for the review, the Executive Committee agreed that the respective executive staffs of the CISET Working Groups on Education, Infrastructure, and Facilities; and Bilateral and Multinational Activities (chaired, respectively, by the National Science Foundation and the Commerce and State Departments) should organize a Joint Task Force on U.S.-Japan Science and Technology Relations with expert representatives from all relevant agencies. The charge to the Task Force was to: (a) identify specific scientific and technological areas in which enhanced cooperative research would be beneficial to both the United States and Japan; (b) explore feasible means for involving U.S. industry in setting the requirements for cooperative research under the bilateral agreement and

[1] The term "Joint" used here gives the impression that it was perhaps a joint U.S. and Japan Task Force. Actually the Task Force consisted only of U.S. Government officials. The intent of the report was to create a plan of action for use in bilateral discussions with the GOJ.

making more effective use of the results obtained from that research; and (c) identify likely opportunities for resolving generic barriers to enhanced bilateral cooperation between the U.S. and Japan.

The detailed Plan of Action that follows is intended to serve as an agenda for technical discussions with the Japanese prior to formal renewal of the non--energy agreement during the spring of 1987. It is based on the results of meetings of the Joint Task Force (held on July 22, August 11 and September 9), on written submissions from participating agencies on both proposed cooperative research areas and generic problems affecting the U.S.-Japan relationship, and on inputs from a September 12 meeting of industry representatives convened by the Office of Science and Technology Policy (OSTP).

The following five procedural recommendations are intended as guides to discussions with the Japanese prior to formal renewal of the non-energy agreement:

1. A formal organization structure should be established by OSTP to plan and execute further cooperation with Japan under the terms of the non-energy agreement. A principal feature of that structure would be a permanent interagency executive committee established at a senior level and constituted in such a way as to insure that both foreign policy objectives are met and mutual scientific benefits optimized.

2. To emphasize the scientific importance the U.S. Government places on the non-energy agreement, any new bilateral science and technology initiative or program identified for cooperation with Japan should be considered for inclusion under the umbrella of that agreement.

3. OSTP should present the proposed management structure to the Japanese along with the proposal that a similar structure be adopted by the Japanese Government to insure increased scientific benefit and improved administrative and technical efficiency.

4. OSTP should present a short list of proposed areas for expanded cooperation developed by the Task Force to the Japanese for acceptance in principle, and make it clear that the U.S. welcomes similar proposals from the Japanese side. Detailed, specific project proposals would be put forward by a designated lead U.S. agency after agreement on a particular area is obtained.

5. To insure that the interests of industry are taken into account, each U.S. Government agency assigned lead responsibility for implementing projects within a designated scientific or technical area under the non-energy agreement should establish an industry advisory group consisting of representatives with technical capabilities in that area.

The principal functions of the executive committee proposed under item 2 [sic, should be 1] would include:

- Assessment of current activities conducted under the non-energy agreement to determine which are consistent with the Presidential status of the agreement, and which would be carried out more appropriately under the auspices of the some other existing bilateral agreement;
- Periodic review of progress made in new initiatives included under the non-- energy agreement and resolution of problems that may arise;
- Selection of appropriate scientific personnel to participate in periodic meetings held with Japanese counterparts, as called for by the non-energy agreement; and
- Provision of mechanisms to insure that other existing bilateral or multilateral science and technology agreements involving Japan are recognized and protected.

The principal substantive areas the Task Force proposes for inclusion under the agreement are:

- Materials science and engineering (particularly electronic materials).
- Life sciences, particularly biotechnology.
- Information science and technology.
- Artificial intelligence (including software development).
- Global geosciences.
- Joint database development (in several fields).
- Joint standardization and nomenclature development (in several fields).

Plan of Action for Proposed Renewal of the U.S. -Japan Agreement for Cooperation in Research and Development in Science and Technology

Introduction

At a meeting of the coordinators of the CISET Joint Task Force on U.S.-Japan Bilateral Science and Technology Relations held at the National Science Foundation on September 30, 1986, at which submissions from participating agencies were reviewed, agreement was reached on a course of action to be presented to the Office of Science and Technology Policy (OSTP) concerning the following matters:

1. Administrative procedures to be followed leading to negotiations with the Japanese Government on renewal of the U.S.-Japan Agreement for Cooperation in Research and Development in Science and Technology [the non-energy agreement];

2. Organization of U.S. Government of the agreement;
3. Identification of several broad scientific and technological areas to be proposed for coverage under the agreement;
4. Procedures for more effective implementation of the agreement; and
5. Involvement of U.S. industry in setting requirements and making use of information received.

These subjects were addressed in considerable depth because of a consensus among the Task Force coordinators that the non-energy agreement hold s the promise of being an important tool in the bilateral relationship with Japan if it can be revitalized and if certain significant change can be made in U.S. (and Japanese) Government policies and procedures.

Negotiations with Japan on Renewal of the Agreement

The Task Force noted that representative of OSTP and several participating Federal agencies plan to visit Tokyo early in 1987 to hold preliminary talks with the Ministry of Foreign Affair and other Japanese officials concerning the future of the agreement. The Task Force agreed that it should recommend to OSTP that a positive, constructive attitude should be shown to the Japanese side. This is based on the judgment that renewal of the agreement offers a valuable opportunity to strengthen the mutuality of benefit derivable from enhanced cooperation provided steps are taken to strengthen policy oversight in both the U.S. and Japan and provided a new management approach to implementing the agreement can be instituted.

Management of the Agreement

In the current political situation, demands are being made in both the public and private sectors for increasing U.S. competitiveness in high technology. The non--energy agreement appears to be potentially useful device for gaining legitimate access to the results of Japanese research and development, and thus helping satisfy these demands. This is particularly true in those fields of science and technology where Japan is among the world leaders (or where current Japanese R&D indicates that Japan may emerge as a future leader), and when these fields are not already within the province of other existing bilateral agreements. New ventures under the aegis of the non-energy agreement should be conducted in a strong spirit of cooperation through balanced, reciprocal exchange of information and personnel and through the joint use of research facilities. The U.S. is the country that has been both the largest supplier of technical information to Japan and the one with whom Japan has enjoyed the longest and closest scientific and technological ties. There now is a great deal of evidence to the effect that the current policy of the Japanese Government is to expand its role in international cooperation in science and technology. The U.S. Government's position should that it intends to take advantage of this new Japanese policy.

Implementation of the non-energy agreement has suffered from lack of an overall strategic plan and limited communication and coordination among involved agencies on both sides. Changes in monitoring personnel in the various agencies, as well as in OSTP, also have contributed to ineffectiveness in coordination. In order to emphasize the importance the U.S. Government places on the non-energy agreement as a means for increasing the effectiveness of its science and technology relations with Japan, the Task Force recommends that a fundamental principle for further implementation should be that any new bilateral science and technology initiative or program identified for cooperation with Japan should be considered for inclusion under its umbrella. However, given the political significance of the non-energy agreement as a President initiative, only those of high scientific and technological priority ought actually to be included.

Two additional problems with the way the non-energy agreement has been implemented should also be noted:

First, the U.S. Government (particularly in the early stages of the agreement) failed to promote projects at the leading edge of science and technology that would have been appropriate for a Presidential level initiative. This failure on the part of the U.S. led to the suspicion, on the Japanese side, that a principal U.S. objective was to induce Japan to provide the bulk of the funding for joint research projects. A related problem was the missed opportunity to use the non-energy agreement for cooperation in areas where Japan is among the world leaders. The Task Force believes that substantial Japanese investment in U.S. R&D project should be sought where appropriate, but that this should not be an overriding objective. Counter-investment in Japanese projects is also worthy of consideration.

Second, there was a failure to solicit or take into account the needs or views of private industry in implementing the agreement. As initial, productive step to rectify this deficiency was taken by OSTP, in cooperation with the Task Force, by convening the September 12 [1986] meeting to assess the viewpoint of industry. But means must be established for insuring that the dialogue between the U.S. Government and the private sector continues.

Finally, the Task Force noted several generic issues affecting overall U.S.-Japan bilateral science and technology relations, as suggested by the participating agencies. These issues are associated with:

- The need for encouraging and facilitating more long-term visits to Japanese laboratories by U.S. scientists and engineers. The Task Force recognized that this will require language training, and suggested that the possibility of access to technical language courses developed by the Japanese for designated numbers of

U.S. scientists and engineers in specific fields should be raised informally with the Japanese Government.
- Obtaining equitable access to Japanese data and laboratories. The Task Force recognized that the access issue is complex and needs further clarification. U.S. scientists have had good access under terms of the bilateral agreement managed by NSF, and a recent survey conducted by NSF's Tokyo Office indicates that several Japanese laboratories (including industrial laboratories) would welcome foreign participation.
- Opportunities for main better use of underutilized facilities, instrumentation in both countries.
- Opportunities for closer cooperation that could be achieved though data base development, standardization and exchange.
- The desirability of raising and attempting to resolve differences in laws and practices in the areas of regulation, liability, and intellectual property protection.

Taking these various points into consideration and recognizing the high level political significance of the non-energy agreement, the Task Force recommends that the substantive thrust of its implementation be limited to a few high priority areas of science and technology. Since continuous involvement of the principal operating agencies on both sides will be essential to effective management of the agreement, the Task Force further recommends that a new, formal organization be established by OSTP to plan and execute further cooperation with Japan under the terms of the Agreement and that a counterpart organization in Japan modeled roughly along the same lines be proposed to the Japanese side during the course of negotiations.

OSTP should remain the nominal focal point of any new organizational structure. However, it is evident that OSTP has neither the time nor the resources to engage in day-to-day planning and administration of an effective agreement. This can best be done by a consortium of the agencies involved. Accordingly, the Task Force recommends that a permanent interagency management or executive committee be established by OSTP at a senior technical level (i.e. deputy assistant secretary or assistant director level). The Committee should be formed in such a way as to ensure that scientific objectives are optimized, that foreign policy objectives are met, and that technology transfer concerns are adequately addressed. Departments or agencies to be represented on the committee would include:

Departments of Agriculture, Commerce (including the National Bureau of Standards and the National Oceanic and Atmospheric Administration), Defense, Energy, Health and Human Services (Public Health Service), State, Environmental Protection Agency, National Aeronautics and Space Administration, National Science Foundation, U.S. Geological Survey.

In addition, representatives of other departments and agencies would be invited to participate in ex officio capacities and/or when appropriate.

The Task Force suggests that the functions of the interagency Executive Committee should include the following:

1. To assess current activities conducted under the non-energy agreement to determine which are consistent with the Presidential status of the agreement, and which would be carried out more appropriately under the auspices of some other existing bilateral agreement.
2. To review periodically the progress made under the agreement and to resolve problems that may arise. (This function alone will raise the level of importance of the agreement and the attention paid to it; it does not exist at the present time.)
3. To develop Government-wide policy proposals for presentation to OSTP, and to respond to OSTP initiatives.
4. To act as the official, acknowledged representative of the U.S. Government in establishing and maintain coordination with industry associations having interests in Japanese science and technology. This function should be applicable to all science and technology agreements with Japan, not just the non-energy agreement.
5. To provide scientific representation to periodic meetings held with Japanese counterparts, as called for by the agreement.
6. To determine which department or agency will take the lead in those technical fields or disciplines which cut across agency responsibilities.
7. To ensure that other existing bilateral or multilateral science and technology agreements involving Japan are recognized and protected. (The Task Force emphasizes that the continued viability of these agreements is important to the overall political and scientific relationship. However, agencies should bring proposed new initiatives to the committee for consideration for inclusion under the non-energy agreement.)

Since it cannot be expected that officials at the deputy assistant secretary level will be able to devote significant fractions of their time to one agreement, their role should be to furnish the imprimatur of authority and concrete evidence of interagency coordination and responsibility. The Task Force visualizes that each member of the Executive Committee would appoint a person within his department or agency who would provide day-to-day coordination at the working level. These persons ideally would be those who have great familiarity with science and technology in Japan and with the operation of the non-energy and other bilateral and multilateral agreements with Japan.

These coordinators collectively are visualized as a perpetuation of the current CISET Task Force, meeting as needed. In addition, they could be given another important function that will be identified later. In any event, the use of coordinators at the working level for an agreement with Japan involving a number of agencies has been demonstrably successful in the case of the U.S.-Japan Natural

Resources agreement. Part of the success of this agreement can be attributed to the dedication and interest of the coordinators on both the U.S. and Japanese sides.

If the OSTP agrees with the essence of this management approach and enacts it, the Task Force recommends that OSTP present the new management structure to the Japanese side (perhaps during a preliminary visit to Japan), along with the proposal that a similar structure be adopted by the Japanese Government to ensure increased administrative and technical efficiency.

Priority Scientific and Technological Areas of Cooperation

A number of Federal agencies responded positively and in depth to the CISET Task Force solicitation of ideas for expansion of cooperation with Japan. Others either were unable to respond with specific ideas or held negative views toward further cooperation in fields judged to be sensitive or at the leading edge in the U.S. A summary of agency responses is attached, as are selected follow-up comments from industry representatives who participated in the September 12 meeting convened by OSTP. These have been incorporated into the synthesis which follows.

It is not surprising that the priority areas of science and technology proposed for expanded cooperation by the U.S. side coincide well with the priorities for R&D that have been established by the Japanese Government. With some variability in definition, they are:

- materials science and engineering (particularly electronic materials).
- life sciences, particularly biotechnology.
- information science and technology.
- automation and process control.
- artificial intelligence (including software development).
- global geosciences.
- joint database development (in several fields).
- joint standardization and nomenclature development (in several fields).

Since none of these subjects falls within the purview of a single U.S. government agency, coordination of the kind postulated earlier will be required to execute any cooperative activity with Japan.

Japanese Government has established formal programs covering a number of these subjects, and in some cases has invited international participation -- to which the U.S. has not responded. Potentially the most important of these is the Human Frontiers Program initiated by MITI. Although this program was discussed in private at the Tokyo Summit in June 1986, the discussion proved to be premature and the Prime Minister returned the concept to a government committee for further study. It will probably surface again. The Human Frontiers Program is concerned almost exclusively with the life sciences and foreign participation would not be limited to the U.S. Other advanced Japanese research programs that are open to foreign participation include the ERATO and Frontier Research programs of the Science and Technology Agency and MITI's Japan Trust. Additionally, programs not nominally open to foreign participation could form the basis for negotiations with the Japanese [EXAMPLES]. Since the U.S. usually does not have formal, discipline-focused programs that act as logical counterparts to those in effect in Japan, devices for matching the two national efforts need to be developed.

Task Force recommends that OSTP present this short list of priority areas for expanded cooperation to the Japanese for acceptance in principle. If this acceptance is obtained, the U.S. would be in a position to propose detailed, specific projects. It should be made explicitly clear that the U.S. would welcome similar proposals from the Japanese side, including proposals for cooperation in general disciplinary fields not covered by the list or by other agreements.

Systems for More Effective Implementation of Agreement

In addition to the more intensive and extensive policy, oversight and interagency management structure proposed, a new level of Government-sponsored involvement in the non-energy agreement is required as a means of accelerating the transfer of advance Japanese technical information to private industry. To this, the Task Force recommends that the U.S. Government agency assigned lead responsibility for each agreed upon area of cooperation under the next phase of the non-energy agreement to establish an industry advisory group consisting of representatives with technical capabilities in that area.

In some areas covered under the non-energy agreement, implementation could also be made more effective by assigning program, disciplinary, or project responsibilities to Government contractor organizations through directives specifying that information obtained from Japan must be further disseminated to interested parties. The same philosophy would be applicable to government research organizations, such as the National Institutes of Health, National Bureau of Standards [now known as National Institute of Standards and Technology], and Agricultural Research Service laboratories, and National Aeronautics and Space Administration and National Oceanic and Atmospheric Administration centers. Part of this process is already in place. For example, National Laboratories administered under contract with the Department Energy already participate in nuclear and non-nuclear energy R&D agreements with Japan, but they are not specifically charged with providing information to the private sector or in factoring into their work with Japan the needs of the private sector. National Bureau of Standards scientists perform similarly.

The Task Force visualizes that the agency coordinators assigned to the agreement could manage this tier of interaction with contractors under the respective agencies' jurisdiction, but it would be the function of the Executive

Committee or the coordinators in concert to decide which contractors or government laboratories should be given government-wide responsibilities for the various initiatives and programs.

Advantages of transferring some responsibility to the working technical level include: (a) better technical awareness of advances being made in Japan, (b) the potential for bringing greater amounts of manpower to bear if needed, (c) greater flexibility in covering costs (including foreign travel, holding conferences, etc), and (d) usually better administrative resources and facilities. Another advantage is that contractor organizations are distributed geographically throughout the U.S. instead of being concentrated in Washington. This should enhance the effectiveness of day-to-day interactions with the public.

This decentralization approach fits the system being employed in Japan quite well. For example, most of MITI's technical expertise is found in the research institutes of MITI's Agency of Industrial Science and Technology, nine of which are located in Tsukuba Science City and seven of which are collected around the country. Likewise, the Ministry of Education sponsors discipline-oriented, inter-university research institutes. The Japanese side could be encouraged to establish direct technical relations by these institutes with designated organizations in the U.S., thereby cutting red tape and increasing the efficiency of information transfer.

The Office of Naval Research (ONR) in its Far East Program has adopted the scheme outlined above in one instance and has found it to be very successful, Pennsylvania State University has been designated (and funded) by ONR to act as the center for interacting with Japan on piezoelectric research. Periodic seminars are held to which both Japanese and American scientists are invited. A condition of attendance is that each participant present a paper giving the latest results of his research. No proprietary information is given or sought. ONR then insures that other American scientists working in this field are informed of the information obtained.

An example of a similar approach can be found in the JTECH studies performed by Commerce, NSF and DARPA [ARPA] through a private contractor, Science Applications International Corporation [since 1989 this contract was transferred to Loyola College, Baltimore MD]. Japanese capabilities in six technical fields have been surveyed over a period of three years by American panels drawn largely from universities and private corporations. A possible problem with this approach is that there is a considerable time lag before the information obtained reaches the public.

Increasing Broad Non-Governmental Involvement in Bilateral Relations

Industrial participants at the September 12 meeting convened by OSTP stressed the desirability of involving U.S. industry in identifying the types of technical information that should be obtained from Japan and in establishing a system for disseminating information obtained to industry. While the decentralization scheme described above would be a marked improvement over current practice in meeting the latter objective, there is little question that the Government has not been effective in taking into account the needs and concerns of U.S. industry. One participant proposed that a joint commission of Japanese and American industrial executives be established for this purpose but limited to "... the application of science to human needs." Another participant proposed that one or more R&D consortia made up of American companies be set up to interact with Japanese counterpart consortia. Both ideas have merit.

The Task Force observes that government-industry interaction for both technical and policy purposes is already well developed in Japan but remains a missing ingredient in the U.S. half of the U.S.-Japan cooperation equation. Beyond using the non-energy agreement to facilitate technology transfer to the private, a means for bringing the best available policy-level thinking to bear on long term science and technology relations between the two countries is required. The Task Force therefore urges OSTP to establish, as an integral component of the U.S. Government's management plan for the non-energy agreement, a blue-ribbon, non-governmental policy oversight panel with representation from both industry and universities whose principal charge would be to develop and assess long-range directions for bilateral relations between the two countries. Obviously, the effectiveness of such a panel would be enhanced considerably if the Japanese were to establish a parallel body. The Task Force is convinced that the mutual benefits that will accrue to both countries would be enhanced significantly by such an arrangement.

Appendix I CISET Joint Task Force on U.S.-Japan Bilateral Science and Technology Relations

[The original appendix contains a list of U.S. government organizations and the names of the persons from those agencies which participated in the Task Force report. However, only the names of the government agencies will be listed below.]

Departments of Agriculture, Commerce, Defense, Energy and State, Central Intelligence Survey, Environmental Protection Agency, Geological Survey, National Aeronautics and Space Administration, National Bureau of Standards, National Oceanic and Atmospheric Administration (Part of DOC), National Science Foundation, Public Health Service (part of the Department of Health and Human Services).

Appendix 4

Meetings of the JWLC, JHLC and JHLAP

Cycle #	Year	JWLC	JHLAP	JHLC
1	1988	9/14/15/1988 Washington, DC	1/17-18/89 Tokyo	10/11-12/1988 Tokyo
2	1989-1990	4/25-26/1989 Tokyo	2/13-14/1990 Irvine, CA	5/4/1990 Washington, DC
3	1990-1991	4/30-5/1/1990 Washington, DC 7/18-19/1991 Tokyo	4/18-19/1991 Hakone	10/17-18/1991 Tokyo
4	1991-1993	9/30-10/1/1992 Washington, DC	4/29-30/1993 Detroit	5/3/1993 Washington, DC
5	1992-1995	2/14/1994 Washington, DC	10/26-27/1994 Gotemba	1/12/1995 Tokyo
6	1996-1997	2/22/1996 Washington, DC	10/19-20/1995 Seattle	5/2/1996 Washington, DC
7	1997	10/23/1997 Tokyo	11/24-27/1996	10/24/1997 Tokyo
8	1998	None	4/9-10/1998a Philadelphia	None
9	1999	Up to July None	None	None

Appendix 5

U.S. and Japanese Delegations

The following were derived from State Department lists when the latter was preparing for negotiating sessions with the GOJ. Each delegations were larger when the meetings were held in their home country. The core for the U.S. delegation came from State, DOC, OSTP; the principal actors in the Japanese delegation came from MOFA, MITI, S&TA and MOE. These lists are not necessarily final lists but the best available and they do show what level, what offices and what governmental departments/ministries were represented.

U.S. Delegation

Chairman: Ambassador Peter Jon de Vos, Deputy Assistant Secretary for Science and Technology Affairs, OES, Department of State

Mr. Philip Agress	Office of Japan, ITA, DOC.
Ms Susan Biniaz	Office of Legal Adviser, State Department.
Mr Arthur Fajans	Deputy Director, International Security Programs, DOD.
Mr. Glen Fukushima	Director for Japan Affairs, USTR.
Ms Judy Goans	Office of Legislation & International Affairs, PTO, DOC.
Ms Susan J. Koch	Strategic Defense & Space Arms Control Policy, DOD.
Dr. Sharon Libicki	OES, State Department.
Mr. Kevin Maher,	Office of Japan Affairs, State Department.
Mr John Masterson	Office of the Chief Counsel for International Trade, DOC.
Mr Shellyn McCaffrey	Deputy Executive Secretary, EPC, White House.
Mr Edward G. Murphy	Office of Economic Policy, Department of Treasury.
Mr Steven D. Needle	Office of General Counsel, DOC.
Mr Martin Prochnik	Director, Office of Cooperative S&T Programs, OES, State Department.
Mr Daniel A. Reifsnyder	Deputy Director, Office of Cooperative S&T Programs, OES, State Department.
Mr Emory Simon	Director, IPR, USTR.
Ms Maureen Smith	Director, Office of Japan, ITA, DOC.
Mr Benjamin Weakley	Office of International Affairs, DOE.
Ms Deborah Wince	Assistant Director for International Affairs, OSTP.
Mr Bill Yee	Office of the Deputy General Counsel, DOC.

Observer:

Dr. Charles T. Owens	Head, Information & Analysis Section, Div of International Programs, NSF.

Japanese Delegation

Chairman:

1st session: Hyōdō Nagao	Minister, Embassy of Japan, Washington, DC.
2nd - 6th sessions:	Ambassador Endō Tetsuya, S&T Affairs Bureau, MOFA.
7th session: Nomura Issei	Minister, Embassy of Japan, Washington, DC.

Mr M. Hamada	Assistant Director, Scientific Affairs Div, MOFA.
Mr. Hinata Seigi,	Director, Scientific Affairs Div, MOFA.
Mr Ikeda Kaname,	Science Counselor, Embassy of Japan, Washington, DC.
Mr Inaba Kenji	Director, International Research Cooperation Div., MITI.
Mr Ishii Masafumi	Treaty Bureau, MOFA.
Mr Kusahara Katsuhide	Director, International Academic Activities Div, MOE.
Mr Miyabayashi Masayoshi	Director, Planning Office, S&TA.
Mr T. Nakahara	Assistant Director, International Affairs Div, S&T Policy Bureau, S&TA.
Mr Toshikage Masakazu	Science Office, Embassy of Japan, Washington, DC.
Mr Shigeda Toyoei	Scientific Affairs, Div, MOFA.

Appendix 6

Selected Bibliography on U.S.-Japan Science and Technology

There have been a few articles on the 1988 U.S.-Japan S&T Agreement in U.S. newspapers and magazines, with somewhat better coverage in the Japanese media. There are, of course, no analytical essays on this Agreement. The following is, therefore, a list of books, articles and reports on various aspects of U.S.-Japanese science and technology from about 1980 to the present. It appeared that from about this time, more serious attention was beginning to be paid to Japanese science and technology and its potential and presumed challenge to U.S. pre-eminence and competitive position. This is a selected bibliography and is not intended to be an exhaustive list. Furthermore, many items that discuss the American competitive position and what to do about it are deliberately omitted since they do not discuss U.S. and Japanese science and technology in relation to each other. This list is also intended to provide a backdrop against which the S&T Agreement was proposed and negotiated.

The first section on U.S. Government reports includes some which might be regarded as ordinary commercial publications. They are included here because they were prepared at the behest of the government and provide another indication of the kind of attention being focused on Japanese S&T.

There are hundreds of articles and many books on Japanese science and technology. The items included are those which are regarded as useful in understanding the background against which the 1988 Agreement was created.

In regard to the Japanese books, it can said quite candidly that there are many more popular books on "high tech" matters in Tokyo bookstores than were available in the U.S. Only a selected few of these books have been listed.

---

1. U.S. Government Reports and Reports for the Government

Bloom, Justin. *Japan as a Scientific and Technological Superpower*. DOC/NTIS. 1990. 194 p.

Defense Science Board. *Task Force Report on Industry-to-Industry International Armaments Cooperation: Phase II-Japan*. 1984. 142 p.

Defense Science Board. *Task Force Report on Defense Semiconductor Dependency*. 1987. 103 p.

Executive Order 02591, April 10, 1987. *Facilitating Access to Science and Technology*.

Foreign Broadcast Information Service. *JPRS Report: Science and Technology*. Weekly, mostly unclassified, some marked "For Official Use only" and available from NTIS.

Gerstenfeld, Arthur, ed. *Science Policy Perspectives: USA-Japan*. Academic Press. New York. 1982. 363 p. Report on the 2nd seminar on science policy under the U.S.-Japan Cooperative Science Program.

Japanese Scientific and Technical Information in the United States. 1983 Workshop Proceedings. National Technical Information Service. 165 p.

Library of Congress. Congressional Research Service. *Japanese Science and Technology: Some recent efforts to improve U.S. monitoring*. 1986. 26 p. Written by Nancy R. Miller.

Library of Congress. Congressional Research Service. *Japanese Technical Information Opportunities to Improve U.S. Access*. 1987. 55 p. Written by Christopher T. Hill.

National Academy of Sciences. *Senior-Level Panel Calls for "Symmetrical Access" to U.S./Japan High-Tech Resources*. 1986. 14 p.

National Bureau of Standards (U.S. Department of Commerce). *U.S. Access to Japanese Technical Literature: Electronics and Electrical Engineering*. Proceedings of a Seminar at NBS. June 1985. 159 p. NBS Special Publication 710.

National Research Council. *Advanced Processing of Electronic Materials in the United States and Japan*. National Academy Press. 1986. 42 p.

National Research Council. *High-Technology Ceramics in Japan*. National Academy Press. 1984. 64 p.

National Research Council. *International Developments in Computer Science*. Report by the Standing Panel to Survey International Developments in Computer Science. National Academy Press. 1982. 98 p.

National Science Board. *Science and Engineering Indicators*. Issued annually.

National Science Foundation. *International Science and Technology Data Update*. Issued annually.

National Science Foundation. *Report Memorandum*. Reports issued periodically during the year from NSF's Tokyo Office on Japan's S&T programs and budgets.

National Science Foundation and the Japan Society for the Promotion of Science. *United States-Japan Cooperative Science Program*. Issued jointly every year to cover the period April 1 to March 31.

Office of Naval Research (U.S. Department of the Navy). *Scientific Bulletin*. Quarterly report from ONR's Tokyo Office. Each issue about 100 pages on specific and scientific and technical developments.

Office of Science and Technology Policy in Cooperation with the National Science Foundation. *Biennial Science and Technology Report to the Congress*. Issued annually.

Office of Technology Assessment. *Commercializing High-Temperature Superconductivity*. 1988. 171 p.

Office of Technology Assessment. *Commercial Biotechnology*. 1984. 612 p.

Science Applications International Corporation. *JTech Panel Report on Advanced Computing in Japan* (1987. 60p),*Advanced Materials* (1986. 100 p.), *Biotechnology* (1985), *Computer Science* (1984. 80 p.), *Mechatronics* (1985. 79 p.), *Opto*

& Microelectronics (1985. 145 p.), *Telecommunications Technology* (1986. 190 p.).
Science, Technology and American Diplomacy. Annual reports (1980+) submitted to the Congress by the President.
U.S. Congress, 99th, *Federal Technology Transfer Act of 1986*. Jan. 21, 1986.
U.S. Congress, 99th, *Japanese Technical Literature Act of 1986*. Aug. 14, 1986. Public Law 99-382.
U.S. Department of Commerce. *Activities of the Federal Government to Collect, Abstract, Translate and Distribute Declassified Japanese Scientific and Technical Information, 1987-1988*. Report to Congress. 1988. 83 p.
U.S. Department of Commerce. *Ceramic and Semiconductor Sciences in Japan*. 1987. 32 p.
U.S. Department of Commerce. *Factory Automation in Japan: Key Trends and Innovations*. 1988.
U.S. Department of Commerce. *Survey of Supply/Demand Relationships for Japanese Technical Information in the United States: The Field of Advanced Ceramics Research and Development*. 1988. 120 p.
U.S. Department of Defense. *Electro-Optics Millimeter/Microwave Technology in Japan*. 1985, 1987 (Final Report).
U.S. Department of Defense. *Japanese Military Technology: Procedure for Transfer to the United States*. 1986.
U.S. Department of Energy. Office of Energy Research. *Basic Research in Ceramic and Semiconductor Science at Selected Japanese Laboratories*. DOE/ER-0314. 1987. 93 p.
U.S. Department of Navy. Office of Naval Research. *Monitoring Foreign Science and Technology for Enhanced International Competitiveness: Defining U.S. needs*. Washington, DC. 1986. 93 p.
U.S. House of Representatives, Committee on Science and Technology. *Background Readings on Science, Technology and Energy R&D in Japan and China*. 1981. 499 p.
U.S. House of Representatives, Committee on Science and Technology. *Science, Technology and Energy Development: Report*. May 1981. 69 p.
U.S. House of Representatives, Committee on Ways and Means. *High Technology and Japanese Industrial Policy: A strategy for U.S. Policymakers*. 1980. 73 p. A study by Dr. Julian Gresser.
U.S. House of Representatives. *Federal Technology Transfer Act of 1986: Conference Report*. Report 99-953. 21 p.
U.S. House of Representatives, Subcommittee on Science, Research and Technology. *The Availability of Japanese Science and Technical Information in the United States* (Committee Report), 1984. 29 p.
U.S. House of Representatives, Subcommittee on Science, Research and Technology. *The Availability of Japanese Scientific and Technical Information in the United States: Hearings*. March 1984. 407 p.

U.S. House of Representatives, Subcommittee on Science, Research and Technology. *Japanese Technological Advances and Possible United States Responses Using Research Joint Ventures: Hearings*. 1983. 476 p.
U.S. House of Representatives, Subcommittee on Science, Research and Technology. *The Role of Technical Information in U.S. Competitiveness with Japan: Hearings*. June 1985. 295 p.
U.S. Senate, Subcommittee on Science, Technology and Space. *U.S.-Japan Science and Technology Agreement: Hearing*. 1987. 38 p.
U.S.-Japan Advisory Commission. *Challenges in United States-Japan Relations*. Washington, DC. 1984. 109 p. Also published in Japanese as *Nichi-Bei Shimon Iinkai Hōkoku*. 127 p.
U.S.-Japan Advisory Commission. *Stabilization and Expansion of Long Term Scientific and Technical Cooperation between the United States and Japan*. 1984. 37 p. Prepared by Justin L. Bloom and Taizō Yakushiji.
U.S.-Japan Advisory Commission. *U.S.-Japan Security Relations in the 1980's and Beyond*. 1984. 40 p. Prepared by Ellen S. Frost.
World Technology Research Center, Loyola College. WTEC(formerly JTEC) Panel Reports on selected technologies comparing U.S. with Japan (1989-1994) and with other countries (1994+ which are not listed here) :
High Temperature Superconductivity (11/89), Space Propulsion (10/89), Nuclear Power (10/90)
Advanced Computing (10/90)
Space Robotics(1/91)
High Definition [TV] Systems (2/91)
Advanced Composites (3/91)
Construction Space Technologies (6/91)
X-Ray Lithography (10/91)
Data Base Use and Technology (4/92)
Bioprocess Engineer ing (5/92)
Material Handling (2/93)
Separation Technology (3/93), Knowledge Based Systems (5/93)
Satellite Communications (7/93)
Advanced Manufacturing Technology for Polymer Composite Structures (4/94), Microelectromechanical Systems (9/94)
Biodegradable Polymers and Plastics (3/95)
Electronic Manufacturing and Packaging (2/95)
Human-Computer Interaction Technologies (3/96)
Optoelectronics (2/96)
ERATO and PRESTOBasic Research (9/96)
Rapid Prototyping in Japan and Europe (3/97)
Adanced Casting in Japan and Europe (3/97)
Power Applicationss of Superconductivity in Japan and Germany (9/97)
Nanoparticles, Nanostructured materials and nanodevices (1/98)

2. Books

Aron, Paul H. *The Robot Scene in Japan: The Second Update.* Daiwa Securities America, Inc. New York. 1985. 69 p. Paul Aron Report 28.

Bartholomew, James R. *The Formation of Science in Japan.* New Haven. Yale University Press. 1989. 392 p.

Bloom, Justin L. *Japan's Research and Development System* undated. 55 p.

Brown, Harold. *U.S.-Japan Relations: Technology, Economics and Security.* The George Washington University. 1987. 32 p.

DiFilippo, Anthony. *Cracks in the Alliance: Science, technology and the evolution of U.S.-Japan Relations.* Aldershot (UK), Ashgate. 1997. 291 p.

Feigenbaum, Edward A. and Pamela McCorduck. *The Fifth Generation: Artificial Intelligence and Japan's Computer Challenge to the World.* Addison-Wesley Publishing Reading, MA. 1983. 275 p.

Fujii, Haruo. *Nihon no Kokka Kimitsu.* Tokyo. Gendai Hyōronsha. 1972. 323 p.

Gamota, George and Wendy Frieman. *Gaining Ground: Japan Strives in science and technology.* 1988. Ballinger Press. 180 p. (Summary of the JTech reports mentioned under Science Application International Corporation under Government reports).

Gibson, Robert W. and Barbara K. Kunkel. *Japanese Science and Technical Literature: A Subject Guide.* Greenwood Press. Westport CT. 1981. 560 p.

Glazer, Herbert. *The Japanese Optoelectronics Industry and Its Relationship to the Strategic Defense Initiative.* 1986. 61 p.

Green, Anthony T. *U.S.-Japan Technology Transfer: Accommodating Different Interests.* Center for Information Policy Research. Harvard University. Cambridge, MA. 1986. 128 p.

Growth and Competitiveness in High-Technology Industries: A Japan-U.S. Comparative Conference. Japan Society. New York. 1984. 38 p.

Haitekku Senryaku Kenkyūkai, comp. *Beikoku no Gijutsu Senryaku: Tsuyoi Beikoku no Saisei o mezashite.* Mita Shuppankai. 1988. 329 p.

Haitekku Senryaku Kenkyūkai, comp. *Nihon no Gijutsu Senryaku.* Mita Shuppankai. 1988. Parts 1, 2 and 3.

Heiduk, Gunter and Yamamoto Kōzō, eds. *Technological Competition anInterdependence: the search for policy in the United States, West Germany and Japan.* Seattle. University of Washington Press. University of Tokyo Press. 1990. 255 p.

Hutchinson, R.A. et al. Information Flow from Japan to U.S. Researchers in Applied and Basic Energy Fields in *Journal of Technology Transfer.* 10:1 (1-7) 1985.

Kimura, Shigeru. *Japanese Science Edge.* University Press of America/The Wilson Center. 1985. 164 p.

Kuehn, Thomas, et al. *The Global Challenge in High Technology Trade: Indicators of Foreign Technology and Economic Power.* Georgia Institute of Technology. Atlantica, GA. 1988. 2 v.

Lynn, Leonard H. *How Japan Innovates: A comparison with the U.S. in the case of*

Oxygen Steelmaking. Westview Press. 1982. 211 p.
Morse, Ronald A., ed. *Option 2000: Politics and High Technology in Japan's Defense and Strategic Future*. Woodrow Wilson International Center. [1987]. 121 p.
Morse, Ronald A. and Richard J. Samuels, eds. *Getting America Ready for Japanese Science and Technology*. Woodrow Wilson International Center. Washington, DC. 1985. 183 p.
Nakayama, Tarō. *Starting from Zero: Transformation of Japan by Science and Technology*. ESCAP Regional Centre for Technology Transfer. Bangalore, India. 1985. 140 p.
Nelson, Richard R. *High Technology Policies: A Five Nation Comparison*. American Enterprise Institute for Public Policy Research. 1984. 94 p.
Nihon Sentaku Kenkyūkai. [MITI study group.] Nyūgurōbarizumu e no Kōken to Shin-Sangyō Bunka Kokka no Sentaku. 1988. 99, 57 p.
Okimoto, Daniel, ed. *Competitive Edge: The Semiconductor Industry in the U.S. and Japan*. Stanford University Press. Stanford, CA. 1984. 275 p.
Patrick, Hugh ed. *Japan's High Technology Industries: Lessons and Limitations of Industrial Policy*. University of Washington Press. Seattle. 1984. 277 p.
Pierre, Andrew J., ed. *A High Technology Gap: Europe, America and Japan*. New York University Press. 1987. 114 p.
SRI International. *Science and Technology as Tools for Foreign Policy*. 1980. Arlington, VA. 244 p.
Uyehara, Cecil H., ed. *U.S.-Japan Science and Technology Exchange: Patterns of Interdependence*. Westview Press. Boulder, CO. 1988. 279 p.
Uyehara, Cecil H., ed. *U.S.-Japan Technological Exchange Symposium, 1981*. University Press of America. Lanham, MD. 1982. 132 p.

3. Selected Articles and Essays from Books

Bloom, Justin. Chapter on Japan (44 p.) in *Performer Organizations and Support Strategies for Fundamental Research in Six Countries*. SRI International. Palo Alto, CA. 1984.
Bloom, Justin L. Bilateral Cooperative Programs: A case study, the United States and Japan in *Journal of the Washington Academy of Sciences*. 77:3 (9/87):87-92.
Bloom, Justin L. The U.S.-Japan Bilateral Science and Technology Relationship: A Personal Evaluation, in *Scientific Technological Cooperation among Industrialized Countries: The role of the United States*. Mitchell B. Wallerstein, ed. National Academy Press. 1984. p. 84-110.
Japanese Technology Today, in *Scientific American*, published annually since 1980-1991.
JEI Report. Published weekly by the Japan Economic Institute occasionally includes reports on U.S.-Japan competition in various technical fields, such as pharmaceuticals, robots, etc.

Morris-Suzuki, Tessa. *The Technological Transformation of Japan: from the 17th Century to 21st* Century.Cambridge. Cambridge University Press. 1994. 304 p.
Peck, Merton J. Technology in *Asia's New Giant: How the Japanese Economy Works*. Brookings Institution. Washington, DC, 1975. p. 529-585.
Reich, Robert B. The Rise of Techno-Nationalism in *The Atlantic Monthly*, 5/87, p. 63-69.
Rise of techno-nationalism: Science is the new battle ground for U.S. Japanese tensions in *Far East Economic Review*, 3/31/86:58-65.
Science in Japan, in *Science*, 233:4761 (7/18/86):267-304.

4. Books and Government reports published in Japan

Ishii, Takemochi. *Nihon no Sentan Gijutsu* (Japan's leading edge technology). Nihon Hōsō Shuppan Kyōkai. Tokyo. 1985. 281 p.
Japanese Technology: Bulls and Bears in *Japan Echo*. Vol. X (1983) (Special Issue). Tokyo.
Kagaku Gijutsuchō. *Kagaku Gijutsu Hakusho* (Science White Paper). Issued annually.
Kagaku Gijutsucho. Kagaku Gijutsu Seisakukyoku. *Waga kuni ni okeru gijutsu hatten no hōkō ni kansuru chōsa -gijutsu yosoku hōkokusho.* 1987. 121 p.
Makino, Noboru and Shimura Yukio. *Nichi-Bei Gijutsu Sensō* (The Japan-U.S.A. conflicts on high technology). Nihon Keizai Shimbunsha. Tokyo. 1984. 278 p.
Ministry of International Trade and Industry, Agency of Industrial Science and Technology. *International QuantitativeComparison of Japanese Industrial Technology*. JETRO. 1983. 46 p.
Moritani, Masanori. *Nichi-Bei-Ō: Gijutsu kaihatsu sensō*(Japan-U.S.-Europe: Technology Development War). Tōyō Keizai Shimpōsha. Tokyo. 1981. 228 p.
Nagasaki Sentan Gijutsu Kaihatsu Kyōgikai. *Nihon no Sentan Gijutsu: 21 Seiki no tembō* (Japan's leading edge technology: outlook for the 21st century). Nikkei Saiensusha. Tokyo. 1985. 239 p.

5. Selected Case Studies

The following selective list would suggest that studies have been made about the decision-making process in Japan in a variety of areas. I compiled this list in an attempt to find out if a study in USJ negotiations and policy-making in the S&T area had been done. It appears from this list that no such study has been made.

Abelson, Donald S. Experiencing the Japanese4 Nagotiating Style, in *Toward a Better Understanding: U.S.-Japan Relations.* Washington, DC. Foreign Service Institute. 1986. p. 39-42.
Akaha, Tsuneo. *Japan in Global Ocean Politics*. Honolulu. University of Hawaii Press. 1985. 224 p.

Allison, Gary D. *Public Servants and Public Interests in Contemporary Japan*. 1980. 36 p. Unpublished paper given to the Washington & Southeast Regional Seminar on Japan, 2/16/80.

Angel, Robert C. *Japan's Prime Minister: The Individual and the Office*. Unpublished paper given at the Southern Political Science Association. November, 1987.

Angel, Robert C. *Supra-Ministerial Policy Coordination in Japan*. Unpublished paper given at the American Political Science Association Convention, 1987.

Angel, Robert C. *U.S.-Japan Economic Relations: Lessons from the 1971 Yen Revaluation Crisis*. 1978. 38 p. Unpublished paper given to Columbia University Seminar on Modern Japan, April 14, 1978.

Apter, David E. and Nagayo Sawa. *Against the State: Politics and Social Protest in Japan*. Cambridge. Harvard University Press. 1984. 271 p.

Baerwald, Hans H. Nikkan Kokkai: The Japan-Korea Treaty Diet, *Cases in Contemporary Politics: Asia*. Ed: Lucian W. Pye. Little Brown. Boston. 1970. p. 19-57.

Bailey, Stephen Kemp. *Congress Makes a Law: The Story behind the Employment Act of 1946*. New York. Columbia University Press. 1950. 282 p.

Blaker, Michael K. *Japanese International Negotiating Style*. New York. Columbia University Press. 1977. 253 p.

Blaker, Michael K. Probe, Push and Panic: The Japanese Tactical Style in International Negotiations, in *The Foreign Policy of Modern Japan*, University of California Press, 1977. Ed. Robert A. Scalapino. p. 55-101.

Campbell, John Creighton. *Contemporary Japanese Budget Politics*.Berkeley. University of California Press. 1987. 308 p.

Campbell, John Creighton. *Entrepreneurial Bureaucrats and Programs for Old People in Japan*. 1978. Paper delivered at the 1978 Annual Meeting of the American American Political Science Association.

Campbell, John Creighton. The Old People Boom and Japanese Policy Making, *Journal of Japanese Studies* 5:2(Summer 79):321-357.

Campbell, John Creighton. *Old People Problem: The career of an issue*.1980. 38 p. Unpublished paper given to the Columbia University Seminar on Modern Japan, 2/13/80.

Campbell, John Creighton. *Progressive Local Government, Policy Innovation, and the Diffusion of Programs for the Agenda in Japan*. 1981. 14 p. Unpublished paper for the Washington & Southeast Regional Seminar on Japan, 4/18/81.

Carlberg, Eileen. *The Legislative Process Surrounding Women's Labor Legislation in Japan* [1972 Working Women's Welfare Law]. Phd Dissertation.

Cho, Paul Bylung-chan. *Japan's Ratification of the Nuclear Non-Proliferation Treaty A study of the Diffusion of Policymaking Power and Consensus Politics in Japanese Foreign Policy making*. PhD dissertation (West Virginia Univ, 1981. 455 p.

Cohen, Bernard C.*The Political Process and Foreign Policy: the making of the Japanese Peace Settlement*. Princeton, NJ, Princeton University. 1957. 293 p.

Destler, I.M. Country Expertise and U.S. Foreign Policymaking: The Case of Japan, in Morton A. Kaplan and Kinhide Mushakoji, eds. *Japan, America and the Future World Order*. New York. Free Press. 1976. p. 125-144.

Destler, I.M. & Hideo Sato, Priscilla Clapp, Haruhiro Fukui. *Managing An Alliance: the Politics of U.S.-Japanese Relations.* Washington, DC. The Brookings Institution. 1976. 209 p.

Emmerson, J.K. & Michael Blaker. Japanese International Negotiating Style. *Political Science Quarterly* 93:2(1978):364-365.

Flanagan, Scott C. *Pathologies of Decision-Making in Prewar Japan.* 1973. 119 p. Unpublished paper given to Japan Studies Seminar at Georgetown University.

Fukui, Haruhiro. *Foreign Policy Making in Japan: Case studies for empirical theory.* 1974. 99 p. Unpublished paper given at 1974 AAS convention.

Fukui, Haruhiro. The Gatt Tokyo Round: The Bureaucratic politics of Multilateral Diplomacy. Paper prepared for "U.S.-Japanese Relations in Multilateral Diplomacy". Project of the East Asia Institute. Columbia University. 1977.

Fukui, Haruhiro. Policy-making in the Japanese Foreign Ministry, *The Foreign Policy of Modern Japan.* Robert A. Scalapino, ed. Berkeley. University of California Press. 1977. p. 3-35.

Fukui, Haruhiro. *The Textile Wrangle: Conflict in Japanese-American Relatio, 1969-1971.* Ithaca, NY. Cornell University Press. 1979. 394 p.

Fukushima, Glen. The Liberalization of Legal Services in Japan: A U.S. Government Negotiator's Perspective, *Law in Japan* 21:5 (1988):5-17.

Fukushima, Glen. The Nature of United States-Japan Government Negotiations, *United States/Japan Commercial Law and Trade*. Valerie Kusuda Smick, ed. Transnational Juris Publications, Inc. 1990. p. 660-667.

Glazer, Herbert. Understanding Japanese Decision-making, *Toward a Better U.S.-Japan Relations.* Washington, DC. Foreign Service Institute. 1986. p. 39-42.

Gross, Bertram M. *The Legislative Struggle: A study in social combat.* New York. McGraw-Hill Book Co. 1953. 472 p.

Hahn, E.J. Negotiating with the Japanese, *California Lawyer* 2(1982):21-59.

Harari, Ehud. Japanese Politics of Advice in Comparative Perspective: A Framework for Analysis and a Case Study, *Public Policy* 22:4 (Fall 74):537-577.

Hirose, Michisada. Pressure Groups in Japanese Politics, *Japan Echo* 11:4(84):61-67.

Hitchcock, David I. Joint Development of Siberia: Decision-Making in Japanese-Soviet Soviet Relations, *Asian Survey.* (3/71).

Holdsworth, Richard. Japanese Peace Treaty Negotiations with the Soviet Union and China, January to April 1975. *Millenium* 5:1(1976):41-57.

Hosoya, Chihiro. *Case Study of Japanese Foreign Policymaking Process: Japan's Response to German-Soviet War, 1941.* Unpublished paper presented to Columbia University Seminar System: Modern Japan. 5/11/1963.

Hosoya, Chihiro. Characteristics of the Foreign Policy Decision Making System in Japan, *World Politics* 26:3(4/1973):353-369.

Hosoya, Chihiro. Japan's Decisionmaking System as a Determining Factor in Japan-U.S. relations, Morton A. Kaplan and Kinhide Mushakoji, eds, *Japan, America and the Future World Order*. New York. Pree Press. 1976. p. 117-124.

Huff, Rodney Louis. *Political Decision Making in the Japanese Civilian Atomic Energy Program*. PhD dissertation. The George Washington University. 1973. 474 p.

Kimura, H. Sovie t and Japanese Negotiating Behavior: The Spring 1977 Fisheries Talks, *Orbis* 24(1980):43-67.

Kobayashi, Naoki. Interest Groups in the Legislative Process, *Japanese Politics: An Inside View*. Ithaca, NY. Cornell University Press. 1973. p. 68-87.

Kobayashi, Naoki. The Small and Medium-sized Enterprises Organization Law, Japanese, Japanese Politics: An Inside View. Hiroshi Itoh, ed. Ithaca, NY. Cornell University Press. 1973. p. 49-67.

Koh, B.C. *Japan's Administrative Elite*. Berkeley. University of California Press. 1991. 312 p.

Kraus, Ellis S. et al, eds. Conflict in Japan. Honolulu. University of Hawaii Press. 1984. 417 p.

Lincoln, Edward J. *The Japanese National Railways: Interaction with the Government*. 1979. 28 p.Unpublished paper given to Washington & Southeast Regional Seminar on Japan, 4/28/79.

Marcot, Neal Abel. *The Japanese Foreign Policymaking Process: A Case Study, Okinawa Reversion*. PhD dissertation. Georgetown University. 1981. 396 p.

McCreary, Don R. and Robert A. Blanchfield. The Art of Japanese Negotiation, Languages in International Perspective. Norwood, NJ. Ablex. 1986. p. 301-350.

McCreary, Don R. *Communicative Strategies in Japanese-American Negotiations*. PhD Dissertation. Newwark, DE. University of Delaware. 1984. 211 p.

McKean, Margaret A. *Energy Conservation Policy in Japan Through Two Crises*. 1982. 20 p. Unpublished paper given to the Washington & Southeast Regional Seminar Seminar on Japan, 2/82.

McKean, Margaret A. Pollution and Policy-making, *Policy Making in Contemporary Japan*. T.J. Pempel, ed. Ithaca, NY. Cornell University Press. 1987. p. 201-238.

Misawa, Shigeo. An Outline of the Policymaking Process in Japan, *Japanese Politics: An Inside View*. Hiroshi Itoh, ed. Ithaca, NY. Cornell University Press. 1973. p. 12-48.

Mochizuki, Mike Masato. *Managing and Influencing the Japanese Legislative Process: The role of parties and the National Diet*. Cambridge. Harvard University.1982. 555 p. PhD dissertation.

Moser, Leo J. Negotiating Style: American and Japanese, *Toward a Better Understanding*. Bendahmane and Moser, eds. Foreign Service Institute. 1986. p. 45-51.

Muramatsu, Michio. *Political Parties and the Bureaucracy in the Japanese Policy Process*. Prepared for colloquium sponsored by Woodrow Wilson International Center for Scholars. 1981. 29 p.

Mushokoji, Kinhide. *The Strategies of Negotiation: An American-Japanese Comparison*. Tokyo. Sophia University. 1970. 42 p.

Ōtsuki, Shinji and Honda Masaru. *Nichi-Bei FSX Sensō: Nichi-Bei Dōmei o yurugasu*

gijutsu masatsu. Ronsōsha. 1991. 365 p.
Overholt, William H. The Central Council for Education, Organized Business and Politics of Education Policy-making in Japan. *Comparative Education Review* 19:2(6/1975).
Packard, George R. *Protest in Tokyo: The Security Treaty: Crisis of 1960.* Princeton. Princeton University Press. 1966. 423 p.
Park, Yung H. The Politics of Japan's China Decision, *Orbis* 19:2(Summer 1975):-562-590.
Park, Yung H. The Tanaka Government and the Mechanics of the China Decision, *China and Japan: A Search for Balance since World War I.* Hilary Conroy and Alvin D. Cox., eds. ABC-Clio Books. Santa Barbara, CA. 1978. p. 387-397.
Pempel, T.J. Japan's Political Institutions and the Decision Making Process, *Japanese Role in the Asia-Pacific Region*, Berkeley. Institute of East Asian Studies. 1990.
Pempel, T.J. Patterns of Japanese Policymaking: Experiences from Higher Education. Boulder, CO. Westview Press. 1978. 248 p.
Pempel, T.J., ed. *Policymaking in Contemporary Japan.* Ithaca. Cornell University Press. 1987. 345 p.
Pempel, T. J. *The Tar Baby Target: "Reform" of the Japanese Bureaucracy under the American Occupation.* Paper for Allied Occupation of Japan Conference, Hawaii. 1978. 65 p.
Redman, Eric. *The Dance of Legislation*. New York. Simon & Schuster. 1973. 319 p.
Reed, Steven R. *Japanese Prefectures and Policymaking.* Pittsburgh, University of Pittsburgh Press. 1986. 208 p.
Rix, Alan G. *Foreign Aid and Policymaking in Japan*. PhD Dissertation. Australian National University. 1977.
Saito, Motohide. *The "Highly Crucial" Decision making Model for Postwar Japan and Prime Minister Hatoyama's Policy toward the U.S.S.R.* New York. Columbia University. 1986. 346 p. PhD dissertation.
Shiels, Frederick L. *The American Experience in Okinawa: A Case Study in Foreign Policymaking Theory.* PhD Dissertation. Ithaca, NY. Cornell University. 1976.
Silberman, Bernard S. Bureaucratic Development and the Structure of Decision-Making in the Meiji Period: The case of the Genro, *Journal of Asian Studies* 27:1 (11/67):81-94.
Steslicke, William E. *National Health Insurance: The Japanese experience.* 1974. 20 p. Unpublished paper given at 1974 AAS Convention.
Thayer, Nathaniel B. and Stephen E. Weiss. Japan: The Changing Logic of a Former Minor Power, *National Negotiating Style.* Washington, DC. Foreign Service Institute. 1987. p. 45-74.
Thurston, Donald R. *Teachers and Politics in Japan.* Princeton. Princeton University Press. 1973. 357 p.
Totten, George O. III. *Ocean Policy of Japan: Rationalization of the Japanese Ship-building Industry.* Public Policy Decision of 1978.

Tsuji, Kiyoaki. Decision-making in the Japanese Government: A Study of Ringisei, *Political Development in Modern Japan.* Robert E. Ward, ed. Princeton. Princeton University Press. 1968. p. 457-475.

U.S. House of Representatives, Committee on Foreign Affairs. *Government Decision making in Japan: Implications for the United States.* U.S. Government Printing Office. 1982. 147 p.

Van de Velde, James R. The Influence of Culture on Japanese-American Negotiation, *Fletcher Forum 7:2 (1983):395-400.*

Vogel, Ezra F. *Modern Japanese Organization and Decision Making.* Berkeley. University of California Press. 1974. 340 p.

Wang, Robert S. Talking Turkey with Tokyo: To have Successful Discussions with the Japanese, the United States must pay more Attention to its Ally's Negotiating Style, *Foreign Service Journal* 62(11/85):34-37.

Watanabe, Akio. *The Okinawa Problem: A chapter in Japan-U.S. Relations.* Melbourne, Australia. Melbourne University Press. 1970.

Witham, Wallace Fernald, Jr. *A Study of Meiji Japan's Foreign Policy Decision-making Structure: Japanese Attitudes Pertaining to the Expansion of the U.S. into the Pacific, 1888-1898.*PhD Dissertation. University of Minnesota. 1974. 337 p.

Yoshitsu, Micheal M. *Japan and the San Francisco Peace Settlem*ent. New York. Columbia University Press. 1982. 124 p.

Appendix 7

List of Interviewees

Almost fifty people in and out of the U.S. and Japanese Governments in Washington, DC, and in Tokyo were interviewed for this study. Most of the interviews took place in 1988-89 and then later in Tokyo in the 1990s. The title mentioned is the known title of position at the time of the interview.

1. The U.S. Government

Department of Agriculture

Tallent, William H. PhD, Assistant Administrator, Cooperative Interactions, Agricultural Extension Service.

Department of Commerce

Agress, Philip, International Trade Specialist, Office of Japan.
Allen, Joseph P., Director, Office of Federal Technology Management.
Brumley, Robert, General Counsel.
Lipsky, Susan Miller, Acting Director, International Operations, Office of Productivity, Technology.
Masterson, John, Office of General Counsel.
McCaffrey, Shellyn G., Associate Deputy Secretary.
Merrifield, Bruce, Assistant Secretary of Commerce, Productivity, Technology & Innovation.
Mullins, Richard H., Senior Policy Adviser, Office of Federal Technology Management.
Smith, Maureen R, Director, Japan Division.
Underwood, G.T., Director, Office of Metric Programs.

Department of Defense

Fajans, Arthur, Deputy Director, International Security Programs.
Koch, Dr. Susan J., Analyst for Strategic Defense & Space Arms Control Policy, Office of the Assistant Secretary, ISA.
Sullivan, Peter M., Assistant Deputy Under Secretary of Defense, Trade Security Policy.

Department of Energy

Hill, Billy.
Smith, Marilyn S., Assistant for International to the Director of the Office Energy Research.
Weakley, Benjamin L. Program Office, Office of International Affairs.
Hill, Billy.

Department of State

Biniaz, Sue, Legal Office.
Chern, Kenneth, Office for Japanese Affairs, EA/J.
Clark, William, Jr., Deputy Assistant Secretary of State, EA.
Crowley, R.W., Office of Science and Technology Cooperation, OES.
Derham, James M., Office of Japanese Affairs, EA/J.
de Vos, Peter, DASS/OES, Chairman, U.S. Delegation for STA negotiations.
Getzinger, Dr. Richard W., Science Counselor, U.S. Embassy, Tokyo, 1986-1990.
Libicki, Shari B., PhD, OES/SCT, AAAS Science Engineering and Diplomacy Fellow.
Maher, Kevin K. Office for Japanese Affairs, EA/J.
McPherson, William, OES/SCT.
Piez, William, Deputy Assistant Secretary of State, EA.
Reifsnyder, Deputy Director, Office of Cooperative Science & Technology Programs, OES/SCT.
Reis, Robert C., Jr., Deputy Director, Office of Investment Affairs.
Robertson, Charles E., Environmental and Scientific Affairs, OES.
Sherman, William, Ambassador, United Nations, formerly DCM American
Embassy, Tokyo.
Sigur, Gaston, Assistant Secretary, Asia and Pacific Affairs, EA.

Department of Treasury

Murphy, Edward E., Office of Economic Policy.

Economic Policy Council

Maroni, William J. Special Assistant to the President & Executive Secretary.

National Institutes of Health

Fairlee, Coralie, Fogarty International Center.
Goodman, Chanty, Fogarty International Center.
Schambra, Philip, Fogarty International Center.

Schmidt, Jack R., PhD, Chief, International Coordination & Liaison Branch, Fogarty International Center.
Stepanian, Alexandra, Program Office for Asia and the Soviet Union, International Coordination and Liaison Branch, Fogarty International Center.
Wallace, Craig K., MD, Associate Director for International Research, Fogarty International Center.

National Science Foundation

Boright, John P., Director, International Programs.
De Angelis, Alexander. NSF representative in American Embassy, Tokyo.
Moore, John, Deputy Director.
Owens, Charles T., International Programs.
Wallace, Charles.

Office of Management and Budget, Executive Office of the President

Bostock, Judith, Nuclear Energy.
Kitti, Carole E., Economist, Special Studies Div, National Security & International Affairs.
Noonan, Norine, Acting Branch Chief, Science and Special Projects.

Office of Science and Technology Policy, Executive Office of the President (The White House)

Bromley, Alan, Science Adviser to the President.
Graham, Dr William, Science Adviser to the President.
Schweitzer, Jeff, PhD, Senior Policy Analyst for Policy & International Affairs.
Wince, Deborah, Assistant Director.

Office of the U.S. Trade Representative, Executive Office of the President

Emory, Simon, General Counsel.
Field, Catherine R., Associate General Counsel.
Fukushima, Glen S. Director for Japanese Affairs.

U.S. Patent and Trademark Office

Cage, Kenneth.
Goans, Judy Winegar, Office of Legislation and International Affairs.

U.S. Senate

McGaffigan, Edward, Jr., Senator Jeff Bingaman's Office.

2. The Japanese Government

Embassy of Japan in Washington, DC

Ikeda, Kaname, Science Counselor (from Science and Technology Agency).
Kisaka, Takashi, Science Counselor (from Science and Technology Agency).
Kuramochi, Tadao, Science Counselor (from Science and Technology Agency).
Nomura, Issei, Minister, MOFA.
Watanabe, Taizō, Minister, MOFA.

House of Representatives

Yoshimura, Harumitsu, Specialist, Research Div., Scientific and Technology Committee, formerly with Science and Technology Agency.

Ministry of Foreign Affairs

Endō, Tetsuya, Ambassador Plenipotentiary, Chairman, Japanese Delegation.
Ishii,Masafumi, Assistant Director, Treaty Div, Treaty Bureau.
Ishikawa, Masamichi, Deputy Director, First International Organizations Division, Economic Affairs Bureau.
Hinata, Seigi, Fellow, Center for International Affairs, Harvard University, Cambridge, MA, was Director of Scientific Affairs Division.
Ōta, Hiroshi, Director-General Scientific and Technological Affairs.
Takahashi, Shūhei, Deputy Director, Scientific Affairs Division.

Ministry of Education

Kusahara, Katsuhiko, Director, Higher Education Divison.

Science and Technology Agency

Miyabayashi, Masayoshi, Director, Planning Division.

3. Other (U.S.)

Blommer, Michael W., Executive Director, American Intellectual Property Law Association, Arlington, VA.
Bloom, Justin, consultant, former Minister-counselor (science), U.S. Embassy, Tokyo.

Buchsbaum, S.J., Executive Vice President, Customer Systems, AT&T Bell Laboratories, Holmdel, NJ, formerly member of the White House Science Committee.

Entin, Stephen J., Institute for Research on the Economics of Taxation, Washington, DC.

Frost, Ellen, PhD, Director, Government Programs, U.S.-Japan Relations, Westinghouse Electric Corporation.

Gaffney, Frank J., Jr., Senior Fellow, Hudson Institute. Was Deputy Assistant Secretary of Defense.

Helfrich, Gerard F., Senior International Affairs Specialist, Meridian Corporation, Alexandria, VA.

Huberman, Benjamin, Vice President, The Consultants International Group, Inc., Washington, DC.

Latker, Norman, patent attorney.

Marcum, John M., President, European Institute of Technology, Paris, France.

O'Rourke, C. Larry, Finnegan, Henderson, Farabow, Garret & Dunner, Washington, DC.

Owczarski, William A., Director of External Technology Development, United Technologies (formerly with OSTP).

Rivers, Lee W., President, Technology Transfer Initiatives, Washington, DC.

Rosner, William R., attorney, Jones, Day, Cleveland, OH, was with OES for a while.

Smith, Michael B., President, SJS Advanced Strategies, Washington, DC, was Deputy USTR.

Trivelpiece, Alvin W., Executive Officer, American Association for the Advancement of Science, Washington, DC.

Other (Japan and other countries)

Inose, Dr. Hiroshi, Director General, National Center for Science Information Systems, Tokyo.

Kondō, Dr. Jirō, President, Science Council of Japan, Tokyo.

Kramer, Dr. Bernd, Science Counselor, Embassy of the Federal Republic of Germany.

Index

About the Author

Cecil H. Uyehara, President of Uyehara International Associates, is a consultant in the Washington, D.C. area on U.S.-Japanese relations (science and technology). He served in the U.S. government for almost 25 years, with the Air Force (weapons systems planning), the Office of Management and Budget (military assistance) and the Agency for International Development (AID, U.S. Department of State). He has published books and articles on Japanese politics, scientific advice and public policy, Japanese calligraphy and Afghan philately. His published works in politics and S&T are: *Socialist Parties in Postwar Japan*, Yale University Press, 1966, co-authored with Alan Cole and George O. Totten; *Future Potential Science and Technology Cooperation between the United States and Japan*, with Shigeo Yoshikawa. Uyehara International Associates. 1996. 117 p.;"JTEC Studies -- A Common Thread" in Preprints *3rd Japanese Science & Technology Information Conference*, Nancy France, May 15-18, 1991; "Appraising Japanese Science and Technology" in *Japan's Economic Challenge*, U.S. Congress, Joint Economic Committee. 1990. p. 289-307; "U.S. Responses to the Japanese Science and Technology Challenge", Conference on Japanese Science, Technology & Commerce. Amsterdam. IOS Press. 1990. p. 87-111; *U.S.-Japan Science and Technology Exchange: Patterns of Interdependence,* editor, Westview Press, 1988; *Technological Exchange: The U.S.-Japanese Experience,* editor, University Press of America, 1982;"The Nuclear Test Ban Treaty and Scientific Advice" in *Knowledge and Power*, The Free Press, 1966; *Leftwing Social Movements in Japan, An Annotated Bibliography,* Charles E. Tuttle Co., 1959;The Social Democratic Movement, *The Annals of the American Academy of Political and Social Science.* 308(11/1956):54-62; *Japanese Ministry of Foreign Affairs Archives, 1868-1945*, Library of Congress, 1954. He organized the first U.S. Congressional hearings on Japanese science and technology information 1984.

He graduated from Keio University (B.A.), the University of Minnesota (M.A.) both in political economy, and received awards/grants from the Ford Foundation, American Philosophical Society, University of Minnesota (Shevlin Fellowship) and the National Institute of Public Affairs Fellowship to attend Harvard University, 1963-1964 as a National Institute of Public Affairs (NIPA) Fellow . He was born in London, England; lived in London, Shanghai, China, Tokyo, Japan, the Washington, DC area and several cities in the United States and in Kabul, Afghanistan.